International Geologic Congress Index Maps to the Field Trips

Physiographic Map of North America with field trip routes in Alaska, Hawaii and Mexico

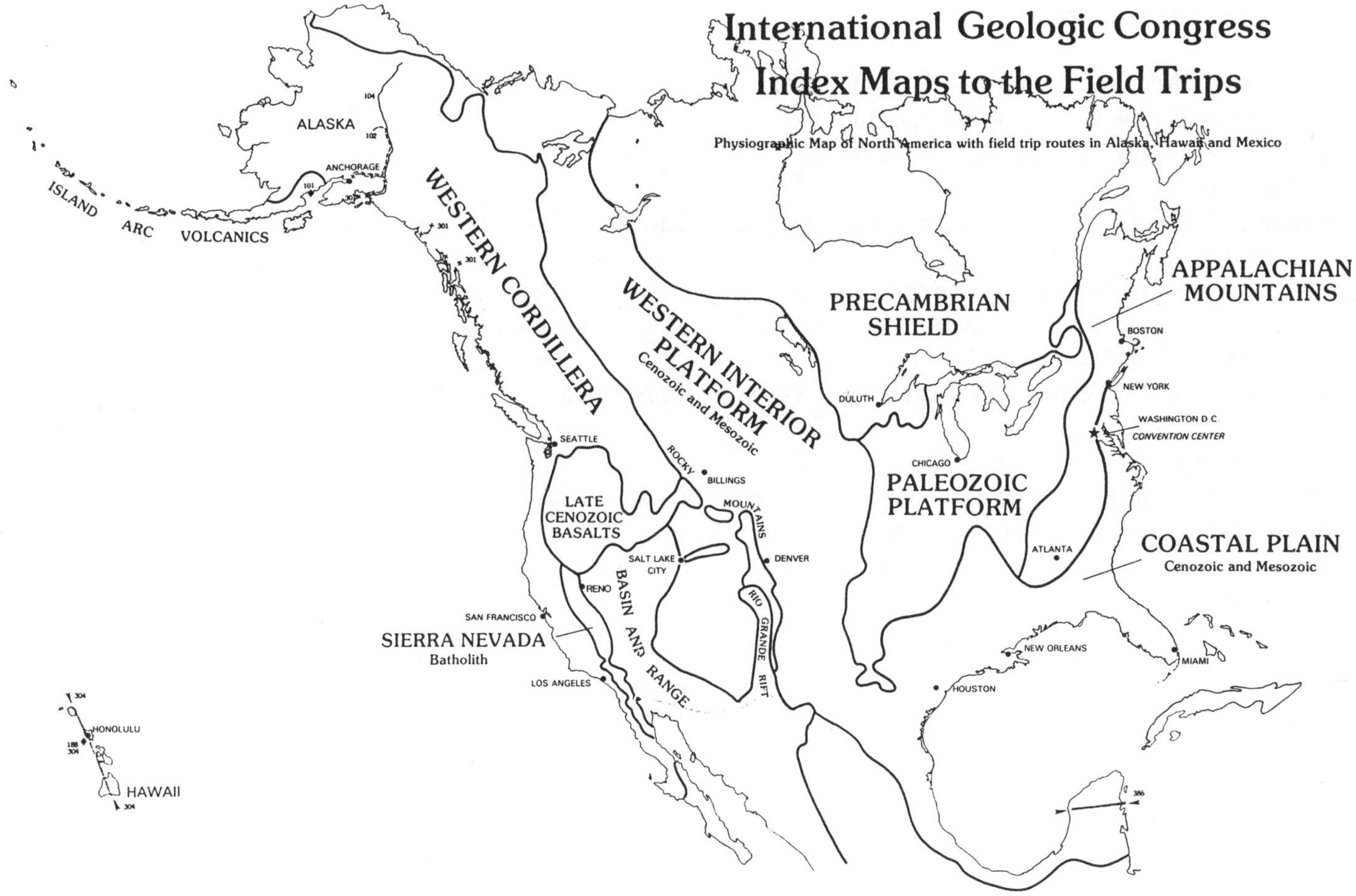

Legend

Field Trip Maps

Cities

175 328 Field trip numbers (refer to text)

Specific routes; line type for differentiating field trip routes; limited to area of geologic interest

Field trip routes which extend over a broad geographic area

Field trip routes which focus around a single area

TITLES IN THE IGC SERIES

Coal and Hydrocarbon Resources of North America *2 volumes*
Coastal and Marine Geology of the United States *1 volume*
Environmental, Engineering, and Urban Geology in the United States *2 volumes*
Geology of Grand Canyon, Northern Arizona (with Colorado River Guides) *1 volume*
Glacial Geology and Geomorphology of North America *2 volumes*
Mesozoic/Cenozoic Vertebrate Paleontology: Classic Localities, Contemporary Approaches *1 volume*
Metamorphism and Tectonics of Eastern and Central North America *3 volumes*
Mineral Deposits of North America *2 volumes*
Sedimentation and Basin Analysis in Siliciclastic Rock Sequences *3 volumes*
Sedimentation and Stratigraphy of Carbonate Rock Sequences *2 volumes*
Sedimentation and Tectonics of Western North America *3 volumes*
Tectonics of the Scotia Arc, Antarctica *1 volume*
Volcanism and Plutonism of Western North America *2 volumes*

PENELOPE M. HANSHAW, *Field Trip Series Editor*

Volcanism and Plutonism of Western North America

Volume 1

American Geophysical Union, Washington, D.C.

Library of Congress Cataloging-in-Publication Data

Volcanism and plutonism of western North America.

1. Volcanism—West (U.S.)—Guide-books. 2. Intrusions (Geology)—West (U.S.)—Guide-books. 3. West (U.S.)—Description and travel—1981- —Guide-books. I. American Geophysical Union.
QE524.V64 1989 551.2'1'0978 89-15054
ISBN 0-87590-681-8

Printed in the United States of America.

TABLE OF CONTENTS

PREFACE

Virtually every geologic province in the United States is encountered in these combined volumes of the field trip guides from the 28th International Geological Congress. Some guides extend into adjoining parts of Canada, Mexico, the Bahamas and the Caribbean. One provides an extensive view of the Scotia Arc, Antarctica. The length of trips and their complexity reflect the great variety of terranes of the North American continent.

These volumes combine trips with similar geography and geology. The main divisions are geographical. Within them an attempt has been made thematically to group various geologic topics.

Field trips for the 28th International Geological Congress (IGC) were conceived and developed with the cooperation of all segments of the Earth Science community in the United States. The Field Trip Committee solicited proposals for trips from all academic institutions, from State and Federal geological surveys, and through a number of widely circulated professional journals. About 250 proposals were received, of which the Committee selected 188 for listing in the IGC First Circular; 100,000 copies of this Circular were distributed in January 1987. Responses indicated that about 140 of these trips were of sufficient interest to go ahead with planning and preparation of guidebooks. Of these, 126 have been published in these volumes and as separate guidebooks.

We feel these guidebooks represent a valuable compendium of the current status of knowledge of the regional geology of the United States. We hope that you, the reader, will use and enjoy these volumes. We believe that publication of the guidebooks is a fitting tribute to the leaders who gave so generously of their time and talents in planning trips and preparing guidebooks. It has been our pleasure to work with them.

For the 28th International Geological Congress

John C. Reed, Jr., *Vice President, Field Trips*
Juergen Reinhardt, *Co-Chairman, Field Trip Committee*
Penelope M. Hanshaw, *Co-Chairman, Field Trip Committee*
Robert M. Mixon, *Chairman, Local Field Trips*
Tom Freeman
Robert N. Ginsburg
A. R. Palmer
J. Keith Rigby
John Rodgers
P. K. Sims

Geological Field Guide to the Hawaiian Islands

Hilo to Honolulu, Hawaii
July 1–7, 1989
July 21–27, 1989

Field Trip Guidebook T188/304

Leaders:
David A. Clague *Richard W. Hazlett*

American Geophysical Union, Washington, D.C.

2000 Florida Ave., N.W., Washington, D.C. 20009

ISBN: 0-87590-562-5

Printed in the United States of America

COVER Space Shuttle image of the Hawaiian Islands looking towards the east. Kauai (center left) and Oahu (center) are partially obscured by clouds, but Niihau (lower left) Molokai, Lanai, Maui, Kahoolawe, and Hawaii (upper right) are mainly clear.

Leaders:

David A. Clague
U.S. Geological Survey
345 Middlefield Rd.
Menlo Park, CA 94025

Richard W. Hazlett
Department of Geology
Pomona College
Claremont, CA 91711

IGC FIELD TRIP T188/304: GEOLOGICAL FIELD GUIDE TO THE HAWAIIAN ISLANDS

David A. Clague
U.S. Geological Survey, Menlo Park, California

Richard W. Hazlett
Department of Geology, Pomona College, Claremont, California

GEOLOGIC SETTING

The Hawaiian Islands form the youngest part of a chain of volcanoes that stretches nearly 6000 km across the north Pacific Ocean (Fig. 1). This unique geological feature consists of at least 107 individual volcanoes having a combined volume greater than 1 million km^3 (Bargar and Jackson, 1974). The chain is age progressive with still-active volcanoes at the southeast end whereas those at the northwest end have ages of about 75-80 Ma. The volcanoes of the chain were erupted onto the seafloor without regard for the age or structure of the underlying ocean crust. The Hawaiian volcanic ridge is surrounded by a symmetrical depression, the Hawaiian Deep, as much as 0.7 km deeper than the adjacent seafloor. The Hawaiian Deep is in turn surrounded by the broad Hawaiian Arch. Clague and Dalrymple (1987) review the setting and age relations of the volcanoes that comprise the chain.

At the southeast end of the chain lie the eight principal Hawaiian Islands (Fig. 2). The Island of Hawaii includes the active volcanoes of Kilauea and Mauna Loa. Loihi Seamount, located about 30 km off the southeast coast of Hawaii, is also active and is considered to be an embryonic Hawaiian volcano (Moore et al., 1982). Hualalai Volcano on Hawaii and East Maui Volcano on Maui have erupted in historical times. West of Niihau, only a few of the volcanoes rise above sealevel as small volcanic islets and coral atolls. West of Kure Island, the volcanoes are entirely submerged beneath the sea. The chain has a pronounced bend in the middle that was caused by a change in the direction of motion of the Pacific plate at 43.1±1.4 Ma. Northwest of this bend the chain is called the Emperor Seamounts.

It is now clear that this remarkable feature was formed during the last 75-80 m.y. as the Pacific lithospheric plate moved north and then west-northwest relative to a melting anomaly that is located in the asthenosphere, and is commonly called the Hawaiian hotspot. According to the hot-spot hypothesis, a trail of volcanoes was formed and left on the ocean floor as each new volcano was progressively cut off from its source and a new volcano was born behind it.

Hawaiian volcanoes erupt lavas of distinct chemical compositions during four different stages in their evolution and growth. The three later stages are well studied and documented (Stearns, 1940a; Macdonald and Katsura, 1964; Macdonald, 1968), but the first stage, which includes the early submarine history of the volcano, has only been examined recently (Moore et al., 1982). This earliest stage, the alkalic preshield stage, is characterized by submarine eruptions of alkalic basalt and basanite followed by eruption of transitional and tholeiitic basalt. These lava types probably are interbedded during this stage, but the trend is toward eruption of less alkalic lavas with time. The second stage, the tholeiitic shield stage, is characterized by eruption of an enormous volume of tholeiitic basalt and tholeiitic picrite that forms the shield volcanoes. The third stage, the alkalic postshield stage, occurs after the construction of the shield and consists of a cap of alkalic basalt and related differentiated lavas (ankaramite, hawaiite, mugearite, benmoreite, and trachyte) that erupt from the summit and from vents commonly aligned along the rift zones developed during the preceeding shield stage. The final stage, the alkalic rejuvenated stage, follows a period of volcanic quiescence lasting from 2.5 to <0.4 Ma and is characterized by infrequent small-volume eruptions of alkalic basalt, basanite, nephelinite, and melilitite from vents unassociated with preexisting structures. All the volcanoes had a shield stage and are inferred to have had a preshield stage, but many did not undergo both postshield and rejuvenated stage activity. The younger volcanoes may not yet have had a chance to develop to the postshield stage (Kilauea and Mauna Loa) or to the rejuvenated stage (Kilauea, Mauna Loa, Hualalai, Mauna Kea, and Kohala). These stages are shown schematically in Fig. 3 and the age relations of the lavas of the different stages from the different volcanoes are shown in Fig. 4. The relation between lithostratigraphic units and eruptive stages are shown in Fig. 5. Table 1 summarizes the volume, elevation, and surface area of the different volcanoes and islands.

In a general way, the volumes of lava produced during each stage reflect the magma supply rates during the sequential stages. Thus, Hawaiian volcanism commences with infrequent eruptions of small-volume alkalic preshield-stage lava. The tholeiitic shield stage consists of frequent eruptions of large volume. The alkalic postshield stage again consists of infrequent eruptions of small volume. The strongly alkalic rejuvenated stage consists of very infrequent eruptions of very small volume. These changes in magma supply rate, and therefore of heat transfer from the mantle to within the volcano, result in an evolution of the magma storage system beneath the volcano (Clague, 1987a).

This fieldguide emphasizes the evolution of

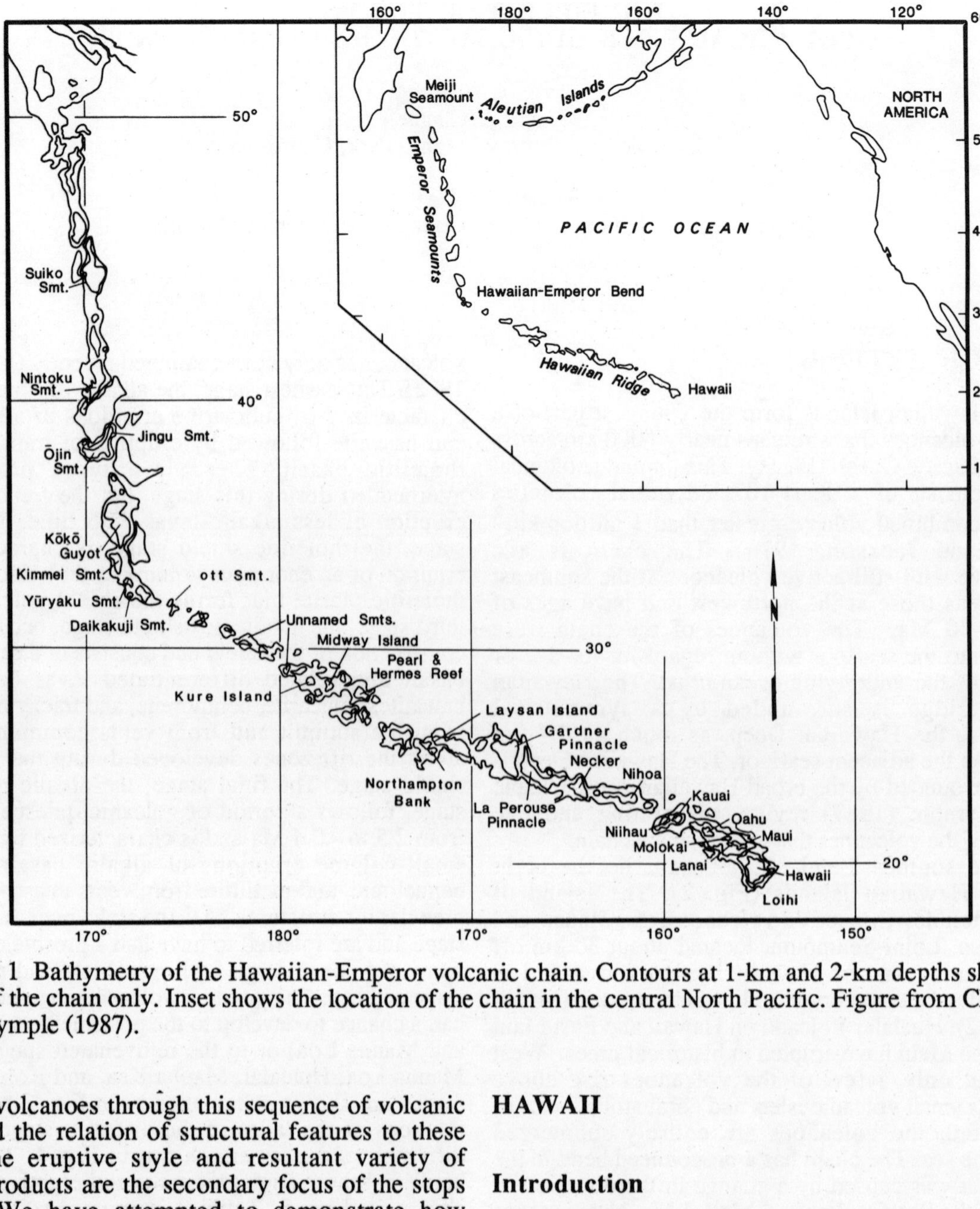

FIGURE 1 Bathymetry of the Hawaiian-Emperor volcanic chain. Contours at 1-km and 2-km depths shown in area of the chain only. Inset shows the location of the chain in the central North Pacific. Figure from Clague and Dalrymple (1987).

Hawaiian volcanoes through this sequence of volcanic stages, and the relation of structural features to these stages. The eruptive style and resultant variety of volcanic products are the secondary focus of the stops selected. We have attempted to demonstrate how Hawaiian volcanoes are similar to one another in their development and chemistry, and how they differ. We begin at Kilauea which is in the tholeiitic shield stage; it is the most active volcano in Hawaii. We then goes to Mauna Loa, the other Hawaiian volcano in the tholeiitic shield stage, and then proceed to examine the alkalic postshield stage at Mauna Kea, Kohala, and Hualalai volcanoes. An intra-island flight is used to highlight the geology of Maui, Molokai, Lanai, Kahoolawe, and Niihau. The shield stage and development of calderas are highlighted on Kauai as is the alkalic rejuvenated stage. The trip concludes with a look deep into the dikeswarm of Koolau Volcano on Oahu, and examination of ash deposits and their enclosed xenoliths from phreatomagmatic eruptions of the rejuvenated stage Honolulu Volcanics, also on Koolau Volcano.

HAWAII

Introduction

The island of Hawaii consists of five volcanoes- Kilauea, Mauna Loa, Mauna Kea, Kohala, and Hualalai (Fig. 6). Two additional volcanoes are part of the Hawaii volcanic complex- Loihi Seamount off the southeast coast is an embryonic Hawaiian volcano, and Mahukona Volcano located to the west of Kohala Volcano is an extinct volcano. The summit of Mahukona volcano was subaerial at one time but has now subsided beneath the sea. Kilauea and Mauna Loa volcanoes are in the tholeiitic shield stage and erupt frequently, Hualalai and Mauna Kea Volcanoes are in the alkalic postshield stage and erupt every few hundred to few thousand years, and Kohala is dormant, having passed through the alkalic postshield stage. Kohala last erupted about 120 ka, Mauna Kea about 3.6 ka, and Hualalai in 1800-1801. Representative analyses of lavas from the five volcanoes

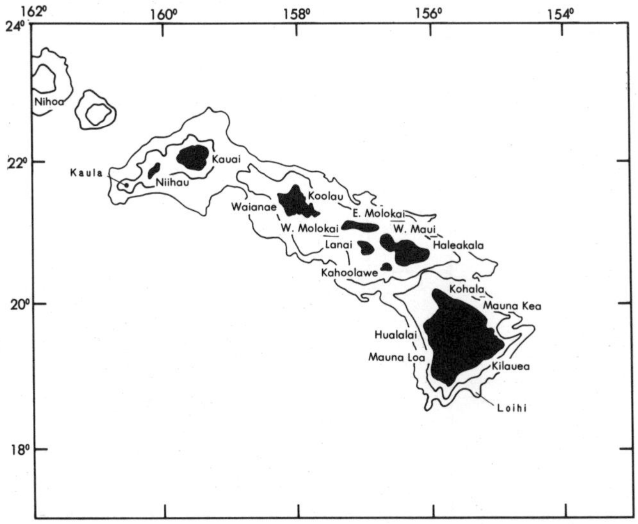

FIGURE 2 The principal islands and volcanoes of the Hawaiian Islands are shown with 1-km and 2-km isobaths. Loihi is a submarine nascent Hawaiian volcano.

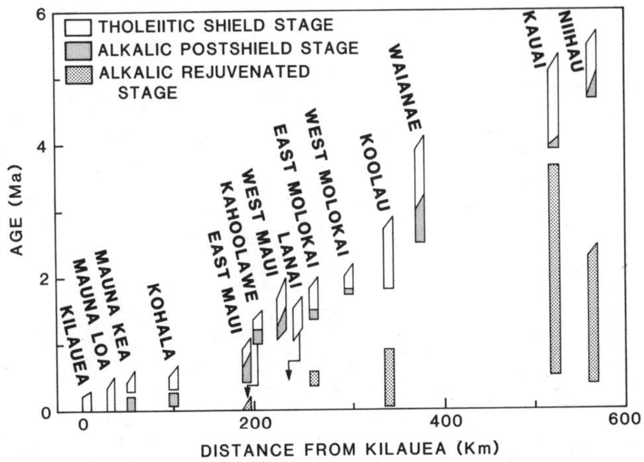

FIGURE 4 Age-distance plot for the Hawaiian Islands. The age data are subdivided according to eruptive stage. The Kahoolawe and Lanai data are shifted along the axis as indicated by the arrows to avoid overlap on the plot. Figure from Clague and Dalrymple (1988).

that make up Hawaii are presented in Table 2.

Kilauea Volcano

Kilauea Volcano has been historically the most active and intensively studied Hawaiian volcano. The subaerial portion of Kilauea covers approximately 1500 km^2. About 90% of the surface is composed of flows <1100 years old (Holcomb, 1987). Kilauea is in the shield stage and all the lavas are tholeiitic basalt, olivine tholeiitic basalt, and picritic tholeiitic basalt. The lavas are subdivided into the older Hilina Basalt and the younger Puna Basalt which are separated by the Pahala Ash.

A small caldera is located just south of the summit and two rift zones radiate seaward from the caldera towards the southwest and the east (Fig. 7). Five eruptions have occurred in the southwest rift zone historically, whereas 26 have occurred in the east rift zone. There have also been over 30 historic eruptions in or near the caldera and periods of sustained lava lake activity at Halemaumau, the principal vent inside Kilauea caldera.

Magma originates in a zone of partial melting at least as deep as 40 km beneath Kilauea (Fig. 8). The magma accumulates in a sponge-like magma chamber extending from 2 to 7 km below the southern portion of the caldera floor (Ryan et al., 1981). From this chamber, magma is supplied to eruptions in the caldera area and in the rift

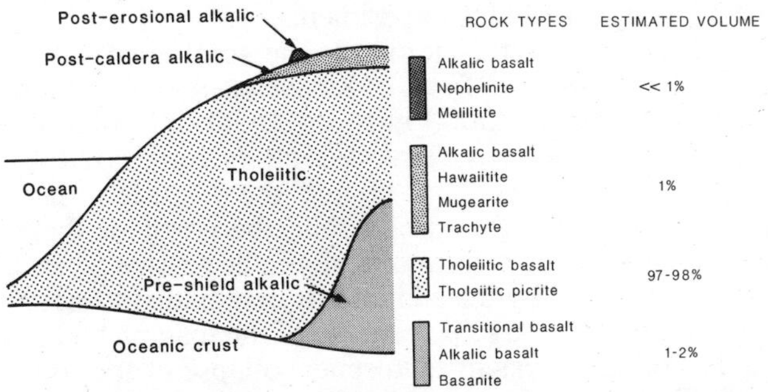

FIGURE 3 Cross-section through a hypothetical Hawaiian volcano showing the age and volume relations of the four eruptive stages. The main rock types and their relative volumes are also shown. Figure from Clague (1987b).

TABLE 1 Sizes of Hawaiian Volcanoes

Volcano	Volume (x10^3 km)	Elevation (feet)	Elevation (km)	Area (mi^2)
Hawaii				4038.0
Kilauea	19.4	4090	1.25	
Mauna Loa	42.5	13677	4.17	
Mauna Kea	24.8	13796		4.20
Hualalai	12.4	8271	2.52	
Kohala	14.0	5480	1.67	
Maui				728.8
East Maui	29.3	10023	3.05	
West Maui	6.4	5788	1.76	
Kahoolawe	8.8	1477	0.45	45.0
Lanai	12.0	3370	1.03	139.5
Molokai				261.1
East Molokai	16.6	4970	1.51	
West Molokai	17.0	1381	0.42	
Oahu				607.7
Koolau	20.9	3150	0.96	
Waianae	25.0	4040	1.23	
Kauai	29.4	5243	1.60	553.3
Niihau	17.0	1281	0.39	73.0

Volume data from Bargar and Jackson (1974), elevations and areas from Atlas of Hawaii (1973).

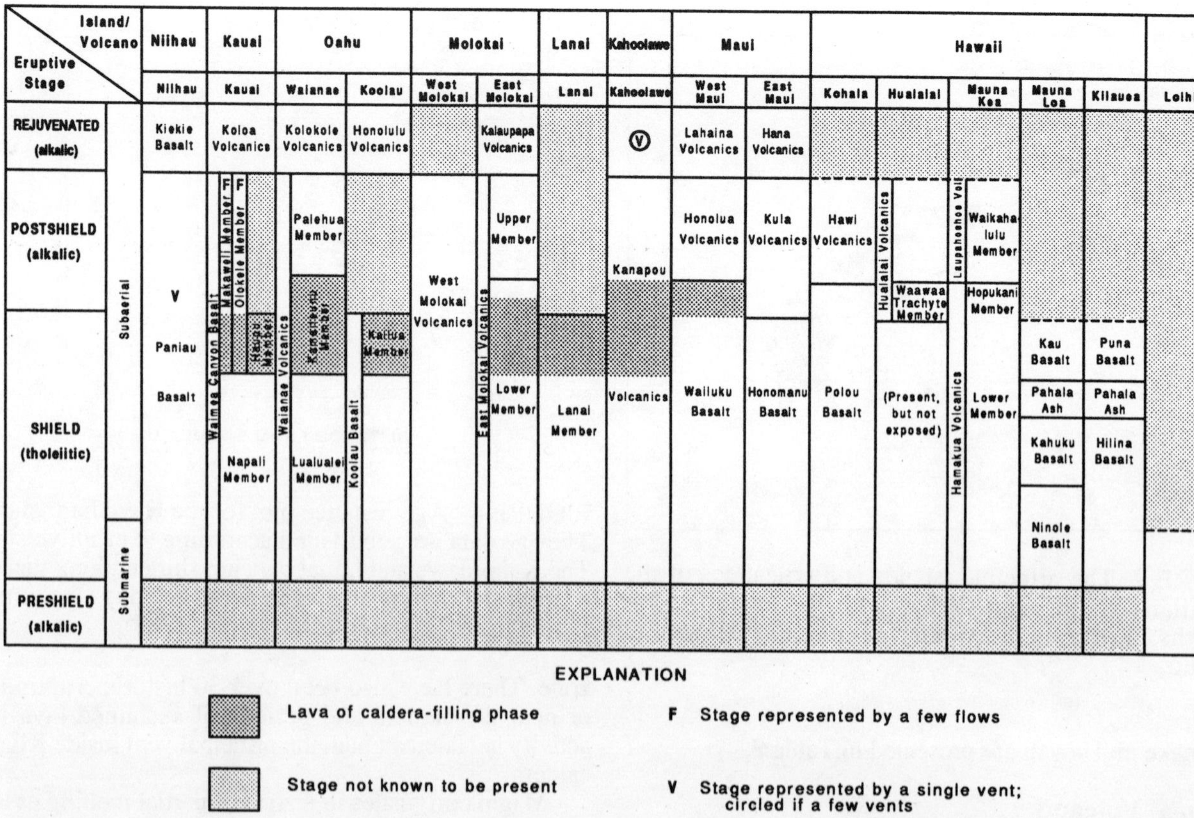

FIGURE 5 Chart showing relations of eruptive stages to volcanoes and stratigraphic units from the southeastern Hawaiian Islands, arranged approximately from northwest (left) to southeast (right). No age correlation between units on different volcanoes is implied. Dashed lines indicate that stage may not be complete. Base of oldest unit on each volcano assumed to be the base of the shield stage (except on Loihi). Figure from Langenheim and Clague (1987).

zones. Dzurisin et al. (1983) conclude that the magma supply rate from the mantle is nearly constant and that about 50% of the magma never reaches the surface but remains entrapped in storage as intrusions. Wright and Helz (1987) review recent work on the melting process, and the nature of the magma source in the mantle beneath Kilauea.

Most of the eruptions at Kilauea are characterized by mild fissure-fed lava fountaining, the type "Hawaiian" activity. Such eruptions are usually brief, lasting only days or weeks. Occasionally, a fissure eruption evolves into sustained lava lake activity, concentrated at one location along the original fissure. A lava lake may remain active for months or years, forming broad lava fields and building up a broad mound, or lava shield, around itself due to repeated short-term overflows. Rarely, phreatomagmatic eruptions occur at Kilauea. The Keanakakoi Ash Member of the Puna Basalt, a pyroclastic deposit rimming the southern margin of the Kilauea caldera, is the product of one such eruption in 1790.

Several prominent fault systems cut the flanks of Kilauea. The Koae fault system is a series of en-echelon, down-to-the-north normal faults linking the southwest with the east rift zones (Fig. 7). Swanson et al. (1976) attribute the formation of the Koae fault system to forceful injection of dikes into the east rift zone. The dikes displace the south flank of the volcano in a S15E direction. Because the summit region is not directly tied to this strain, but remains relatively stationary, a tear-away structure--the Koae faults--has developed, influenced by the shape of the bend in the east rift zone. Duffield (1975) modelled the formation of the Koae fault system using clay-block experiments.

The Hilina fault system cuts the southern flank of Kilauea. The southern flank of Kilauea descends without lateral support over 5000 m to the seafloor, unlike the more stable northern flank which is buttressed by neighboring Mauna Loa. Hence the southern flank undergoes more severe strain than the northern flank, occasionally failing catastrophically in large earthquakes. Strain is imposed by the addition of magmatic intrusions which represent a large percentage of the magma budget. The Hilina fault system represents collapse of the south flank. The faults are normal, with mostly down-to-the-south displacements. Some escarpments rise as much as 300 m. At depth, the largest faults may have a listric geometry, flattening seaward. A low angle shear zone may underlie most of the southern flank of the volcano near the base of the volcanic pile, accomodating its

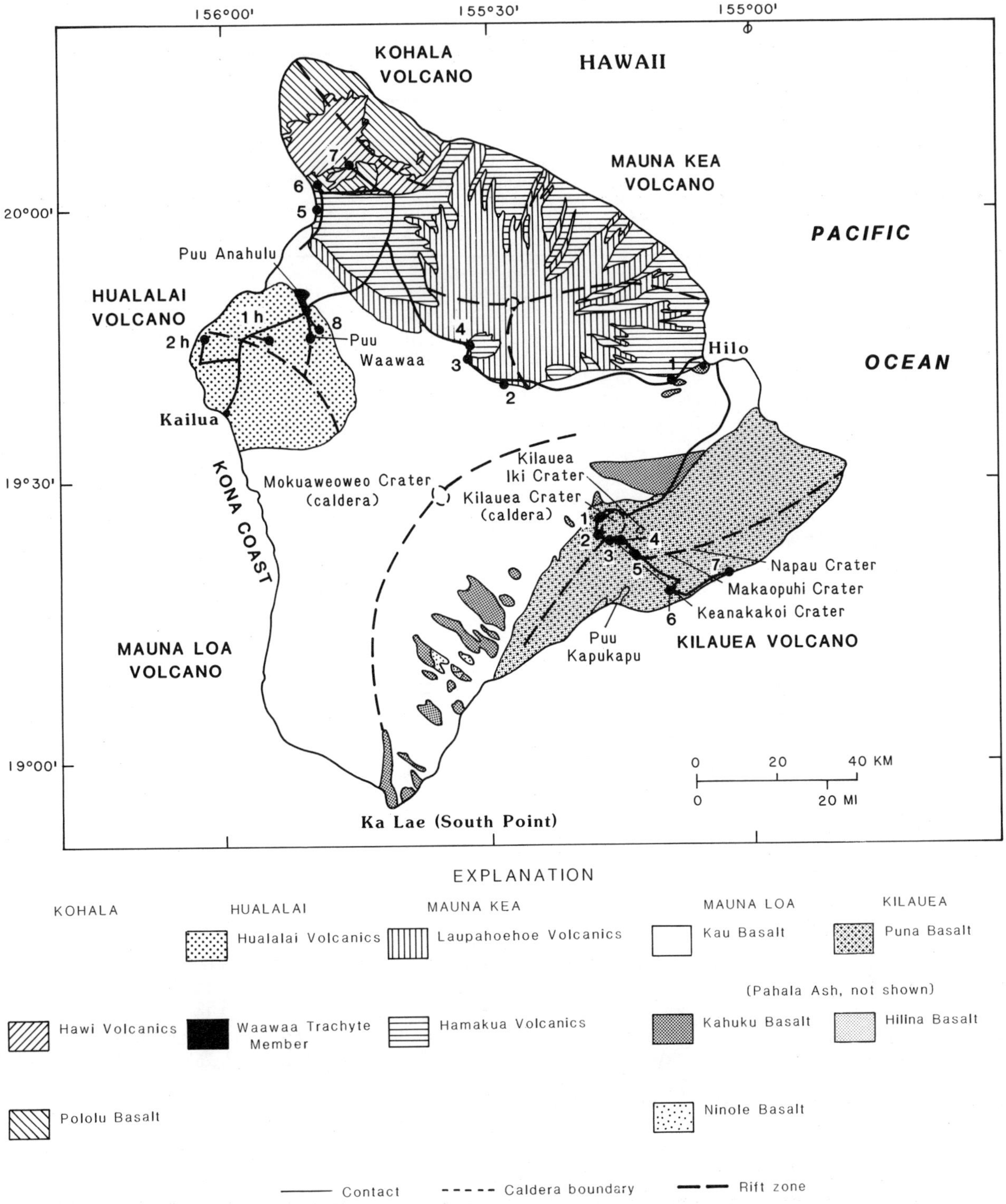

FIGURE 6 Generalized geological map of the Island of Hawaii. Figure from Langenheim and Clague (1987) which should be consulted for primary references.

movement away from the rift zones (Crosson and Endo, 1982).

Six field stops have been selected to provide a general reconnaisance of the major features of Kilauea, as outlined above. Site descriptions given below are extracted and updated from a more comprehensive field guide to the geology of Kilauea by Hazlett (1987).

Roadlog to Kilauea Volcano, Hawaii - Day 1.

Stop 1. Hawaiian Volcano Observatory, Jagger Museum of Volcanology, and Caldera Overlook. The Hawaiian Volcano Observatory was established in 1912 by Thomas A. Jagger, who came to Hawaii from the Massachusetts Institute of Technology

TABLE 2 Chemical analyses of lavas from Hawaii.

Volcano	Kilauea		Mauna Loa		Hualalai			
Unit	Puna Basalt		Kau Basalt		Hualalai Volcanics			
Stage	Shield Stage		Shield Stage		Shield Stage		Postshield Stage	
Lava Type	Tholeiitic Basalt	Picritic Tholeiitic Basalt	Tholeiitic Basalt	Picritic Tholeiitic Basalt	Tholeiitic Basalt	Picritic Tholeiitic Basalt	Alkalic Basalt	Trachyte
ID#	HM68-15	OXO78-0	1949ML	1868ML	KK-14	KK7812	65HU-10	67-101A
SiO_2	50.25	47.97	52.04	46.75	50.6	46.8	46.62	63.10
Al_2O_3	13.61	10.48	13.94	8.40	13.4	8.48	14.72	17.79
Fe_2O_3	1.49	1.67	1.58	2.29	1.99	1.29	2.00	2.24
FeO	9.86	10.04	9.69	9.27	9.10	10.4	10.71	1.35
MgO	7.63	17.03	7.14	23.58	7.78	23.0	9.10	0.52
CaO	11.14	8.30	10.63	6.21	10.6	6.49	10.04	0.79
Na_2O	2.32	1.67	2.25	1.32	2.15	1.36	2.82	7.63
K_2O	0.55	0.31	0.33	0.26	0.34	0.20	0.91	4.97
TiO_2	2.62	1.98	2.05	1.34	2.07	1.33	2.37	0.50
P_2O_5	0.26	0.18	0.26	0.14	0.19	0.14	0.27	0.12
MnO	0.17	0.17	0.18	0.17	0.17	0.12	0.19	0.52
CO_2	0.00	0.01	0.00	0.01	0.08	0.04	0.01	0.00
H_2O+	0.06	0.14	0.01	0.00	0.32	0.20	0.00	0.26
H_2O-	0.00	0.04	0.02	0.02	0.16	0.08	0.04	0.04
Total	100.00	100.02	100.13	99.78	99.0	100.0	99.88	99.71

Volcano	Mauna Kea				Kohala				
Unit	Hamakua V.	Laupahoehoe V.			Pololu Basalt	Hawi Volcanics			
Stage	Shield	Postshield			Shield	Postshield			
Lava Type	Tholeiitic Basalt	Alkalic Basalt	Ankaramite	Hawaiite	Tholeiitic	Alkalic Basalt	Hawaiite Basalt	Mugearite	Benmoreite
ID#	20	C-77	C-67	C-79	W9	KH11	MG9a	MG4b	KH1
SiO_2	47.94	46.65	44.54	49.79	50.32	47.99	51.13	53.18	59.04
Al_2O_3	10.43	14.72	13.52	18.37	12.73	15.97	13.61	18.15	18.37
Fe_2O_3	2.03	3.63	4.47	6.03	-	-	-	-	-
FeO	9.54	8.98	7.76	4.82	11.00	12.31	13.11	9.66	6.26
MgO	17.34	7.96	11.92	4.01	9.37	6.49	4.45	2.89	1.73
CaO	7.42	10.14	10.71	6.68	10.08	10.70	8.19	5.40	3.39
Na_2O	1.99	2.53	1.69	4.58	1.99	2.80	3.02	4.04	6.07
K_2O	0.39	0.80	0.59	2.08	0.30	0.80	1.10	2.11	2.71
TiO_2	1.98	3.22	2.60	2.51	2.15	3.08	3.50	2.14	1.39
P_2O_5	0.24	0.33	0.32	0.56	0.22	0.42	0.66	1.48	0.59
MnO	0.17	0.18	0.18	0.22	0.18	0.20	0.26	0.26	0.23
CO_2	0.02	-	-	-	-	-	-	-	-
H_2O+	0.49	0.63	0.52	0.32	-	-	-	-	-
H_2O-	0.09	0.51	0.77	0.24	-	-	-	-	-
Total	100.17	100.3	99.6	100.2	98.3	100.8	99.0	99.3	99.8

Mauna Kea tholeiitic basalt analysis from Moore (1966), Mauna Kea alkalic lava analyses from Macdonald (1968), all Kohala analyses from Feigenson et al. (1983).

attracted by the unique opportunities for volcanological research made available by the safety of access, and convenience of study of Kilauea's frequent eruptions. The Observatory has been directed by the U. S. Geological Survey since 1948. Twenty-five to thirty scientists and technicians assigned to the observatory carry out a broad range of geophysical, geochemical, and geological studies. Heliker et al. (1986) describes the history and operations of the observatory and Wright and Swanson (1987) describe the current volcanological studies underway at Kilauea. A valuable reference to older research is the collection of Volcano Letters reprinted by the Smithsonian Institution (Fiske et al., 1987). Popular publications about the volcano are for sale in the Jagger Museum, together with exhibits illustrating the techniques used by the Observatory in its studies of Kilauea.

From the terrace besides the Jagger Museum, the

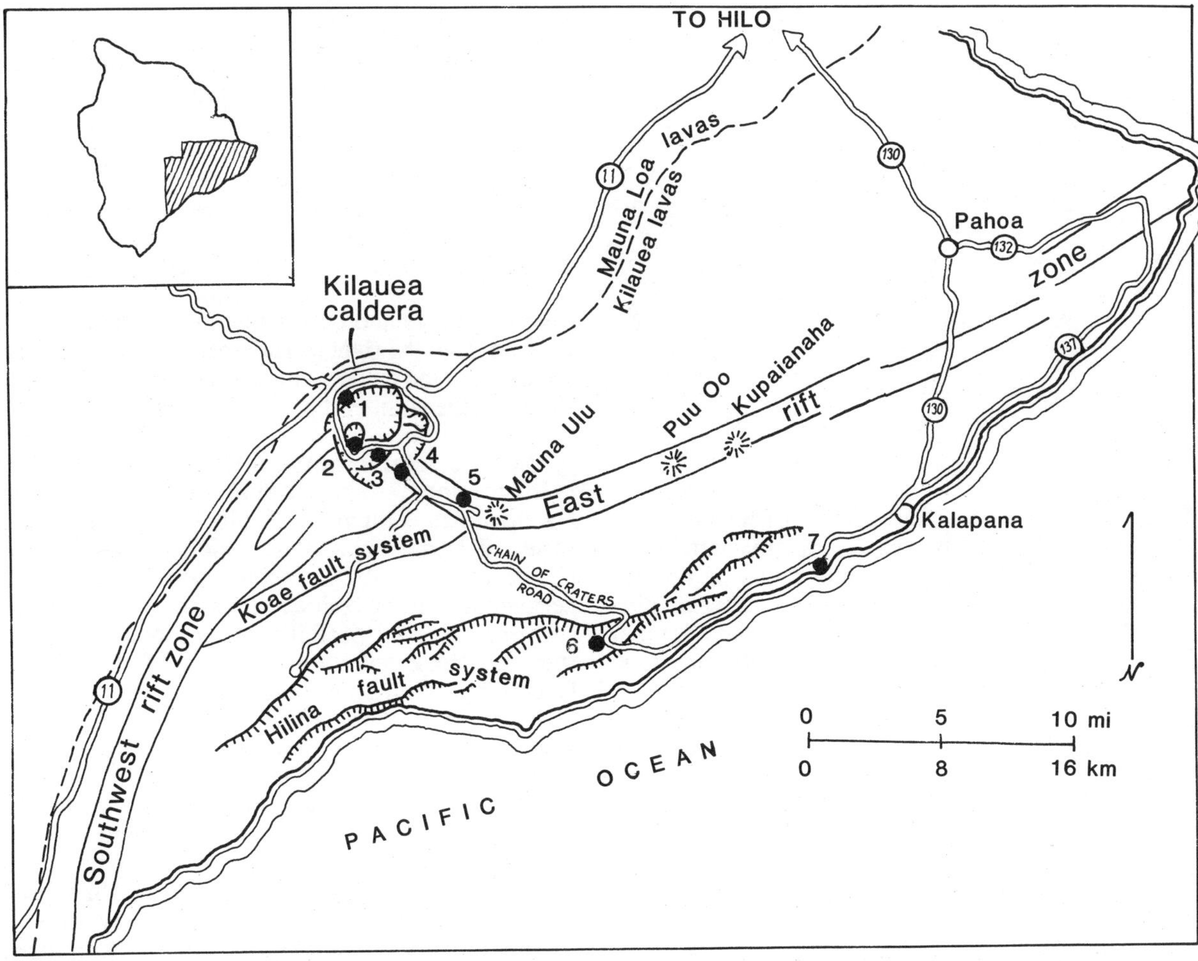

FIGURE 7 Kilauea location map showing the locations of the stops discussed in the roadlog (black dots), and major structural elements of Kilauea Volcano.

view of Kilauea caldera and Halemaumau Crater is superb. Kilauea caldera measure 3.25 by 4.75 km, and is breached on the southwest side. On the east side of the caldera, near the Volcano House, vertical walls rise 120 m from floor to rim. The portion of Kilauea caldera you see may have formed shortly before the catastrophic eruption of 1790, the source of the Keanakakoi Ash Member. Since that large eruption, the history of the caldera has been one of rapid infilling, interupted by intervals of renewed collapse. Upward development of the caldera floor proceeded most rapidly in the 19th century, primarily by sustained lava lake overflows (Brigham, 1909). When the party of Reverend William Ellis viewed the caldera in 1823, most of the caldera floor was over twice as deep as it is today. An inner sink, about 90 m deep, was separated from the crater walls by a mezzanine termed the "black ledge." Within the sink, especially toward its southern end, were numerous active spatter cones and a set of lava lakes (Ellis, 1827, reprinted 1963, Fig. 9). By the end of 1831, the inner sink was completely filled. However, by September 1832, the caldera floor had again collapsed, forming a new black ledge.

The new sink refilled slowly, supplied by lava from lakes at the southwestern and northwestern caldera floor. In April 1868, a powerful (M7.5?) earthquake disrupted the magma chamber underlying the caldera and the floor once again collapsed. Dr. William Hildebrand was among the first to report this event (Brigham, 1909, p. 106): "More than 2/3 of the old floor of Kilauea has caved in, and sunk from one hundred to three hundred feet below the level of the remaining floor. . . .The caving in of the floor seemed to be still in progression, for twice during our exploration of the crater, our nerves were disturbed by a prolonged heavy rumbling and rattling noise, as from a distant platoon-file of musketry."

By 1871, the new sink was being refilled from lava lakes in the Halemaumau area. Over the next five years, the sink was filled to the brim, and a large, gently sloping lava shield, with Halemaumau at its summit, grew rapidly over the entire caldera floor. The top of the shield rose above the southwestern rim of the caldera by 1894, but in July of that year, the summit of the shield collapsed. Except for brief eruptions within the sulphurous, funnel-shaped collapse pit at Halemaumau, lava lake activity did not resume until November 1905 (Brigham, 1909).

From 1905-1924, Halemaumau crater almost continuously contained a churning lava lake as much as

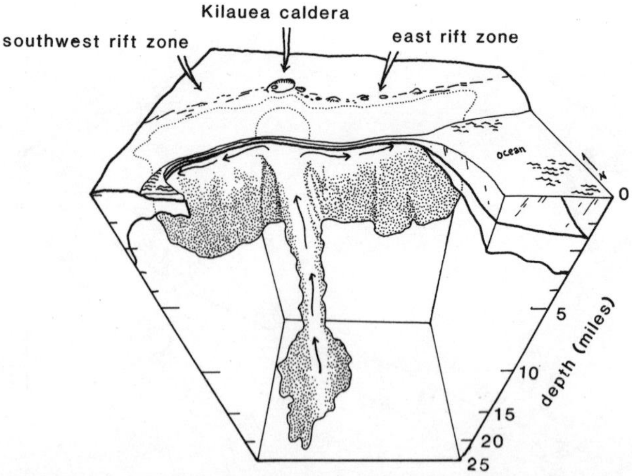

FIGURE 8 A schematic cross-section of Kilauea. The stippled pattern highlights areas of magma intrusions. Arrows in this field show principal paths of magma movement: first from the mantle to a shallow magma chamber beneath the summit, then from the chamber laterally into the rift zones. Figure is modified from Ryan (1987).

500 m across. Occasionally the lake would overflow onto the surrounding caldera floor, but most of the time it remained confined by levees within the crater. At its highest level, the lava lake reached an elevation equal to that of the Volcano House Hotel (Howard A. Powers, pers. commun., 1986).

In 1924, the lava lake drained away. A series of steam blasts ensued, doubling the diameter of Halemaumau to about its present size. Except for a seven-month interlude in 1967-1968, the Halemaumau lava lake has not returned, nor has the level of the surrounding caldera floor changed significantly. Following a dormant interval lasting from 1934 to 1952, the principal eruptive activity of Kilauea shifted from its caldera into the east rift zone.

A series of small, brief fissure eruptions from 1954-1982 covered most of the caldera floor with a new veneer of lava, and constructed several low spatter ramparts near the rim of Halemaumau (Fig. 9).

Near the rim of Kilauea caldera are two satellitic pit craters, Keanakakoi and Kilauea Iki, and an oxidized cinder and spatter cone, Puu Puai, that formed downwind from the 1959 Kilauea Iki eruptive vent (Fig. 8). The southwest rift zone is visible as a line of steaming vents and cones extending beyond the rim of the caldera to the right of Halemaumau. The east rift zone trends from the caldera rim towards Mauna Ulu, the lava shield on the southeast horizon.

Stop 2. Halemaumau Crater and the central caldera floor. The ground alongside the trail to Halemaumau is littered with blocks deposited during the series of steam explosions which jarred Halemaumau Crater in May, 1924. Ejected rocks may be divided petrographically into sintered, thermally metamorphosed, and occasionally partially remelted lava, and coarse intrusive rock analagous to that of the Uwekahuna laccolith, an intrusive body which we shall see later. Sparse blocks of tholeiitic basalt with phenocrysts of augite, and rare biotite-bearing and two-pyroxene gabbro also occur. Macdonald (1944) and Chapman (1947) describe the petrography of these samples in detail.

As described at Stop 1, Halemaumau has been the historical focus of summit eruptive activity. It was the principal vent of Kilauea volcano from 1823-1924, and represents the shortest, and frequently most direct conduit to the summit magma reservoir. Prolonged lava lake activity in Halemaumau ended with the 1924 steam blast eruptions, though 18 eruptions, ranging from less than a day to eight months, have since occurred in the crater. In May 1924, after the first steam blast, the floor of Halemaumau was over 400 m deep. Gradually the floor has been built up by repeated accumulation of new

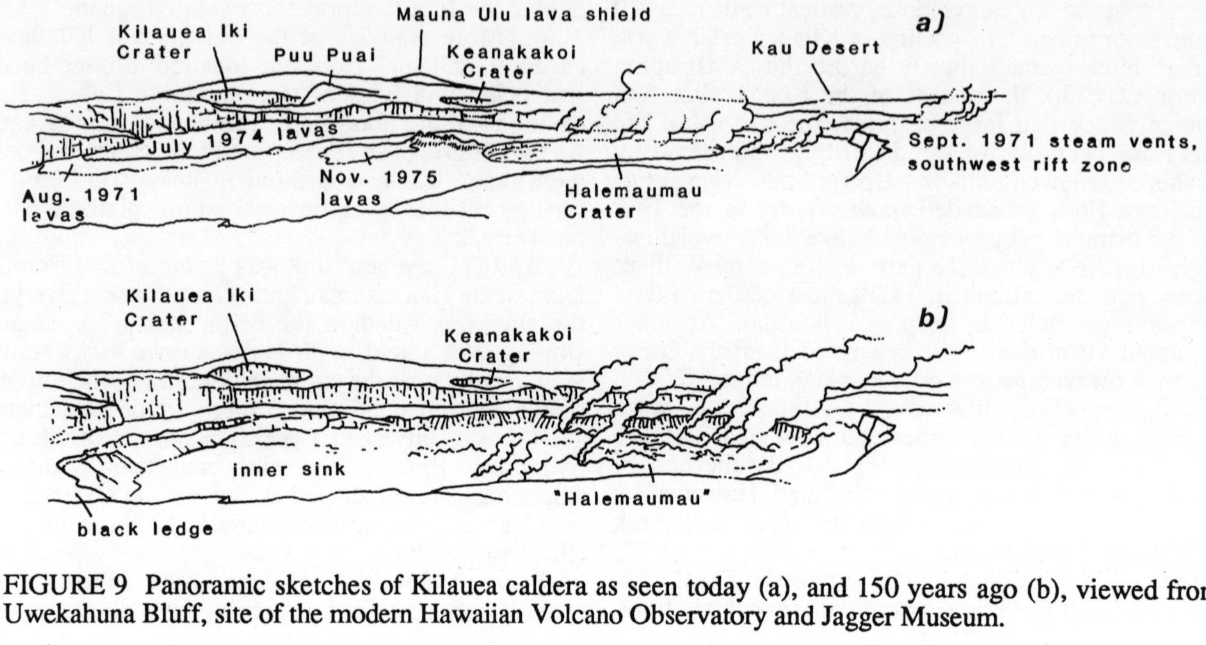

FIGURE 9 Panoramic sketches of Kilauea caldera as seen today (a), and 150 years ago (b), viewed from Uwekahuna Bluff, site of the modern Hawaiian Volcano Observatory and Jagger Museum.

flows, interspersed with four episodes of collapse triggered by sudden draining of the underlying magma chamber. Today, the floor of Halemaumau lies 85 m beneath its rim.

Most of the lava making up the floor of Halemaumau was erupted in September 1974. Fissures that fed this eruption are discernable on the opposite (northwest) wall of the crater. The broken rim around the margin of the crater floor indicates drainback of lava after the eruption. The prominent circumferential ledge visible on the wall is the remnant of a former crater floor (representing a high stand of a lava lake active in 1967-1968) that collapsed 45 m in September 1971.

Exhalation of gases in Halemaumau commonly occurs in fractures in the weakened crater walls rather than through the floor, which is plugged by relatively impervious lava.

A few hundred meters beyond the Halemaumau Overlook, following the Halemaumau Trail towards Volcano House, one encounters a series of four parallel, nearly overlapping spatter ramparts related to eruptions occurring between 1954 and 1982. The oldest rampart, formed in May-June 1954, is yellow-brown due to intensive palagonitization. Vents stand out as oxidized, reddish cavities. Where overhung with spatter, vent walls bristle with stalactites. From the vantage point of these ramparts, looking northward, one may see in cross-section, exposed in the crater wall, the shallow internal structure of the summit of Kilauea. Numerous thin basalt flows are punctuated by fractures and intrusive bodies.

The prehistoric Uwekahuna laccolith is exposed in the caldera wall as two light-colored lenticular bodies a short distance to the right and beneath the Volcano Observatory, near the caldera floor (Fig. 10). The laccolith measures 25 by 295 m, and is partly buried by landslide talus dislodged by an earthquake in 1983. Although superjacent flow bedding is arched up over the intrusive body, the bedding is sharply broken elsewhere (Macdonald et al., 1983). Murata and Richter (1961) showed that the lava in the laccolith differentiated into tholeiitic picrite, tholeiitic olivine gabbro, and an aphanitic rock approaching quartz basalt in composition. Mechanisms involved were an initial settling of olivine and a final filter pressing of the residual liquid. The range of compositions is as great as that of all previously analyzed lavas from Kilauea.

Stop 3. Keanakakoi Ash Member and lava wash at Keanakakoi Crater. Near Keanakakoi Crater, a stream channel cut in ash of the Keanakakoi Ash Member became a spillway for a stream of very fluid pahoehoe erupted at the head of the channel in July 1974. As this lava flowed toward the floor of Kilauea caldera, it washed up outer walls around bends in the channel. The level of the flow through the channel rapidly subsided as the volume of lava diminished, and as the flow exited the channel onto the caldera floor below and spread out laterally. Flow structures preserved in the channel walls attest to the low viscosity and high speed of freshly erupted Hawaiian lava moving down a steep slope.

A few hundred meters south of the lava wash, excellent exposures of the Keanakakoi Ash Member crop out in the low southern walls of the caldera. Most of the layering in these exposures appears to be planar from a distance, but closer inspection reveals the presence of gentle dune structures along some horizons, indicative of rapid lateral transport rather than airfall emplacement. The deposit here represents a mixture of predominantly pyroclastic surge with pyroclastic flow and airfall tephra (Fig. 11, Swanson and Christiansen, 1973).

Precursors of large pyroclastic eruptions at Kilauea have not been monitored or observed historically. However, oral accounts from survivors of the 1790 eruption indicate that several days of intense earthquake activity preceeded the eruption (Dibble, 1843; Brigham, 1909). Stratigraphy of the Keanakakoi Ash Member in the Kau desert also shows that an interval of high fountaining, accompanied by heavy tephra fall, occurred immediately prior to the explosive discharge. The pyroclastic surges and flows spread over an area of about 200 km^2 (Malin et al., 1983; Swanson and Christiansen, 1973), mostly to the southwest of the summit, laying down a simmering blanket of vitric to lithic ash and cinder, with a maximum thickness of 5-10 m near the source vent (Halemaumau?). The uppermost layers of the deposit are rich in lithic fragments, possibly representing collapse of magma chamber or vent walls, or erosive coring of the vent during the eruptions (Fig. 11). Excellent examples of bomb and block sags may also be found near the top of this deposit.

In addition to the Keanakakoi Ash Member, one other large pyroclastic deposit is known from the stratigraphy of the Kilauea summit area. Underlying the caldera floor, and exposed locally beneath the Uwekahuna laccolith, is the Uwekahuna Ash Member,

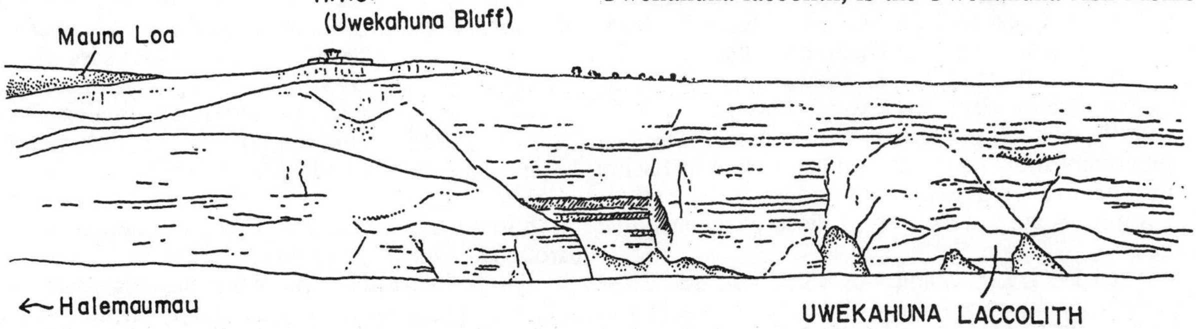

FIGURE 10 Location of the Uwekahuna laccolith, looking northwest from the central caldera floor.

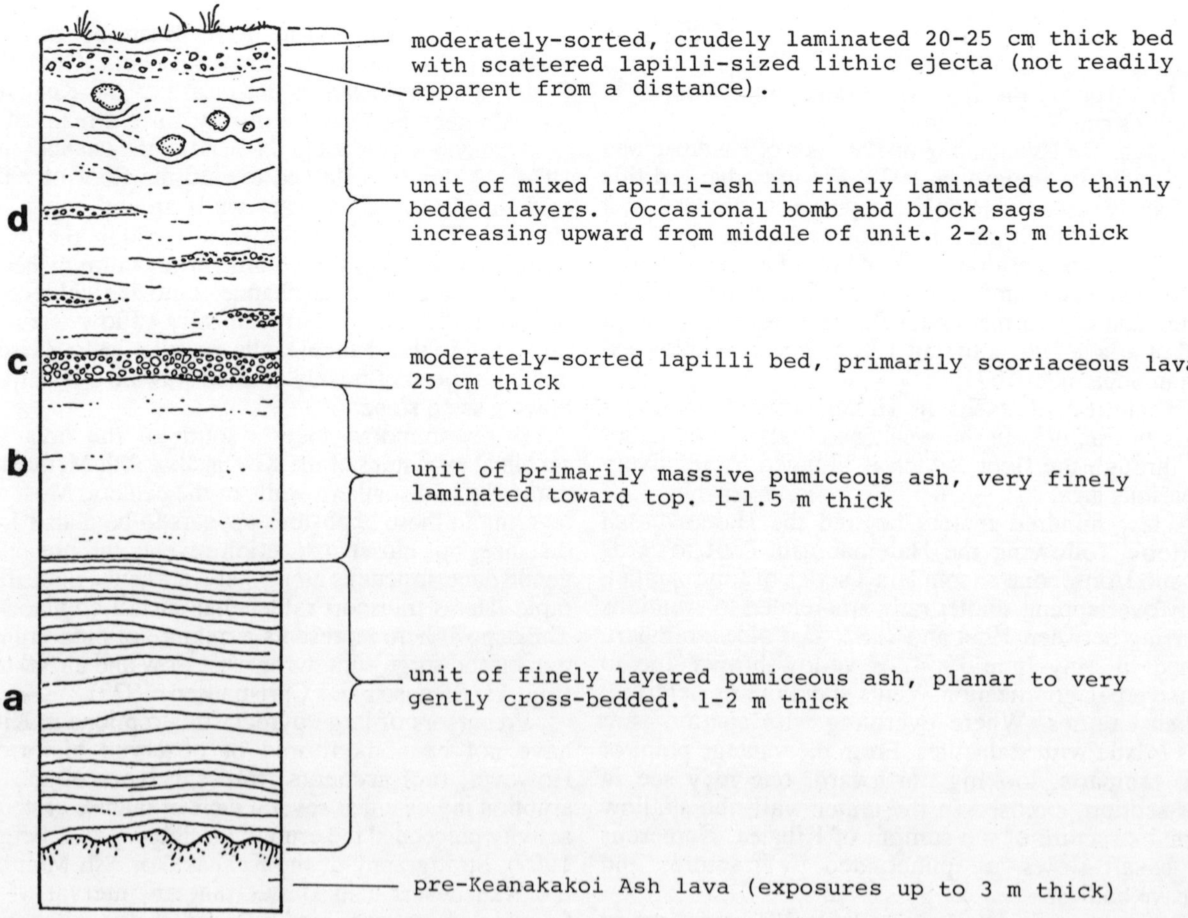

FIGURE 11 Stratigraphy of the Keanakakoi pyroclastic deposit, as seen in the southern wall of Kilauea caldera near Keanakakoi Crater. These exposures are typical of Keanakakoi Ash Member found elsewhere, though most exposures also include thin beds of pumice, Pele's hair, and Pele's tears both at the base and top of the deposit, indicating that high fountaining occurred both shortly before and after the pyroclastic eruption.

withcarbon-14 dates that fall in the range 2200-1000 y bp (Holcomb, 1987).

Stop 4. Lua Manu Crater and lava trees. Lua Manu is the first in a series of pit craters encountered as one travels down the Chain of Craters Road. These collapse pits mark the upper east rift zone. Holcomb (1987) suggests that most of the pit craters are collapse features that formed in conjunction with the 1790 eruption of the Keanakakoi Ash Member, due to sudden draining of magma stored in the upper east rift zone and summit magma reservoir. Some pits, with flat, lava-veneered floors such as Lua Manu, have been partially filled with lava since collapse, often from vents some kilometers distant. Eruptive fissures may open on the floor, walls, or even at the rim of a pit crater. The fact that pit craters show no control on the pattern of fissure development during eruptions indicates that their structural roots are not as deep as those of the eruptive fissures.

Prior to 1974, Lua Manu was a shallow, tree-filled depression, its presence largely obscured by vegetation. On July 19, 1974, the crater partially filled with lava that poured in from fissures which opened just northwest and east of the present visitor overlook. Some of the lava pooling in the crater drained back into the earth through fissures in the floor of the crater, leaving a high-level mark, or "bathtub ring" of lava in the form of a veneer on the crater walls.

During the eruption of July 19, 1974, numerous lava trees formed, as the area of the present lava flow was then heavily forested. A short walk westward several hundred meters across the 1974 flow leads one to the finest locality for examining these structures. Figure 12 illustrates the process by which the lava trees formed. Drainback textures are exposed on the edges of some of the lava trees where subsiding blocks of flow crust scraped by surfaces still hot and pasty. In some molds, charcoal from burned vegetation still remains. The native ohia trees (Metrosideros collina var. polymorpha) are moist hardwoods and lava is quickly chilled when it flows around them. For this reason there are many unburned trunks on this flow, toppled or protruding from bases enclosed in lava.

From Lua Manu, the trip continues down Chain of Craters Road. Near Hiiaka Crater, on the west (right) side of the road there is a large fresh-appearing fault scarp that is related to the Koae fault system.

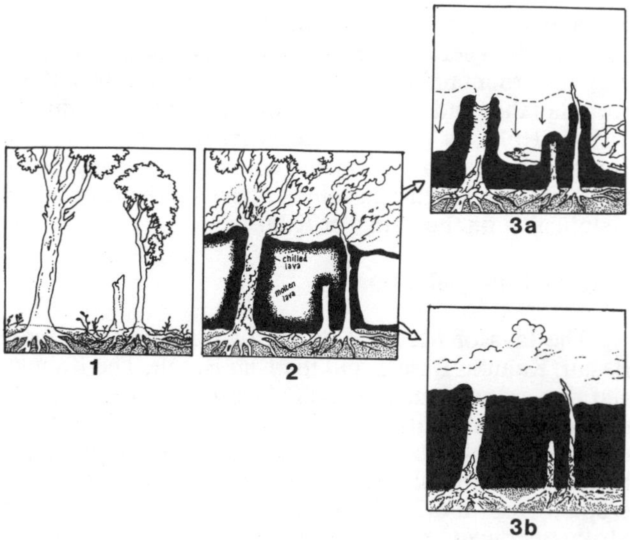

FIGURE 12 A cartoon depicting the formation of lava trees and tree molds. A flow of pahoehoe enters a forest (panels 1 and 2). The lava is chilled solid locally around tree trunks (panel 2). If the level of the flow subsides (as it frequently does near vents), the chilled lava around the tree trunk is left standing as pillar-like lava trees (panel 3a). If the level of the lava does not subside, well-like tree molds form instead (panel 3b).

Stop 5. Pauahi Crater, Mauna Ulu Overlook. Pauahi Crater (Fig. 13) is an example of a composite pit crater; at least three stages of collapse are represented in its form. The crater is approximately 500 m long and 110 m deep. It has been the site of three historical eruptions. The first was on May 5, 1973 when the earliest of the vents opened on the floor of Pauahi Crater. Activity quickly ceased here, however, and shifted upslope to the vicinity of Hiiaka Crater. The small amount of lava extruded onto the floor of Pauahi Crater was quickly buried during the second historical eruption six months later. The eruption of November 10, 1973 began in the early evening when an eruptive fissure opened on the floor of Pauahi Crater, just below the present visitor viewing platform. The eruption followed by several hours the sudden draining of a lava lake at Mauna Ulu, the broad lava shield with the oxidized summit on the southern horizon, suggesting that the plumbing system feeding Mauna Ulu had been tapped and diverted. Initially lava fountaining was confined to the bottom of Pauahi Crater, but activity soon spread up both the eastern and western crater walls and into the adjacent forest (Fig. 13). Within a matter of hours, a 3-km long WSW striking en echelon fissure system was venting lava from the vicinity of Puu Huluhulu, the forested cinder cone next to Mauna Ulu, to just west of Pauahi Crater. During the climax of the eruption, lava fountaining from fissures in the walls and along the rim of Pauahi Crater cascaded as much as 100 m to the floor of the crater, feeding two lava lakes, one in the western and the other in the central or main pit (Fig. 13). The lakes were separated by a septum which was soon overtopped, allowing lava to cascade from the central pit into the western one. Huge whirlpools churned in the combined lakes, suggesting that lava was draining through the floor of the crater, even as it fountained from adjacent wall and rim fissures. The draining lava may have been mixed with new magma to be reerupted (Donald W. Peterson, pers. commun., 1982).

One of the fissures active during this eruption, clearly visible from the observer platform, extends as a dark line down the far northeast wall of the crater, in the direction of Puu Huluhulu. The other may be viewed by carefully walking behind a small spatter mound just north of the wooden viewing platform. Small lava trees, interesting textures of spatter accumulation, iridescent glasses, and lava drainback structure occur around the oxidized fissure mouth.

Most activity of the November 10 eruption ceased in about 10 hours, but a small amount of new lava continued to well up through the floor of the western pit until December 9. The molten lava of the central pit drained rapidly back into underlying fissures, leaving a high-standing lava bathtub ring which veneers the southern crater walls (Peterson et al., 1976). During the final phase of the eruption, the lava lake in the western pit also partially subsided.

The last historical eruption in Pauahi Crater occurred after six years of quiescence. The outbreak of November 16, 1979 was preceeded by a seismic swarm of more than 11 hours duration. Venting began in the western pit of the crater, at the foot of the cliff beneath the visitor outlook. Then a fissure opened westward outside of the crater, cutting Chain of Craters Road and feeding a lava flow which advanced 500 m into the forest. The eruption was over within 24 hours.

Stop 6. Holei Pali and the Mauna Ulu lava field. After leaving the East Rift Zone, Chain of Craters Road passes Mauna Ulu lava shield and descends the block-faulted southern flank of Kilauea to the coastal platform below. The road crosses the 300-400 m-high escarpment of Holei Pali, the eastern end of the Hilina fault system, which is draped with numerous lava flows of the 1969/1974 Mauna Ulu eruptions. Pahoehoe flows are glassy and reflective, hence have a light gray color. Aa flows are much darker than pahoehoe.

From the Holei Pali Pullout, one can see from the color differences that the cascades of lava on the face of Holei Pali comprise a chaotic pattern of aa and pahoehoe. Donald W. Peterson (written commun., see also Peterson and Tilling, 1980) explained how this pattern may develop: "When the front of the advancing lava streams encounters a steep slope, the lava frequently changes to aa because of the internal turbulence induced by the more rapid movement (Macdonald, 1953). Thus much of the lava exposed on pali slopes is in the form of aa. However, as the flow continues, lava tubes near the source extend downstream, and lava tube lava, insulated for a longer period, is hotter and less viscous when it reaches the pali slope (Swanson, 1973; Peterson and Swanson, 1974). This new lava retains its pahoehoe form as it descends the slope, and it partly covers the earlier aa. At a later time, newly formed flows may convert to aa as they descend the slope, partly covering both pahoehoe and earlier aa. Throughout the course of the Mauna Ulu eruptions, these processes were repeated many times, resulting in intricate mixtures of aa and pahoehoe on pali slopes."

The pahoehoe field at the Holei Pali Pullout consists of flows erupted from the east rift zone in 1971 and 1972. Where these and other flows from the Mauna Ulu eruptions reached the sea, about six hectares of new land were added to the island.

Stop 7. Wahaula area and Kupaianaha lava flows. The Kilauea fieldtrip ends near the National Park Visitor Center at Wahaula Heiau, a 13th century Hawaiian temple complex on the coast. A short distance east of Wahaula Heiau, lava flows from Kupaianaha, a young lava shield in the east rift zone, have covered the highway, destroyed numerous houses, added new land to the coast, and provided material for the creation of several spectacular new black sand beaches. Depending on the state of eruptive activity, we shall explore the coastline to examine geological features which illustrate how new land is added when pahoehoe flows enter the sea.

Kupaianaha resembles Mauna Ulu, a feature you have already seen. The lava shield lies about 8 km inland from the Wahaula area. It began to grow in July 1986, and has been almost continuously active up to the time of this writing (June, 1988). A short distance west of Kupaianaha rises a 250 m-tall cinder cone named Puu Oo, built in a series of eruptive phases characterized by high lava fountaining between June 1983 and July 1986. Kupaianaha and Puu Oo were constructed during different phases of the same eruption. This eruption, beginning in January 1983 and continuing to the present, represents the longest sustained activity to have occurred historically on the flanks of Kilauea.

Mauna Loa Volcano

The lava of Mauna Loa is divided into the Ninole Basalt, Kahuku Basalt, and the Kau Basalt. The Kahuku and Kau are separated by the Pahala Ash and are thought to be coeval with the Hilina and Puna Basalt, respectively, of Kilauea. The exposed lava is all tholeiitic basalt and picritic tholeiitic basalt of the shield stage that erupted from the 2.5 by 4 km summit caldera, named Mokuaweoweo, and two rift zones. The rift zones extend southwest and east-northeast. The southeastern and the southwestern slopes of Mauna Loa have been steepened by downfaulting along the Kaoiki and Kealakakua fault systems. Possible remnants of two earlier shields are exposed in the Ninole Hills (Ninole Basalt) and in the area around Kulani. Neither of these earlier shields is well documented and both may be merely deeper parts of the Mauna Loa shield uplifted

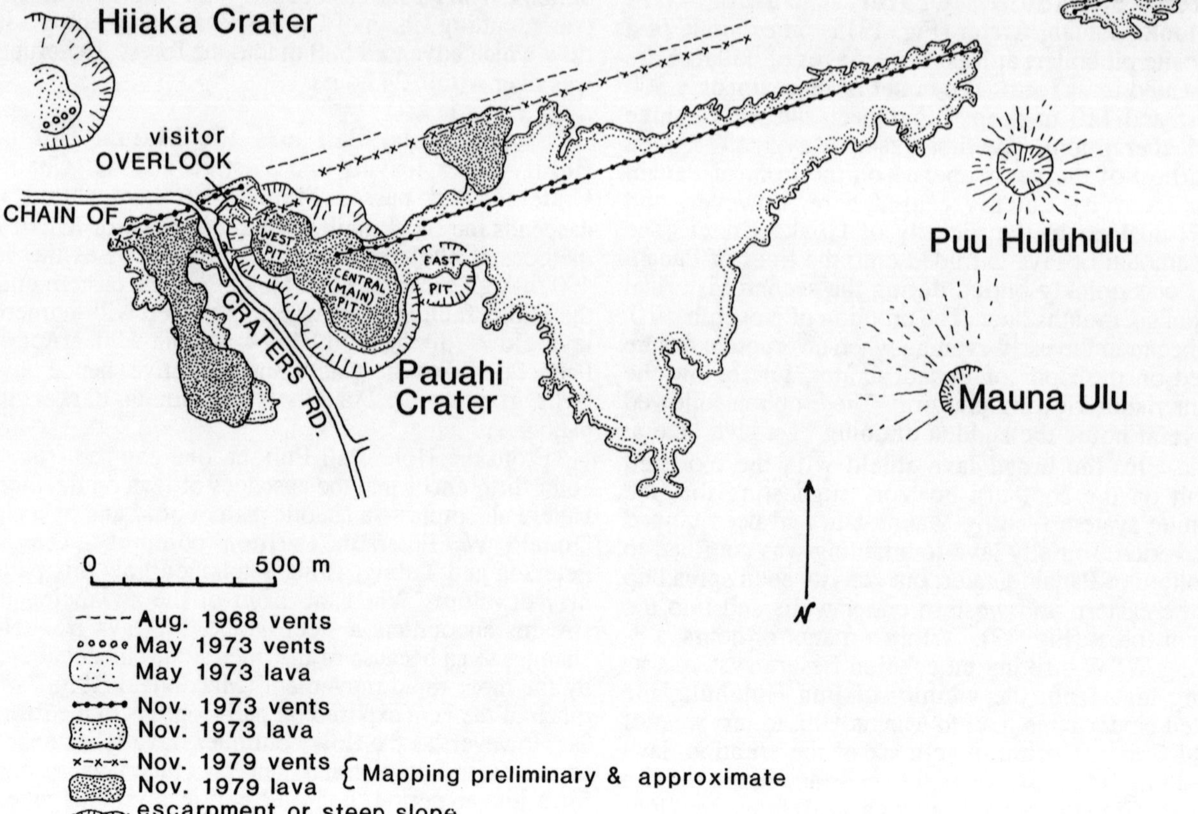

FIGURE 13 Pauahi Crater area, showing distribution of vents and lava flows of eruptions which have broken out historically in Pauahi Crater. Field trip participants will view Pauahi Crater from the visitor overlook.

along normal faults in a manner analogous to that in the Hilina fault system near Puu Kapukapu. Numerous radial eruptive fissures occur on the north and northwest flank of the volcano. Detailed geochemical studies of Mauna Loa include studies by Wright (1971), Tilling et al. (1987), and Rhodes (1988).

Mauna Kea Volcano

The volcano has a well exposed section of shield lava capped by postshield alkalic lava. The rocks are divided into the older Hamakua Volcanics and the younger Laupahoehoe Volcanics. The Hamakua Volcanics consists of the shield stage tholeiitic basalt and picritic tholeiitic basalt, and overlying postshield-stage alkalic basalt, ankaramite, and hawaiite. There is no clear boundary between the shield and postshield lavas; tholeiitic and alkalic lavas are intercalated near the boundary (Frey et al., 1984). Much of the surface is blanketed by the younger Laupahoehoe Volcanics which is mostly hawaiite, with much less ankaramite and alkalic basalt (West and Garcia, 1982). The Hamakua Volcanics is exposed only in deep erosional canyons. The summit region was glaciated during the Pleistocene (Porter, 1979a).

The rift zones are not well-defined, but cinder cones are roughly aligned in westerly and southerly directions from the summit. A nearly buried east rift zone is well delineated by a submarine ridge extending nearly 40 km to sea. It is uncertain if a summit caldera existed, but the crude arcuate alignment of some of the cinder cones, coupled with a large gravity high just south of the summit, indicate that a former caldera may be buried beneath the Laupahoehoe Volcanics.

Kohala Volcano

Kohala Volcano is composed of the older Pololu Basalt and the younger Hawi Volcanics. The Pololu Basalt is a succession of thin flows of tholeiitic basalt, olivine tholeiitic basalt, and picritic tholeiitic basalt of the shield stage except near the top of the section, where alkalic basalt of the alkalic postshield stage occurs. The younger Hawi Volcanics formed during the alkalic postshield stage and is separated from the Pololu Basalt by an unconformity. Most of the Hawi lava is mugearite, but hawaiite and benmoreite are also present. Cinder cones of the Hawi align along two rift zones trending northwest and southeast. Arcuate faults near the summit indicate that a caldera probably formed during eruption of Pololu lava, but later Hawi lava buried it. A gravity high located southeast of the summit may correspond to the approximate location of this inferred caldera. Hawi lava is absent from an 11-km section of the northwest side of the volcano. Flows are deflected from this zone by a series of fault scarps in the summit region that bound a northwest trending graben 10 km long and 2-5 km wide. Lava erupted inside the graben, filled it, and overflowed most of the southwest rim. Detailed geochemical studies of Kohala lavas have been done by Feigenson et al. (1983), Lanphere and Frey (1987), and Hofmann et al. (1987).

Hualalai Volcano

Hualalai Volcano was last active in 1800-1801 when two major and three minor flows issued from a series of vents on the northwest rift zone. With the exception of a large trachyte cone and flow at Puu Waawaa (the Waawaa Trachyte Member) the entire surface consists of flows of alkalic basalt of the alkalic postshield stage. All subaerial rocks of Hualalai are called the Hualalai Volcanics. A detailed study of the volcano (Moore et al., 1987) shows that nearly all the lava is alkalic basalt, with only a few flows that are transitional to hawaiite. There are two well-defined and one poorly defined rift zones; the well-defined ones are to the northwest and the south, the poorly defined one to the north. It is uncertain if a summit caldera existed in the past, although Clague (1987a) infers the existence of a shallow magma reservoir until the trachyte formed at 105 ka. Tholeiitic lavas have been recovered from the submarine northwest rift zone (Clague, 1982) and from the bottoms of several water wells (Moore et al., 1987). The tholeiitic lavas demonstrate that Hualalai had a shield stage characterized by eruption of tholeiitic basalt and that the cap of alkalic postshield stage lava is thin. Geochemical studies of the lavas are given by Clague et al. (1980) and Moore et al.(1987).

Roadlog of Mauna Loa, Mauna Kea, Kohala, and Hualalai Volcanoes - Day 2. This roadlog begins in downtown Hilo and ends in downtown Kailua on the Kona coast. The starting point is the intersection of Kamehameha Avenue and Waianuenue Avenue in the center of Hilo. This location is milepost 0 on State Highway 200 (the Saddle Road). The route is shown on Fig. 6. The first part of this guide has been selected from a more complete fieldguide by Porter et al. (1987).

miles (km)

0 (0) Intersection of Kamehameha Avenue and Waianuenue Avenue in Hilo. Proceed to the west on State Highway 20.

1.3 (2.1)Bear left onto Kaumana Drive

3.5 (5.6) Cross onto the Mauna Loa 1880-1881 flow. The eruption feeding this flow continued for 280 days, erupted 2.3 x10^8 m^3 of lava, and the flow advanced to within 2 km of Hilo harbor (Lockwood and Lipman, 1987).

4.2 (6.8) **Stop 1.** Pull off to right side of road at Kaumana Cave Park. Kaumana cave is a lava tube in the 1880-1881 Mauna Loa flow. During the eruption, this tube was a continuous channel for 45 km and extended from the source vent on the northeast rift zone at an altitude of 3080m. Well developed tube systems form during sustained eruptions characterized by little or no fountaining. The source vent is an elongate crack with no spatter. Kaumana Cave can be traversed for about 1 km from this entrance. Inside the tube there are several lava shelves and smaller tubes developed late in the eruption as the lava levels inside the tube subsided. Greeley (1974) describes the interior of the tube in a fieldguide.

9.0 (14.5) Contact between Mauna Loa flows dating from 1880-1881 and 1855-1856. Both flows are aphyric and have been mapped based on their trace element characteristics (Rhodes et al., 1982). The 1855-1856 eruption lasted 450 days and was characterized by periods of high fountaining; much of the lava is aa. The flows have a volume of 2.8 x 10^8 m^3, but stopped 10 km from Hilo harbor because no well-developed tube system formed. The road follows this flow for the next 13 miles (21 km).

21.8 (35.1) Contact with the 1935-1936 Mauna Loa flow. At this location the flow is mostly aa but the complex transition from pahoehoe occurs within the next 5 miles.

28.0 (45.1) **Stop 2.** Turn left onto pullout adjacent to Puu Huluhulu, a Pleistocene cinder cone of the Laupahoehoe Volcanics of Mauna Kea. The cone is surrounded by Mauna Loa flows, mainly the 1935-1936 flow. A cinder quarry on the west side of the cone expose the internal structure of the cinder cone including a late steeply dipping flow and a dike near the top of the quarry. The cone is hawaiite in composition and formed during the alkalic postshield-stage.

This stop also provides an excellent view of the summit of Mauna Kea Volcano to the north. Most of the cinder cones near the summit are hawaiite of the Laupahoehoe Volcanics and are late Pleistocene in age. A few of the Laupahoehoe vents are Holocene; these are mainly located low on the flank of the volcano towards the northeast. The summit region was glaciated in the Pleistocene and several glacial drifts are interstratified with lava flows (Porter, 1979a). These glacial drifts can be seen from this viewpoint as light-colored material near the summit. The youngest of the glacial drifts define an area of about 70 km^2 that was covered by ice. On the south side of the mountain, moraines formed along two gulches at elevations as low as 10,650 ft (3250 m). Porter (1987) describes six pyroclastic cones and their associated flows that erupted subglacially.

30.1 (45.1) 1843 Mauna Loa flow. This is a paired flow consisting of early aa and late pahoehoe. This relationship is frequently observed and is caused by the development of lava channels and tubes later in the eruption and the change from early high fountaining to later sustained output of pahoehoe with little or no fountaining. The Puu O'o eruptions of Kilauea followed this same pattern.

33.2 (53.5) **Stop 3.** This stop provides a panoramic view of Mauna Loa Volcano. The low shield shape is formed by repeated subaerial eruption of fluid tholeiitic basalt. The vast majority of Mauna Loa eruptions occur in the summit caldera, named Mokuaweoweo, and along the two rift zones which extend towards the northeast and southwest. However, some eruptions have issued from radial vents located on the north and northwest flanks of Mauna Loa (Fig. 14). To the north of this stop, extending roughly 2 km along the highway, is a spatter rampart that is roughly 3000 years old. The flow is olivine-rich and can be traced about 11 km to the west. This Mauna Loa vent cuts Mauna Kea lavas. Other radial vents from Mauna Loa cut Hualalai lavas to the south of the Hualalai summit. The large 1859 Mauna Loa flow, which we will cross later in the trip, also issued from a radial vent, in this case located just northwest of the summit. This stop is located on the 1843 Mauna Loa

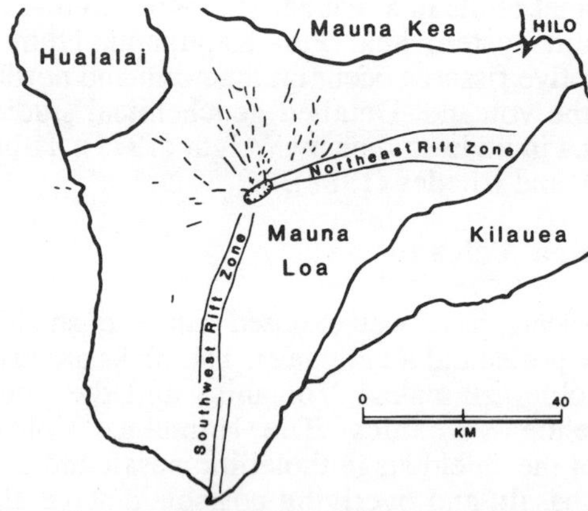

FIGURE 14 Simplified map showing the distribution of Mauna Loa radial vents on the north and northwest flanks of the volcano. Figure from Porter et al. (1987).

flow which was fed from a radial vent at 2990 m elevation. The eruption had begun at the summit a few days earlier, but activity shifted to the north flank and continued for about three months. The flow consists of about 2 x 10^8 m^3 of tholeiitic basalt that is plagioclase phyric; it is the only historic Mauna Loa flow that contains plagioclase as a major phenocryst phase.

35.1 (56.5) **Stop 4.** Mauna Kea State Park (rest stop). Pohakuloa Gulch to the north has eroded through the Waihu Glacial Member of the Laupahoehoe Volcanics, through a few aa flows of the Laupahoehoe Volcanics, and into the thin-bedded flows of the Hamakua Volcanics. The Waihu Glacial Member consists mainly of conglomerate deposited by outburst floods generated by subglacial eruptions.

49.7 (80.0) Kohala Volcano is in view straight ahead. The cinder and spatter cones near the summit are part of the Hawi Volcanics and are mostly composed of hawaiite and mugearite.

53.7 (86.5) Junction with State Highway 190; turn left towards Kailua.

58.4 (94.0) Junction with Waikoloa Road; turn right.

62.9 (101.3) Puu Hinai, a hawaiite vent of the Hamukua Volcanics of Mauna Kea. This is one of the most distal of the Mauna Kea surface vents.

70.5 (113.5) Junction with State Highway 19; turn right towards Kawaihae.

73.0 (117.5) The highway crosses a series of flows of the Hamakua Volcanics. The low, flat, tree covered area towards the coast consists of younger Mauna Loa lavas that have spread out upon entering the ocean. Recent work on submerged coral terraces offshore suggests that Mauna Loa lavas began entering the ocean in this area between about 125 and 13 ka.

76.4 (123.0) **Stop 5.** (Lunch Stop) Turn left into Hapuna Beach State Park parking area. Hapuna Beach is one of only a few coral sand beaches on Hawaii. The sand has accumulated between two thick flows of the Hamakua Volcanics. The flow to the north is ankaramite whereas the flow to the south is a weakly porphyritic alkalic basalt. Three older flows are exposed between these two flows. The coral reef offshore is no longer alive so the sand is not being actively replenished by new

reef growth. Further offshore along the entire coast north of Kailua, large quantities of coral sand occur on top of the submerged reef terraces in water depths as shallow as 100 m.
77.5 (124.9) Return to Highway 19 and turn left.
79.7 (128.3) Junction with State Highway 270; turn left towards Kawaihae.
80.4 (129.4) **Stop 6.** Kawaihae Quarry. This quarry was the source of the large lava blocks used to construct the breakwater that protects Kawaihae Harbor. The sequence of flows exposed in the quarry include a basal transitional plagioclase-phyric pahoehoe flow of the Pololu Basalt of Kohala, an olivine plus augite phyric alkalic basalt aa flow of the Hamakua Volcanics of Mauna Kea, and an upper aphyric mugearite aa flow from Puu Loa of the Hawi Volcanics of Kohala. Between the Pololu Basalt flow and the Hamakua Volcanics flow are 30 cm of reddish-brown soil and 8-10 cm of hawaiite tephra from Puu Kawaiwai, an early vent of the Hawi Volcanics. K-Ar ages on these flows show the basal Pololu flow to be about 320 ka and both the Hamakua and the Hawi flows to be about 120 ka (Porter et al., 1987). This Hawi flow is one of the youngest flows on Kohala Volcano. The succession of flows demonstrate that the Hamakua Volcanics of Mauna Kea and the Hawi Volcanics of Kohala erupted at the same time. Retrace your route along Highway 270.
81.1 (130.6) Junction with State Highway 19; turn left towards Waimea.
89.2 (143.6) Junction with State Highway 250; turn left towards Hawi.
94.6 (152.3) **Stop 7.** Puu Makela benmoreite dome. Park in scenic overlook on lefthand side of the road. This is one of only a few volcanic domes in Hawaii. This lava was apparently viscous enough to form a dome because of its strongly fractionated composition. The benmoreite is nearly aphyric but rare phenocrysts of plagioclase, amphibole, and fayalitic olivine occur with microphenocrysts of apatite and titanomagnetite. Small rare xenoliths of gabbro and dunite are seen occasionally. The highway roadcut provides a cross section through the dome and shows several meters of rubble beneath the dome surface. The interior of the dome is massive and somewhat altered. A flow issued from the northeast side of the dome; both the dome and flow are surrounded by the mugearite flow from Puu Loa that was examined at the Kawaihae Quarry stop. Return along State Highway 250 towards Waimea.
97.3 (156.7) On the right hand side of the road is Puu Kawaiwai, the source of the hawaiite tephra observed in the Kawaihae Quarry. Most of the tephra fell close to the vent, indicating relatively low fountaining. Numerous volcanic bombs occur in the tephra.
100.1 (161.2) Junction with State Highway 19; turn left towards Waimea.
102.1 (164.4) Junction with State Highway 190; turn right towards Kailua.
108.5 (174.7) Junction with State Highway 200 (Saddle Road); continue on State Highway 190.
113.2 (182.3) Junction with Waikoloa Road; continue on State Highway 190.
116.4 (187.4) Contact of first Mauna Loa tholeiitic basalt flows after crossing numerous Mauna Kea flows.
121.0 (194.8) Contact of the 1859 Mauna Loa flow.
123.2 (198.4) Steep scarp marks the edge of the Puu Waawaa trachyte flow form Hualalai Volcano. Mauna Loa and Hualalai flows have ramped in and overflowed the east side of the trachyte flow whereas the west side of the flow is a steep escarpment. Some of these flows have cascaded over the western escarpment. The trachyte flows consist of several flow lobes each 250-500 feet thick.The trachyte pumice cone and flow are about 105 ka (Clague, 1987a), whereas the surface flows of alkalic basalt on Hualalai are less than about 15 ka (Moore et al., 1987).
123.8 (199.3) Turn left into Puu Waawaa Ranch road and take left fork in road. This road and the quarry are on private land and permission to pass should be obtained.
125.9 (202.7) **Stop 8.** Puu Waawaa Quarry. The quarry exposes the interior of the trachyte pumice cone. The cone is fluted on the surface, possibly because of deposition by glowing avalanches. The pumice deposits include fragments of trachyte obsidian. Extremely rare dunite xenoliths occur in the trachyte; they have reacted with the enclosing trachyte to form a light- colored mica. Drillholes in the area demonstrate that the pumice fall was quite widespread. The trachyte represents an extreme differentiate of alkalic basalt magma and apparently formed as the sub-caldera magma chamber solidified soon after the transition from tholeiitic to alkalic volcanism (Clague, 1987a). Return back to Highway 190.
128.0 (206.1) Turn left onto State Highway 190 towards Kailua.
132.8 (213.8) The 1800 Kaupulehu flow crosses the highway for the next 1.8 miles (2.9 km). This flow erupted from one of five active vents in 1800-1801. The flows are all alkalic basalt of the postshield stage and represent the most recent activity on Hualalai Volcano. Two stops on the day 3 itinerary examine xenoliths and lava tubes in flows erupted in 1800-1801.
145.9 (234.9) Junction of Palani Road and Kuakini Highway in Kailua.

Roadlog of Hualalai Volcano, Hawaii - Day 3. This half day roadlog will examine two alkalic basalt flows erupted in 1800-1801 on Hualalai Volcano. The access to the first site and the site itself are on private land and permission to pass must be obtained from Hualalai Ranch or the Bishop Estate.
miles (km)
0 (0) Junction of Palani Road and Kuakini Highway in Kailua where the previous day trip ended. Head east on Highway 190 (up the hill).
0.3 (0.5) Junction of Palani Road (State Highway 190) and State Highway 19. Continue on Highway 190 to the north.
3.8 (6.1) Junction with State Highway 180; continue on Highway 190.
10.2 (16.4) Turn right onto unpaved road just past milepost 29.
10.4 (16.7) Locked gate.
12.2 (19.6) **Stop 1h.** Telephone relay station. The area

around the relay station is part of the 1800 Kaupulehu flow which erupted from a vent at 1880 m elevation on the northwest rift zone of Hualalai Volcano. The site is remarkable for the tens of thousands of xenoliths exposed here. A detailed plane-table map and descriptions of the xenoliths are presented by Jackson and Clague (1982) and Jackson et al. (1981), respectively. The site is shown in a simpified map in Fig.15. The xenoliths are dominantly dunite and wehrlite, but gabbro, websterite, and anorthosite also occur; lherzolite is absent. This locality was first described by Richter and Murata (1961), but the mechanism by which the xenoliths were concentrated has remained controversial. We favor a model in which the xenoliths were entrained in the magma at depths up to about 20 km, erupted during a late vigorous phase of the eruption, carried downslope in a tube system, and deposited downslope when the tube system clogged. The xenoliths apparently formed a hydraulic dam as they accumulated and eventually the lava created skylights in the tube system through which the lava escaped, carrying many xenoliths even further downslope. These skylights are surrounded by spatter and look similar to small vents. Some of the gabbro xenoliths are fragments of the ocean crust underlying Hualalai. The largest xenoliths are about 35 cm in diameter and had to be transported to the surface quickly in order to stay in suspension. This observation indicates that Hualalai alkalic basalt rises from depths of about 20 km extremely rapidly; volcanic hazards posed by potential future eruptions are increased due to the inferred rapid magma migration. Return to Highway 190 along the same road.

14.2 (22.9) Turn left onto State Highway 190.

18.5 (29.8) Turn right at Kona Palisades Estates onto Ka'imi nani Drive.

22.2 (35.7) Junction with State Highway 19; turn right towards Kawaihae.

24.9 (40.1) **Stop 2h.** Lunch stop at the1801 alkalic basalt flow. Park on shoulder on right side of road. The large tube was part of the distribution system of the 1801 tube-fed pahoehoe flows. The vent is located at an elevation of about 500 m and consists of a large spatter rampart oriented upslope; the flows reached the sea over a wide front. The walls of the tube, which were exposed by highway construction, contain numerous small xenoliths of mainly dunite, wehrlite, and gabbro. Xenoliths had been searched for in this flow by the late E. D. Jackson with no success. The roadcut demonstrates that the xenoliths settled to the base of the flow where they could not be seen. Detailed mapping on Hualalai Volcano (Moore et al., 1987) has identified numerous other flows that contain abundant xenoliths; all the flows where xenoliths have been seen in natural outcrop are aa. Xenoliths have also been seen in roadcuts of a few pahoehoe flows.

The tube system displays many of the features seen at Kamauna Cave, but they are better preserved here because of the low precipitation. Upslope a hundred or so meters the tube contains a large island, apparently a section of the tube wall that was rafted downstream. Turn around on Highway 19 and head towards Kailua.

27.1 (43.6) Turn right on airport road and proceed to the terminal. The next part of the fieldtrip will be an aerial tour of the islands of Maui, Kahoolawe, Lanai, Molokai, and Niihau. We will also examine the summits of Mauna Kea and Kohala, and parts of Oahu and Kauai. The flight will emphasize the volcanoes that we will not tour on the ground. The brief summaries of the geology of these volcanoes are from Clague and Dalrymple (1987).

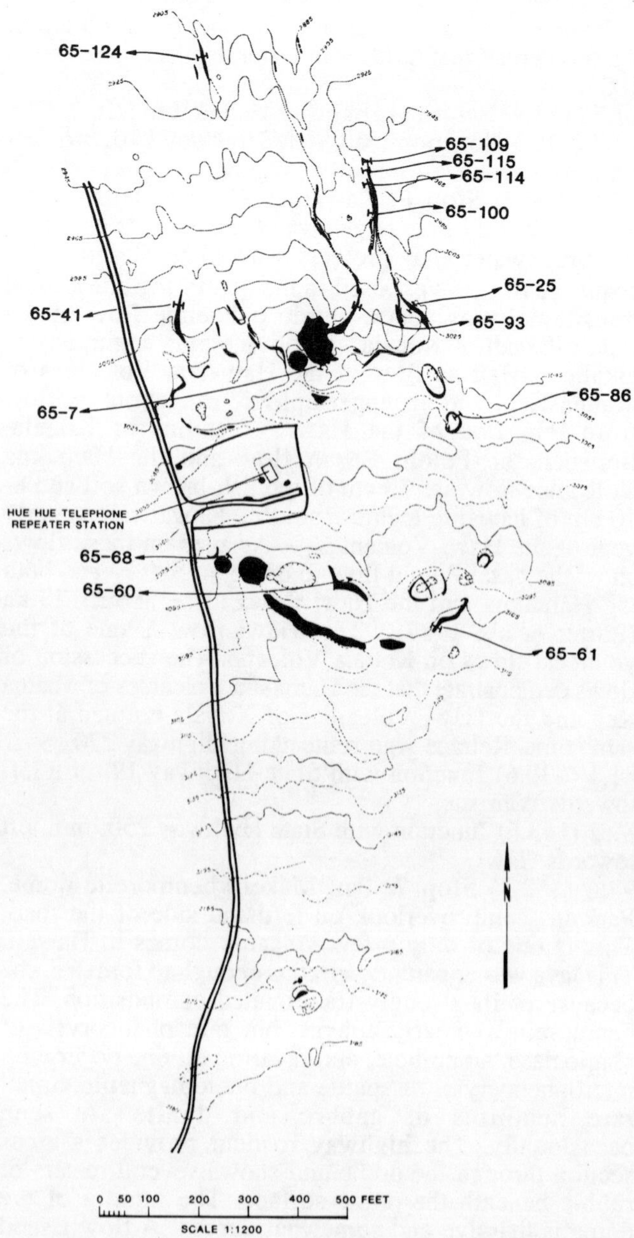

FIGURE 15 Simplified geologic map showing the distribution of xenolith beds in the Hualalai 1800 Kaupulehu flow near the Huehue telephone repeater station. The numbers indicate locations where xenoliths were systematically counted and measured by Jackson et al. (1981). The map is modified from Jackson and Clague (1982).

MAUI

Maui consists of two volcanoes, East Maui or Haleakala Volcano and West Maui Volcano (Fig. 16). East Maui Volcano is the youngest volcano in the Hawaiian Islands that has had alkalic rejuvenated-stage volcanism. The volcano was last active in about 1790 (Oostdam, 1965). The oldest unit is the Honomanu Basalt, a series of tholeiitic basalt, olivine tholeiitic basalt, picritic tholeiitic basalt flows that represent the shield stage. Above sea level these flows are almost completely buried beneath those of the overlying Kula and Hana Volcanics. The Kula Volcanics is composed mostly of hawaiite with some ankaramite and alkalic basalt and represents the alkalic postshield stage. There is little evidence for extensive erosion between the shield and postshield lavas. The Hana Volcanics is composed mostly of the same rock types as the Kula Volcanics, but it erupted after an erosional period; it represents the alkalic rejuvenated stage of volcanism. Three rift zones are delineated by the location of vents for the Kula and Hana Volcanics. The east and southwest rifts have vents of both units, whereas the north rift has only vents of the Kula Volcanics. The summit area has a large depression that is similar to a caldera, but actually formed by erosion. The Hana Volcanics is unique among rejuvenated stage volcanic rocks in that its vents are aligned along the preexisting rifts, the duration of the erosional period is short (<0.4 Ma), and ankaramite and hawaiite are present. Chen and Frey (1983, 1985) present detailed geochemistry of lavas from all three stages.

The volcanic rocks of West Maui are subdived into the Wailuku Basalt, Honolua Volcanics, and Lahaina Volcanics.The Wailuku Basalt consists of tholeiitic basalt, olivine tholeiitic basalt, and picritic tholeiitic

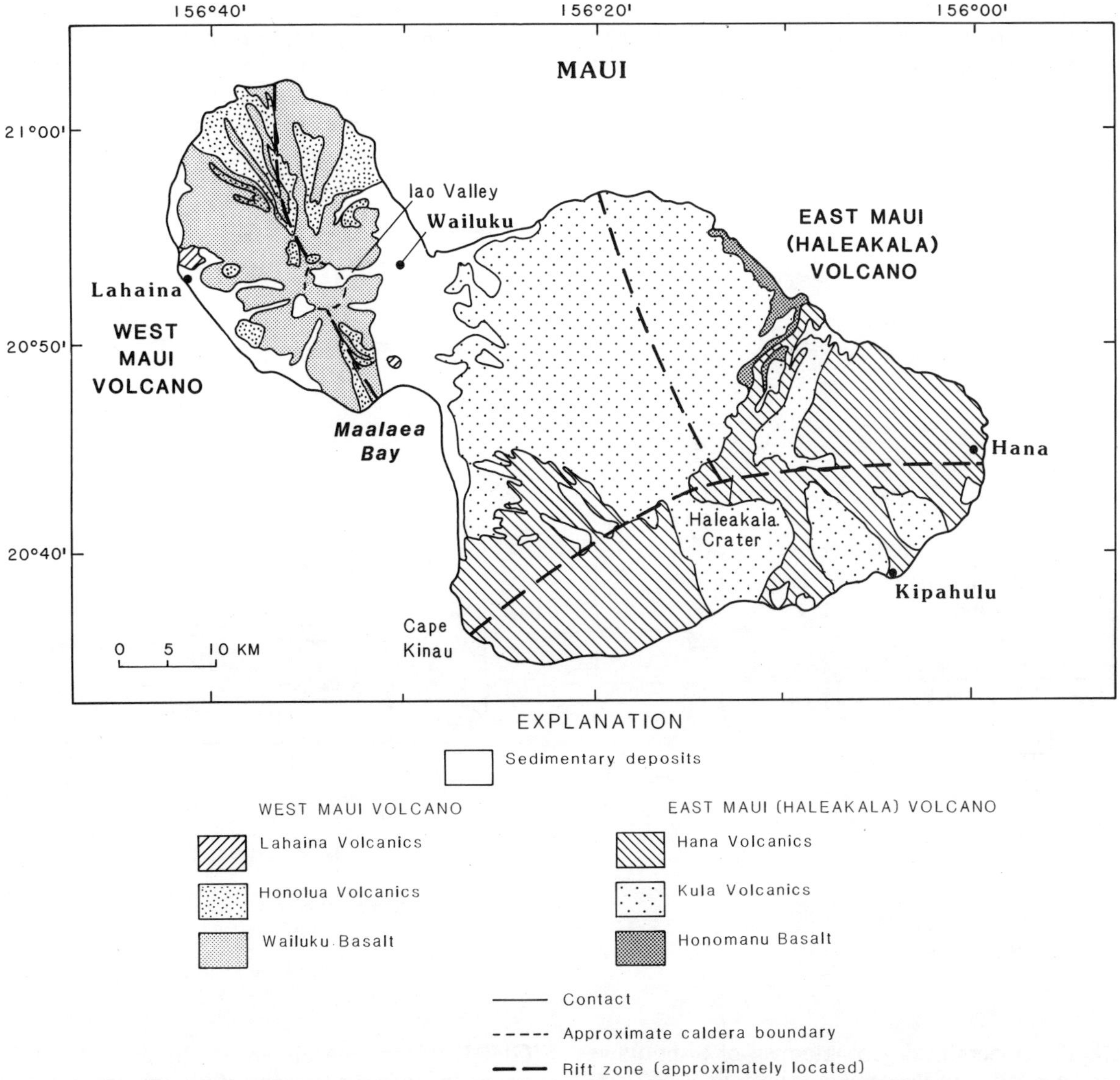

FIGURE 16 Generalized geologic map of Maui showing the distribution of units for East and West Maui Volcanoes. Figure from Langenheim and Clague (1987) which should be consulted for primary references.

basalt of the shield stage and of alkalic basalt of the caldera-filling phase. The Honolua Volcanics, which represent the alkalic postshield stage, consists of a thin, discontinuous cap of mugearite and less trachyte and hawaiite. The Lahaina Volcanics followed a long period of erosion and consists of the cones and flows of four small eruptions of basanite and olivine-rich basanite. The Lahaina Volcanics represents the alkalic rejuvenated stage. The volcano has poorly-defined rift zones delineated by dike swarms that trend northeast and south and by vents of the Honolua Volcanics that trend north and south from the small central caldera. The caldera-filling lava is severely altered by late gases; erosion has preferentially removed these altered rocks in Iao Valley.

KAHOOLAWE

The volcanic rocks of Kahoolawe have not been subdivided. The only formation, the Kanapou Volcanics, includes tholeiitic basalt and olivine tholeiitic basalt of the shield stage, tholeiitic and alkalic basalt of the caldera-filling phase, and alkalic basalt and hawaiite of the alkalic postshield stage (Fig. 17). Five vents along the seacliff in Kanapou Bay erupted alkalic basalt following an extended period of volcanic quiescence. These vents presumably represent an alkalic rejuvenated stage. The volcano was built by eruptions along a prominent west-southwest rift zone and two less pronounced rifts trending north and east. Most of the vents have been removed by erosion, but remnants of about six vents remain. The caldera of 5 km diameter lies at the eastern end of the island and has been breached by the sea.

LANAI

Only tholeiitic lava was erupted during the shield stage and caldera-collapse phase of Lanai Volcano; no later alkalic lavas are known. The Lanai Basalt consists of tholeiitic basalt, olivine tholeiitic basalt, and picritic tholeiitic basalt that erupted from the northwest, southwest, and southeast rift zones and from the summit caldera underlying the present-day Palawai Basin (Fig.18). An extensive dike swarm marking the southwest rift crops out along the Kaholo Pali. Most of the dikes are nearly vertical and are about 30 cm thick.

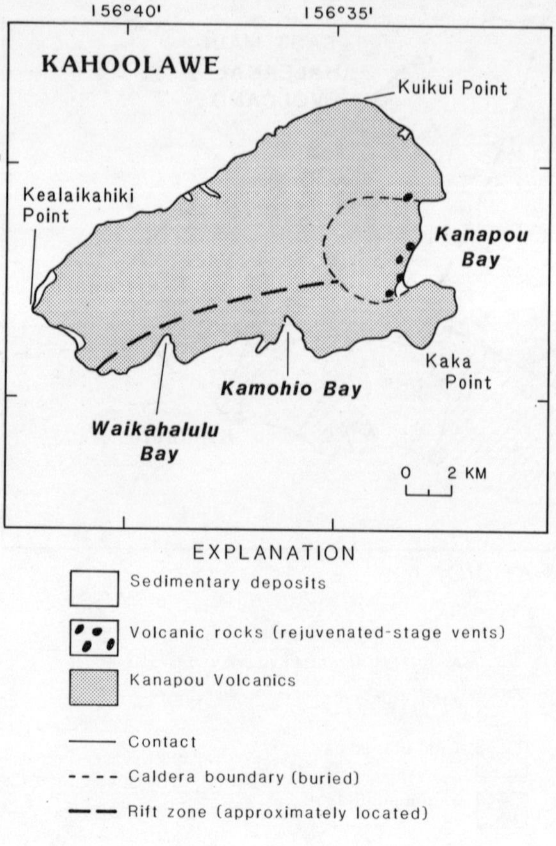

FIGURE 17 Generalized geologic map of Kahoolawe showing the location of the caldera and rift zones. Figure from Langenheim and Clague (1987) which should be consulted for primary references.

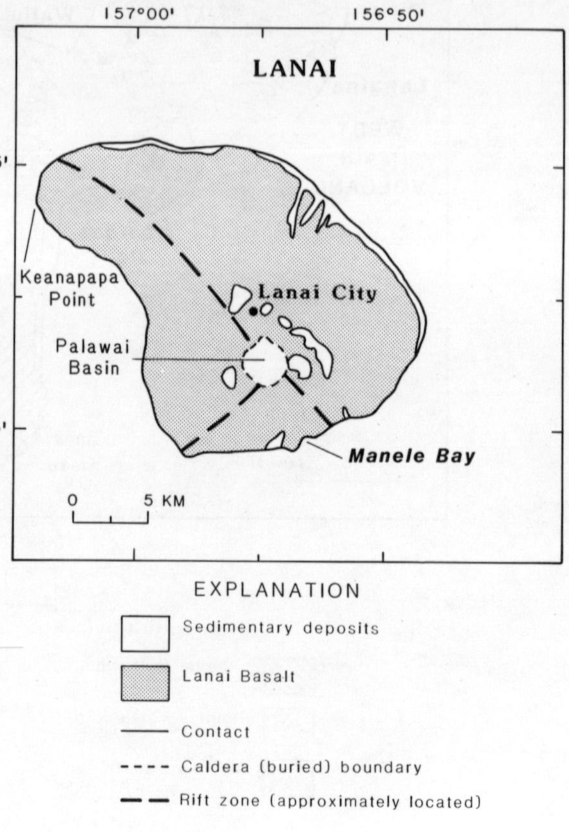

FIGURE 18 Generalized geologic map of Lanai showing the inferred location of the caldera and the three rift zones. Figure from Langenheim and Clague (1987) which should be consulted for primary references.

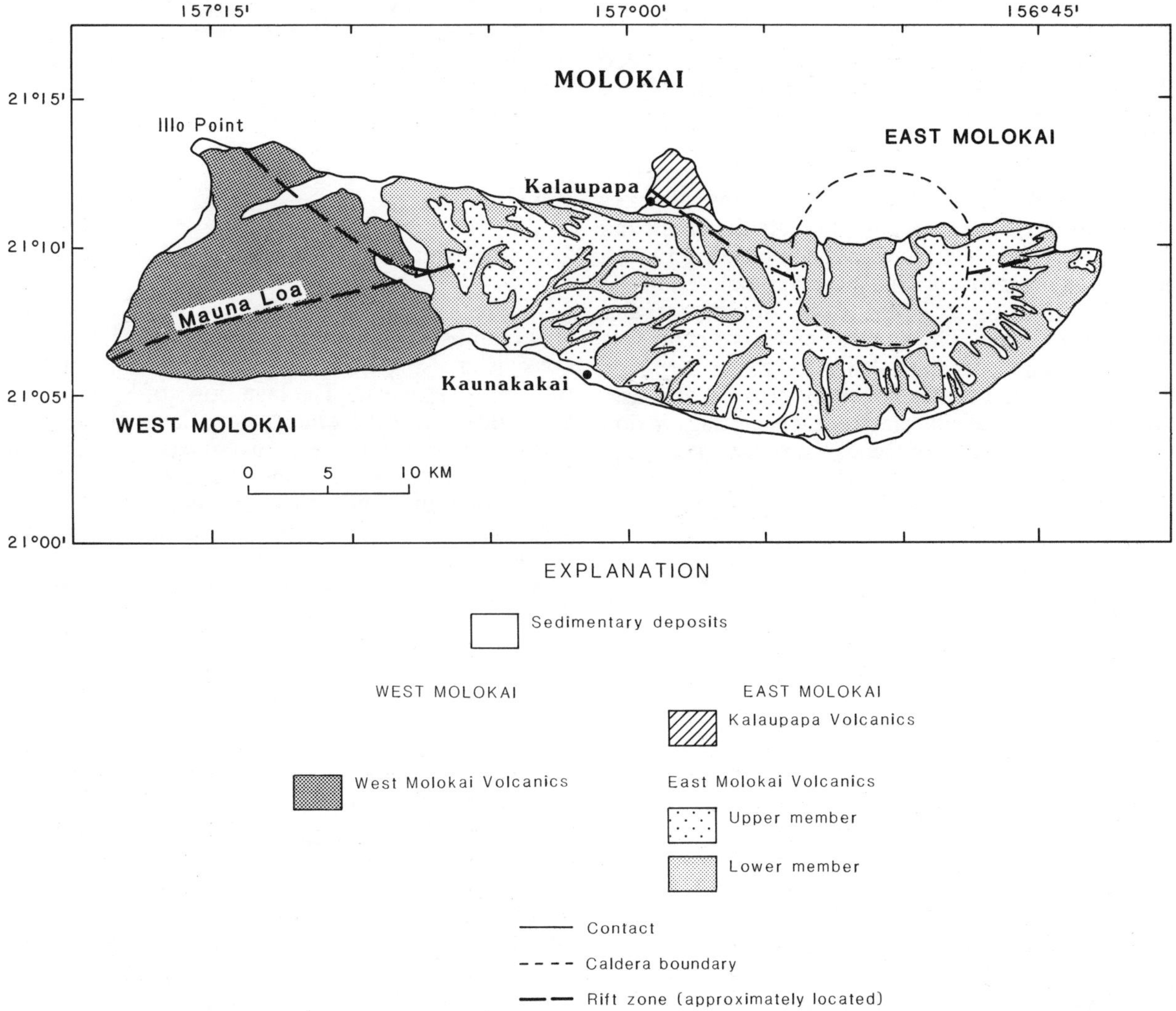

FIGURE 19 Generalized geologic map of Molokai showing the location of the caldera on East Molokai and the rift zones on both volcanoes. Figure from Langenheim and Clague (1987) which should be consulted for primary references.

MOLOKAI

Molokai consists of two large shield volcanoes, East and West Molokai Volcanoes (Fig. 19). East Molokai lavas are informally subdivided into upper and a lower members of the East Molokai Volcanics. The lower member consists of tholeiitic basalt, olivine tholeiitic basalt, and picritic tholeiitic basalt of the shield stage, but alkalic basalt of the alkalic postshield stage occurs in the upper part of the lower member. Beeson (1976) and Clague and Beeson (1980) have shown that tholeiitic and alkalic basalt are intercalated in the upper part of the lower member. The upper member consists predominantly of mugearite, with smaller amounts of hawaiite and trachyte and represents the alkalic postshield stage. Following an extensive erosional period, during which the 1200-m cliffs formed on the north side of the island, alkalic basalt and basanite of the alkalic rejuvenated stage erupted to form the Kalaupapa Peninsula and Mokuhooniki Island, a tuff cone located offshore the east end of the island. East Molokai Volcano was constructed by eruptions along east and west-northwest trending rift zones and from a summit caldera about 2.5 x 7 km in size. The caldera-filling lavas are similar to tthose of the lower member except that they are horizontal, more massive, and more extensively altered.

West Molokai lavas are assigned to the West Molokai Volcanics. Most of the exposed lava is tholeiitic basalt of the shield stage. This lava erupted from an east-northeast rift zone that crosses the summit area. A less pronounced rift trends towards the northwest. There is no evidence of a summit caldera. Alkalic lava of the alkalic postshield stage, predominantly hawaiite with subordinate alkalic basalt, erupted from a series of cinder and spatter cones located mainly on the northwest rift. No alkalic rejuvenated stage lava is known. Longitudinal sand dunes parallel the direction of the trade winds near the west end of the island.

NIIHAU

The island of Niihau consists of a deeply eroded shield volcano mantled by lava of the alkalic rejuvenated stage on the north, west, and south sides (Fig. 20). The Paniau Basalt consists of tholeiitic basalt and olivine tholeiitic baslt of the shield stage and the remnants of a single alkalic postshield vent at Kaeo. Several dikes exposed near the eastern coast are also alkalic basalt and presumably fed vents that have been completely removed by erosion. A magnificent dike swarm is exposed in the eastern seacliff; these dikes trend southwest and represent a rift zone. The summit of the volcano was northeast of the present-day island, and the eastern side of the volcano has been removed by erosion or mass-wasting.The period of volcanic quiescence and erosion was followed by eruption of the alkalic rejuvenated stage

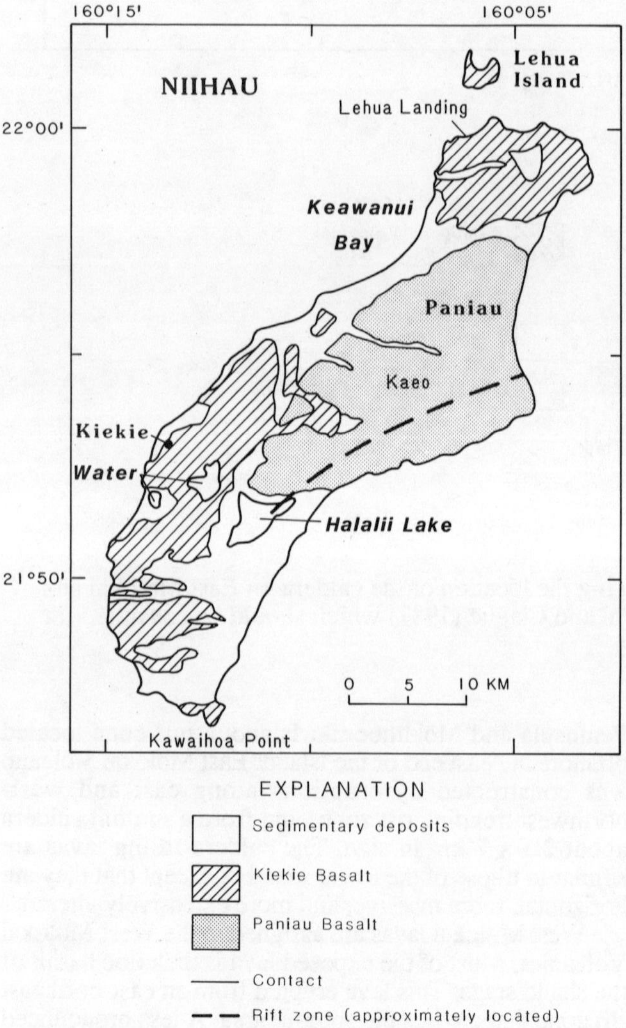

FIGURE 20 Generalized geologic map of Niihau showing the distribution of the shield stage Paniau Basalt and the rejuvenated stage Kiekie Basalt. The rift zone is spectacularly displayed where it intersects the seacliff on the east coast. Figure from Langenheim and Clague (1987) which should be consulted for primary references.

Kiekie Basalt, which is entirely alkalic basalt. Lehua Island off the north shore is a breached tuff cone of the Kiekie Basalt. As on West Molokai, longitudinal sand dunes that parallel the trade winds occur near the southern end of the island.

KAUAI

Introduction

Kauai and its nearest neighbor island, Niihau, are the westernmost of the principal Hawaiian Islands. Kauai consists of a single shield volcano that is about 40 km in diameter at sea level. The lava flows of Kauai have been subdivided into the shield-stage lava of the Waimea Canyon Basalt and the rejuvenated-stage lava of the Koloa Volcanics. The Waimea Canyon Basalt is subdivided into the following units, from oldest to youngest: the Napali Member, which constitutes the main shield of the volcano; the Olokele Member, which fills the enormous (17-19 km diameter) summit caldera; the Haupu Member, which fills a small, elongate flank caldera in the southeastern part of the island, and the Makaweli Member, which fills a 6.4 km wide graben extending southward from the summit (Fig. 21). The lava flows of the Waimea Canyon Basalt consist mostly of tholeiitic basalt and tholeiitic picrite, except for rare flows of hawaiite and mugearite that occur in the uppermost parts of the Makaweli and Olokele Members. These rare flows constitute the alkalic postshield stage on Kauai. The flows are so rare and of such limited areal extent that they have not been subdivided as a separate unit. The Koloa Volcanics consists of vents and flows of alkalic basalt, basanite, nephelinite, and nepheline melilitite. Some of the vents along the coast are tuff cones in which the ash is generally altered to palagonite.

The shield-stage flows of the Napali Member dip outward 6-12° from the center of the island and are generally between 1 and 4.5 m thick. In Waimea Canyon, about 80% of the exposed flows is pahoehoe, whereas the remaining 20% is aa. Kauai, unlike most other Hawaiian volcanoes, has no well defined rift zones; dikes radiate outward from the summit caldera in all directions, but are most concentrated trending northeastward and west-southwestward from the summit. The summit caldera, now filled by the ponded flows of the Olokele Member, was unusually large compared to other Hawaiian volcanoes.The graben, now containing the ponded flows of the Makaweli Member, is a feature not seen on any of the other volcanoes. Flank calderas, one now filled with the ponded lavas of the Haupu Member and a second that was only partially filled by tholeiitic lavas, the Lihue Depression, are unique to Kauai (Macdonald et al., 1960).

The tholeiitic basalt flows of the Waimea Canyon Basalt have been dated by K-Ar techniques and yield ages indicating that the tholeiitic shield stage lasted from about 5.1 to 4.3 Ma (McDougall, 1979). The rare hawaiite and mugearite flows of the alkalic postshield stage erupted about 3.92 Ma (Clague and

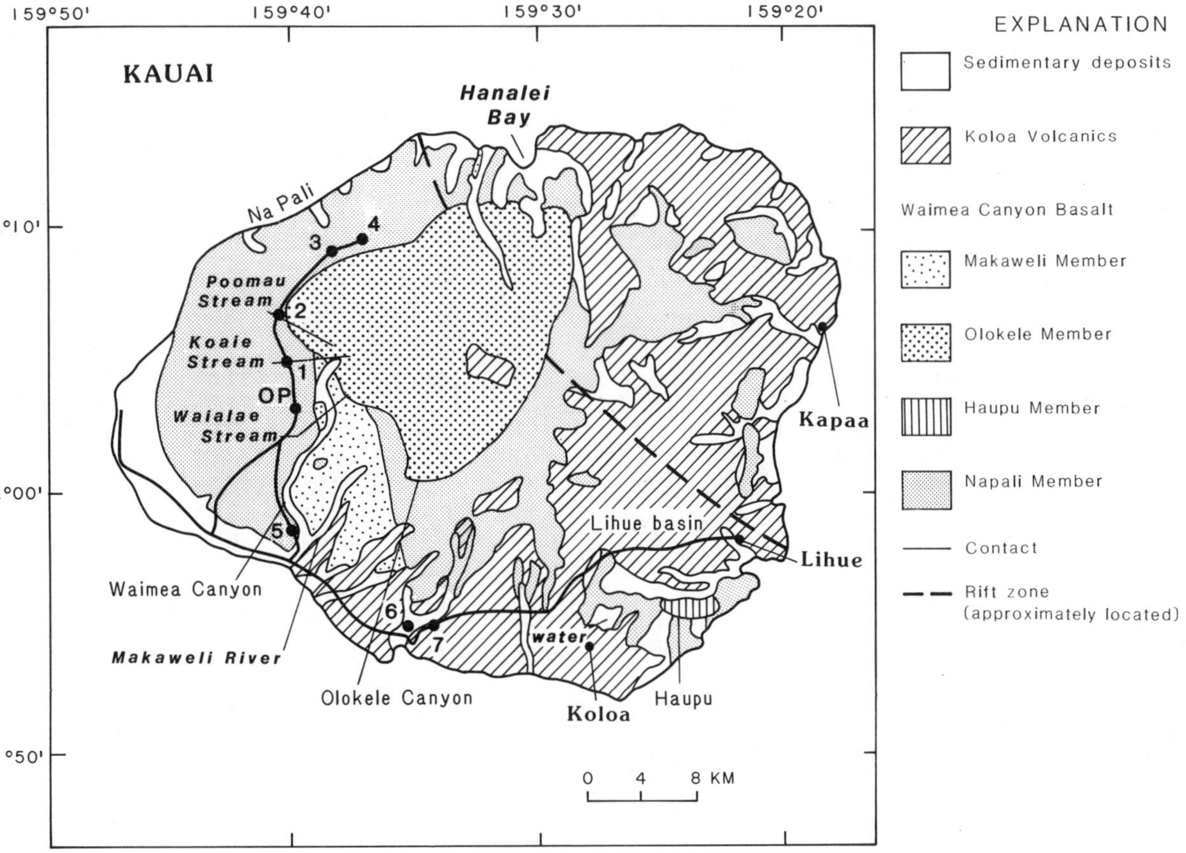

FIGURE 21 Generalized geologic map of Kauai showing the main lithologic units and the route and stops of the roadlog. Waimea Canyon is defined by stream drainages shown in left center of map. Figure from Clague and Bohrson (1987).

Dalrymple,1988). The flows and vents that comprise the rejuvenated-stage Koloa Volcanics erupted from 3.65 to 0.52 Ma from vents mainly scattered over the eastern half of the island (Clague and Dalrymple, 1988). The various lithologies have no spatial or chronological pattern. However, the flows of the Koloa Volcanics older than 1.7 Ma occur in the west-northwestern half of the island whereas those younger than 1.5 Ma occur in the east-southeastern half. All but two of the 21 dated samples are between 0.52 Ma and 2.01 Ma. A distinct peak in eruptive activity occurred at 1.2-1.3 Ma.

A number of analyses of Kauai tholeiitic lavas have been published, most in Macdonald and Katsura (1964) and Feigenson (1984). Table 3 lists representative analyses of the Waimea Canyon Basalt including analyses of tholeiitic basalt from the Napali, Makaweli, Haupu, and Olokele Members and of the rare hawaiite and mugearite flows that occur at the top of the Makaweli and Olokele Members. Many of the analyses of Kauai tholeiitic basalt have low K_2O and Rb, apparently caused by groundwater leaching. The dense samples of dikes define an MgO variation trendline for K_2O that suggests that many of the flow samples have lost half or even more of their original K_2O. The Kauai tholeiitic lavas have rare-earth patterns similar to those of other Hawaiian tholeiitic lavas that are characterized by relatively unfractionated middle- to light-rare-earth elements and strongly fractionated middle- to heavy-rare-earth elements (Clague and Frey, 1980; Feigenson, 1984). The tholeiitic lavas have relatively radiogenic Sr isotopic ratios that are also variable (ratios range from 0.70384 to 0.70343, Feigenson,1984; Stille et al., 1986, and White and Hofmann, 1982) The postshield-stage flows of hawaiite and mugearite have less radiogenic Sr isotopic ratios than the tholeiitic lavas (ratios range from 0.70337 to 0.70331, Clague and Dalrymple, 1988).

The lavas of the Koloa Volcanics cover much of the southeastern half of the island where they erupted from at least 40 vents. Some of these vents are crudely aligned along North-trending rifts; fewer are aligned along rifts oriented N60E. The locations of a few vents are inferred based on the location of small gabbroic bosses (Macdonald et al., 1960); one such boss, consisting of ijolite and alkali gabbro, has been studied in detail (Johnston and Stout, 1984; Johnston et al., 1985). Koloa lavas reach a thickness of 650 m in the east wall of Hanalei Valley. Many flows of the Koloa Volcanics contain sparse to abundant xenoliths of lherzolite and less common dunite. Three flows containing particularly abundant xenoliths crop out in the west wall of Hanapepe Valley in Hanapepe, in Grove Quarry, and in the seacliff on the east side of Kalihiwai Bay. Analyses

TABLE 3 Chemical analyses of lavas from Kauai.

Unit	Waimea Canyon Basalt				Koloa Volcanics			
Sage	Shield Stage		Postshield Stage		Rejuvenated Stage			
Lava Type	Tholeiitic Basalt	Picritic Tholeiitic Basalt	Hawaiite	Mugearite	Melilitite	Nephelinite	Basanite	Alkalic Basalt
ID#	74WAI11	73NP1	86KA5	86KA4	65Ka2	76K9	74K5	74K12
SiO_2	49.5	44.1	47.7	48.1	38.3	40.6	41.8	45.1
Al_2O_3	13.4	8.33	15.4	18.5	9.90	10.6	11.6	12.2
Fe_2O_3	2.37	3.00	5.64	8.20	5.39	8.74	5.47	4.13
FeO	9.74	9.43	6.17	3.33	9.28	5.27	7.86	7.96
MgO	7.54	23.5	5.62	2.55	14.0	12.6	12.6	12.5
CaO	10.3	5.72	8.39	4.88	12.5	12.3	11.3	11.0
Na_2O	2.45	1.22	3.57	4.08	2.50	3.13	2.90	2.46
K2O	0.40	0.07	1.46	2.09	1.22	0.75	0.66	0.63
TiO_2	2.68	1.56	3.28	2.82	3.29	2.98	2.54	1.67
P_2O_5	0.25	0.16	0.81	1.15	0.76	0.64	0.39	0.33
MnO	0.17	0.16	0.16	0.15	0.21	0.20	0.18	0.17
CO_2	-	-	<0.02	0.30	0.02	0.01	-	-
H_2O+	0.34	1.08	0.19	1.85	1.20	0.73	0.98	0.47
H_2O-	0.38	1.05	0.93	1.50	0.90	1.13	0.70	0.50
TOTAL	99.5	99.4	99.3	99.5	99.5	99.7	98.9	99.2

Shield stage lava analyses are unpublished data of D. A. Clague, the rest of the analyses are from Clague and Dalrymple (1988).

of Koloa lavas have been published by numerous workers (see summary in Clague and Dalrymple, 1988), but the new data of Clague and Dalrymple (1988) for 50 lavas is the most extensive and internally consistent to date. This data set includes analyses for 4 flows of nepheline melilitite, 1 sample of the ijolite boss, 18 flows of nephelinite, 7 flows of basanite, and 20 flows of alkalic basalt. Nepheline melilitite flows are more common than indicated by the number of analyzed flows but these flows alter more readily than the others and few are fresh enough to analyze or date by K-Ar. Representative analyses of the various lithologies are presented in Table 3.

The lavas of the Koloa Volcanics have strongly light-rare-earth element enriched patterns similar to those of the Honolulu Volcanics on Oahu (Kay and Gast, 1972; Feigenson,1984; Clague and Dalrymple, 1988). The Koloa lavas also have relatively unradiogenic Sr isotopic ratios (range 0.70299 to 0.70324, (O'Nions et al., 1977; Feigenson, 1984; Clague and Dalrymple, 1988). Detailed petrogenetic models of the source compositions of the Koloa Volcanics are presented by Feigenson (1984) and Clague and Dalrymple (1988).

The roadlog emphasizes the geological evolution of Kauai starting with the formation of the shield, the collapse and infilling of the summit caldera and then the Makaweli graben, and, finally, following a period of volcanic quiescence, the eruption of the rejuvenated-stage lavas of the Koloa Volcanics. The roadlog is a shortened version of a guide to Waimea Canyon by Clague and Bohrson (1987).

Roadlog of Kauai - Day 4.

Miles (km)

0 (0) Intersection of Hawaii 50 and 550 in Kekaha. Turn inland at sign to Waimea Canyon. The coastal plain beneath Kekaha consists of uplifted lagoonal deposits.

1.1 (1.8) Thin bedded flows of tholeiitic basalt and olivine tholeiitic basalt of the Napali Member of Waimea Canyon Basalt. These thin-bedded flows dip 6-12° radially outward from the center of the island and erupted during the shield stage about 5.1-4.3 m.y. ago. The cliff separating the coastal plain from the lava exposures is an ancient seacliff eroded at the time the lagoonal deposits formed.

3.7 (5.9) The roadway climbs up onto the deeply weathered original flow surface. Sugar cane is grown in the thick soils developed on this dip-slope.

10.15 (16.2) Dike on east side of road. Other examples occur at 10.25, 10.5, and 11.15 mi (16.4, 16.8, and 17.8 km), all on the east side of the road. These dikes are part of the poorly defined dike swarm that radiates from the central caldera. The dikes were feeders for overlying fissure eruptions. Most other Hawaiian volcanoes have well-defined dike swarms aligned along prominent rift zones.

11.2 (17.9) **Stop 1.** Turnout to Waimea Canyon Lookout. Short walk from parking area to lookout. The photomosaic in Fig. 22 outlines the geology of the Waimea Canyon Basalt. The Napali Member is located to the west of the canyon. Numerous dikes cross-cut these outward dipping flows. To the east and northeast across the canyon are thick, ponded flows of the Olokele Member. These flows ponded within the summit caldera and are commonly 50 ft (15m) thick and flat-lying. To the east and southeast across the canyon are thick, ponded flows of the Makaweli Member. These flows postdate those filling the caldera and spilled out of the caldera and ponded within a structural graben on the south flank of the shield. The lava of the Makaweli

FIGURE 22 Photomosaic taken from Waimea Canyon Lookout showing major volcanic members of the Waimea Canyon Basalt. Right edge is looking southeast and left edge is looking north-northeast. Figure from Clague and Bohrson (1987).

Member dips gently to the south. Looking towards the east-southeast, one can see the contact between the higher elevation Olokele Member to the north and the lower elevation Makaweli Member to the south.

12.2 (19.5) Puu Ka Pele is the small hill immediately to the south-southeast. It is a small pit crater filled with thick ponded flows that are more resistant to weathering than the thin-bedded surrounding flows.

14.9 (23.8) **Stop 2.** Turnout to Puu Hinahina Viewpoint. There are two lookout points here. The first is a view of Waimea Canyon looking southeast down the canyon. Olokele and Napali Members occur on east and west sides of the valley, respectively. The base of the canyon roughly follows the contact, which defines the caldera-bounding fault. In the distance to the southeast, the lower elevation block is the Makaweli Member. The second lookout provides a view of Niihau Island where the eroded tholeiitic shield (Paniau Basalt) and the low-lying plains of rejuvenated stage alkalic basalt (Kiekie Basalt) can be seen on a clear day. The cone north of Niihau is Lehua Island, a rejuvenated stage tuff cone. A similar tuff cone occurs at the southern point of Niihau Island.

19.35 (31) **Stop 3.** Kalalau Lookout. Bear left into parking area. Thin-bedded flows of the Napali Member have been eroded by heavy runoff, forming "knife-edge" ridges. Some of these ridges are aligned along more resistant dikes. The amphitheatre shape of the valley is similar to the heads of all the larger canyons. The steep slopes plunging to the sea are probably wave-eroded cliffs. This valley is accessible by trail from the northeast end of Hawaii 56 near Haena.

20.55 (32.9) **Stop 4.** Puu O Kila Lookout. A different view into Kalalau Valley. Return along same road.

31.0 (49.6) **Optional stop.** Kukui Trailhead. A short walk along a dirt trail takes one to a viewpoint into the canyon. The thin, outward dipping flows directly beneath the viewpoint are shield-stage tholeiitic lavas of the Napali Member. Across the canyon to the east, the lower block consists of lava ponded in the Makaweli graben, whereas the more distant higher elevation area consists of lava ponded inside the caldera. The Makaweli graben formed after the caldera had been filled to overflowing. Lava ponded in the graben generally erupted within the caldera and flowed southward into the graben The brown ridge just west of the canyon floor is a conglomerate unit deposited by the ancestral Waimea River. The conglomerate is capped by a flow of rejuvenated stage Koloa Volcanics. This flow apparently erupted within the boundaries of the caldera and flowed down the river canyon, covering the conglomerate. This flow is the oldest dated flow of the Koloa Volcanics and is 3.65 Ma. An optional hike down the Kukui trail allows the opportunity to examine the flows of the Napali and Makaweli Members, as well as the conglomerate and overlying alkalic basalt flow of the Koloa Volcanics. The hike to the canyon floor can be completed in 3-4 hr; the trail is steep, strenuous, and slippery when wet.

32.95 (52.7) Intersection of Hawaii 550 and Waimea Canyon Road. Turn left onto Waimea Canyon road. Drive carefully as the road is steep, curvy, and rough.

37.4 (59.8) **Stop 5.** Stop with view of Olokele Canyon to the northeast. Looking southeast, the areas covered with sugar cane are rejuvenated flows of the Koloa Volcanics. These flows weather to bright orange soils.

38.9 (62.2) Steep grade down the ancient seacliff.

39.7 (63.5) Intersection of Waimea Canyon Road and Hawaii 50. Turn left into town of Waimea.

46.2 (73.9) **Stop 6.** Turn left into Hanapepe and continue along west side of Hanapepe River for 0.2 mi, just past bridge on right side of road . The basanite flow exposed in the cliff contains common xenoliths of lherzolite up to about 3 inches in diameter.

48 (76.8) **Stop 7.** Overlook of the Hanapepe River valley. The flows exposed in the canyon walls are all of the Koloa Volcanics. Some of the Koloa flows in this valley flowed down canyons cut in previous Koloa flows thereby indicating the sporadic nature of rejuvenated stage activity. The flows exposed in the

canyon walls are thick mainly because they ponded within the erosional valley. At other locations, young Koloa flows rest on soils developed on earlier Koloa flows, again indicating the duration of rejuvenated stage activity.

OAHU

Introduction

The island of Oahu consists of the eroded remnants of two shield volcanoes- - the younger Koolau Volcano to the east and Waianae Volcano to the west (Fig. 23). Both volcanoes have lost their original shield outline and are now northwest trending ridges shaped mainly by erosion. Each volcano had a summit caldera and well-defined rift zones now identified by dike swarms. The island is roughly 30 by 44 miles .

The lava flows of Waianae Volcano have been subdivided into the shield and postshield stage Waianae Volcanics and the rejuvenated stage Kolekole Volcanics (Sinton, 1986). The Waianae Volcanics is subdivided into the shield stage lavas of the Lualualei Member, the late shield to caldera-filling tholeiitic and alkalic lavas of the Kamaileunu Member, and the alkalic postshield lavas, mainly hawaiite, of the Palehua Member. The Kolekole Volcanics consists of roughly six cinder and spatter cones and their associated flows (Sinton, 1986); these lavas contain xenoliths of mainly dunite. The caldera was about 7 km in diameter and centered on Lualualei Valley. The lavas of the Lualualei Member have been K-Ar dated at 3.8 to 3.0 Ma with an average of 3.62±0.11 Ma (McDougall, 1964; Doell and Dalrymple, 1973). Lavas of the Kamaileunu Member range in age from 3.46 to 2.95 Ma and of the Palehua Member from 2.95 to 2.4 Ma (McDougall, 1964; Funkhouser et al., 1968; Doell and Dalrymple, 1973). The lavas of the Kolekole Volcanics have not been dated but are inferred to be Pleistocene in age (Sinton, 1986).

The lava flows of Koolau Volcano have been subdivided into the shield-stage tholeiitic lavas of the Koolau Basalt and the alkalic rejuvenated stage lavas of the Honolulu Volcanics. The Koolau Basalt includes the Kailua Member which consists of the flows which ponded inside the 13 km wide caldera. The lavas consist mostly of tholeiitic basalt and olivine tholeiitic basalt; picritic tholeiitic basalt is rare. Many of these flows contain phenocrysts and microphenocrysts of orthopyroxene, a rare mineral in lavas from other Hawaiian volcanoes. Some flows near the top of the Koolau Basalt are transitional in composition between tholeiitic and alkalic basalt. The lavas of the Kailua Member are hydrothermally altered; secondary minerals include chlorite, quartz, epidote, zeolites, opal, chalcedony, and clays (Stearns and Vaksvik, 1935).The rift zones are marked by the presence of an intense dike swarm that contains an estimated 7800 near-vertical sub-parallel dikes ranging in width from 10 cm to 3 m(Walker, 1987a).The shield-stage flows from east of the caldera and part of the caldera have been removed by erosion or mass-wasting from a major landslide (Moore,

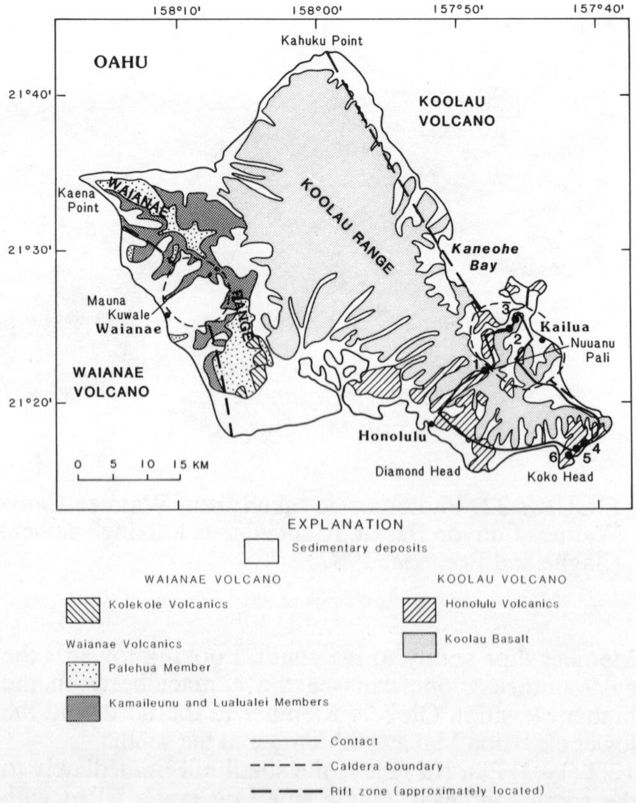

FIGURE 23 Generalized geologic map of Oahu. The roadlog covers the southeastern corner of the island. Figure from Langenheim and Clague (1987) which should be consulted for primary references.

1964). Lavas of the Koolau Basalt have been dated by K-Ar and yield ages of 1.8-2.7 Ma (Doell and Dalrymple, 1973).

The original surface of the Koolau shield is exposed locally, but erosion has carved canyons as deep as 600 m and has exposed the caldera complex and dike swarm to a depth of 1 km. This dissection of the shield, caldera, and dike swarm provide a rare opportunity to examine the upper part of the underground plumbing system of a Hawaiian volcano. The volcano is not dissected deeply enough to expose the intrusive rocks thought to occur beneath the caldera, such as occurs on many of the Society Islands (McBirney and Aoki, 1968). This period of erosion was followed by eruption of the alkalic rejuvenated-stage lavas of the Honolulu Volcanics.

The Honolulu Volcanics consist of the eruptive products from 37 vents or groups of vents on the southeastern part of the Koolau shield. Eruptive vents for the tuff and flows of the unit were mapped by Stearns and Vaksvik (1935), Stearns (1939, 1940a), and Winchell (1947). These vents are arranged along rifts perpendicular to the rift system of the underlying Koolau shield (Jackson and Wright, 1970). The total volume of the Honolulu Volcanics is no more than a small fraction (<1%) of the underlying Koolau shield. Many of the flows of the Honolulu Volcanics fill valleys eroded in the older Koolau Basalt thereby indicating eruption following a prolonged period of volcanic quiescence.

The earliest eruptions of the Honolulu Volcanics occurred in the vicinity of the old Koolau caldera, whereas many of the more recent eruptions occurred along the Koko rift along the southeast coast of Oahu. Table 4 lists these eruptions sequentially by vent (Jackson and Wright, 1970, modified slightly from Winchell, 1947), and includes K-Ar ages. The lavas and tuffs of the Honolulu Volcanics are younger than 1 Ma (Lanphere and Dalrymple, 1980), although most of the flows are only poorly dated due to problems apparently related to the presence of excess argon.

The eruptive products of the Honolulu Volcanics include nepheline melilitite, nephelinite, basanite, alkali olivine basalt, and alkalic basalt (nomenclature of Clague and Frey, 1982).The major lava type erupted from each vent is listed in Table 4. There is no clear relation between lava type and the sequence of eruption; nepheline melilitie and nephelinite were erupted throughout the eruptive sequence. Tuffs and flows from 18 of the vents contain rare to abundant coarse-grained inclusions of dunite, lherzolite, wehrlite, and garnet peridotite (White, 1966; Jackson and Wright,1970; Sen and Presnall, 1986). These xenoliths are zonally distributed with lavas erupted near the caldera containing mostly dunite, whereas farther from the caldera, they contain mostly lherzolite and , and finally, garnet peridotite. However, the areal distribution of lava types proposed by Jackson and Wright (1970) is far less systematic than they proposed (Clague and Frey, 1982).

Vents of the Honolulu Volcanics near the coast experienced phreatomagmatic eruptions which produced wide tuff-rings. Much of the ash generated was deposited from base-surges and commonly displays dune bedding. Most of the ashes are now lithified palagonite tuffs. Many of these vents experienced periods of strombolian activity in the final stages when the vent became sealed to incursion of seawater (Walker, 1987b). Several stops on the fieldtrip will examine these tuff rings.

Many analyses of tholeiitic basalt of the Koolau Basalt have been published, most in Macdonald and Katsura (1964), Macdonald (1968), Jackson and Wright (1970), and Roden et al. (1984). Table 5 lists a few representative analyses of the Koolau Basalt that show the high-SiO_2 and low-FeO nature of Koolau tholeiitic basalt. The Koolau tholeiitic lavas have rare-earth patterns similar to to those of other Hawaiian tholeiitic lavas that are characterized by relatively unfractionated middle- to light-rare-earth elements and strongly fractionated middle- to heavy-rare-earth elements. The Koolau tholeiitic lavas are distinctive isotopically in that they have the highest $^{87}Sr/^{88}Sr$ ratios and the lowest $^{143}Nd/^{144}Nd$ ratios of any Hawaiian lavas, and are isotopically heterogeneous (Roden et al., 1984; Stille et al., 1983). The highest Sr isotopic ratio reported is 0.70455 (Roden et al. ,1984).

Analyses of the Honolulu Volcanics have been published by many workers, but a particularly complete suite of analyses (41 samples) is presented by Clague and Frey (1982). Representative analyses of lavas of the Honolulu Volcanics from this paper are presented in

TABLE 4 Vents for the Honolulu Volcanics.

No.	Vent Name	Rock Type	Xenoliths	K-Ar Age
1	Puu Hawaii Loa	M	Y	-
2	Pali Kilo	N	N	-
3	Pyramid Rock	N	Y	0.71
4	Moku Manu	N	N	-
5	Ulupau	N,T	Y	-
6	Mokolea (Mokapu South)	M,T	N	-
7	Kalihi	M	Y	0.46, 0.58
8	Haiku	N(M)	Y	-
9	Rocky Hill Group	M	N	-
10	Aliamanu Group	M?,T	Y	-
11	Kaneohe Group	M	Y	0.70
12	Luakaha	M,N	Y	0.42, 0.36
13	Makuku	N	N	-
14	Pali Group	B	Y	-
15	Makawao	BR	N	-
16	Kaau	M	Y	0.65
17	Mauumae	N	Y	0.43
18	Salt Lake Group	B,T	Y	-
19	Makalapa	T	Y	-
20	Ainoni	N	N	2.01
21	Castle	N	Y	0.85, 1.01
22	Maunawili	N	Y	-
23	Training School	M	Y	1.13
24	Diamond Head	N?,T	N	-
25	Kaimuki	B	N	0.28
26	Black Point	B	N	0.29, 0.41
27	Kamanaiki	N	N	-
28	Punch Bowl	N	Y	0.30, 0.53
29	Manana	M	N	-
30	Kahauloa	AB	N	-
31	Kalama	AB	N	0.03
32	Koko Head Group	AB	N	-
33	Koko Crater Group	AB	N	0.04
34	Kaupo	AB	N	0.03, 0.32
35	Kaohikaipu	AB	N	-
36	Tantalus Group	M	N	-
37	Sugarloaf Group	M	N	0.07

Rock Types: M, melilitite; N, nephelinite; B, basanite; AB, alkalic basalt; BR, breccia; T, tuff.
Xenoliths: Y, present; N, absent.
Table modified from Clague and Frey (1982).

Table 5. Most of the lavas are primary or near-primary melts generated by variable small percentages of partial melting of a carbon-bearing, isotopically nearly homogeneous, garnet-lherzolite source (Clague and Frey, 1982). The lavas of the Honolulu Volcanics have strongly light-rare-earth enriched patterns and relatively unradiogenic Sr isotopic ratios (0.70317 to 0.70339, Stille et al., 1983; Roden et al., 1984). The nepheline melilitite lavas were generated by about 2% partial melting whereas the alkalic basalt was generated by about 11% partial melting (Clague and Frey, 1982). These low percentages of partial melting failed to deplete the source of minor residual phases, probably including phlogopite, amphibole, and another Ti-rich phase (an oxide?), but not apatite. None of the xenoliths appear to have the modal mineralogy of the inferred mantle source or residue.

The xenoliths in the Honolulu Volcanics have been the focus of numerous studies (e.g. White, 1966;

TABLE 5 Chemical Analyses of lavas from Koolau Volcano, Oahu.

Unit	Koolau Basalt			Honolulu Volcanics			
Stage	Shield Stage			Rejuvenated Stage			
Lava Type	Tholeiitic Basalt	Ol-rich Tholeiitic Basalt	Glass Rim on Dikes	Melilitite	Nephelinite	Basanite	Alkalic Basalt
ID #	69MAK2	WW9948	-	65MOIL2	68BP2	65PAL1	65KAL1
SiO_2	52.76	51.27	52.8	35.52	40.15	42.27	45.14
Al_2O_3	14.32	13.10	14.15	11.01	12.01	12.11	13.37
Fe_2O_3	3.46	2.49	-	9.82	5.80	4.25	3.85
FeO	6.78	8.02	10.15	7.13	7.20	9.17	9.00
MgO	6.62	11.68	6.74	11.46	13.27	12.96	11.46
CaO	9.47	8.16	9.60	12.17	12.73	10.57	10.39
Na_2O	2.71	2.51	2.80	4.84	4.13	3.62	2.94
K_2O	0.50	0.14	0.47	1.72	1.06	0.87	0.65
TiO_2	2.05	1.54	2.00	2.78	1.90	2.42	1.97
P_2O_5	0.28	0.20	0.31	1.05	0.98	0.69	0.35
MnO	0.17	0.16	0.16	0.25	0.23	0.21	0.20
CO_2	0.02	0.01	-	0.26	0.03	0.01	0.05
H_2O+	0.33	0.31	-	0.58	0.14	0.33	0.19
H_2O-	0.24	0.34	-	0.63	0.14	0.31	0.15
TOTAL	99.83	100.08	99.2	99.33	99.95	99.91	99.86

69MAK2 and WW9948 from Roden et al.(1984), Koolau Basalt glass rim on dikes is an average of 6 unpublished microprobe analyses of D. A. Clague, all Honolulu Volcanics data from Clague and Frey (1982).

Beeson and Jackson, 1970; Jackson and Wright,1970; Frey, 1980; Leeman et al., 1980; Sen, 1983, Sen and Presnall, 1986; Goto and Yokoyama, 1988). Table 6 lists major element analyses and mineralogy of most of the major xenolith types found in the Honolulu Volcanics. Most of these xenoliths are from Salt Lake Crater, one of the stops on the fieldtrip.

The roadlog emphasizes features of the geology of the Koolau shield and of the Honolulu Volcanics. Many of these features occur on other volcanoes as well, but are either poorly exposed,not well developed, or inaccessible.The stops selected were choosen from more complete guides to Oahu by Garcia (1979), Walker and Garcia (1987), Walker (1987b), and Moberly and Walker (1987)

Roadlog of Koolau Volcano, Oahu - Day 5

Miles (km)

0 (0) The roadlog begins at the intersection of the Pali Highway (State Highway 61) and the H-1 Freeway at offramp 21B. Begin going northeast on the Pali Highway up Nuuanu Valley.

5.9 (9.5) **Stop 1.** Exit to right for Pali Lookout and Nuuanu Pali State Park. Proceed to parking area.The panoramic view of the Pali (cliffs) and the windward side of the Koolau Range includes most of the caldera complex. The Pali is roughly parallel to the southwest boundary of the caldera, but appears to be caused by stream erosion rather than faulting. The thin-bedded tholeiitic basalt flows of the Koolau shield are exposed along the old Pali Highway that decends east as a trail from the lookout. The flows, half pahoehoe and half aa, are cut by rare dikes that comprise less than 5% of the outcrops. Some of these dikes have glassy margins and jointing perpendicular to the margins. Few of these dikes are vertical. Further along the trail, the Koolau lavas are unconformably overlain by a weathered cinder cone and lava flow of the Honolulu Volcanics. The nephelinite flow contains abundant weathered inclusions of lherzolite as large as 10 cm. The Pali vent of the Honolulu Volcanics erupted on the crest of the Koolau Range and the scoria deposits drape the northeast and southwest sides of the Range. More detailed descriptions of this stop are given by Garcia (1979), Walker (1987b), and Walker and Garcia (1987). Return to the Pali Highway headed northeast, pass through two tunnels.

8.4 (13.5) Turn left at crossroads at traffic signals near bottom of the grade onto State Highway 83 heading west towards Kaneohe.

10.8 (17.4)Turn right onto Kaneohe Bay Drive (Highway 65). Proceed for 2 miles up Mokapu Saddle Road. Turn right onto on-ramp to H3 Highway just before the bridge, proceed on northbound lane towards Kaneohe MCAS.

13.8 (22.2) **Stop 2.** First deep roadcut beyond the bridge displays roughly 90 dikes. The dikes at this location dip NE at 70-80° and the direction of dike opening plunges southwest at about 17°. The dikes are narrow; they average only 74 cm wide and the widest is 4.5 m. Many dikes are injected along the margins of earlier dikes; some are injected up the center of an earlier dike. Some dikes step from one margin of the earlier dike to the other margin. Many of the dikes have thin

TABLE 6 Chemical Analyses and Modes of Xenoliths, Salt Lake Crater.

Rock Type ID#	Lherzolite 68SAL3	Garnet Lherzolite 66SAL1	Garnet Clinopyroxenite 68SAL11	Garnet Websterite 68SAL26	68SAL6
SiO_2	44.03	44.82	44.57	48.42	50.20
Al_2O_3	2.11	8.21	13.6	9.46	8.37
FeO*	8.06	9.77	12.24	10.20	7.14
MgO	42.84	26.53	13.34	16.85	19.00
CaO	1.57	8.12	11.42	12.15	13.28
Na_2O	0.28	0.89	1.69	1.41	0.87
K_2O	0.01	0.03	0.02	0.01	0.02
TiO_2	0.07	0.52	1.76	0.70	0.51
P_2O_5	0.02	0.04	0.02	0.03	0.02
MnO	0.12	0.19	0.21	0.17	0.16
Modal Proportions					
Olivine	76.1	28.7	0.4	-	-
Orthopyroxene	17.5	12.1	-	16.2	22.0
Clinopyroxene	5.5	33.5	61.2	65.6	70.5
Garnet	-	22.1	36.5	16.2	7.3
Oxides	0.9	3.6	1.4	1.4	tr
Phlogopite	-	-	tr	0.5	0.2

All data from Jackson and Wright (1970).

glassy margins, now mainly altered to orange palagonite or a green chlorite-like mineral. Some of the dikes have lineated outer surfaces which indicate the flow direction. The details of this and other outcrops of the dikeswarm are discussed and illustrated by Walker (1987a). A key feature of the outcrop is that the country rock is a coarse breccia having a maximum clast size of about 1 m. The breccia consists of fragments of older dikes and flows and apparently formed during caldera subsidence. The breccia was then lithified by a hydrothermal or low-grade metamorphic event so that it behaved rigidly during injection of younger dikes. Some of the older dikes are amygdaloidal which suggests shallow emplacement whereas the younger dikes are dense, indicating emplacement at greater depth. Continue on H3 for 0.7 miles to parking lot at Kaneohe Viewpoint.

14.5 (23.3) The outcrops display relatively rare dense dikes cutting massive flows. The peak to the northeast is on the axis of the Koolau dike complex; lavas dip away from the axis. Continue on H3 for another 0.5 miles to next deep roadcut just before off-ramp to Kaneohe Bay Drive.

15.0 (24.1) **Stop 3.** Another deep roadcut on H3 that displays at least 150 dikes that comprise about 50% of the outcrop. The countryrock at this and the previous two stops includes massive intracaldera lavas. At this and the first stop in the dikeswarm, older amygdaloidal dikes occur as well as younger massive dikes. The older dikes are absent and there are fewer of the younger dikes at the second stop at the Kaneohe Viewpoint. Figure 24 from Walker (1987a) shows the relations of the three stops and illustrates the development of the dike complex. Exit H3 and turn left onto Kaneohe Bay Drive. Cross Mokapu Boulevard and proceed on Kalaheo Avenue. The canal drains the Kawainui swamp to the south which marks the location of the center of a large positive Bouger gravity anomaly (320 mgals) and the location where rocks having P-wave velocity of 7.7 km/sec occur as shallow as 2 km (Adams and

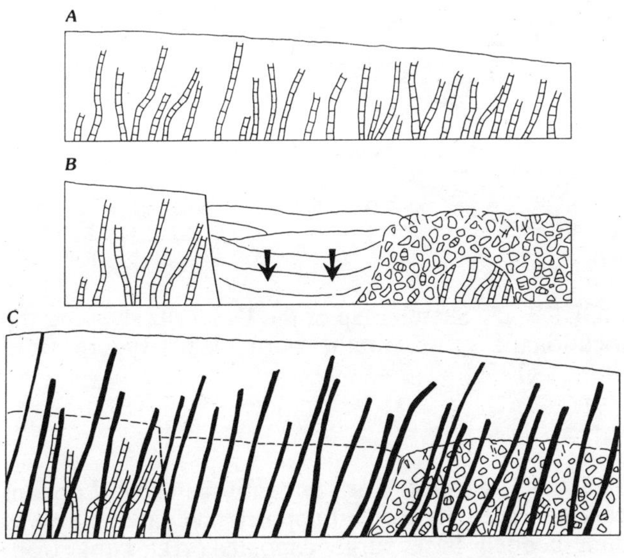

FIGURE 24 Relations of Koolau dikeswarm at three locations discussed in the roadlog. The sites show the development of the dikeswarm starting with injection of dikes near the surface followed by caldera subsidence and local brecciation of the dike complex. The last phase is metamorphism of the complex, lithification of the breccia, and injection of younger dikes at considerable depth. Figure and interpretation from Walker (1987a).

Furumoto,1965). Continue on Kaleheo Ave. for 1.7 miles (2.7 km) and then turn right onto Kuulei Road/Kailua Road. The center of the Koolau caldera is located near a point roughly 0.9 miles along Kailua Road. The Koolau caldera is much larger than those of the active volcanoes Kilauea and Mauna Loa, but still considerably smaller than that on Kauai. The ponded lavas inside Koolau caldera dip towards the center of the caldera; there are relatively few dikes near the center of the caldera. Continue another 1 mile on Kailua Road and turn left onto Highway 72, the Kalanianaole Highway.

19.9 (32.0) Intersection of Kailua Road and Kalanianaole Highway. Proceed towards the southeast around Makapuu Head Lookout (at 32.2 miles, 51.8 km). The Manana (Rabbit)and Kaohikaipu islands to the north of Makapuu Head Lookout are part of the Koko Rift which extends towards the southwest and includes at least 11 vents of the Honolulu Volcanics (Fig.25). Many of these vents are nested tuff rings.

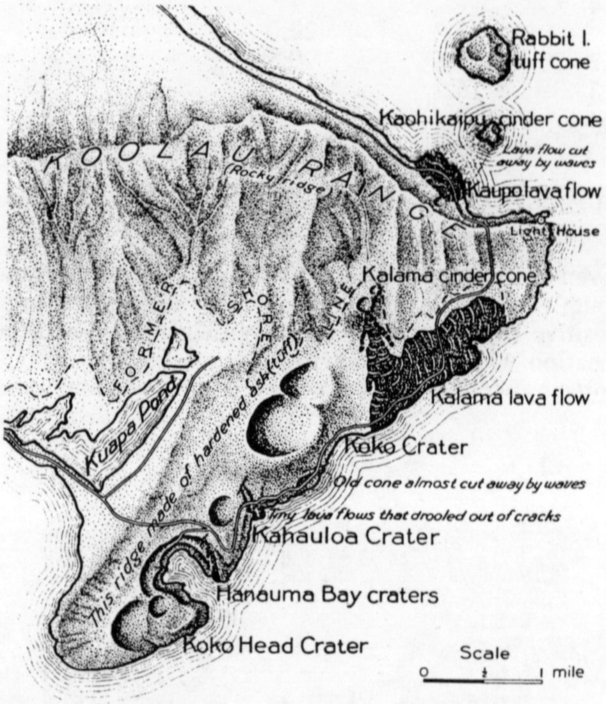

FIGURE 25 Sketch map of the Koko rift showing the location of vents, southeastern Oahu. Figure from Stearns (1939).

35.8 (57.7) **Stop 4.** Turn left into the Lanai Lookout parking area. The roadcut opposite the parking area is dune-bedded base surge deposits originating from Kahauloa Crater to the west. The yellow-brown to gray ash is rich in limestone fragments. Deposition of the ash is predominantly on the steeper lee side of the dunes and can be used to infer flow direction. The tuffs between the parking area and the shore contain well-preserved accretionary lapilli and have preserved imprints of raindrops on several bedding plane surfaces. These features indicate subaerial deposition of the tuff. Reef limestone is exposed on the shore platform below the tuff. A roadcut 60 m northeast of the parking area shows a thin aa flow from Koko Crater on stongly folded ash. The deformed ash is a slumpsheet along the seaward base of Koko Crater. Much of the uppermost part of the tuff is altered to palagonite which forms a boundary that is discordant to bedding. Hay and Iijima (1968) described the sequence of alteration and paragenesis of zeolites in these tuffs. A more complete description of this location is given by Moberly and Walker (1987). Continue southwest on Highway 72.

37.2 (59.9) **Stop 5.** Park on the left side of the road just before right bend in road at top of the hill. The road at the bend crosses the rim of a crater. The walk from the road down the rim of the crater to the coast is on the crest of a constructional antiform. The moat to the southwest separates this crater rim from the rim of the crater surrounding Hanauma Bay. Many ballistic bombs of Koolau Basalt, reef limestone, and an earlier tuff pockmark the bedding plane surface and create impact sags in cross-section. Some blocks apparently originated from Hanauma Bay and others from Kahauloa Crater, indicating that the two craters erupted simultaneously. Return to the road by way of the moat between the nested craters, which contains a thin flow of basanite. A more detailed description of this site is given by Walker (1987b). Continue along Highway 72 towards the southeast.

37.6 (60.5) **Stop 6.** Exit to left on side road to Hanuma Bay. The view from the rim shows a beautiful pocket beach protected by a fringing reef that spans the breach in the tuff ring. This is a state park famous for its marine life. A detailed description of the volcanic geology and a description of the limestone reef is given by Moberly and Walker (1987). Return to the highway, turn left and return to Honolulu.

Roadlog of Salt Lake Crater, Oahu - Day 6.

A second excursion starts in Honolulu and goes to Salt Lake Crater, a tuff ring famous for its abundant ultramafic inclusions. The cluster of tuff rings (Fig. 26) includes Aliamanu and Salt Lake which contain garnet-bearing xenoliths. All the tuff rings except Makalapa, which has few xenoliths, could have formed during the same eruption. The basal deposits consist of pyroclastic scoria and ash. Lithic fragments increase in abundance upsection and were deposited by base surges. Dune bedded surge deposits are generally separated by thin layers of airfall containing accretionary lapilli. The lithic fragments consist mainly of blocks of Koolau tholeiitic basalt that are as large as 1.5 m and are commonly coated by a thin layer of nephelinite. The vents closest to the sea also contain abundant fragments of reef limestone. Aliamanu and Salt Lake contain abundant ultramafic xenoliths as large as 50 cm although most are in the 2-10 cm range. The xenoliths consist mainly of pyroxenite and lherzolite, less abundant garnet-bearing pyroxenite, and rare hornblende and spinel rich xenoliths. Clinopyroxene megacrysts as large as 5 cm also occur. The mileage log begins at the intersection of Highway 78 (the westward continuation of H1) and Puuloa Road, just north of the Honolulu Airport.

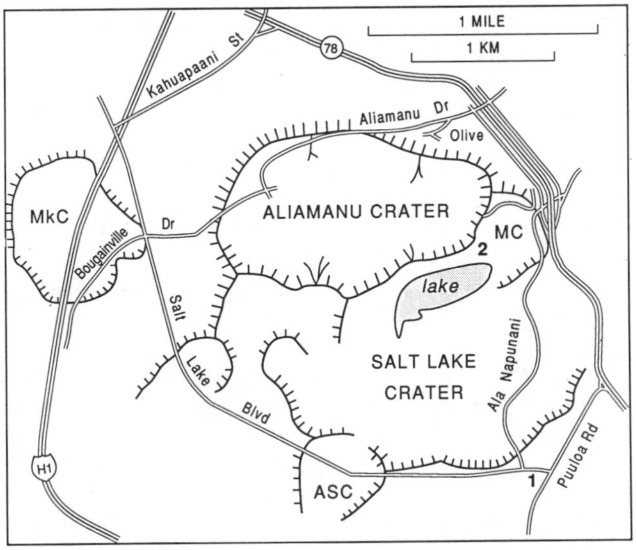

FIGURE 26 Map of the cluster of nested tuff rings around Salt Lake Crater. MkC is Makalapa Crater, ASC is Aliamanu School Crater, and MC is Moanaloa Crater. The stops discussed in the roadlog are indicated by dots. Figure from Walker (1987b).

0.4 (0.7) **Stop 1s.** Proceed south on Puuloa Road towards the airport to the intersection with Salt Lake Boulevard, park beyond the intersection on the right side of the road. About 200 m south of the intersection, the roadcut exposes basal fall deposits from Salt Lake Crater on a thick soil. The basal ash contains some slender tree molds still in their growth positions. Nearer the intersection is upsection in the pyroclastic sequence. A 1 m thick deposit near the base is probably a pyroclastic flow that contains horizontal tree molds that aim away from the crater. This flow is overlain by dune bedded base surge deposits representing at least 25 separate base surges; most are separated by thin airfall layers. Ballistic blocks as large as 70 cm produced impact sags in the surge deposits. Walker (1987b) has described this and other locations in the Salt Lake Crater cluster. Retrace and enter Salt Lake crater on Salt Lake Boulevard . **Stop 2s.** Turn right on Ala Napunani. The crater is 1.5 miles in diameter and flat-floored. The rim is less than 20 m high suggesting that the crater has been infilled by slumping of the rim into the vent. Xenoliths may be collected anywhere within the crater. Please refrain from attempting to collect xenoliths from outcrops of the tuff as they are so friable that they invariably break and destroy the outcrop.

EPILOGUE

This ends the fieldtrip through the Hawaiian Islands. The volcanic history of this area is still being written with submarine preshield to shield stage eruptions occurring frequently at Loihi Seamount, shield stage eruptions occurring frequently on Mauna Loa and Kilauea, postshield stage eruptions likely within the next 200 years on Hualalai and possible on Mauna Kea within a timeframe of several thousand years. Rejuvenated stage eruptions have occurred very infrequently in the past so the liklihood of an eruption is small. However, Niihau, Kauai, Waianae and Koolau on Oahu, West and East Molokai, Lanai, Kahoolawe, and West and East Maui have the potential to have rejuvenated stage eruptions within the next 10,000 to 100,000 years. Of these, all have already had some rejuvenated stage eruptions except West Molokai and Lanai. The most likely to experience rejuvenated stage activity is East Maui, which last erupted in 1790.

Acknowledgments

This fieldguide is based in large part on previous guides prepared by others. In particular, we would like to acknowledge the work of Mike Garcia, Jack Lockwood, Steven Porter, George Walker, and Bill Wise whose fine guides to Mauna Loa, Mauna Kea, Kohala, and the Honolulu area were difficult to improve on. This, and previous, guides are heavily based on the pioneering work by Harold Stearns and Gordon Macdonald. The manuscript was sharpened by careful reviews by Mel Beeson and Alice Davis.

REFERENCES

Adams, W.M., and Furumoto, A.S., 1965, A seismic refraction study of the Koolau volcanic plug: *Pacific Science*, v. 19, p. 296-305.

Atlas of Hawaii, 1973, edited by R. Warwick Armstrong, The University Press of Hawaii, 222 p.

Bargar, K.E., and Jackson, E.D., 1974, Calculated volumes of individual shield volcanoes along the Hawaiian-Emperor Chain: *U.S. Geological Survey Journal of Research*, v. 2, p. 545-550.

Beeson, M.H., 1976, Petrology, mineralogy and geochemistry of the East Molokai Volcanic Series, Hawaii: *U.S. Geological Survey Professional Paper* 961, 53 p.

Beeson, M.H., and Jackson, E.D., 1970, Origin of garnet pyroxenite xenoliths at Salt Lake Crater, Oahu: *Mineralogical Society of America Special Paper* 3, p. 95-112.

Brigham, W.T., 1909, The volcanoes of Kilauea and Mauna Loa: *Memoirs of the Bernice Pauahi Bishop Museum of Polynesian Ethnology and Natural History*, v. 2, no. 4: Honolulu, Hawaii, Bishop Museum Press, 222 p.

Chapman, P.W., 1947, Crystallization phenomena in volcanic ejecta from Kilauea, Hawaii: *American Mineralogist*, v. 32, p. 105-110.

Chen, C.-Y., and Frey, F.A., 1983, Origin of Hawaiian tholeiites and alkalic basalt: *Nature*, v. 302, p. 785-789.

Chen, C.-Y., and Frey, F.A., 1985, Trace element and isotope geochemistry of lavas from Haleakala Volcano, East Maui: Implications for the origin of Hawaiian basalts: *Journal of Geophysical Research*, v. 90, p. 8743-8768.

Clague, D.A., 1982, Petrology of tholeiitic basalt dredged from Hualalai Volcano, Hawaii: *Eos, Transactions, American Geophysical Union*, v. 63 p. 1138.

Clague, D.A., 1987a, Hawaiian xenolith populations, magma supply rates, and development of magma chambers: *Bulletin of Volcanology*, v. 49, p. 577-587.

Clague, D.A., 1987b, Hawaiian alkaline volcanism: *Journal of the Geological Society of London*, Special Publication No. 30, p. 227-252.

Clague, D.A., and Beeson, M.H., 1980, Trace element geochemistry of the East Molokai Volcanic Series, Hawaii: *American Journal of Science*, v. 280-A, p. 820-844.

Clague, D.A., and Bohrson, W., 1987, Geologic field guide to Waimea Canyon, Kauai, Hawaii: *Geological Society of America Centennial Field Guide: Cordelleran Section*, p. 1-4.

Clague, D.A., and Dalrymple, G.B., 1987, The Hawaiian-Emperor volcanic chain Part 1. Geologic evolution: in Decker, R.W., Wright, T.L., and Stauffer, P.H., eds., Volcanism in Hawaii,*U.S. Geological Survey Professional Paper* 1350, p. 5-54.

Clague, D.A., and Dalrymple, G.B., 1988, Age and petrology of alkalic postshield and rejuvenated stage lava from Kauai, Hawaii: *Contributions Mineralogy Petrology*, v. 99, p. 202-218.

Clague, D.A., and Frey, F.A., 1980, Trace element geochemistry of tholeiitic basalts from site 433c, Suiko Seamount, in Jackson, E.D., Koisumi, I., and others, *Initial Reports of the Deep Sea Drilling Project* v. 55: Washington, U.S. Government Printing Office, p. 559-569.

Clague, D.A., and Frey, F.A., 1982, Petrology and trace element geochemistry of the Honolulu Volcanics, Oahu: Implications of the oceanic mantle beneath Hawaii: *Journal of Petrology*, v.23,447-504.

Clague, D.A., Jackson, E.D., and Wright, T.L., 1980, Petrology of Hualalai volcano, Hawaii: Implication for mantle composition: *Bulletin Volcanologique*, v. 43-4, p. 641-656.

Crossan, R.S., and Endo, E.T., 1982, Focal Mechanisms and locations of earthquakes in the vicinity of the 1975 Kalapana earthquake aftershock zone 1970-1979: Implications for tectonics of the south flank of Kilauea Volcano, Island of Hawaii: *Tectonics*, v. 1, p. 495-542.

Dibble, S., 1843, *History of the Sandwich Islands*: Lahainaluna, Hawaii, 65 p.

Doell, R.R., and Dalrymple, G.B., 1973, Potassium-argon ages and paleomagnetism of the Waianae and Koolau Volcanic Series, Oahu, Hawaii: *Geological Society of America Bulletin*, v. 84 p. 1217-1242.

Duffield, W.A., 1975, Structure and origin of the Koae Fault system, Kilauea Volcano, Hawaii: *U. S. Geological Survey Professional Paper* 856, 12 p.

Dzurisin, D., Koyanagii, R.Y., and English, T., 1983, Magma supply at Kilauea Volcano, Hawaii, 1958-83: *Eos, Transactions of the American Geophysical Union*, v. 64, p. 900-901.

Ellis, W., 1963, *Journal of William Ellis; narrative of a tour of Hawaii: or Owhyee; with remarks on the history, traditions, manners, customs, and language of the inhabitants of the Sandwich Islands* (reprint of the London 1827 edition and the Hawaii 1917 edition): Honolulu, Hawaii, Advertiser Publishing Company, Ltd., 342 p.

Feigenson, M.D., 1984, Geochemistry of Kauai Volcanics and a mixing model for the origin of Hawaiian alkalic basalts: *Contributions to Mineralogy and Petrology*, v. 87, p. 109-119.

Feigenson, M.D., Hofmann, A.W. and Spera, F.J., 1983, Case studies on the origin of basalt II. The transition from tholeiitic to alkalic volcanism on Kohala Volcano, Hawaii: *Contributions to Mineralogy and Petrology,* v. 84, p. 390-405.

Fiske, R.S., Simkin, T., and Nielson, E.A., eds., 1987, *The Volcano Letter*, Smithsonian Institution Press.

Frey, F.A., 1980, The origin of pyroxenites and garnet pyroxenites from Salt Lake Crater, Oahu, Hawaii: *American Journal of Science*, v. 280-A, p. 427-449.

Frey, F.A., Kennedy, A., Garcia, M.O., West, H.B., Wise, W.S. and Kwon, S.-T., 1984, Mauna Kea Volcano, Hawaii: Lava compositions and their evolution: *Eos, Transations, American Geophysical Union,* v. 65, p. 300.

Funkhouser, J.G., Barnes, I. L., and Naughton, J.J., 1968, The determination of a series of ages of Hawaiian volcanoes by the potassium-argon method: *Pacific Science*, v. 22. p 369-372.

Garcia, M., 1979, Road log of volcanic rocks and features of part of Oahu: in *Field Trip Guide to the Hawaiian Islands*, Hawaii Institute of Geophysics Special Publicaion, p. 67-80.

Goto, A., and Yokoyama, K., 1988, Lherzolite inclusions in olivine nephelinite tuff from Salt Lake Crater, Hawaii: *Lithos*, v. 21, p. 67-80.

Greeley, R., 1974, Kaumana lava tube, in Greeley, R. (ed.), *Hawaiian Planetology Conference*: NASA Special Publication TMX62632, p. 233-238.

Hay, R.L., and Iijima, A., 1968, Nature and origin of palagonite tuffs of the Honolulu Group on Oahu, Hawaii: *Memoir Geological Society of America*, v. 44, p. 219-230.

Hazlett, R.W., 1987, *Geological Field Guide, Kilauea Volcano*: Hawaii Natural History Association, 117 p.

Hekiler, C., Griggs, J.D., Takahashi, T.J., and Wright, T.L., 1986, Volcano monitoring at the U. S. Geological Survey's Hawaiian Volcano Observatory: *Earthquakes and Volcanoes*, v. 18, no. 1, 69 p.

Hofmann., A.W., Feigenson, M.D., and Raczek, I., 1987, Kohala revisited: *Contributions to Mineralogy and Petrology,* v. 95, p. 114-122.

Holcomb, R.T., 1987, Eruptive history and long-term behavior of Kilauea Volcano, a selective photographic compilation, in Decker, R.W., Wright, T.L., and Stauffer, P.H., eds., Volcanism in Hawaii: *U. S. Geological Survey Professional Paper* 1350, p. 261-350.

Jackson, E.D., and Clague, D.A., 1982, The nodule beds of 1800-1801 Kaupulehu flow, Hualalai

Volcano, Hawaii: *U.S. Geological Survey Miscellaneous Field Studies Map* MF-1355.

Jackson, E.D., and Wright, T.L., 1970, Xenoliths in the Honolulu Volcanic Series, Hawaii: *Journal of Petrology*, v. 11, p. 405-430.

Jackson, E.D., Clague, D.A., Engleman. E., Friesen, W.B., and Norton, D., 1981, Xenoliths in the alkalic basalt flows from Hualalai Volcano, Hawaii: *U.S. Geological Survey Open-File Report* 81-1031, p.33

Johnston, A.D., and Stout, J.M., 1984, A highly oxidized ferrian salite-kennedyite-, forsterite-, and rhonite-bearing alkalic gabbro from Kauai, Hawaii and its mantle xenoliths: *American Mineralogist*, v. 69, p. 57-58.

Johnston, A.D., Stout, J.H., and Murthy, V.R., 1985, Geochemistry and origin of some unusually oxidized alkaline rocks from Kauai, Hawaii: *Journal of Volcanology and Geothermal Resources*, v. 25, p.225-248.

Kay, R.W., and Gast, P.W., 1973, The rare-earth content and origin of alkali-rich basalts: *Journal of Geology*, v. 81, p. 653-682.

Langenheim, V.A.M., and Clague, D.A., 1987, The Hawaiian-Emperor Volcanic chain Part II. Stratigraphic framework of volcanic rocks of the Hawaiian Islands: in Decker, R.W., Wright, T.L., and Stauffer, P.H., eds., Volcanism in Hawaii , *U.S. Geological Survey Professional Paper* 1350, p. 55-84.

Lanphere, M.A., and Dalrymple, G.B., 1980, Age and strontium isotopic composition of the Honolulu Volcanic Series, Oahu, Hawaii: *American Journal of Science*, v. 280-A (Jackson Volume), p. 736-751.

Lanphere, M.A., and Frey, F.A., 1987, Geochemical Evolution of Kohala Volcano, Hawaii: *Contributions to Mineralogy and Petrology*, v. 95, p. 100-113.

Leeman, W.P., Budahn, J.R., Gerlach, D.C., Smith, D.R., and Powell, B.N., 1980, Origin of Hawaiian tholeiites: trace element constraints: *American Journal of Science*, v. 280-A, p. 794-819.

Lockwood, J.P., and Lipman P.W., 1987, Holocene eruptive history of Mauna Loa Volcano, in Decker, R.W., Wright, T.L., and Stauffer, P.H. (eds.), Volcanism in Hawaii: *U.S. Geological Survey Professional Paper* 1350, ch. 18, p. 509-535.

Macdonald, G.A., 1944, Unusual features in ejected blocks at Kilauea Volcano: *American Journal of Science*, v. 242, p. 322-326.

Macdonald, G.A., 1953, Pahoehoe, aa, and block lava: *American Journal of Science*, v. 251, p. 169-191.

Macdonald, G.A., 1968, Composition and origin of Hawaiian lavas, in Coats, R.E., Hay, R.L., and Anderson, C.A., eds., Studies in volcanology: *Geological Society of America Memoir*, v. 116, p. 477-522.

Macdonald, G.A., Abbott, A.T., and Cox, D.C., 1983, *Volcanoes in the Sea*: Honolulu, University of Hawaii Press, 517 p.

Macdonald, G.A., and Katsura, T., 1964, Chemical composition of Hawaiian lavas: *Journal of Petrology*, v.5, p. 82-133.

Macdonald, G.A., Davis, D.A., and Cox, D.C., 1960, *Geology and groundwater resources of the Island of Kauai, Hawaii* (geologic map of Kauai enclosed): Hawaii Division of Hydrography, Bulletin 13, 207 p.

Malin, M.C., Dzurisin, D., and Sharp, R.P., 1983, Stripping of Keanakakoi tephra on Kilauea Volcano, Hawaii: *Geological Society of America Bulletin*, v. 94, p. 1148-1158.

McBirney, A.R., and Aoki, K., 1968, Petrology of the Island of Tahiti: *Geological Society America Memoir*, no. 116, p. 523-556.

McDougall, I., 1964, Potassium-argon ages from lavas of the Hawaiian Islands: *Geological Society of America Bulletin*, v.75, p. 107-128.

McDougall, I., 1979, Age of shield-building volcanism of Kauai and linear migration of volcanism in the Hawaiian Island chain: *Earth and Planetary Science Letters*, v. 46, p. 31-42.

Moberly, R., and Walker, G.P.L., 1987, Coastal and volcanic geology of the Hanauma Bay area, Oahu, Hawaii: *Geological Society of America Centennial Field Guide, Cordilleran Section*, p. 5-10.

Moore R.B., Clague, D.A., Rubin, M., and Bohrson, W.A., 1987, Hualalai Volcano: A preliminary summary of geologic, petrologic, and geophysical data: in Decker, R.W., Wright, T.L., and Stauffer, P.H., eds., Volcanism in Hawaii, *U.S. Geological Survey Professional Paper* 1350, p.571-586.

Moore, J.G., 1964, Giant submarine landslides on the Hawaiian Ridge: *U.S. Geological Survey Professional Paper* 501-D.

Moore, J.G., 1966, Rate of palagonitization of submarine basalt adjacent to Hawaii: *U.S. Geological Survey Professional Paper* 550-D, p. D163-D171.

Moore, J.G., Clague, D.A., and Normark, W.R., 1982, Diverse basalt types from Loihi Seamount, Hawaii: *Geology*, v. 10 p. 88-92.

Murata, K.J., and Richter, D.H., 1961, Magmatic differentiation in the Uwekahuna laccolith, Kilauea caldera, Hawaii: *Journal of Petrology*, v. 2, pt. 3, p. 424-437.

O'Nions, R.K., Hamilton, P.J., and Evensen, N.M., 1977, Variations in $^{143}Nd/^{144}Nd$ and $^{87}Sr/^{86}Sr$ ratios in oceanic basalts: *Earth and Planetary Sciences Letter* v.34, p. 13-22.

Oostdam, B.L., 1965, Age of lava flows on Haleakala, Maui, Hawaii: *Geological Society of America Bulletin*, v. 76, p. 393-394.

Peterson, D.W., and Swanson, D.A., 1974, Observed formation of lava tubes: *Studies in Speleology*, v. 2, pt. 6, p. 209-223.

Peterson, D.W., and Tilling, R.I., 1980, Transition of basaltic lava from pahoehoe to aa, Kilauea Volcano, Hawaii: field observations and key factors: *Journal of Volcanology and Geothermal Research*, v. 7, no. 3/4, p. 339-355.

Peterson, D.W., Christiansen, R.L., Duffield, W.A., Holcomb, R.T., and Tilling, R.I., 1976, Recent activity of Kilauea Volcano, Hawaii, in Gonzales-Ferran, o., ed., *Symposium on Andean and Antarctic Volcanological Problems*, Santiago, Chile, September 1974, Proceedings: Rome, Italy, International Association of Volcanology and Chemistry of the

Earth's Interior, p. 646-656.
Porter, S.C., 1979, Quaternary stratigraphy and chronology of Mauna Kea, Hawaii; a 380,000 year record of mid-Pacific volcanism and ice-cap glaciation: *Geological Society of America Bulletin*, v. 90, p. 980-1093.
Porter, S.C., 1987, Pleistocene subglacial eruptions on Mauna Kea, in Decker, R. Wright, T.L., and Stauffer, P. (eds.), Volcanism in Hawaii: *U.S. Geological Survey Professional Paper* 1350, p. 587-598.
Porter, S.C., Garcia, M.O., Lockwood, J.P., and Wise, W.S., 1987, *Guidebook for Mauna Loa--Mauna Kea--Kohala Field Trip*: Hawaii Symposium on How Volcanoes Work, Jan. 19-25, 1987, Hilo, Hawaii, 40p.
Rhodes, J.M., 1988, Geochemistry of the 1984 Mauna Loa eruption: implications for magma storage and supply: *Journal of Geophysical Research*, v. 93, no. B-5, p. 4453-4466.
Rhodes, J.M., Lockwood, J.P., and Lipman, P.W., 1982, Generation of major basalt types: *Abstract Volume, IAVCEI Scientific Assembly*, Reykjavik, Iceland, Abstract 43.
Richter, D.H., and Murata, K.J., 1961, Xenolithic nodules in the 1800-1801 Kaupulehu flow of Hualalai Volcano: *U.S. Geological Survey Professional Paper* 424-B, p. 215-217.
Roden, M.F., Frey, F.A., and Clague, D.A., 1984, Geochemistry of tholeiitic and alkalic lavas from the Koolau Range, Oahu, Hawaii: Implications for Hawaiian Volcanism: *Earth and Planetary Science Letters*, v. 69, p.141-158.
Ryan, M.P., 1987, Elasticity and contractancy of Hawaiian olivine tholeiite and its role in the stability and structural evolution of subcaldera magma reservoirs and rift systems: in Decker, R.W., Wright, T.L., and Stauffer, P.H., eds., Volcanism in Hawaii: *U. S. Geological Survey Professional Paper* 1350, p. 1395-1447.
Ryan, M.P., Koyanagi, R.Y., and Fiske, R.S., 1981, Modeling the three-dimensional structure of macroscopic magma transport systems: application to Kilauea Volcano, Hawaii: *Journal of Geophysical Research*, v. 86, p. 7111-7129.
Sen, G., 1983, A petrologic model for the constitution of the upper mantle and crust of the Koolau Shield, Oahu, Hawaii, and Hawaiian magmatism: *Earth and Planetary Science Letters*, v. 62, p.215-228.
Sen, G., and Presnall, D.C., 1986, Petrogenesis of dunite xenoliths from Koolau Volcano, Oahu, Hawaii: Implications for Hawaiian Volcanism: *Journal of Petrology*, v. 27, p. 197-217
Sinton, J.M., 1986, Revision of Stratigraphic Nomenclature of Waianae Volcano, Oahu, Hawaii: U.S. Geological Survey Bulletin 1775-A, p. A9-A15.
Stearns, H.T., 1939, *Geologic map and guide of the Island of Oahu*, Hawaii (geologic map of Oahu enclosed): Hawaiian Division of Hydrography, Bulletin 2, 75 p.
Stearns, H.T., 1940a, Four-phase volcanism in Hawaii [abs.]: *Geological Society of America Bulletin*, v. 51, p. 1947-1948.
Stearns, H.T., 1940b, *Supplement to the geology and ground-water resources of the Island of Oahu, Hawaii* (Includes chapters on geophysical investigations by J.H. Swartz, and petrography by G.A. MacDonald): Hawaii Division of Hydrography, Bulletin 5, 164 p.
Stearns, H.T., and Vaksvik, K.N., 1935, *Geology and ground-water resources of the Island of Oahu, Hawaii*: Hawaii Division of Hydrography, Bulletin 1, 479 p.
Stille, P., Unruh, D.M., and Tatsumoto, M., 1983, Pb, Sr, Nd, and Hf isotopic evidence of multiple sources for Oahu, Hawaii basalts: *Nature*, v. 304, p. 25-29.
Stille, P., Unruh, D.M., and Tatsumoto, M., 1986, Pb, Sr, Nd, and Hf isotopic constratins on the origin of Hawaiian basalts and evidence for a unique mantle source: *Geochimica Cosmochimica Acta*, v. 50, P. 2303-2319.
Swanson, D.A., 1973, Pahoehoe flows from the 1969-1971 Mauna Ulu eruption, Kilauea Volcano, Hawaii: *Geological Society of America Bulletin* , v. 84, p. 615-626.
Swanson, D.A., and Christiansen, R.L., 1973, Tragic base surge in 1790 at Kilauea Volcano: *Geology*, v.1, p. 83-86.
Swanson, D.A., Duffield, W.A., and Fiske, R.S., 1976, Displacement of the south flank of Kilauea Volcano: the result of forceful intrusion of magma into the rift zones: *U.S. Geological Survey Professional Paper* 1056, 55p.
Tilling, R.L., Wright, T.L., and Millard, H.T. Jr , 1987, Trace-element chemistry of Kilauea and Mauna Loa in space and time: a reconnaissance: in Decker, R.W., Wright, T.L., and Stauffer, P.H., eds., Volcanism in Hawaii: *U. S. Geological Survey Professional Paper* 1350, p. 641-689.
Walker, G.P.L., 1987a, The dike complex of the Koolau volcano, Oahu: internal structure of a Hawaiian rift zone: in Decker, R.W., Wright, T.L., and Stauffer, P.H., eds., Volcanism in Hawaii: *U. S. Geological Survey Professional Paper* 1350, p. 961-993.
Walker, G.P.L., 1987b, Volcanic geology of the Honolulu Group, Oahu, Hawaii: *Geological Society of America Field Trip Guide, Cordilleran Section Meeting*, May 1987, 15p.
Walker, G.P.L., and Garcia, M.O., 1987, Volcanic geology of the Honolulu Group, Oahu, Hawaii: *Geological Society of America Field Trip Guide, Cordilleran Section Meeting*, May 1987, 15p.
West, H.B., and Garcia, M.O., 1982, Composition and evolution of the alkalic cap stage lavas from Mauna Kea Volcano, Hawaii: *Eos, Transactions, American Geophysical Union*, v. 63, p. 1138.
White, R.W., 1966, Ultramafic inclusions in basaltic rocks from Hawaii: *Mineralogie und Petrographie Mitt.*, v. 12 p. 245-314.
White, W.M., and Hofmann, A.W., 1982, Sr and Nd isotope geochemistry of oceanic basalts and mantle evolution: *Nature*, v. 296, p. 821-825.
Winchell, H., 1947, Honolulu Series, Oahu, Hawaii: *Bulletin of the Geological Society of America*, v. 58, p. 1-48.

Wright, T.L., 1971, Chemistry of Kilauea and Mauna Loa in space and time: *U. S. Geological Survey Professional Paper* 735, 40p.

Wright, T.L., and Helz, R.T., 1987, Recent advances in Hawaiian petrology and geochemistry, in Decker, R.W., Wright, T.L., and Stauffer, P.H., eds., Volcanism in Hawaii: *U. S. Geological Survey Professional Paper* 1350, p. 625-640.

Wright, T.L., and Swanson, D.A., 1987, The significance of observations at active volcanoes, in Mysen, B.O., ed., *Magmatic processes: physiochemical principles*: The Geochemical Society, Special Publication No. 1.

South Cascades Arc Volcanism, California and Southern Oregon

Red Bluff to Redding, California
July 20–26, 1989

Field Trip Guidebook T312

Leader: *L. J. Patrick Muffler*

Associate Leaders:
Charles R. Bacon *Robert L. Christiansen* *Michael A. Clynne*
J.M. Donnelly-Nolan *D.R. Sherrod* *J.C. Smith*

American Geophysical Union, Washington, D.C.

Published 1989 by American Geophysical Union

2000 Florida Ave., N.W., Washington, D.C. 20009

ISBN: 0-87590-563-3

Printed in the United States of America

Cover: Comparison of the effects of the eruptions of Lassen Peak on the 19th and 22nd of May, 1915. Both pictures were taken from the same location by B.F. Loomis, a local photographer, and are from the archives of Lassen Volcanic National Park. The upper picture was taken on the morning of May 22nd and shows the path of the avalanche and debris flow of May 19th. The path of the avalanche was controlled by topography and its edge is sharp and clearly discernable. Trees were uprooted and incorporated into the avalanche. The large rock in the foreground is a piece of a lava dome that was disrupted by the explosion that initiated the avalanche. Note the tree-covered covered slopes on the left and right sides, the splash marks on the snow on Lassen Peak, and the lava flow just to the right of the summit of Lassen Peak.

The lower picture was taken later in the summer of 1915. It shows additional area devastated by the pyroclastic flow of May 22nd. Trees that were still standing on the morning of May 22nd were blown down or snapped off above the ground. Their trunks were left lying on the ground pointing away from Lassen Peak. The lava flow below the summit is gone, and additional debris flows and airfall pumice smoothed the surface of the avalanche deposit in the middle ground. See Stop 5 of the Lassen Peak section of this field trip guide for additional description of the 1915 eruptions of Lassen Peak.

IGC FIELD TRIP T312:
SOUTH CASCADES ARC VOLCANISM,
CALIFORNIA AND SOUTHERN OREGON

TABLE OF CONTENTS

Leader:

L. J. Patrick Muffler
U.S. Geological Survey, MS 910
345 Middlefield Rd., Menlo Park, CA 94025

Associate Leaders:

Charles R. Bacon, Robert L. Christiansen,
Michael A. Clynne, and J. M. Donnelly-Nolan
U.S. Geological Survey, MS 910
345 Middlefield Rd., Menlo Park, CA 94025

D. R. Sherrod
U.S. Geological Survey, MS 901
345 Middlefield Rd., Menlo Park, CA 94025

J. C. Sherrod
U.S. Geological Survey, MS 941
345 Middlefield Rd., Menlo Park, CA 94025

IGC FIELD TRIP T312: SOUTH CASCADES ARC VOLCANISM, CALIFORNIA AND SOUTHERN OREGON

L.J.P. Muffler[1], C.R. Bacon[1], R.L. Christiansen[1], M.A. Clynne[1], J.M. Donnelly-Nolan[1], C.D. Miller[2], D.R. Sherrod[1], and J.G. Smith[1]

INTRODUCTION

The Cascade Range is a late Tertiary and Quaternary volcanic arc that extends north from northeastern California through Oregon and Washington into British Columbia. The volcanic arc lies above an easterly dipping active subduction zone along which the Juan de Fuca, Gorda, and Explorer plates are thrust beneath the North American plate (Riddihough, 1984). In addition to the major composite volcanoes that have erupted andesites, dacites, and even rhyolites, there are many smaller, commonly monogenetic vents that erupted primarily calc-alkaline basalt and basaltic andesite throughout the history of the arc (McBirney, 1978; Luedke and Smith, 1981, 1982; Smith and Luedke, 1984; Guffanti and Weaver, 1988). The volcanic arc is active, with certain historic eruptions at Lassen Peak (1914-17) and Mount St. Helens (mid-1800's; 1980-86), and possible historic eruptions at Mt. Shasta (1786), Mt. Baker (mid-1800's), Mt. Hood (mid-1800's), Mt. Rainier (mid-1800's) and Cinder Cone (east of Lassen Peak; 1851?).

This field guide is designed as a 6-day introduction to the volcanic geology of the Cascade Range in northern California and southern Oregon (**Figure 1**). Emphasis is placed on four major Quaternary volcanic centers of the High Cascades:

- **The Lassen volcanic system** - an andesitic stratovolcano with a large, young silicic domefield on its northeast flank.
- **Medicine Lake volcano** - a largely basaltic shield complex in an extensional environment just east of the main axis of the High Cascades.
- **Crater Lake (Mount Mazama)** - site of a catastrophic pyroclastic eruption of a zoned magma chamber about 6800 years ago.
- **Mount Shasta** - a classic subduction-zone stratocone.

The guide also treats regional Quaternary basaltic and andesitic volcanism of the High Cascades and includes a leg through the Western Cascade Range, the mildly deformed and deeply eroded mid-Tertiary analog of the younger High Cascades.

Most of the basic geologic mapping, petrology, geochemistry, and isotopic dating upon which this guide is based have been developed in the last 10 years as part of the Geothermal Research Program of the U.S. Geological Survey. In turn, these modern investigations rest on the foundations provided by Howel Williams (1932a, 1932b, 1942) and Charles A. Anderson (1933a, 1933b, 1940, 1941) in their pioneering investigations of volcanism in the southern Cascade Range.

Acknowledgments

We acknowledge with gratitude the able assistance of Anne E. Gartner in preparation of the illustrations, as well as helpful technical reviews by Marianne Guffanti and Donald W. Peterson.

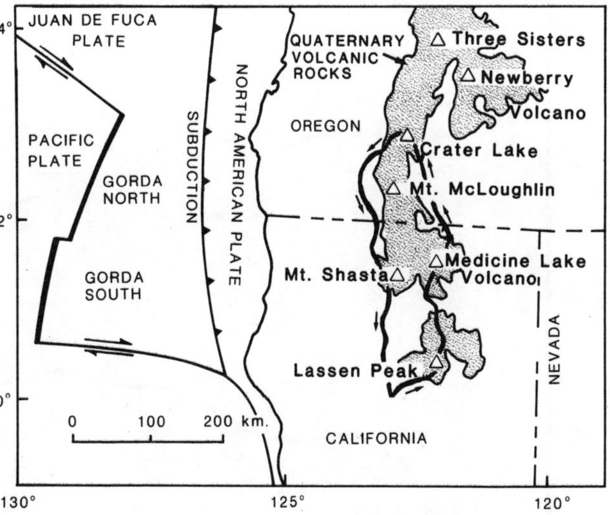

FIGURE 1 Map of the southern Cascade Range showing Quaternary volcanic rocks (pattern) and field trip route.

[1]U.S. Geological Survey, Menlo Park, California
[2]U.S. Geological Survey, Cascades Volcano Observatory, Vancouver, Washington

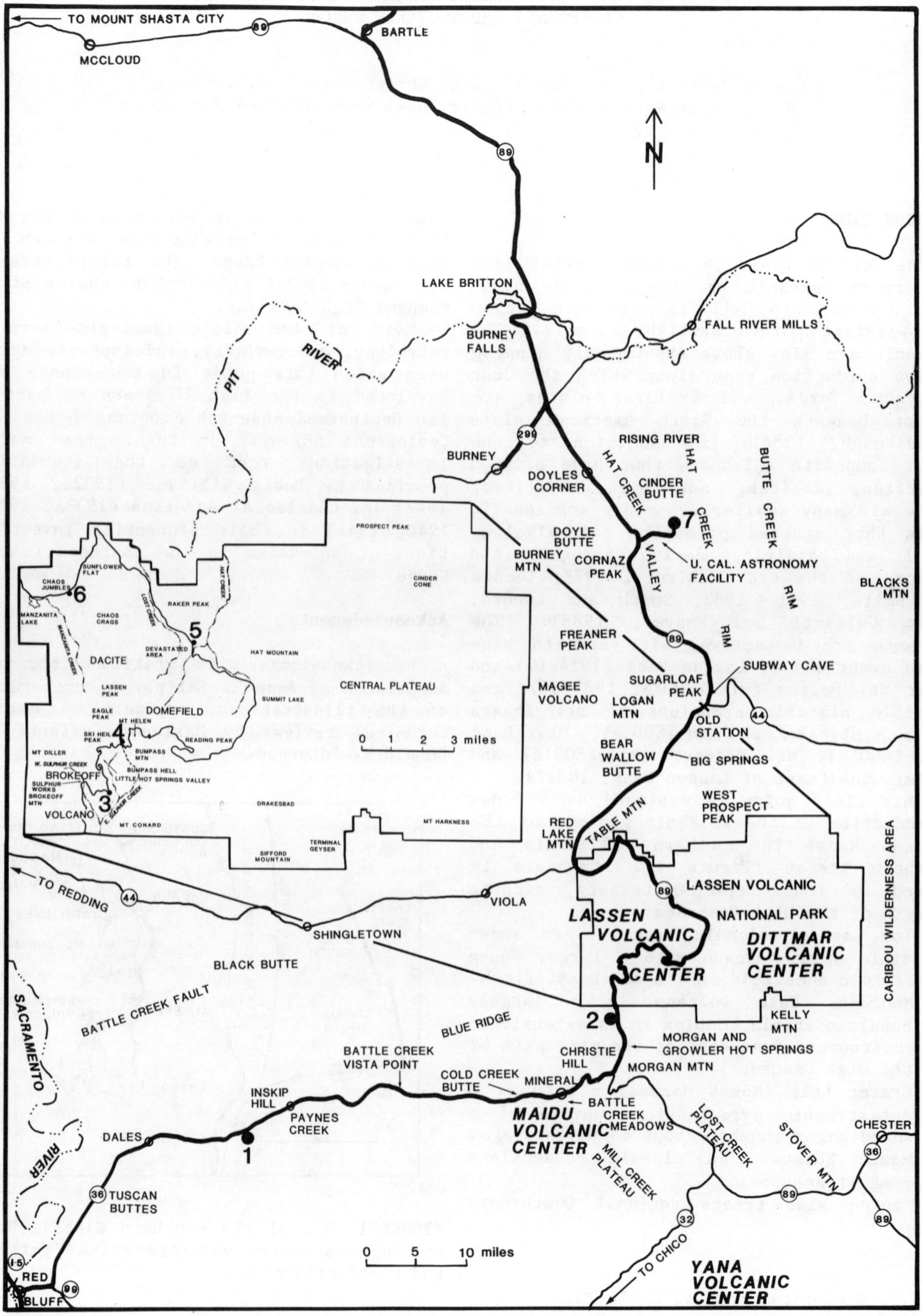

FIGURE 2 Map of the Lassen region showing field trip route.

LASSEN VOLCANIC NATIONAL PARK AND VICINITY

by

M.A. Clynne and L.J.P. Muffler

This leg of the field trip provides an overview of Quaternary volcanic rocks in and around Lassen Volcanic National Park, with a stop at each major stratigraphic unit of the Lassen volcanic center. Additional stops focus on the 1915 eruption of Lassen Peak, the Lassen geothermal system, regional tectonism, and mafic volcanism.

On a regional scale, Quaternary volcanism in the southernmost Cascade Range is predominantly basaltic to andesitic and consists of hundreds of coalescing volcanoes of small to medium volume (10^{-3} to 10^{2} km^3) and relatively short lifetimes (10^0 to 10^3 yr). Superimposed on this regional mafic volcanism are a few long-lived, much larger volcanic centers that have erupted products ranging from basaltic andesite to rhyolite.

Each of the larger centers consists of an andesitic composite cone and flanking silicic domes and flows. Four such centers younger than about 3 Ma have been recognized in the Lassen area (**Figure 2**). Each was built in three stages:

- **Stage I:** Cone-building basaltic andesite and andesite lava flows and pyroclastic rocks.
- **Stage II:** Thick, cone-building lava flows of andesite and silicic andesite.
- **Stage III:** Silicic domes and flows flanking the main cone.

The silicic magma chamber of Stage III provides a heat source for a hydrothermal system that develops within the core of the main cone. Alteration of permeable rocks of the cone facilitates glacial and fluvial erosion of the central part of the volcano. The result is selective preservation of a resistant rim of Stage-II lavas and flanking silicic rocks around a central depression. The three older volcanic centers in the Lassen area, the Dittmar, Maidu, and Yana volcanic centers, have reached this stage, and their hydrothermal systems are extinct. The fourth and youngest center, here called the Lassen volcanic center (LVC), hosts active silicic volcanism and a well developed active hydrothermal system.

Stages I and II of LVC produced the Brokeoff volcano (BV), an 80 km^3 andesitic stratocone (**Figure 3**). The bulk of BV consists of Stage-I deposits (olivine-augite and hypersthene-augite andesite lava flows and stratified pyroclastic deposits) that erupted from a central vent at 600-470 ka. Stage I culminated in eruption of a small volume of hornblende-pyroxene dacite lava. During Stage II, which lasted about 70,000 y, thick flows of porphyritic augite-hypersthene silicic andesite, generally lacking interbedded pyroclastic material, were erupted. Stage III was initiated by eruption of at least 50 km^3 of rhyolitic magma (Rockland air fall and ash flows) at about 400 ka. This eruption is thought to have produced a caldera that is now filled by a silicic domefield. This domefield consists of three groups of rocks totaling about 30-50 km^3:

- **Group 1:** Hornblende-biotite rhyodacite lavas related to the Rockland magma (e.g., Raker Peak).
- **Group 2:** Six domes and flows of 2-pyroxene-hornblende dacite lavas erupted at 250-200 ka (e.g., Bumpass Mt.; Ski Heil Peak; Mt. Helen).
- **Group 3:** Hornblende-biotite dacite and rhyodacite erupted as domes, lava flows, and pyroclastic flows in at least ten episodes, mostly during the past 100 ka (e.g., Lassen Peak; Chaos Crags).

Throughout Stage III, large flows of hybrid andesite totaling 10 km^3 and consisting of thoroughly mixed mafic and silicic magma were erupted peripherally to the silicic domefield, primarily on the Central Plateau.

Porphyritic andesite and dacite with high Al_2O_3, low TiO_2, and medium K_2O contents and FeO/MgO ratios of 1.5-2.0 are the most abundant rock types in LVC. Sparsely porphyritic rhyolite pumice (Rockland) is the single most voluminous unit. Basaltic andesite, rhyodacite, and hybrid andesite are subordinate in abundance. Rocks of LVC resemble other calc-alkaline volcanic rocks emplaced on continental margins overlying sialic crust. Major-element Harker-variation diagrams of LVC show smooth trends from 55 to 75% SiO_2 (Clynne, 1984). The general compositional evolution is from mafic to silicic with time, although the evolution is not strictly sequential (**Figure 4**). Petrographic characteristics and geochemical systematics indicate a complex origin for mafic and intermediate LVC magmas, involving mixing and crystal fractionation of heterogeneous mantle-derived parental melts. Partial melting of young mafic crust may play a role in the origin of the more silicic magmas. Hybrids and ubiquitous quenched mafic inclusions provide abundant evidence for the interaction of mafic and silicic magma. The long span of activity at LVC and the presence of a large hydrothermal system imply an evolving magma chamber.

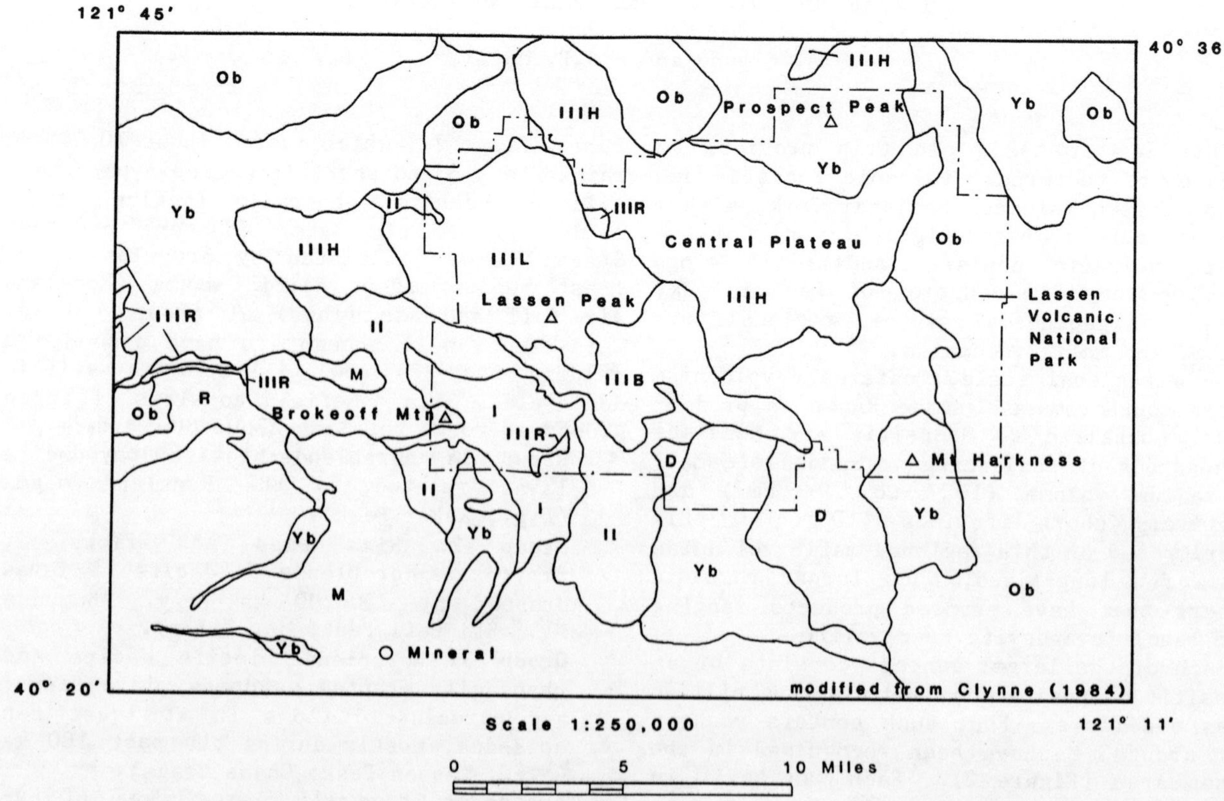

VOLCANIC CENTERS

LASSEN VOLCANIC CENTER

STAGE III

IIIT Hybrid andesites; shields and cones; augite andesite and olivine andesite; 0.30-0 Ma

IIIL Dacite domefield; domes, flows and pyroclastic flows; hornblende-biotite dacite and rhyodacite, mostly 0.1-0 Ma

IIIB Dacite domefield; domes and flows; 2-pyroxene-hornblende dacite, .25-.20 Ma

IIIR Rockland Tephra & associated debris flows, domes and flows; hypersthene-hornblende rhyolite and hornblende-biotite rhyolite approximately .40 Ma

STAGE II

II Brokeoff Volcano; flows; augite-hypersthene andesites; .47-.40 Ma

STAGE I

I Brokeoff Volcano; flows and stratified pyroclastic deposits; olivine-augite andesite and 2-pyroxene andesite, .60-.47 Ma; flows; 2-pyroxene-hornblende dacite; .47 Ma

MAIDU VOLCANIC CENTER

M Composite cone; mafic andesite to andesite; dacite domefield; rhyolite flows; 0.8-2.0 Ma

DITTMAR VOLCANIC CENTER

D Composite cone; mafic andesite to dacite and peripheral silicic rocks; approximately 1.4-2.0+ Ma

REGIONAL VOLCANISM

Yb Younger basalts to andesites; shield volcanoes & cinder cones

Ob Older basalts to andesites; shield volcanoes & cinder cones

FIGURE 3 Generalized geologic map of Lassen Volcanic National Park and vicinity.

LVC differs from most other Cascade volcanoes by having a larger volume of silicic rocks, perhaps due to moderate E-W extension, which favors retention of mafic magma by mixing with more differentiated magma in the crust. The present magma system of LVC can be envisaged (**Figure** 5) as an evolving body of magma 5-8 km in diameter in the middle crust (10-20 km depth) under the NW part of LVC. The upper portion of the chamber contains crystal-rich rhyodacite, which has varied little in composition over at least the last 50 ka years. The system is probably zoned to more mafic compositions at depth. Basalt from depth provides heat and material input to maintain the system in its partially molten state.

Despite the volcanologic and petrologic evidence of a magma system, teleseismic and seismic-refraction studies (Berge and Monfort, 1986; Berge and Stauber, 1987) have failed to detect a magma chamber beneath LVC. A 25-km, oval, 50 mGal negative gravity anomaly is centered on the silicic domefield and the Central Plateau (hybrid andesites) of LVC. This gravity low is probably the expression of low-density volcanic rocks near the surface and Quaternary plutonic rocks at depth. NNW-trending normal faults in the Lassen region reflect the impingement of Basin and Range tectonics on the Cascade arc (Guffanti and Weaver, 1988).

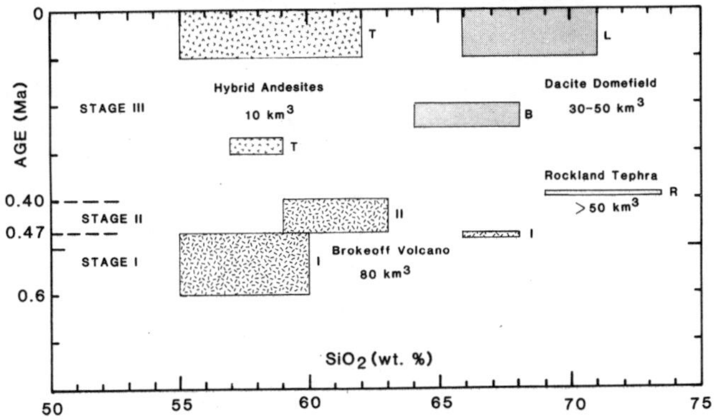

FIGURE 4 Time-composition diagram for rocks of the Lassen Volcanic Center. Symbols as in Figure 2. Some units in Stage III with poorly constrained ages may fall outside the indicated fields. Volume estimates are approximate.

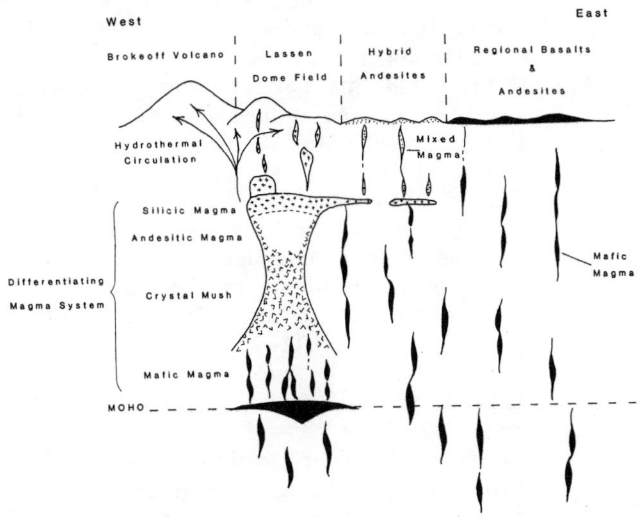

FIGURE 5 Model of Lassen volcanic center.

Ages of Pleistocene rocks in the Lassen region are based on K-Ar dates by G.B. Dalrymple and A.L. Cook, ^{14}C and U-Th dates by D.A. Trimble and S.W. Robinson, and paleomagnetic determinations by D.E. Champion.

Mileage

0.0 At Red Bluff turn E from Interstate 5 onto California Highways 36 and 99. **2.1**

2.1 Turn L on California Highway 36 toward Susanville and Lassen Volcanic National Park. **1.7**

3.8 Roadcuts in Quaternary terrace gravels related to the Sacramento River. **1.1**

4.9 Contact of terrace gravels overlying Tuscan Formation (3.3-2 Ma). The Tuscan Fm. forms a widespread apron W of the southernmost Cascade Range, and consists predominantly of volcanic mudflows (lahars) derived primarily from the Yana and Dittmar volcanic centers. Volcanic sedimentary deposits of fluvial origin are locally significant. Silicic ash-flow and air-fall tuffs, basaltic to andesitic lava flows, and intrusions also occur. The Tuscan Fm. originally covered about 5000 km^2 to a maximum thickness of 500 m and had a volume of 1200 km^3 (Lydon, 1968). Weathering and erosion of debris-flow matrix produces a characteristic "stony plain." **5.2**

10.1 In the distance ahead is the scarp of the E-trending Battle Creek fault. This fault cuts rocks as young as 1 Ma, but lavas slightly younger than 350 ka drape the scarp without offset. **3.2**

13.3 The settlement of Dales and the intersection with county road A16 to the L; continue straight on California 36. For the next mile there are views at 12:00 of the W side of Brokeoff volcano (BV) and of parts of the Maidu volcanic center (MVC). **5.7**

19.0 Roadcuts for the next three miles

alternate between Tuscan Fm. and lava flows from Inskip Hill, a late Pleistocene shield volcano located trenchward of the Cascade arc. Inskip Hill is composed of calc-alkaline basalt that is depleted in incompatible elements relative to the majority of basalts found in the Lassen area. **0.8**

19.8 **STOP 1, debris avalanche in the uppermost part of the Tuscan Fm.** Pull off on broad turnout on R side of road; the debris avalanche is exposed in the roadcut directly opposite the turnout. **Beware of high-speed logging trucks when crossing the highway.** The deposit is probably about 2 Ma, has lost its original hummocky topography, and consists of megablocks of various volcanic rock types in a matrix of crushed material (**Figure** 6). The three most abundant rock types (dark-gray ash-flow tuff with black pumice and sporadic fragments of obsidian; light-gray ash-flow tuff; andesite rich in large green clinopyroxene) can be traced to their source in the Dittmar volcanic center, approximately 48 km E.

Notable features of this debris flow are the heterogeneity of the matrix composition, the invasion of matrix material into fractures and joint surfaces of lava blocks, the jigsaw fit of some blocks, and features indicating laminar flow of matrix material beneath the megablocks at the base of the deposit.

In the roadcut 150 m W, the megabreccia is underlain by an ash-flow tuff of lithology identical to the ash-flow tuff blocks in the megabreccia. Between the ash-flow tuff and the megabreccia is a discontinuous, poorly sorted, channel-filling alluvial deposit.

The roadcut 150 m E exposes thin basalt flows from Inskip Hill, 2.7 km NNE. **11.3**

FIGURE 6 Photograph of roadcut exposing debris flow in the Tuscan Formation (Stop 1).

31.1 Battle Creek Vista Point. View of the Tuscan Fm. overlain by the Blue Ridge Rhyolite (1.2 Ma), a single 15 km^3 flow from the Maidu volcanic center (MVC). Similar huge flows form Lost Creek Plateau, Mill Creek Plateau, and Stover Mtn. on the SE flank of MVC. **0.4**

31.5 White deposits of Rockland ash flow (0.4 Ma) with orange soil are exposed in low roadcuts for the next 2.5 miles. **5.2**

36.7 Roadcuts of andesitic flows and pyroclastic rocks from the MVC. **2.2**

38.9 Quarry in the cinder cone of Cold Creek Butte, an upper Pleistocene olivine-augite calc-alkaline basalt. Flows from this vent dammed Battle Creek to create Battle Creek Meadows. **1.1**

40.0 Bare rubble to R is a basalt flow from Cold Creek Butte. **0.1**

40.1 Cross the South Fork of Battle Creek into Battle Creek Meadows, the center of the hydrothermally altered and deeply eroded core of the andesitic stratocone of MVC; andesites of the stratocone are exposed on the cliffs N of the creek. The stratocone of MVC was active from about 2 to 1.5 Ma. The petrologic evolution of MVC is similar to that of the younger LVC. **1.7**

41.8 Mineral Lodge. To the E, Highway 36 climbs over the resistant rim of the andesitic stratocone of the MVC. The intensity of hydrothermal alteration diminishes abruptly. **2.4**

44.2 White roadcuts for the next two miles are in altered dacites of Christie Hill, one of many dacite domes emplaced at 1.1 Ma as part of Stage III of MVC. **2.1**

46.3 Turn L onto California Highway 89. To the R for the next 1.4 miles is the dacite dome of Morgan Mtn., another of the Stage-III dacite domes of MVC. The valley to the L between Christie Hill and Morgan Mtn. is filled by a 300 ka olivine basaltic andesite flow from a vent high on the flank of Brokeoff volcano (BV). This flow represents a return to regional mafic volcanism after the extinction of BV at 400 ka. **2.3**

FIGURE 7 Photograph illustrating the features of a silicic andesite lava exposed in the Bluff Falls Quarry (Stop 2).

48.6 Exposure in roadcut to L of glacial till of middle Tioga age (approximately 25-18 ka). Glacial deposits in LVNP and vicinity are thought to represent at least five periods of glaciation, which have been tenuously correlated to Sierra Nevadan glacial stratigraphy by Crandell (1972) and Kane (1982). **1.6**

50.2 **STOP 2, Bluff Falls Quarry.** Park on R, walk 100 m N, cross the highway carefully, and climb up the slope just N of the small stream. The quarry exposes nearly the entire thickness of a typical silicic andesite lava flow from Stage II of BV. The base of the flow (not exposed here) consists of thin reddened scoriaceous flow breccia, which is overlain by a massively jointed, faintly flow-banded, glass-rich zone (**Figure** 7). This zone grades upward into a thick zone of thin platy jointing. In the flow interior, platy jointing is nearly horizontal to slightly wavy, but at flow edges it can be ramped upward or highly contorted. On the glaciated flanks of BV, the orientation of the platy jointing can be used to map flow edges and flow direction. The platy interior of the flow grades upward into a zone with columnar joints superimposed on the platy jointing. The upper surface of the flow is glassy and vesicular.

This flow is a porphyritic, 60%-SiO_2 rock containing abundant plagioclase, hypersthene, and augite phenocrysts in a cryptocrystalline groundmass. Sparse olivine xenocrysts are derived from disaggregation of abundant glomeroporphyritic clots that have gabbroic composition and texture and consist of the same minerals as the phenocrysts plus olivine. Patchy recystallization of the plagioclase, vague exsolution lamallae in the pyroxenes, and solid-state replacement of olivine by orthopyroxene suggest that the clots represent cumulate material removed from an andesite magma similar to the andesite of Bluff Falls Quarry and incorporated into its present host. These clots are abundant in all Stage-II lava flows. **1.6**

51.8 Entrance station for Lassen Volcanic National Park. Just beyond is the Lassen Chalet, with a restaurant, rest rooms, and Park information. The entrance station and the buildings beyond are located on a large landslide that 3310 y ago moved 7 km down Mill Creek from the cliff at the base of Brokeoff Mtn.

Note that **collecting or disturbing rock, mineral, or other natural specimens in Lassen Volcanic National Park is prohibited except by special permit. 1.0**

52.8 Sulphur Works, the most accessible of the thermal areas in LVNP. A short boardwalk guides visitors safely through an area of drowned fumaroles, boiling pools, and steaming ground. **1.4**

54.2 **STOP 3, Diamond Peak in the center**

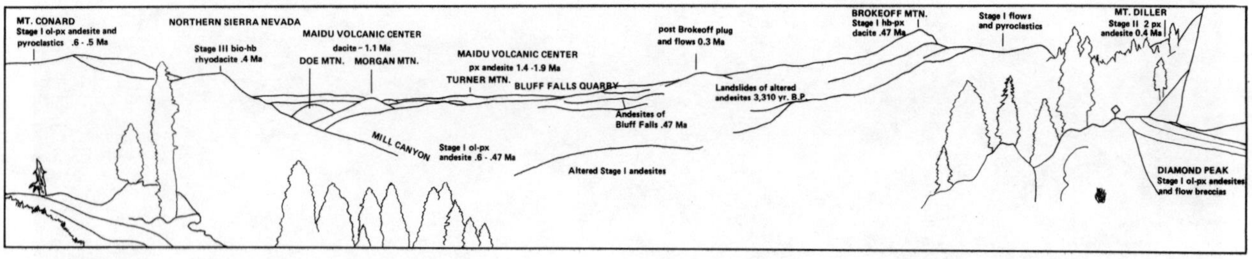

FIGURE 8 Panorama of Brokeoff volcano from Diamond Peak (Stop 3).

of Brokeoff volcano (BV). Most of the BV rocks that can be seen from this viewpoint (**Figure 8**) are at least incipiently altered, and most areas in the core of the cone are strongly to totally altered. Flows and breccias on Diamond Peak dip steeply eastward, indicating that the Stage-I summit vent of BV was nearby to the W. To the S, in the glacial valley of Mill Creek, the base of BV is exposed where it overlies rocks of MVC. Mill Canyon and the flanks of Brokeoff Mtn. expose nearly the entire BV stratigraphy. Thick flank flows of olivine-augite and hypersthene-augite andesite can be correlated across Mill Canyon with no offset by faulting. The triangle-shaped peak of Brokeoff Mtn. to the W is resistant to erosion by virtue of a thick flow of dacite that ended Stage I. To the E, Little Hot Springs Valley displays numerous thermal features. Cliffs on the E side of the valley expose thin lava flows of Stage I intercalated with hydrothermally altered flow breccias and pyroclastic deposits. The cliffs are capped by Stage-II flows. To the SE, Stage-I flows and breccias form most of Mt. Conard. The cliff forming the prominent shoulder above Mill Canyon is a single Stage-II lava flow overlain by a thick rhyodacite lava flow of Stage III, Group 1. Both flows followed the course of a canyon cut in Stage-I rocks of BV. The rhyodacite flow is about 400 ka and illustrates the amount of erosion in the core of BV since that time.

Williams (1932a) attributed the amphitheater in the central area of BV to a caldera-forming pyroclastic eruption. Our detailed mapping, however, has failed to reveal caldera-bounding faults. On the contrary, although the cone is deeply eroded, the stratigraphy is coherent. The Rockland tephra, originally thought to be older than BV, has been shown to be younger than the cone and contains BV rocks as lithic fragments; its source appears to have been the N flank of the BV. The central depression of BV thus was probably caused by fluvial erosion greatly enhanced by hydrothermal alteration and glaciation. Valley walls of altered rock oversteepened by glacial erosion were further destabilized by rapid fluvial downcutting. During interglacial times, landslides were a significant mass-wasting process. A large proportion of the terrain below Brokeoff Mtn. and Mt. Diller consists of landslide deposits, and the valley of Sulphur Creek below Sulphur Works is filled by reworked landslide material. **2.6**

56.8 White, kaolinite-rich altered Stage-I andesite of BV at sharp bend in road. **0.6**

57.4 Just before the S end of Emerald Lake the road crosses the contact between the uppermost preserved rocks of BV and the 250 ka pyroxene-hornblende dacite dome of Ski Heil Peak (Group 2 of Stage III). Straight ahead, Eagle Peak is a 57 ka rhyodacite dome (part of Group 3) that was preceded by a flow (the cliffs ringing the mountain), air-fall pumice (that mantles Ski Heil Peak), and a pyroclastic flow. **0.4**

57.8 **STOP 4, 4-km round-trip walk to Bumpass Hell.** First, walk out to the S end of the parking lot for excellent views of all three stages of the LVC. The parking lot is built on the contact between Stage-II BV flows and Stage-III dacites of Group 2. Note the glacial striations and glacial boulders; the prominent

2-m boulder is pyroxene dacite from Bumpass Mtn.

The gentle, level trail to Bumpass Hell winds along the contact between Stage II and Stage III of LVC. The massive dacite of Bumpass dome (Group 2 of Stage III) is marked by well preserved glacial straiae. All the dacite domes of the Group 2 have suffered extensive glacial erosion. Although their shape has not been greatly modified, no trace of an outer glassy or pumiceous carapace or talus mantle is preserved. The middle part of the trail has been blasted from the massively jointed devitrified interior of the dome. At the viewpoint where the trail turns E, 0.85 km from the parking lot, Stage-II andesite is exposed. The upper flow surface of the andesite was eroded before deposition of the overlying dacite breccia.

The Lassen geothermal system (**Figure 9**) consists of a central vapor-dominated reservoir at a temperature of 235°C underlain by a reservoir of hot water (Muffler et al., 1982; Ingebritsen and Sorey, 1985). The focus and major thermal upflow is at Bumpass Hell, which is located at the vent for Bumpass Mtn., along the contact between BV and the Stage-III silicic domefield. Natural discharge from the deep hot-water part of the Lassen geothermal system occurs only at Morgan Hot Springs and Growler Hot Spring, both located in the canyon of Mill Creek nearly 1000 m below Bumpass Hell. These springs, which discharge near-neutral chloride water and deposit silica, occur at the contact between BV rocks and rocks of the underlying MVC. Part of the deep hot water also flows laterally to the SE, where it is encountered in well Walker "O" No. 1 at Terminal Geyser.

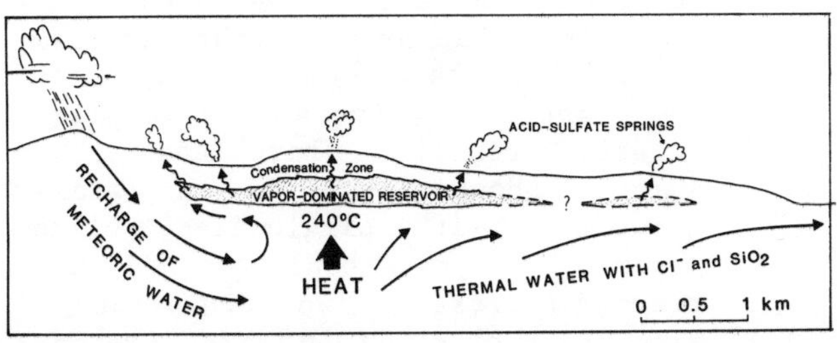

FIGURE 9 Schematic cross-section of the Lassen geothermal system (from Muffler et al., 1982).

Bumpass Hell (Muffler et al., 1983) contains numerous superheated fumaroles, one of which had a temperature in the summer of 1988 of 161.4°C (C.J. Janik, oral commun.). Approximately 75 major fumaroles, acid-sulfate hot springs, and mudpots plus myriad smaller features occur in an area of approximately 0.13 km^2 that is intensely altered to an aggregate of opal and kaolinite. Much of the surface of the active part of Bumpass Hell is covered with orange and yellow sulfates. Pyrite is common in many of the hot springs as linings of the vents and discharge channels, as scum floating on the surface of pools, and as dispersions in gray or black mudpots.

The acid-sulfate water from Bumpass Hell is typical of hot springs related to a vapor-dominated reservoir in having low pH, high sulfate, and no significant Cl. Most other thermal areas in LVNP have thermal features and chemistry similar to Bumpass Hell and are also manifestations of the vapor-dominated reservoir. Some springs near Sulphur Works, in Little Hot Springs Valley, and at Drakesbad are relatively rich in HCO^{-3} and deposit travertine. These springs are interpreted to be surface discharge from the zone of steam condensate that overlies the vapor-dominated reservoir (Muffler et al., 1982). **1.3**

59.1 The large parking lot marks the start of a good trail that gains 600 m in elevation as it switchbacks for 4 km to the summit of Lassen Peak, a dacite dome complex that we think was emplaced between 29,000 and 18,000 years ago. At the summit one can see the 1915-17 eruption craters and deposits and on a clear day have good views of the Sierra Nevada, Sacramento Valley, Coast Ranges, Klamath Mountains, Mount Shasta, and the Medicine Lake Highlands. **0.2**

59.3 Summit of road at 8512 ft. To the R is the Mt. Helen dome, rich in mafic blobs, and on the L is the Kings Creek flow, a 30 ka hornblende-biotite rhyodacite flow. This flow and associated pyroclastic flows issued from a vent now covered by Lassen Peak. Much of the pumiceous glassy carapace and structure of the flow is preserved because it has not been heavily glaciated. Directly ahead, the next roadcut exposes thick breccia at the base of the flow. To the R beyond Mt. Helen are the 230 ka pyroxene-hornblende dacite dome and flow complex of Bumpass Mtn. and a short stubby pyroxene-hornblende dacite flow from Mt. Helen. **0.7**

60.0 View of Reading Peak, a deeply glaciated, 212 ka pyroxene-hornblende dacite dome complex of Group 2. To the R and behind Reading Peak are Mt. Harkness (a late Pleistocene basaltic andesite shield), Kelly Mtn. (a remnant of the Dittmar volcanic center), and Sifford Mtn. (another late Pleistocene basaltic andesite shield). **2.5**

62.5 Excellent view of Lassen Peak at 12:00, Mt. Helen at 10:30, and Bumpass Mtn. at 9:30. The shoulder to the L of Bumpass Mtn. is Stage-II andesite of BV. Further to the L is Mt. Conard, composed of Stage-I andesites. **2.0**

64.5 Cross Kings Creek. To the L is the dacite dome of Reading Peak. To the R is the large dacite flow which vented at Bumpass Mtn.. **0.8**

65.3 The skyline to the R, visible through the trees for the next 2.7 miles, is in the Caribou Wilderness Area, an intensely glaciated area of many small late Pleistocene basalt and andesite vents. **1.3**

66.6 Elevation 7000 ft. For the next 3/4 mi, volcanic vents of the Central Plateau are visible to the R through the trees. Rocks of the Central Plateau are all hybrid andesites. The youngest is Cinder Cone, which definitely erupted 450 years ago and may have erupted in 1851. At 12:00 is Hat Mtn., dated at approximately 25 ka. Exposed along the road is monolithologic till derived from Reading Peak. **1.5**

68.1 Summit Lake. For the next three miles, the road passes through till with sporadic outcrops of Stage-III rhyodacite. **3.3**

71.4 **STOP 5, Devastated Area parking lot.** New field studies by Christiansen and Clynne (1986 and unpublished mapping) and a reanalysis of previous work provide a geologic map at 1:24,000 and a revised scenario of the May 1915 eruptions of Lassen Peak. After a year of intermittent steam blasts and formation of a summit crater 350 m in diameter, in mid-May of 1915 lava welled into the crater to form a small dome of black glassy dacite. The dome grew for 5 or 6 days and on the night of May 19-20 was disrupted by a single explosion. No juvenile material was ejected by this explosion, but fragments of the still-hot dome fell on the snow-covered upper flanks of Lassen Peak, generating an avalanche that flowed 5 km down a 0.8 km-wide path. Fragmentation of the avalanching lava blocks melted enough snow to generate a large debris flow that went 15 km further down Lost Creek. Also on the night of May 19-20, dacite lava erupted from the vent opened by the dome-disrupting explosion, spilled over the low W and NW crater rims, and flowed approximately 300 m down the steep slopes.

After two quiet days, an explosive subplinian eruption on the afternoon of May 22 produced a new crater and erupted compositionally layered (banded) dacite-andesite pumice and unlayered dacite to rhyodacite pumice. Fallback onto the still partly snow-covered E slope of Lassen Peak generated a pyroclastic flow that incorported snow. Melting of this snow transformed the pyroclastic flow into a highly fluid debris flow which continued down Lost Creek. Continued eruption of pumice deposited a fallout lobe to the E and produced several smaller viscous debris flows on the slopes of Lassen Peak.

In the May 1915 eruptions, four types of volcanic rock were produced from Lassen Peak: (1) a black lava dome (64-65% SiO_2) formed between May 14 and 19 and a dacitic lava flow (64-65% SiO_2) emplaced on May 19, (2) dark bands in banded pumice (60-61% SiO_2), (3) light bands in banded pumice (64-69% SiO_2) and unlayered light pumice erupted on May 22, and (4) quenched inclusions of andesitic magma (57-60% SiO_2) that occur in the lavas and light pumice.

The black lava dome and flow and the light pumice contain large plagioclase, biotite, hornblende, and quartz phenocrysts in a glassy groundmass with plagioclase, clinopyroxene, hypersthene, and oxide microphenocrysts. Two populations of each phenocryst type are present: one that is unresorbed, and one that is strongly resorbed. Olivine phenocrysts (Fo_{84}, 2400 ppm Ni) are present in all the rock types.

The dark pumice bands have small olivine phenocrysts in a dark glassy groundmass with tiny pyroxene and plagioclase microlites. All degrees and scales of mixing between the light and dark bands can be found.

Quenched inclusions up to 50 cm comprise up to 5% of the black lavas and light pumice bands. The gray to black inclusions have phenocrysts of olivine in a microvesicular network of acicular plagioclase, clinopyroxene, and hypersthene microphenocrysts, with abundant interstitial light- to dark-brown glass. The dark pumice bands and the quenched inclusions also contain resorbed felsic phenocrysts.

All the 1915 rock types are mixtures. End members are hornblende-biotite rhyodacite (70% SiO_2) and olivine basalt or basaltic andesite (52-54% SiO_2). The quenched andesitic inclusions contain about 35% silicic components. The black lavas were formed by disaggregation of quenched inclusions in rhyodacite and contain about 30-40% mafic components. Dark bands in banded pumice are similar in composition to quenched andesitic inclusions, and light bands to black lava.

Leaving the Devastated Area parking lot, the cliff ahead and to the R is the rhyolite of Raker Peak. This rhyolite belongs to Group 1 of Stage-III and has a composition identical to the most silicic Rockland tephra. The glaciated front face of the dome may have been a caldera-bounding fault related to eruption of the Rockland magma. Note the internal flow structure of the dome revealed by truncation. For the next 1.8 miles, the road crosses 1915 mudflows, then a Tioga moraine, and then drops down onto the mudflows again. **2.5**

73.9 Lost Creek. 1915 debris flows overlie

TABLE 1 Summary of events in the formation of Chaos Crags. Unpublished data of R.L. Christiansen and M.A. Clynne with ^{14}C and tree-ring correlation ages from the USGS Menlo Park Radiocarbon Laboratory.

--3 cold rockfall avalanches from dome 2 formed Chaos Jumbles, ^{14}C age 275 ± 25 years (weighted average of 3 samples), tree-ring correlation age 1619 ± 63 A.D.

hiatus of approximately 700 years

--hot dome-collapse pyroclastic flow from dome 4
--emplacement of dome 4
--emplacement of dome 3b
--warm dome-collapse avalanche from 3a
--emplacement of dome 3a
--hot dome-collapse pyroclastic flow from dome 2
--emplacement of dome 2

hiatus ??

--emplacement of dome 1
--explosive disruption of dome 0 by eruption of pyroclastic flow C, accompanied by a coignimbrite fall deposit and formation of a tuff cone, ^{14}C age 1062 ± 14 years (weighted average of 7 samples), tree-ring correlation age 984 ± 15 A.D.

hiatus of approximately 75 years

--emplacement of dome 0,
--eruption of two column-collapse pyroclastic flows, A and B, ^{14}C age 1124 ± 15 years (weighted average of 7 samples), tree-ring correlation age 911 ± 27 A.D.
--initial vent opening, air-fall pumice and lithic deposit and formation of a tuff cone

pyroclastic flows from the 1000-year-old Chaos Crags eruptions; excellent exposures of both can be seen in quarries 1.5 km upstream.

For the next two miles, the road passes through Tioga till that contains boulders of dacite from Lassen Peak. **2.1**

76.0 To the L are the Sunflower Flat domes, a group of seven 35 ka hornblende-biotite rhyodacite domes preceded by eruption of air-fall pumice and a pyroclastic flow (the flat over which the road passes). Mature forest obscures the relatively youthful morphology. **0.6**

76.6 Two-pyroxene andesite of Table Mtn. on the R came from a vent at 10:30. **1.4**

78.0 Talus on L is from one of the Sunflower Flat dacite domes; the road is once again on pyroclastic flow of Sunflower Flat. **0.8**

78.8 Edge of Chaos Jumbles. **0.6**

79.4 **STOP 6, Chaos Crags, Chaos Jumbles, and quenched mafic inclusions.** Detailed geologic mapping (R.L. Christiansen and M.A. Clynne, unpublished) has added significant new detail to the general stratigraphy of the Chaos Crags given by Crandell et al. (1974). The initial event of the Chaos Crags eruptive sequence 1000 years ago was the eruption of pumice and lithic debris and the formation of a tuff cone (**Table 1**). Eruption of two pyroclastic flows, A and B, quickly followed; these were confined to the valleys of Lost and Manzanita Creeks, where they flowed to a distance of 5 km. Dome 0 grew in the vent of the pyroclastic eruptions and then was partly destroyed by a large pyroclastic eruption that emplaced the widespread pyroclastic flow C. At 3 km from the vent, pyroclastic flow C became confined to the drainages of Lost Creek and Manzanita Creek, where it flowed to a distance of 18 km and 28 km respectively. A coignimbrite fall deposited a lobe of pumice that can be traced for 40 km NE. Near the vent, the coignimbrite fall deposit contains up to meter-sized blocks of dome-0 lava.

A series of domes were then emplaced in numerical order of **Table 1**. Dome 3 of Crandell et al. (1974) is now recognized as two domes: 3a and 3b. Domes 2, 3a, and 4 had hot collapse events that produced short pyroclastic flows. Approximately 300 years ago, the Chaos Jumbles were formed when dome 2 partially collapsed in three cold rockfall avalanches.

All the Chaos Crags units are porphyritic hornblende-biotite dacites or rhyodacites containing 66-70% SiO_2. Dense rocks contain about 40% crystals of plagioclase, biotite, hornblende, quartz, and xenocrystic olivine. The total volume of the Chaos Crags deposits is on the order of 2 km^3.

Quenched inclusions ("blobs") of mafic magma occur in all the rocks but are especially abundant in the later domes. The quenched inclusions consist of a microvesicular network of acicular plagioclase, pyroxene, and hornblende microphenocrysts, with abundant interstitial glass. Euhedral grains of oxide are a conspicuous accessory. The inclusions have a variety of textures from aphyric (with a relatively coarse-grained groundmass) to porphyritic (with a relatively fine-grained groundmass). The phenocrysts in the porphyritic variety are resorbed felsic crystals derived from mixing with the silicic host magma at the time of inclusion formation. Where both aphyric and porphyritic types occur in the

FIGURE 10 Panorama from the Hat Creek fault scarp (Stop 7).

same inclusion, the finer-grained porphyritic material concentrically surrounds the coarser-grained aphyric material. Some quenched inclusions also contain olivine, calcic plagioclase, and clinopyroxene phenocrysts inherited from their basaltic parent.

The Chaos Jumbles consist of three separate rockfall avalanches covering 6.8 km^2 that can be distinguished on the basis of flow margins and grain characteristics. The deposits consist of a monolithologic breccia of Chaos Crags dacite blocks in a matrix of pulverized dacite. The deposits have steep distal and lateral margins 3-5 m in height. The first rockfall avalanche was the largest and traveled 4.5 km with 650 m vertical drop from the breakaway scar on dome 2 of Chaos Crags. The other avalanches were successively smaller and shorter, but thicker. The avalanche paths were controlled by existing topography. The intial direction of each avalanche was WNW toward Table Mtn. The first avalanche deposit rode up onto Table Mtn., where it is found up to 100 m above the valley floor, before being deflected to the W. Groovelike linear features are interpreted by Eppler et al. (1987) to be strike-slip faults caused by compression during flow as the deposit was deflected by Table Mtn. Regularly spaced surface ridges are oriented perpendicular to the direction of flow. The avalanches apparently were emplaced as a high yield-strength material capable of deforming and shearing rather than a plug of nondeforming material being carried on a deforming basal layer (Eppler et al., 1987). **1.6**

81.2 To the L across Manzanita Lake can be seen Loomis Peak and Eagle Peak, both Stage-III rhyodacite domes. The oldest of the three Chaos Jumbles rockfall avalanches crossed and dammed Manzanita Creek. Radiocarbon ages of wood samples collected by D.A. Trimble from trees submerged in the lake that formed behind the dam confirm the conclusion of Crandell et al. (1974) that all three avalanches occurred in quick succession about 300 years ago. **0.6**

81.8 Turn R on Highways 89 and 44. Roadcuts are in pyroclastic flow from Eagle Peak, dated at 57 ka. **1.3**

83.1 Eskimo Summit. Cinder cones to the L comprise the Red Lake Mtn. center, source of basaltic andesite flows to the W and N. For the next 6 miles, the road lies on these calc-alkaline flows. **4.6**

87.7 Flat to the R displays boulders of the 1915 mudflow lying on 1000-year-old pyroclastic flow from Chaos Crags. **1.0**

88.7 To L is the flow front of a young but undated pyroxene andesite from Bearwallow Butte. **0.9**

89.6 The flat plateau at 12:00 is the Hat Creek Rim, uplifted along a normal fault bounding the Hat Creek Valley on the E. The rounded shield volcano on the skyline is Blacks Mtn., an olivine andesite. Good views to the S of Lassen Peak and Chaos Crags. **2.1**

91.7 Big Springs, source of much of the water in Hat Creek, discharge along a prominent fault uplifting lower Pleistocene diktytaxitic basalt, exposed in roadcuts for the next 0.7 mile. **0.7**

92.4 Old Station Post Office, on glacial outwash. **1.5**

93.9 Parking area for the trail to the vents of the Hat Creek Basalt, a diktytaxitic low-K olivine tholeiite (LKOT) flow of probable late Pleistocene age that extends N for 20 miles. **1.6**

95.5 Highway 44 turns R to Susanville. Continue straight on Highway 89. **0.3**

95.8 Turnoff R to Subway Cave, a lava tube in the Hat Creek Basalt. For the next 3.6 miles, the road is on the Hat Creek Basalt. The prominent bare flows on the L are 2-pyroxene andesite from Sugarloaf Peak. **3.7**

99.5 Cross Hat Creek at Bridge Picnic Area, passing onto lower Pleistocene olivine-plagioclase basalt. **3.3**

102.8 Wilcox Road. Roadcuts for 1.5 miles are Hat Creek Basalt, followed by roadcuts in Pleistocene andesites from vents to the W. **4.3**

107.1 Turn R off Highway 89 onto Doty Road at Fireside Village. **1.5**

108.6 Turn R at small white sign for the University of California Radio Astronomy Facility. **1.9**

110.5 Road to R to University of California Radio Astronomy Facility. Continue straight. At 1:30 is a young fault scarp uplifting the Hat Creek Basalt 100 feet along the Hat Creek fault. **0.5**

111.0 End paved road. Turn L on cindered road up fault scarp. 0.3

111.3 **STOP 7, The Cascade Crest (Figure 10), the Hat Creek fault, and basalt magma types.** Walk carefully 100 ft to the N on the tilted flow surface of the Hat Creek Basalt.

To the S is the retreated fault scarp of the Hat Creek Rim and the younger, lower scarp that cuts the Hat Creek Basalt. Rocks exposed on the retreated fault scarp are calc-alkaline basalts and andesites in general similar to the volcanic rocks W of Hat Creek Valley. Previous workers (e.g., Anderson, 1940) have interpreted the younger scarp as a lava slump scarp, produced by the withdrawal of lava from a temporarily dammed part of the Hat Creek Basalt. Our recent geologic mapping, however, shows that the scarp in the Hat Creek Basalt is a true fault scarp. Gravels overlying the Hat Creek Basalt are displaced up to 20 m by the scarp. Furthermore, 0.5 mile N of this stop, a strand of the fault cuts through the Hat Creek Basalt into the underlying basaltic andesite from Cinder Butte. Finally, scarps cutting the Hat Creek Basalt exist only along pre-existing older fault scarps along the E side of the valley; there are no scarps in the Hat Creek Basalt where it lies against nonfaulted volcanic rocks on the W side of the valley.

Near the central parts of individual strands of the Hat Creek fault, the scarps cutting the Hat Creek Basalt are single and vertical, reaching a maximum height of 30 m. Near the end of individual strands, however, the scarps decrease in height and pass into monoclinal folds where the recent movement on the fault has been taken up by adjustment along columnar cooling joints in the basalt flow.

The existence of such a prominent, young fault with no associated modern seismicity suggests that movement on the Hat Creek fault is episodic over periods of perhaps hundreds or thousands of years.

Two types of mafic lavas occur in the area surrounding the Lassen volcanic center. The first is similar to low-K, olivine tholeiite (LKOT) found throughout the NW part of the Great Basin (called HAOT by Hart et al., 1984). LKOT are highly fluid magmas erupted from fissures and producing valley-filling lava flows. The rocks are typically aphyric or contain sparse olivine or plagioclase phenocrysts in a holocrystalline, often diktytaxitic groundmass composed of plagioclase, augite, olivine, and magnetite. LKOT are abundant in the Pliocene and Pleistocene record but rare within the presently active axis of the Cascades, where they are exposed primarily in fault scarps. A wave of LKOT volcanism may have progressed W across the Cascade axis 1-2 million years ago in conjunction with impingement of Basin and Range tectonism on the Lassen portion of the Cascade arc (Guffanti and Weaver, 1988).

LKOT from the NW Great Basin form a homogeneous geochemical group showing little variability (Hart et al., 1984), and LKOT from the Lassen area are chemically equivalent to the average LKOT from the Great Basin. The bulk of these rocks are primitive (Mg# >65; >100 ppm Ni). Distinctive features of LKOT are low SiO_2 (48-50%), low alkali and large-ion lithospheric (LIL) elements, and flat rare-earth element (REE) patterns.

The second magma type is similar to orogenic lavas from continental and mature island arcs and here is designated calc-alkaline basalt and associated andesite (CABA). CABA forms shield volcanoes and lava flows with associated cinder cones. CABA basalts have 49-53% SiO_2, 4-11% MgO, FeO/MgO 0.7-1.7, low TiO_2 (0.5-1.8%) and low K_2O (0.2-1.5%). Modeling suggests that CABA andesites can be derived from basalts by crystal fractionation of observed phenocryst phases, but magma mixing and assimilation also probably occur. Based on abundances and ratios of incompatible trace elements, three groups of CABA can be recognized within a continuum. A **normal** group of medium-K_2O basalts and andesites forms the main trend. A **depleted** group contains lower abundances of TiO_2, K_2O, and incompatible trace elements except for Sr, in which it is often greatly enriched. An **enriched** group contains higher incompatible trace-element and titanium-group element abundances, characteristics normally associated with alkalic rocks.

Preliminary interpretation of isotopic and trace-element systematics of LKOT and CABA suggest that the most primitive samples of the two basalt types could be primary and that they cannot be related by any closed system or assimilation process (T.D. Bullen and M.A. Clynne, unpublished data). We are left with the alternative that the mantle source region and conditions of generation of basaltic magmas are heterogeneous in the Lassen region.

Return down cindered road and paved road to the W. **2.7**

114.0 Turn R at T junction. **0.9**

114.9 Turn R on California Highway 89. Road is still on Hat Creek Basalt. Lower Pleistocene andesites exposed on slopes to W. **4.2**

119.1 Doyles Corner. The most northerly outcrops of the Hat Creek Basalt occur in this meadow. 1 mile to the R is Rising River, an immense group of cold springs discharging groundwater derived from the Lassen highlands 45 km S. The low fault scarp 0.3 miles straight ahead exposes a Pleistocene diktytaxitic LKOT. **4.9**

124.0 Flashing red stop light at junction with California Highway 299. Continue straight. **5.8**

129.8 Turnoff L to Burney Falls. Continue straight. **1.7**

131.5 S end of bridge across Lake Britton in the valley of the Pit River. Roadcuts both N and S of the bridge expose Pliocene diatomite interlayered with diktytaxitic LKOT. **1.1**

132.6 Roadcut on R exposes pillow lava interbedded with lacustrine sedimentary rocks, diatomite, and air-fall tuffs. **8.5**

141.1 Road to R to Dana and MacArthur. Continue straight. **11.4**

152.5 Cross railroad. **1.2**

153.7 Turn R on Harris Spring Road toward Medicine Lake. **0.1**

153.8 Stop sign at railroad tracks. **4.3**

158.1 Intersection with paved road to R.

MEDICINE LAKE VOLCANO CALIFORNIA

by

Julie M. Donnelly-Nolan

Medicine Lake volcano is a Pleistocene and Holocene shield volcano covering about 2500 km^2 with a volume of about 600 km^3. Its lower flanks are dominated by mafic lavas, especially basalt, which is commonly primitive high-alumina basalt. Higher on the volcano, andesitic lavas dominate, and some high-silica lavas are present, notably the late Holocene rhyolites and dacites of Glass Mountain and Little Glass Mountain, and the Medicine dacite flow. Well preserved Holocene flows of a variety of compositions from basalt to rhyolite will be seen on this field trip as well as evidence of tectonic activity in the form of open ground cracks and major normal faults.

The classic geologic reference for Medicine Lake is Anderson (1941), which contains excellent descriptions and the best published geologic mapping of the volcano. Numerous other papers have been written about the volcano, particularly petrologic studies by Mertzman (1977a, 1977b), Grove and Baker (1984), Grove and Donnelly-Nolan (1986), and Grove et al. (1988). These papers and Donnelly-Nolan (1988) include references to earlier work.

A previous field guide contains more logistical information and some different stops (Donnelly-Nolan, 1987). Some of the stops written up here (**Figure 11**) have been taken directly from that guide, although additional information has been added.

Mileage

0.0 Junction of paved roads 4.3 mi N of Highway 89 and Bartle. Turn R and drive NE across Pliocene high-alumina basalt known regionally as the Warner Basalt (Anderson, 1941). Note thickness of red soil, much thicker than on any Medicine Lake lavas. **9.1**

9.1 The road drops over an E-facing fault scarp onto post-glacial basalt erupted from the S flank of Medicine Lake volcano. The road crosses the lava tube that begins at Giant Crater and can be traced at least 14 mi, making it one of the longest known lava tubes (Greeley and Baer, 1971). **8.8**

17.9 Sign for Jot Dean ice cave and parking area. The cave is in Giant Crater lava erupted at Double Hole Crater. Note the deflation that occurred here after the flow formed a "high-water mark" and the lava flowed away to the south. **0.7**

18.6 Turn L onto good graveled road that heads W toward Grasshopper Flat. Half a mile after the turnoff, pass the large spatter vent of Double Hole Crater on your L. Continue W another 1.7 mi to shallow quarry on L. **2.2**

20.8 **STOP 1, Giant Crater event.** Turn into quarry on L and park on mafic air fall from Chimney Crater, one of many vents that erupted basaltic lava during the Giant Crater event about 10.6 ka (D.E. Champion, unpublished carbon-14 data) (**Figure 12**). Walk SW up hill on near-vent agglutinate that has apparently been intruded and uplifted by compositionally identical basaltic lava (52.7% SiO_2) that can be seen in the E wall of Chimney Crater, located about a quarter mile W of the vehicles. This viscous lava is the earliest and most silicic lava erupted during the Giant Crater event, which may have continued for as much as a few tens of years on the basis of paleomagnetic data (D.E. Champion, unpublished data). Successively more

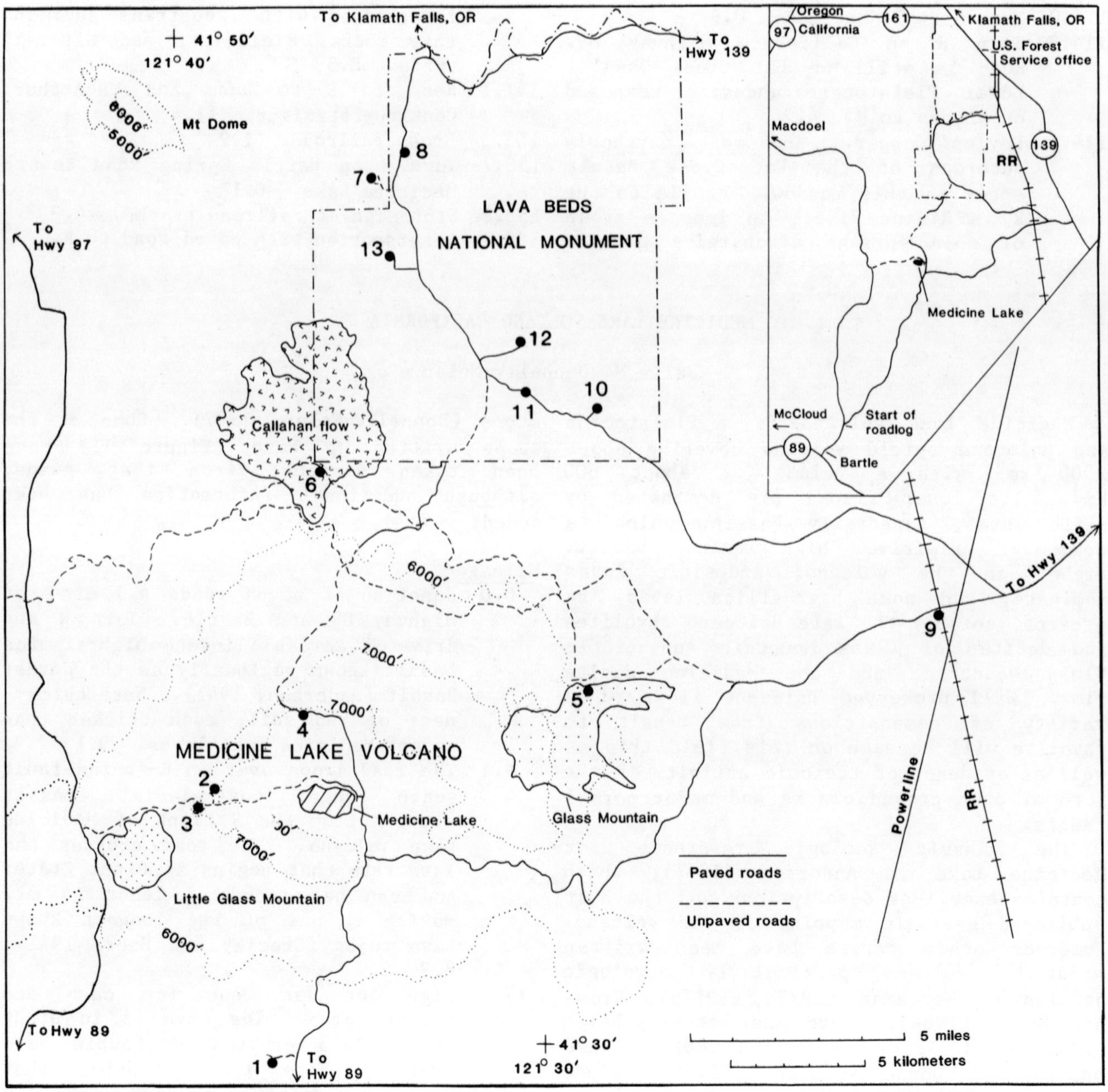

FIGURE 11 Location map of Medicine Lake volcano showing major roads and features and field trip stops.

mafic lava was produced from a number of vents including Chimney Crater, Giant Crater, Double Hole Crater, and several other unnamed vents aligned N 55° E, N 5° E, and N 20° W. Chimney Crater and vents to the N and S aligned N 5° E produced contaminated basaltic lava containing quartz, sodic plagioclase, and iron-rich orthopyroxene xenocrysts (see groups I and II on **Figure 13**). Melted silicic inclusions several cm across have been found at Chimney Crater. Numerous silicic inclusions that have "popped" can be found in the Chimney Crater agglutinate. Look in the mafic spatter between the crater and the vehicles for round void spaces up to several cm across that have a rind of dark-gray rhyolitic-looking glass. Mafic cumulate inclusions can also be found in the agglutinate. Chimney Crater also erupted more mafic, more fluid lava (group III on **Figure 13**) that overflowed to the W and traveled by lava tube in the direction of Giant Crater to the WSW. It is unclear whether there is any connection between the Chimney tube and Giant Crater. Subsequently, lavas of group IV erupted in large volumes, perhaps exceeding 2 km^3 from Giant Crater and Double Hole Crater. Much of the lava transport to the S was by lava tubes formed during this part of the event. The S end of the flow consists mostly of group V lava, basalt with about 0.12% K_2O that was produced from the most primitive composition by fractional crystallization of olivine. Group V lava was transported by tube from Giant Crater although little of the approximately 1.5 km^3 of this material is found near the vent. Most of the group VI lava, the most primitive composition containing only 0.07% K_2O, was transported away from the vents by lava tube. Volume of this diktytaxitic group VI basalt is less than 0.5 km^3. We will not see any of the group IV, V, or VI lava on this trip, but much older lava of nearly identical composition can be seen at Stop 9 of the Lava Beds segment of the field trip. Total volume of the Giant Crater event is estimated at 5 km^3, one of the largest units known at Medicine Lake volcano and 5 times larger than the largest known silicic eruption, Glass Mountain (Donnelly-Nolan, 1988).

Retrace route and return to paved road. **2.3**

23.1 Turn L on paved road and drive N toward Medicine Lake. **6.3**

29.4 Junction of paved roads. Continue straight ahead to the N. Road crosses the poorly defined S rim of the Medicine Lake caldera. **1.6**

31.0 Turn L on paved road toward Medicine Lake. **0.4**

31.4 At stop sign at T junction, turn R and follow paved road around N side of Medicine Lake past end of pavement at entrance to Medicine campground.

FIGURE 12 Aerial view of Giant Crater, Chimney Crater, and other vents of the Giant Crater system (Stop 1).

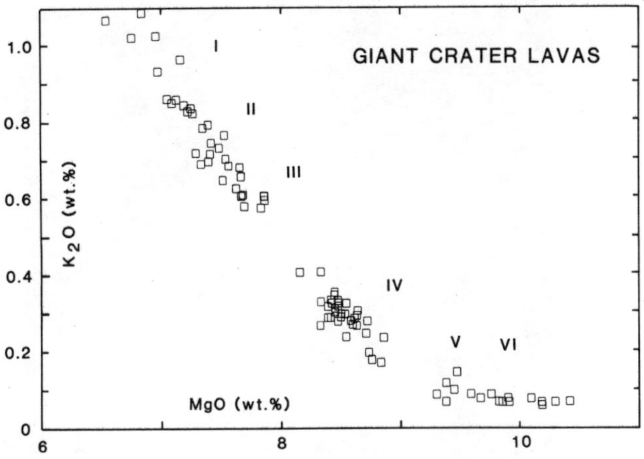

FIGURE 13 Plot of K_2O vs. MgO for Giant Crater lavas (Donnelly-Nolan and T.L. Grove, unpublished data) (Stop 1). Groups I-VI are labeled.

Continue straight at guard station. **1.7**

33.1 Dead-end road to Schonchin Spring on R; continue straight. **0.6**

33.7 Keep R at a Y junction of equally poor roads. **1.6**

35.3 Turn R onto a nearly hidden narrow dirt track marked by a green arrow. Follow it N, then NW to first wide spot on L where several vehicles can be parked. **0.3**

35.6 **STOP 2, Cracks.** Cross the road and walk about N 30° E through area of fallen trees. You will begin to see large cracks in the ground. Continue walking NE about 100 m to the largest crack near the edge of unlogged forest. This crack is as deep as 10 m and about as wide (**Figure 14**). There is evidence of collapse that occurred about 1985. Note that the ragged edges of the cracks can be fit back together in an E-W sense. This set of cracks developed in andesite over a buried rhyolite dike (Fink and Pollard, 1983) during emplacement of the late Holocene Little Glass Mountain and several domes of identical composition on trend farther to the NE. The walls of the cracks expose the interior of a fountain-fed andesite flow. You can readily see the agglutinated texture and angular lithic fragments typical of this type of flow. The andesite was erupted from the glaciated cinder cone just S of the parking spot. Return to vehicles and retrace route to road junction. **0.3**

FIGURE 14 Photo of ground crack that opened during Little Glass Mountain eruptions (Stop 2). Jon Fink for scale. Snow is several feet deep.

35.9 Turn R and drive W to base of Little Mt. Hoffman. **0.2**

36.1 Junction of dirt roads. Proceed straight ahead on left-hand road. **0.7**

36.8 **STOP 3, Little Mt. Hoffman lookout.** Take middle of the three branches of the road as you approach the summit of Little Mt. Hoffman. Located on the W rim of the caldera, this is one of the best viewpoints on the volcano. Spread out below you to the W is the late Holocene rhyolite of Little Glass Mountain. Mt. Shasta dominates the W skyline. Sixteen km E is the dome of Glass Mountain (see Stop 5), on the E side of the caldera. To the S on a clear day you can see Lassen Peak, and to the N, Mt. McLoughlin and sometimes even Mt. Scott on the E rim of Crater Lake caldera. To the N on the lower flank of Medicine Lake volcano is the area where flooding took place following emplacement of andesite tuff (Donnelly-Nolan and Nolan, 1986, and Stop 7 description). The age of Little Mt. Hoffman is unknown, but I think that it was overtopped by glacial ice during the late Pleistocene. Little Mt. Hoffman cinders contain 51.7% SiO_2.

Little Glass Mountain has been the subject of several recent papers (Mertzman and Williams, 1981; Fink, 1983; Grove and Donnelly-Nolan, 1986). This flow and the time- and composition equivalent domes to the NE contain a suite of inclusions varying from granite to hornblende gabbro cumulate to basaltic and andesitic magmatic blobs.

Retrace route E past turnoff to Stop 2. **3.0**

39.8 View of Medicine Lake to R at Guard station. Continue straight. **1.1**

40.9 Pavement resumes. Continue straight. **0.6**

41.5 Turn L away from Medicine Lake. **0.3**

41.8 Turn L onto main paved road from S. Road narrows to N. Follow it around the late Holocene Medicine dacite flow, which lies just N of Medicine Lake on the NW floor of the caldera, and drive W up the S-facing N rim of the caldera. **3.1**

44.9 **STOP 4, N rim of caldera.** Pause at the wide turnout on the L, half-way up the brush-covered slope. This is a brief photo stop for a view of the Medicine dacite flow and Medicine Lake to the S. The lake sits in a bed of glacial clay that prevents the water from draining away into the permeable volcanic rocks. The caldera does not appear to be the result of ash-flow eruptions. The suggestion of Anderson (1941) that it resulted from voluminous eruption of andesite lavas around the rim of the caldera appears to be the best explanation, although previous eruptions early in the history of the volcano may have initiated the summit depression. The andesite of the N rim is very similar to the andesite at Stop 2; both are late Pleistocene in age and both have been glaciated. The muted topography of the caldera is partly a result of glaciation, which has stripped the tops of lava flows and rounded off cinder cones. This rim andesite has been glaciated, and an outcrop displaying good glacial polish and striations can be seen on the L just after the vehicle completes the climb up the rim and turns to the N.

Continue driving N. **0.5**

45.4 Pavement ends just beyond intersection. Continue straight (N) on very rough road. **3.9**

49.3 Intersection of major graveled roads. Turn R on Cougar Butte Road. Continue E on this main graveled road to first major road on R. **6.3**

55.6 Turn R and proceed S. Good dirt road becomes narrow track in pumice. Pumice cover thickens markedly as you approach Glass Mountain. When you arrive in the pumice mining area, drive toward the front of the big rhyolite flow that sparkles ahead of you. Turn L on the major haul road, then R onto a small road around the front of the rhyolite lobe rising steeply above you. Park at the edge of the flow where a big pile of obsidian obstructs a bulldozer track that goes up onto the flow. **2.1**

57.7 **STOP 5, Glass Mountain (Figure 15).** Walk up the bulldozer cut and examine the rhyolite. Pieces range from black aphyric obsidian to pumiceous, lighter-colored samples; some samples have interesting textures that show evidence of breakage, oxidation and flowage while still hot. This rhyolite forms the youngest lobe of Glass Mountain. It contains about 74% SiO_2; USGS rock chemistry standard RGM-1 was collected from this lobe. The bulk of the flow is rhyolite ranging from 73-74% SiO_2, some of which contains basaltic andesite magmatic inclusions, although this youngest lobe contains none. Below you on the E is the earliest part of the flow, consisting of dacite of about 67% SiO_2, which contains abundant magmatic inclusions of basaltic andesite. This flow was studied by Anderson (1933) who pointed out that the rhyolite and dacite are intimately mingled in places, but do not appear to mix into gradational types. My observations suggest that this is not strictly true, and I have collected a few transitional samples. Eichelberger (1975) first explained all the mafic inclusions as lithic fragments but later recognized that they had been injected into the host dacite and rhyolite as liquid blobs (Eichelberger, 1981). Two mixing events are suggested: first a basaltic

FIGURE 15 Aerial view of part of Glass Mountain flow. Location of Stop 5 is shown where youngest rhyolite lobe comes down across earlier dacite lobe. Dome formed by last-erupted rhyolite is in middle distance.

component was mixed into rhyolite magma and homogenization to dacite took place; secondly, basaltic andesite magma was injected into the dacite, forming blobs that chilled against the host silicic magma. Glass Mountain lava erupted from a dike trending N 30° W along which are at least 13 vents, 2 or 3 for the main flow, one forming a small dome to the S, and 10 forming domes to the N. The dome to the south and the 6 northernmost domes contain a varied suite of inclusions in addition to the typical basaltic andesite blobs. Some magmatic inclusions contain abundant olivine and are basaltic. Hornblende gabbro cumulates are present as are granitic fragments. The range of inclusion types is similar to that found in the slightly older Little Glass Mountain flow. The age of these late Holocene flows is uncertain but is estimated to be about 1000 years. Heiken (1978) reviewed the ^{14}C and stratigraphic evidence.

Return to vehicles and retrace the route N from the pumice mining area to the T intersection with Cougar Butte Road. **2.0**

59.7 Turn L onto Cougar Butte Road. **0.3**

60.0 Turn R at first good road. This road proceeds approximately NW down the N flank of the volcano. **5.5**

65.5 Cross paved Medicine Lake-Lava Beds road onto narrow dirt road that was an old railroad grade when logs were carried away by train rather than by truck. Drive W about 3 mi onto late Holocene basaltic andesite flow shown on Anderson's 1941 map as the Callahan flow. Where you initially intersect the flow, its silica content is 52.2%. Just beyond is Cinder Butte, the principal vent (54-55% SiO_2). Silicic inclusions that were partly melted can be found among the cinders that make up the cone. Proceed past the cone to a small quarry in the flow. **3.2**

68.7 **STOP 6, Callahan flow.** Quarrying has exposed the center of a small flow lobe that has been sampled here more than once for paleomagnetism and chemistry. SiO_2 content of the lava here is 54.1%. The paleomagnetic direction (D.E. Champion, unpublished data) is in good agreement with a ^{14}C date of 1110±60 years (Donnelly-Nolan and Champion, 1987) from a dead tree in the edge of the flow. The flow fronts on the N and E edges of the flow are steep and high, indicating viscous lava, and the silica content at these localities is 57-58%. As many as 5 sub-flows of different compositions have been recognized (J.M. Donnelly-Nolan, T.L. Grove, and R.J. Kinzler, unpublished data, and Kinzler, R.J., 1985), such that the earliest lava is most silicic and the latest is more mafic by nearly 6% SiO_2, suggesting a stratified magma reservoir.

Retrace route E to paved road. **3.1**

71.8 Turn L onto paved road. **1.7**

73.5 Pavement ends as you enter Lava Beds National Monument. Note sign for Mammoth Crater as you enter the monument. This pit crater appears to have been the source for several major lava tubes that distributed basalt throughout the monument and beyond. Several additional vents to the N are part of the same event, which produced about 5 km^3 of basaltic lava in a very short time span (D.E. Champion, unpublished paleomagnetic data) in the late Pleistocene. **2.6**

76.1 Turn L on main paved road through Lava Beds National Monument. **2.9**

79.0 Scenic overlook to R. Good view as you drive NW of N-trending fault blocks tilted to W. Mt. McLoughlin is visible in the distance. **3.1**

82.1 Turn L on a paved road that quickly becomes dirt. **0.5**

82.6 Keep R at Y junction onto less frequently used road. **0.1**

82.7 Keep L at Y junction at large juniper trees. **0.1**

82.8 **STOP 7, Andesite tuff.** Walk down into shallow gully on L and examine the brownish-red ash-flow tuff that forms low outcrops on both sides. SiO_2 content of single pumice lumps is about 63%. Estimated thickness here prior to some quarrying was less than 2 m, perhaps less than 1 m. The base of the tuff is exposed at the N end of the gully. This small patch of partially welded andesite tuff is one of many on the N and W sides of Medicine Lake volcano. It is also present in a few small patches on the E side. The spatial distribution of the tuff indicates that it was erupted at or near the center of the volcano. The largest pumice and lithic fragments are found in the one large outcrop in the caldera, suggesting a source within the caldera. The tuff is younger than one of the andesite flows on the NW rim of Medicine Lake volcano (the andesite in the cracks at Stop 2), but that andesite flow is not broken by any caldera-forming faults, implying that eruption of the tuff was not responsible for creating the caldera. The complete absence of this unit anywhere on the caldera rim, together with indications of hydrothermal alteration at the caldera outcrop, indicate that the tuff erupted through an ice cap on the volcano and deposited in the caldera on the only exposed ground surface, where a fumarole had melted the ice. Coarse gravels and anastomosing dry channels cut in the tuff on the NW flank of the volcano indicate that meltwater from the interaction of the ash-flow and the ice formed a catastrophic flood (Donnelly-Nolan and Nolan, 1986). The tuff has not been dated directly, but the indirect argument about the presence of an ice cap, together with evidence from younger lava flows, some of which have been glaciated, point to a late Pleistocene age predating the latest glaciation. Anderson (1941) recognized the andesite tuff but incorrectly interpreted its common exposure in the bottoms of gullies to indicate that it was one of the volcano's oldest units. Stratigraphically, the tuff is of major importance as the volcano's only marker bed.

Retrace route to paved road. **0.7**

83.5 Turn L on paved road. **0.8**

84.3 **STOP 8, Devils Homestead overlook.** Stop at overlook along the Gillem fault and view the basalt of Devils Homestead below you. The basalt was erupted from spatter vents at Fleener Chimneys farther S along the fault. To the S is a panoramic view of Lava Beds National Monument and the Medicine Lake shield, the N flank dotted with cinder cones (**Figure 16**). To the N, Gillem fault continues and forms the W margin of Tule Lake basin. Medicine Lake volcano lies astride the S end of the Klamath graben in a strongly extensional tectonic environment E of the main axis of the Cascades.

Continue N on paved road that turns E near the N edge of the monument. **2.6**

86.9 Turn L (N) on Hill Road and follow it to the N boundary of the monument where the road is graveled for about 7 miles. Gillem fault scarp is on L; Tule Lake (in places reclaimed for farm land) on R. **9.0**

95.9 Headquarters of Klamath Basin National Wildlife Refuges on L after pavement resumes. Continue straight N. **3.9**

99.8 Turn L on Highway 161 (State Line Road) and almost immediately turn R on Malone Road and continue driving N (you are now in Oregon). **1.9**

101.7 Turn L on Oregon Highway 39. **2.2**

103.9 Prominent flagpole with American flag at center of Merrill. End of roadlog.

FIGURE 16 View of Medicine Lake shield from NE.

LAVA BEDS NATIONAL MONUMENT

by

Julie M. Donnelly-Nolan

Samples may not be collected in Lava Beds National Monument. Note also that rattlesnakes may be present. The trip route is shown on **Figure 11.**

Mileage

0.0 Prominent flagpole with American flag at center of Merrill, OR. Drive SE on Oregon Highway 39. **2.6**

2.6 Bear 30° R on Highway 39 toward Tule Lake. **2.4**

5.0 State line. Continue straight on the highway, which becomes California Highway 139. **2.7**

8.7 Flashing yellow light at town of Tulelake. **1.5**

10.2 Modoc County Line. Excellent views to the R of Gillem Bluff, Medicine Lake Highland, and Mt. Shasta in the far distance. **5.8**

16.0 On the R are eroded palagonite tuff rings formed by eruption of basalt into ancient Tule Lake, a basin that has been present for at least 3 Ma and fresh for most of its history despite its lack of a surface outlet (D.P. Adam, unpublished results from 1100-ft drill core). On the L is what little remains of the largest of the World War II Japanese-American relocation camps. **14.3**

30.3 Highway crosses over a railroad at Perez Overhead. Good view to the R of Medicine Lake Highland; the prominent gray area is the Glass Mountain flow, and the white patches are pumice quarries. **2.8**

33.1 Turn R on county highway 97. **2.6**

35.7 To R is National Forest road 10 to Lava Beds National Monument. Continue straight. **1.5**

37.2 **STOP 9, Tionesta high-alumina basalt.** Park at graveled area to the L of road just past railroad tracks. Walk back E across the railroad tracks, then N along a jeep track to a conspicuous pine tree, and then a short distance W to the railroad. Examine diktytaxitic, tholeiitic high-alumina basalt (SiO_2 47.7%, Al_2O_3 18.3%, K_2O 0.07%, TiO_2 0.70%) in walls of railroad cut. Samples from this unit have been studied by Nockolds and Allen (1953), Yoder and Tilley (1962), Hart (1971), and Philpotts et al. (1971), as well as by Anderson (1941), all of whom documented the primitive nature of the lava. Basalts similar to this one can be found all around the lower flanks of Medicine Lake volcano as well as pre-dating the volcano. They have erupted throughout the history of the volcano into post-glacial time (e.g., basalt of Giant Crater) and represent the parental material for the derivative lavas of Medicine Lake volcano by a variety of processes including assimilation, mixing, and fractional crystallization (Grove and Baker, 1984; Donnelly-Nolan, 1988). This basalt was called Warner Basalt by Anderson (1941), but it is much younger than most basalt of the Modoc Plateau that is generally described under this loosely used name. Its surface has been muted by a covering of glacial outwash gravel and late Holocene white pumice from Glass Mountain. The basalt probably erupted from now-buried vents on the E flank of Medicine Lake volcano.

Walk approximately 50 m directly toward a conspicuous snag in line with Lyons Peak (just to L of Glass Mountain) to N-S tensional cracks related to regional extensional tectonics. Follow tensional cracks S to paved road.

Retrace route E along paved road. **1.5**

38.7 Turn L on National Forest road 10 toward Lava Beds National Monument. **9.9**

48.6 Enter Lava Beds National Monument. **2.1**

50.7 Turn R toward Valentine Cave. **0.2**

50.9 **STOP 10, Valentine Cave.** In the words of Aaron Waters (see Donnelly-Nolan et al., 1981, p. 143) who studied and mapped the monument caves for the Park Service:

"Valentine is an interesting, clean and varied cave. Here one can see a lava-tube system which is almost undamaged by post-lava collapse. The cave shows most of the features to be found in lava tubes: among them pahoehoe floors, lava pools, and lava cascades; well developed lavacicles and dripstone on ceilings and walls; extraordinarily interesting lava benches, one of which marks a high-level stand of viscous lava that attempted to crust outward from the walls; and still another kind of bench made by penetration and "bulldozing" away of collapse blocks and rubble as molten lava under hydraulic pressure forced its way through the tube. Large pillars around which the lava stream divided and reunited are present in the upper part of the cave. The central part of the cave shows interesting lava falls and cascades through which the lava stream was transferred from a higher level to a lower level. Downstream from this area of breakdowns, the Valentine tube subdivides into distributaries which are gradually phased out downstream by filling with lava."

The basalt of Valentine Cave has an SiO_2 content of 52.8%, unusually high for a unit with such well developed lava tubes. The spatter vents for this flow lie about 5.5 km to the SW outside the monument. The flow overlies unconsolidated glacial outwash sediments. This is one of three units in the monument that host the nearly 300 lava-tube caves, most of which occur along the half dozen tube systems of the basalt of Mammoth Crater (e.g. Skull Cave, Mushpot Cave).

Return to main road. **0.2**

51.1 Turn R at T junction with main road. **1.7**

52.8 **STOP 11, Visitors Center and Mushpot Cave.** This brief stop will allow you to look at the displays in the Visitors Center and buy books and postcards if you wish, as well as visit the restrooms. Mushpot Cave is located in the center of the parking lot and is lighted.

Return to main paved road and turn L. **1.4**

54.2 Road to L to Mammoth Crater; continue straight. **0.4**

54.6 Turn R onto Lyons Road. **1.1**

55.7 **STOP 12, Skull Cave.** This enormous lava tube transported lava from Modoc Crater, one of the vents of the Mammoth Crater event, around the S and E sides of the preexisting andesite of Schonchin Butte (large cinder cone with lookout, just to the N) to the NE part of the monument and the edge of Tule Lake. At least 3 lava tubes are stacked on top of each other at this locality. As many as 5 levels have been found in at least one cave in the basalt of Mammoth Crater. Note that the law of superposition fails here because the inner lava is youngest and is commonly underneath older lava.

Return to the main monument road. **1.1**

56.8 Turn R at T junction with main road. **3.5**

60.3 Turn L to Fleener Chimneys on paved road that becomes gravel almost immediately. **0.8**

61.1 **STOP 13, Fleener Chimneys.** Climb on and around Holocene spatter vents for Devils Homestead basalt flow. Note that near-vent lava is pahoehoe, not aa. SiO_2 content is 50.9%. The vents are located on the Gillem fault. Vents for basalt at Medicine Lake volcano are typically spatter cones such as these or lava ponds. Cinder cones are rare as vents if silica content is below 52%.

Drive vehicles back to paved road. **0.8**

61.9 Turn L at T junction with main road. Drive N about 2 mi where the road turns NW and crosses the aa basalt flow known as the Devils Homestead, the downstream equivalent of Fleener Chimneys lava. Ahead of you is the fault scarp called Gillem Bluff. The road turns N up the face of the fault. **2.9**

64.8 Devils Homestead Overlook. Continue N on paved road that turns E near the N edge of the monument. **2.6**

67.4 Turn L (N) on Hill Road and follow it to the N boundary of the monument where the road is graveled for about 7 miles. Gillem fault scarp is on L; Tule Lake (in places reclaimed for farm land) on R. **9.0**

76.4 Headquarters of Klamath Basin National Wildlife Refuges on L after pavement resumes. Continue straight N. **3.9**

80.3 Turn L on Highway 161 (State Line Road) and almost immediately turn R on Malone Road and continue driving N (you are now in Oregon). **1.9**

82.2 Turn L on Oregon Highway 39. **2.2**

84.4 Prominent flagpole with American flag at center of Merrill. End of roadlog.

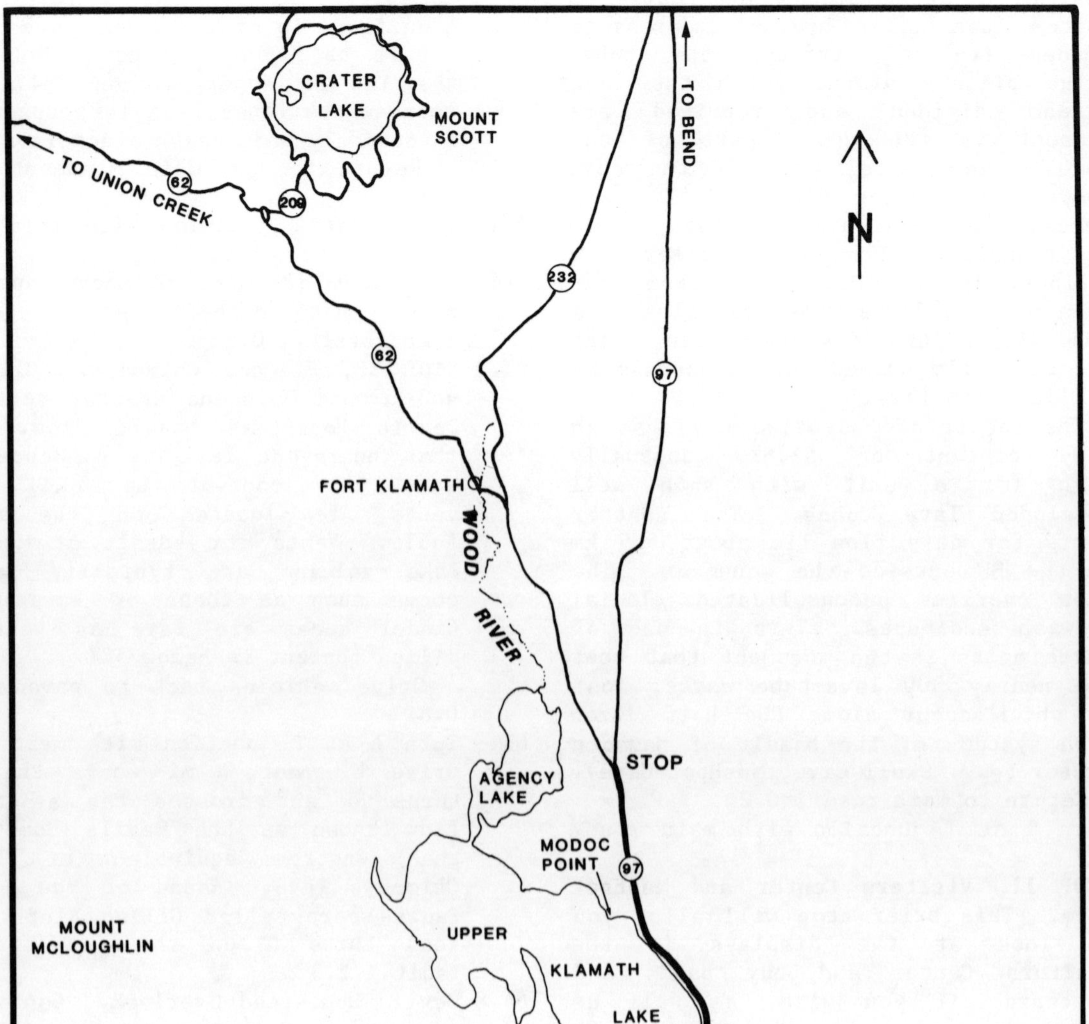

FIGURE 17 Route map for Klamath Basin.

KLAMATH BASIN

by

David R. Sherrod

The Klamath Basin is a composite graben that forms the westernmost structural trough of the Basin-and-Range physiographic province in Oregon and California. The S part of the basin includes Tule Lake and Lower Klamath Lake; the N part contains Upper Klamath Lake and Agency Lake. Bedrock in the Klamath Basin is chiefly late Miocene and Pliocene basalt lava flows and late Miocene to Holocene fluvial and lacustrine sedimentary rocks. Some Quaternary basalt was erupted along the S edge of the basin near Tule Lake, California.

Lower Klamath Lake, now drained and reclaimed as farmland, occupies a simple graben that coincides with a gravity depression of 20 mGal relief relative to adjacent lowlands. To account for this gravity anomaly, the sedimentary fill may be as thick as 2 km (Sammel and Peterson, 1976).

Klamath Falls is built on a sequence of NE-tilted fault blocks that separate the alluviated N and S parts of the Klamath Basin. Homes and businesses in Klamath Falls use thermal water for space heating, tapping approximately 600 wells that produce water at 15°-95°C from depths of 20-550 m (Peterson and McIntyre, 1970; Sammel and Peterson, 1976). The water is heated by deep circulation along range-bounding faults.

Mileage

0.0 Mileage begins at city-center offramp of U.S. 97 **(Figure 17). US 97 in the Klamath Basin is a narrow, high-speed highway with heavy truck traffic. Use caution, particularly in attempting to pull off at any point. 2.3**

2.3 Oregon Institute of Technology (OIT) at 2:00. Both OIT and the adjacent hospital utilize warm geothermal waters for space heating and cooling. **0.8**

3.1 Upper Klamath Lake on L. Diatomite quarry on R. Quarry operations and roadcuts along highway near here expose basalt that invades late Miocene or Pliocene diatomite and sandstone. **2.8**

5.9 Excellent views at 11:00 of N Klamath Basin and Upper Klamath Lake. Upper Klamath Lake is the largest lake in Oregon, covering about 24,000-28,000 hectares (60,000-70,000 acres). The lake, which averages 2-3 m deep, floods a graben that extends nearly to Crater Lake. The E escarpment of the graben has 450 m of structural and topographic relief in the Klamath Falls area. Although topographic relief diminishes toward the N, the structural relief probably persists. Gravity data indicate that 0.4-1.2 km of sedimentary rocks may fill the basin N of Modoc Point (Veen, 1982). **2.1**

8.0 Truck stop at Shady Pine road. Quarry exposes the footwall of a normal graben-bounding fault cutting Pliocene(?) basalts. **1.2**

9.2 Roadcuts to R for 0.3 mile expose Pliocene basaltic andesite lava flows and breccia. A 2-5 m thick bed of dark-gray to orange-brown epiclastic and pyroclastic mafic strata, which locally separates lava flows, is deformed and cut by massive intrusions and brecciated lava. Near the N end of these exposures, the clastic interval contains an angular unconformity, with underlying beds dipping as steeply at 70°-80° E. The tilting is probably the result of deformation along an

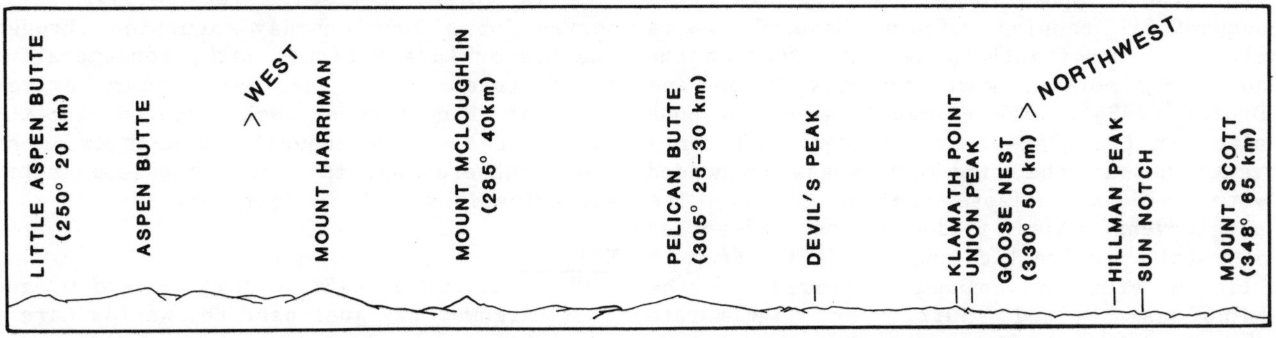

FIGURE 18 Skyline panorama of Cascade Range looking W and NW across Upper Klamath Basin. Distance (in km) and azimuth (in degrees) from junction of Highway 97 and Algoma Road (mile 9.5 in guide book). Crater Lake is in the mountains between Hillman Peak and Mt. Scott.

extensional fault; the abundance of near-vent material in the clastic rocks suggests that volcanism and deformation were synchronous. **0.3**

9.5 Algoma road. At 12:00 is a good view of fault escarpments created by differential displacement along splays of the range-bounding fault system. Basalt at top of escarpment on R has been dated by K-Ar at about 2 Ma.

Many prominent peaks of the Cascade Range from Mt. McLoughlin N to the Crater Lake area can be seen to the W and NW **(Figure 18)**. Mt. McLoughlin is a shield volcano built of olivine- and clinopyroxene-phyric basaltic andesite lavas that surround a vent complex of basaltic andesite cinders and palagonitized lapilli (Maynard, 1974). The precipitous N slopes of the volcano result from late Pleistocene glaciations. Mt. McLoughlin is probably younger than 100 ka, based on its erosional state compared to other dated volcanoes in the Cascade Range. **2.7**

12.2 Hagelstein Park. A U.S. Forest Service road 30 m N of the park proceeds up the escarpment described above. One of the best views of the Klamath Basin and adjacent Cascade Range is obtained by driving part way up this road and taking a short walk to the escarpment edge. **5.1**

17.3 Junction to L with Modoc Point road. Continue N on Highway 97. **6.1**

23.4 Pull into L lane, stop at stop sign, and turn L on Oregon Highway 62 toward Crater Lake National Park.

MOUNT MAZAMA AND CRATER LAKE CALDERA, OREGON

by

Charles R. Bacon

Crater Lake partially fills the caldera within Mt. Mazama at Crater Lake National Park, southern Oregon (**Figure 19**). Mt. Mazama is an andesitic stratovolcano cluster that collapsed 6845±50 yr BP during a catastrophic compositionally zoned pyroclastic eruption of about 50 km^3 of magma. Prior to its climactic eruption, Mt. Mazama was one of the major Quaternary volcanoes of the Cascade Range. Mazama's eruptive history has been delineated by recent mapping of its caldera walls and flanks. Products of the climactic eruption partially fill surrounding valleys and mantle the slopes and caldera rim.

The first detailed account of the geology of the park was USGS Professional Paper 3, by Diller and Patton (1902). Howel Williams showed in his classic monograph (1942) that the caldera formed by collapse and suggested that the compositional zonation reflected sequential tapping of a layered magma chamber. Other authors have reported on the zoned ignimbrite, most recently Bacon and Druitt (1988). The climactic eruption took place in two stages: single-vent and ring-vent phases; the ring-vent phase coincided with caldera collapse (Bacon, 1983). The single-vent phase produced the widespread air fall (climactic pumice fall) from a Plinian eruption column, followed by the Wineglass Welded Tuff, an ignimbrite deposited by ash flows that formed when the column collapsed. Deposits of the ring-vent phase range from pumiceous ignimbrite in the valleys around Mazama to coignimbrite lithic breccia (lag breccia) near the caldera rim (Druitt and Bacon, 1986).

After construction of the stratovolcano complex of Mt. Mazama and prior to the climactic eruption, several rhyodacitic lava flows were emplaced. These lavas apparently were derived from the climactic chamber during various stages of its growth. The andesite and dacite lavas of Mt. Mazama are well exposed in the caldera walls and along the Rim Drive. Mt. Mazama is built upon basaltic to andesitic High Cascade lava flows and, on its E and S sides, on an extensive field of rhyodacite lava flows. Basaltic andesite and lesser amounts of andesite and basalt were erupted from monogenetic vents and small shield volcanoes throughout the Quaternary, and these are visible on the flanks of Mt. Mazama and beyond.

This guide, revised after Bacon (1987), serves for a long one-day excursion through the Crater Lake National Park, concentrating on features near the caldera-rim drive. Ages of Pleistocene rocks quoted in the guide are based on unpublished K-Ar dates by M.A. Lanphere and, in part, on paleomagnetic determinations by D.E. Champion.

Mileage

0.0 Junction of U.S. Highway 97 and Oregon Highway 62, just past the Rapids Cafe. **7.6**

7.6 Small quarry to R as Highway 62 curves L. Near-vent spatter and scoria, along with a feeder dike, are all cut

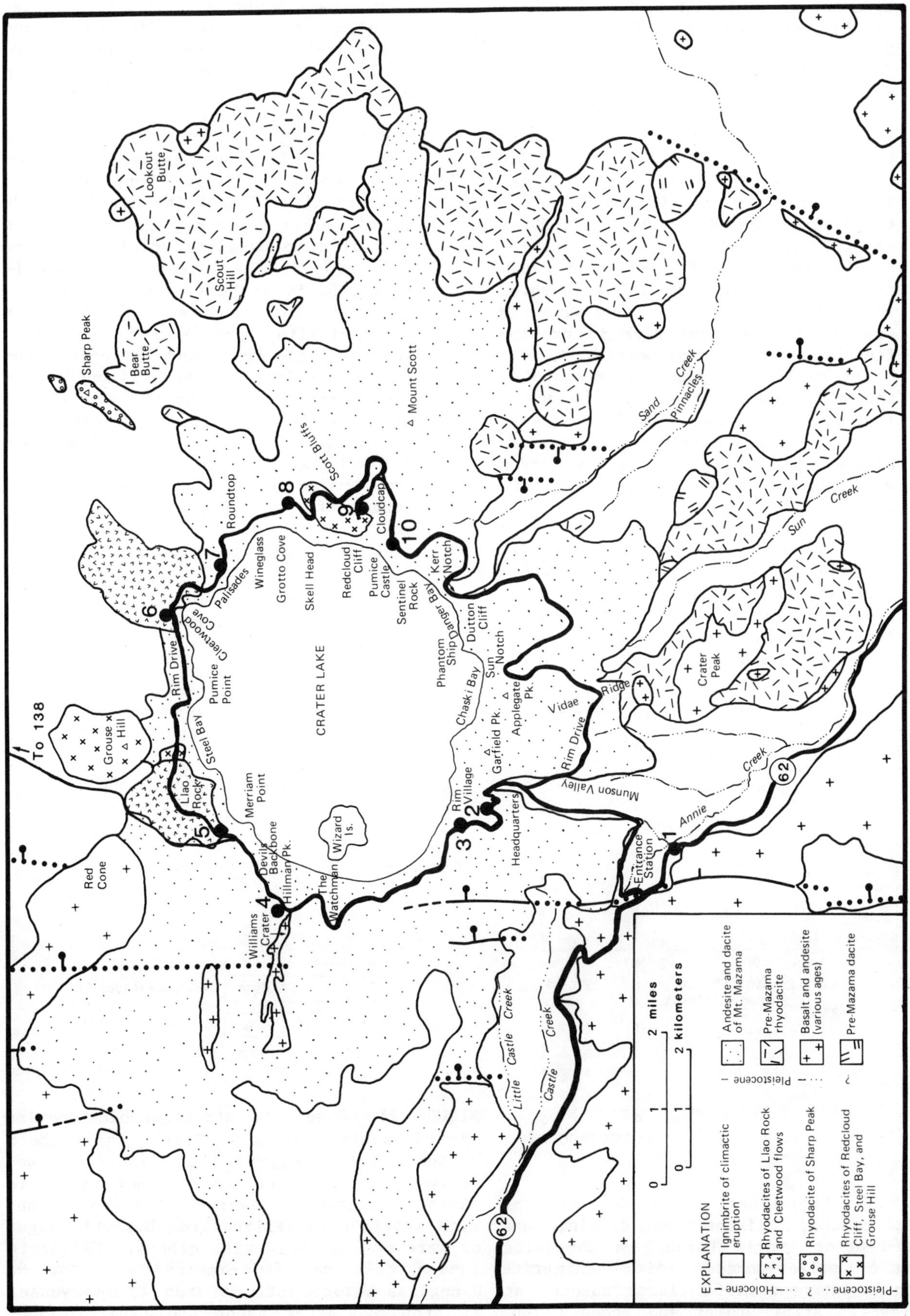

FIGURE 19 Generalized geologic map of Mount Mazama with field trip route and stops indicated.

by a normal fault. **6.6**

14.2 Cattle Crossing Cafe in Fort Klamath. **2.1**

16.3 To the R for the next mile is an excellent view of the peaks on the Cascade crest and the rim of Crater Lake caldera. From W to E, summits to the N are the andesitic cinder cone of Crater Peak, Garfield and Applegate Peaks and Dutton Cliff on the S rim, and Mt. Scott just E of the caldera. **4.1**

20.4 Boundary of Crater Lake National Park. **Note that collecting or disturbing rock or other natural specimens in Crater Lake National Park is prohibited except by special permit.**

The road climbs the gentle slope of the ring-vent-phase ignimbrite where the conifer forest begins. Between here and stop 1 note cuts in the top of the ignimbrite; precipitous canyon walls of Annie Creek are cut into columnar-jointed ignimbrite. High Cascade basaltic andesite and andesite are W of the road; high bluffs to E are pre-Mazama rhyodacite. **8.6**

29.0 **STOP 1, Godfrey Glen; turnout (note stone wall with sign) with view of Annie Creek canyon.** Excellent exposure of valley-filling medial-facies pumiceous ignimbrite of ring-vent phase of climactic eruption (**Figures 20 and 21**; Bacon, 1983, Figs. 6c, 6d, 7, and 9; Druitt and Bacon, 1986; Bacon and Druitt, 1988). Lower half of section is silicic ignimbrite dominated by rhyodacite pumice clasts with 70.4% SiO_2 (all SiO_2 values are recalculated volatile-free). Most of upper half is mixed ignimbrite containing 20 to 80% silicic pumice, the rest being andesitic to basaltic scoria of 61 to 48% SiO_2. The uppermost approximately 30 ft are mafic ignimbrite with <20% silicic pumice. Scoriae in valley-filling ignimbrite generally are hornblende- and plagioclase-phyric; olivine- and clinopyroxene-rich clasts occur near the top of the section. Most scoriae apparently are cumulates. The color change from buff below to gray above reflects increasing emplacement temperature of the matrix of the deposit upward, as much as its composition. Note the bleached zone of fumarolic alteration beneath

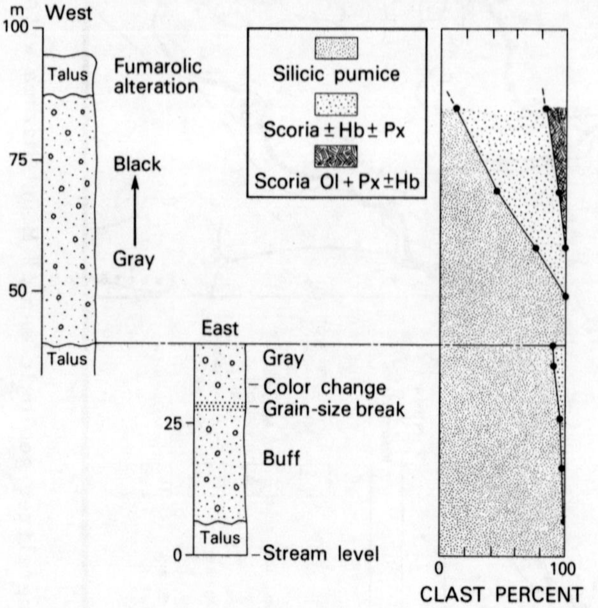

FIGURE 20 Variations in proportions of silicic pumice, olivine-free scoria, and olivine-bearing scoria through a composite section of valley-ponded medial ignimbrite at stop 1. More than 50 clasts counted at each level. After Druitt and Bacon (1986, figure 12).

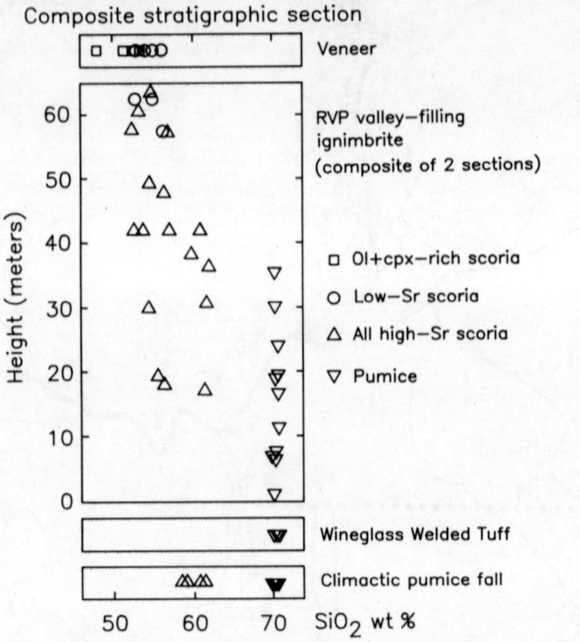

FIGURE 21 Composite stratigraphic section of climactic ejecta. Scoria have been divided into high-Sr and low-Sr types derived from at least two distinct andesitic parental magmas. Olivine- + clinopyroxene-rich scoria crystallized from basaltic magma or were contaminated with gabbro. Climactic pumice fall can be examined at stop 6, Wineglass Welded Tuff at stop 7, and veneer at stop 5. After Bacon and Druitt (1988, figure 6).

the uppermost 3 ft of fine ash. Erosion-resistant pinnacles and coarse columnar joints (near Duwee Falls on tributary on opposite side of Annie Creek) result from vapor-phase induration (Partial welding and columnar jointing of ignimbrite downstream along Annie Creek apparently results from greater glass content of ignimbrite matrix farther from source.). Entire section was deposited without significant breaks, as no sharp grain-size breaks are evident. Note, however, rusty-stained lithic-rich horizon near lower part of deposit. Good exposures of mixed ignimbrite are accessible just beyond the N end of the long turnout. Garfield and Applegate Peaks on the caldera rim are visible on the skyline to the N. Lithic breccia high on the slopes of these peaks and in the adjacent valleys forms the proximal facies of the same eruptive unit exposed here at Annie Creek. **0.7**

29.7 View to L of Arant Point, a subvolcanic intrusion of olivine andesite cut by a normal fault. **0.4**

30.1 Turn R from Oregon Highway 62 onto road to Crater Lake. **0.2**

30.3 Entrance station to Crater Lake National Park. **3.8**

34.1 To R is road to E rim of Crater Lake.

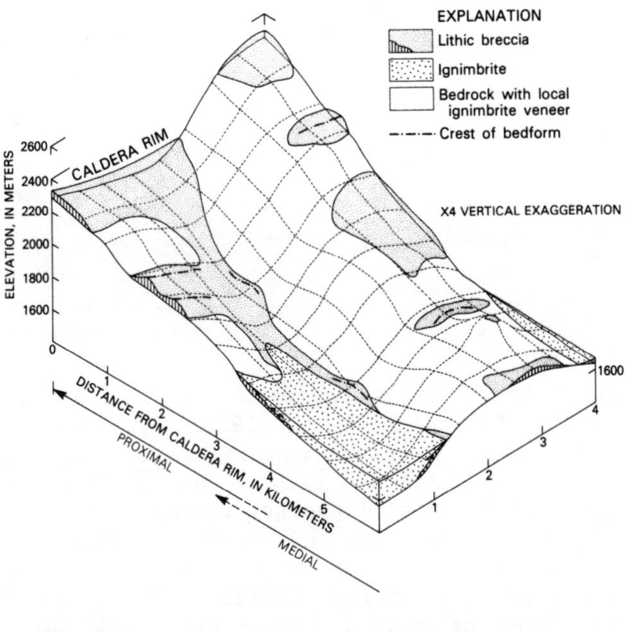

FIGURE 22 Schematic sketch of aerial distributions of lithic breccia and ignimbrite and their relationships to surface topography. After Druitt and Bacon (1986, figure 2).

Continue straight. Information, literature, and restrooms are available at Park Headquarters on L between 8 a.m. and 5 p.m. First cut on R beyond the headquarters exposes coarse lithic breccia of proximal facies of ring-vent-phase pyroclastic-flow deposits of the climactic eruption (Bacon, 1983; Druitt and Bacon, 1986, Fig. 3). Roadcut is in a bedform near the E side of Munson Valley. Similar material underlies floor of Munson Valley in this vicinity. Lag breccia also occurs on summits of all nearby peaks, both in the lee of or upslope from obstacles and locally on the upper slopes of Mt. Mazama (**Figure 22**). Longitudinal and transverse bedforms are common in the lag breccia. **1.3**

35.4 **STOP 2, dacitic fragmental deposits.** Pull off into turnout on S (L) side immediately after hairpin turn from NE to SW. Long straight cut exposes two subunits of approximately 40 ka dacitic fragmental rocks (Bacon, 1983, p. 77-79). Eastern subunit is monolithologic debris-flow or avalanche deposit that apparently records collapse of dome(s) SW of Mazama's summit. Similar material is present in the tree-covered swale high on Garfield Peak to the E. Western subunit is the younger and contains the same dacite, altered dacite, and dense prismatically jointed dacite and pumiceous dacite blocks believed to represent the final pulse of magma responsible for destruction of the dome(s) (Bacon, 1985, Fig. 7). The western subunit is thought to be a lithic-pyroclastic-flow deposit. The dacite contains chilled inclusions of andesite magma (Bacon, 1986, Figs. 7 and 8, Table 3). **1.4**

36.8 **STOP 3, Rim Village and view of caldera.** At the intersection of Rim Drive (L) and road to Rim Village (R), turn R for parking. Morning light is good for viewing N and W caldera walls (**Figure 23**), afternoon for E walls (stop 10; **Figure 24**). Between 400 and about 40 ka, the locus of andesitic volcanism moved around 10 km from E to W during construction of Mt. Mazama. Imposing cliff on N caldera wall is Llao Rock, the vent-filling rhyodacite flow and associated pumice-fall deposit dated at 7015±45 yr BP by radiocarbon (Bacon, 1983). This compositionally zoned silicic eruption (72-70% SiO_2) preceded the climactic eruption by about 100 to 200 years (Bacon, 1983, p. 105). The walls below Llao Rock consist of andesite and dacite flows and minor dacite tephra. As many as five eroded surfaces are present. Lava at lake level here has been dated at about 190 ka. Sheet-like thin flows here and elsewhere in the walls consist of agglutinated spatter topped by rubble. Such fountain-fed lavas grade from bomb beds and agglutinate near source, through streaky lava flows with thick rubbly tops, to distal homogeneous lava with relatively thinner rubble zones.

Eye-shaped cliff at rim E of Llao Rock marks the rhyodacite of Steel Bay (71.6% SiO_2), believed to be one of the earliest products of the climactic magma chamber, emplaced about 30 ka. The ages of this flow and its cousins, Grouse Hill and Redcloud Cliff, are constrained to within the interval of about 20-30 ka by field relations and measurements of paleomagnetic secular variation (D. E. Champion, unpublished data, 1985). E of Llao Rock are Pumice Point, approximately in line with symmetrical flat-topped cone of Timber Crater N of caldera, and Cleetwood Cove, immediately R of tour boat dock (boats are light blue). Cliffs at rim at Cleetwood Cove (stop 6) are in rhyodacitic Cleetwood Flow, identical in composition (volatile-free) to climactic pumice and still hot when caldera collapsed. The dark lava descending to water level is the "backflow", which oozed into the caldera during collapse (Bacon, 1983, p. 100-101, Fig. 11). W of Llao Rock are Devils Backbone dike, Hillman Peak,

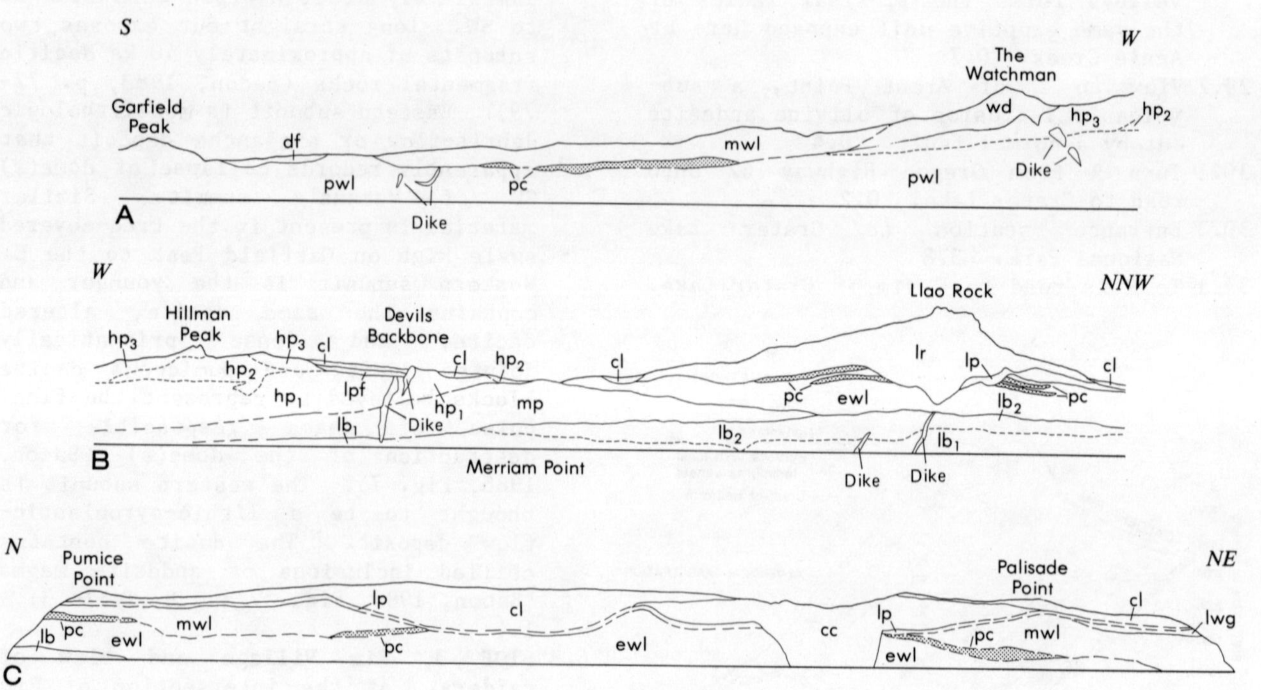

FIGURE 23 Geologic sketch maps of caldera walls; bases from photographic panoramas and not to scale. Thicknesses of most dikes exaggerated. Modified after Bacon (1983, figure 4).

A. SW wall. hp = intrusions, lavas, and fall deposits of Hillman Peak; pc = dacitic ignimbrite; pwl = pre-Wisconsin lava flows; wd = The Watchman and related dacite flows and dike; mwl = middle Wisconsin lava flows; df = dacitic fragmental deposits of stop 2.

B. NW wall. lb = lava sheets of Llao Bay; mp = lava flows and domes and fragmental deposits of Merriam Point; ewl = early Wisconsin lava flows; pc = dacitic tephra, lower unit and deposit in C provisionally correlated with tephra of Pumice Castle; lpf = dacitic lithic-pyroclastic-flow deposits correlated with those of stop 2; lp = Llao Rock pumice fall; lr = Llao Rock rhyodacite flow; cl = undivided deposits of climactic eruption. Most climactic deposits are not visible from this angle.

C. N wall. lwg = late Wisconsin till; cc = Cleetwood rhyodacite flow. Note backflow at Cleetwood Cove (stop 6). Cleetwood pumice fall is included in unit lp from Cleetwood Cove to E.

and The Watchman. Gray vegetated slope just below rim S of Devils Backbone marks lithic-pyroclastic-flow deposit correlated with that of stop 2. Hillman Peak consists of pyroxene andesite, overlain by hornblende andesite (forms gray scree slopes on near-vent fall deposits), overlain by pyroxene andesite with prominent augite and blocky plagioclase. All are about 70 ka. The Watchman flow is about 50 ka. The dike feeding the dacite flow that forms The Watchman can be seen on the caldera wall below the saddle between The Watchman and Hillman Peak. In the SW wall are lavas older than Hillman Peak, an eroded surface overlain by dacitic ignimbrite that weathers orange, and andesite flows as young as about 50 ka (Bacon, 1985, Figs. 6 and 7). Wizard Island is the tip of a thick pile of postcaldera andesite that apparently was erupted when the lake level was about 300 ft lower.

Return to intersection and turn R (NW) on Rim Drive. **2.1**

38.9 Long unpaved turnout on L side of road across from large cut in andesite lava. Younger flow shows glassy brecciated base grading into subtle columns, then upward into platy joints that parallel flow base. Platy zone grades up into coarsely blocky-jointed interior. Original rubbly top has been removed by glaciation. In some flows a second set of platy joints forms steeply dipping sheets high in the flow where this set dominates. Plagioclase phenocrysts in this flow have a distinctive seriate texture. Older flow is more typical of Mazama andesites with plagioclase, augite (green), hypersthene (brown), and titanomagnetite phenocrysts, common glomerocrystic clots, and sparse magmatic inclusions. Olivine (yellow) is a common minor constitutent of many Mazama andesites. **0.4**

39.3 Cut on NE of road exposes crudely bedded lag breccia lying on glaciated surface of dacite lava similar to the Watchman flow (Druitt and Bacon, 1986, Fig. 6). Fumarolic alteration shows that the breccia was deposited hot. Note platy jointing in the Watchman flow between here and stop 7. **0.5**

39.8 The Watchman flow. Paved turnout on L where Rim Drive heads N across the Watchman dacite flow gives good view to S to Mt. McLoughlin (61 km) and Mt. Shasta (200 km). Union Peak is the eroded core of a cinder cone atop a basaltic andesite shield in the middle

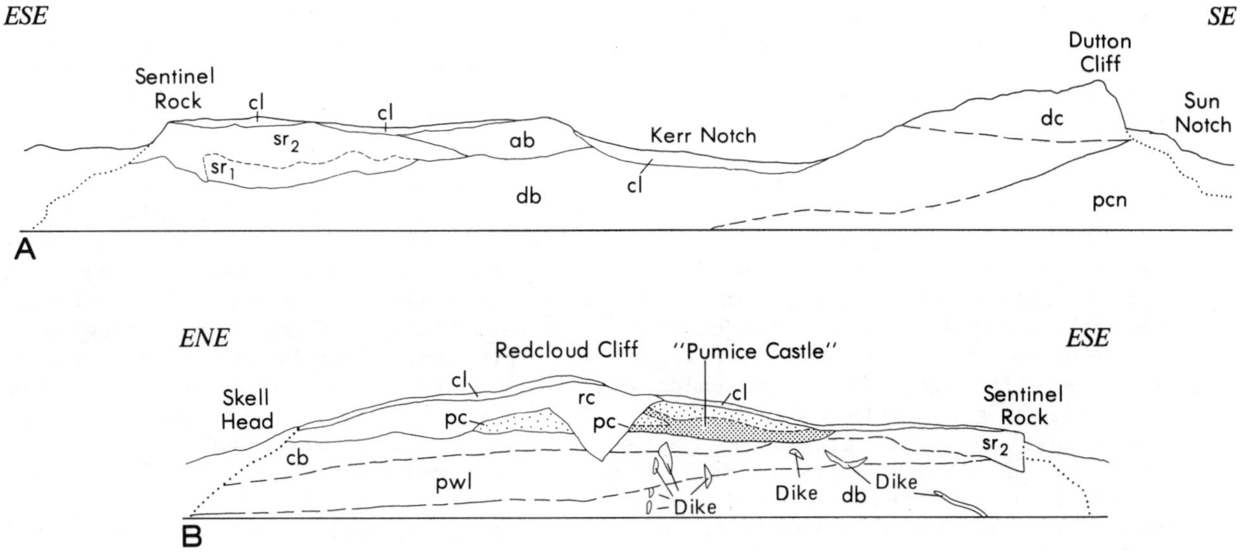

FIGURE 24 Geologic sketch maps of caldera walls; bases from photographic panoramas and not to scale. Modified after Bacon (1983, figure 3).

A. SW wall. pcn = lavas and fragmental deposits of Phantom Cone; db = lavas of Danger Bay; ab = lavas of Anderson Bluffs; dc = lavas of Dutton Cliff; sr = intracanyon lavas of Sentinel Rock. Kerr notch is beheaded glacial valley.

B. E wall. cb = lava sheets of Cloudcap Bay; pc = dacite tephra (heavy stippling) and lava (light stippling) of Pumice Castle; rc = Redcloud Cliff rhyodacite flow. Deposits of climactic eruption not shown where less than a few m thick.

distance. In the roadcut a remnant of the pumiceous blocky carapace of the dacite flow can be examined and seen to grade down through glassy dacite into devitrified rock. This lava has 67.5% SiO_2 and contains rare andesitic magmatic inclusions. Its porphyritic texture and two-pyroxene, plagioclase, and Fe-Ti oxide phenocrysts are typical of Mazama dacites. **1.4**

41.2 **STOP 4, View to N, and Williams Crater.** Large turnout on NW side of road gives view of Red Cone (N) and Bald Crater (NNW) basaltic andesite cinder cones, and part of the S face of Llao Rock. In the middle distance, L to R, are Mt. Bailey, Diamond Peak (beyond Diamond Lake), and Mt. Thielsen. Three Sisters volcanoes may be seen in the far distance on clear days. To reach Williams Crater (formerly Forgotten Crater, Bacon, 1983, p. 80-82) walk S along Rim Drive. In the first cut note lag breccia bleached by fumarolic alteration lying on basaltic lapilli from Williams Crater, which in turn rest on andesite breccia of the youngest unit of the Hillman Peak center. The andesite lava has distinctive blocky plagioclase phenocrysts, hypersthene, and prominent augite. Just past this cut turn W and walk about 400 m to Williams Crater, noting the small dacite dome immediately S of where you left the road. This dome is the easternmost lava of the Williams Crater complex which consists of a basalt flow, cinder cone, and three more flows of commingled andesite and dacite with basalt inclusions. Before reaching Williams Crater note the arcuate bedform of lag breccia banked against it. Most clasts here are variably altered andesite from conduit or vent walls. Systematic variation in clast lithology around the caldera is evidence for multiple vents during this part of the eruption. Note rounding of these clasts due to thermally induced spalling of corners. Some clasts are highly altered, friable andesite; others are partially fused granitoids. Matrix here contains hornblende andesite pumice. At some localities juvenile material zones from rhyodacite pumice at the base upward to mafic scoria.

Williams Crater contains basaltic bombs cored with angular fragments of commingled andesite-dacite lava, themselves bearing inclusions of the basalt. Walk to the S side of the cinder cone and visit the roughly 10-m cliff in commingled lava to see a variety of proportions of andesite and dacite. A basaltic(?) dike evidently intersected the margin of a silicic magma reservoir, venting first basalt, but forcing convective mixing in the reservoir. The basalt is contaminated with ubiquitous gabbroic fragments to a few mm in size, variably disaggregated, which are responsible for a more mafic composition of the lava than that of the original basaltic andesite(?) liquid. This contaminated basalt thoroughly mixed with dacite to form hybrid andesite which then mingled with the dacite. Blobs of basalt occur in both andesite and dacite, and may have been incorporated in dacite before the mixing that produced the andesite. Fragments of the commingled magma were thrown out in the basaltic eruption. Finally, viscous commingled lava oozed from vents on a trend radial to the present caldera. All of these events took place rapidly because paleomagnetic results obtained by D. E. Champion show the same direction of thermoremanent magnetization for all units of Williams Crater. The dacite is chemically most similar to that of the fragmental deposits in Munson Valley (stop 2); tephra from Williams Crater lies on the deposit S of Devils Backbone noted at stop 3. **1.2**

42.4 Below second of two big paved turnouts on R side of road is a glacially striated outcrop of hornblende andesite of Hillman Peak (middle unit). Similar rock crops out about 30 m W of the road. Nearby are mafic blocks in the final veneer deposit of the climactic eruption. These are plagioclase-pyroxene cumulates with variable amounts of olivine or poikilitic hornblende. The veneer locally mantles lag breccia, which to the W occurs in bedforms with clasts to 5 m (Druitt and Bacon, 1986, Fig. 10). Llao Rock can be seen well from this point. The bedded climactic pumice fall present near its summit was eroded from its flanks by ring-vent-phase pyroclastic flows. The dark streak truncating air-fall beds on Llao Rock is scoria-bearing ignimbrite veneer. **0.3**

42.7 Turn R at junction of East Rim Drive and North entrance road. At caldera rim, lag breccia grades down into pumiceous ignimbrite which grades down into a small outcrop of the Wineglass Welded Tuff (stop 7) that lies on bedded fall deposits. **0.3**

43.0 **STOP 5, Llao Rock obsidian and proximal ignimbrite.** Park in very large turnout on L side of road past big cut in obsidian on R. Ignimbrite veneer lies on lag breccia at N end of cut. About 20 cm of pink vitric ash, the ash-cloud deposit or layer 3 of Sparks et al. (1973) caps the veneer. The climactic pumice fall and most of the original blocky pumiceous carapace of the Llao Rock flow have been removed by the pyroclastic flows. Pumiceous rhyodacite grades down into obsidian, which becomes lithophysal and spherulitic deeper into the flow S along the cut. Note abundant andesitic magmatic inclusions (Bacon, 1986, Figs. 5 and 6, Table 2). **1.4**

44.4 For the next half mile there are glimpses NE of Grouse Hill (70-71% SiO_2), previously thought to be Holocene (Bacon 1983, 1985) and now believed to be Pleistocene (ca. 20-30 ka). Grouse Hill is chemically and mineralogically similar to the rhyodacites of Steel Bay and Redcloud Cliff (Bacon and Druitt, 1988). **0.5**

44.9 Cut in the rhyodacite of Steel Bay (hill 7352' of Bacon, 1983, 1985 Fig 8.). Rock contains plagioclase, hornblende, augite, hypersthene, titanomagnetite, and ilmenite phenocrysts, plus rare andesitic inclusions. Cuts between here and Stop 6 are in andesite erupted about 50 ka. **1.1**

46.0 Pumice Point. Long paved turnout on R providing access to Pumice Point by walking about 300 m E along abandoned and overgrown road to rim at Pumice Point. Visible on the steep slope below are the following deposits, from base to top: (1) Faintly orange air-fall pumice erupted from a vent near the E wall of the caldera at Pumice Castle (stop 10) and resting on glaciated andesite; (2) a glaciated andesite lava flow dated at about 50 ka and filling an erosional depression in the pumice of (1); (3) a few feet of light-gray Llao Rock pumice fall forming a gentle slope above the lava; (4) climactic pumice fall, pinkish, in well defined beds with nonwelded pyroclastic-flow interbeds, supporting the steep slope; (5) orange incipiently welded Wineglass Welded Tuff grading upward into (6) lag breccia; (7) ignimbrite veneer; (8) ash-cloud deposit, as at stop 5; (9) crystal and lithic ash, presumably re-ejected from the caldera by phreatic explosions following collapse; and (10) a few feet of wind-reworked material forming a small dune at the caldera rim. Near the base of (4) is a meter-thick, reddish, incipiently welded bed of climactic air-fall pumice that contains rare hornblende andesite scoria and partially fused granitoid blocks. **1.2**

47.2 Parking lot to L for trail to lake and tour boats. Lake tour takes 2 hours, with optional stop at Wizard Island. Highly recommended for a second day. Inquire at Park Headquarters or Crater Lake Lodge for current schedule and fee. Allow 20 minutes to descend and 30 minutes to hike out of caldera. Trail has good exposures of partly welded Wineglass Welded Tuff, and cuts showing climactic air-fall, lithic-rich Cleetwood air fall, and Llao Rock air-fall lying on till. Lava flow below till is silicic andesite dated at about 100 ka. Underlying material may be an avalanche deposit of related blocks that preceded emplacement of the lava flow. **0.3**

47.5 **STOP 6, Cleetwood Cove.** Park at paved turnout on R across road from brick red pumice. Below are cliffs in Cleetwood rhyodacite flow and tongue of lava (backflow) that oozed down caldera wall during ring-vent phase of climactic eruption (Bacon, 1983, Fig. 11). Cliffs E of cove show internal flow structure of lava. Roadcut exposes climactic pumice fall lying on Cleetwood flow. Near their mutual contact these units are highly oxidized. Nowhere but on top of this lava flow is the pumice fall oxidized and sintered. The pumice, which was hot enough when it landed to support steep slopes, fell as a hot blanket on the cooling lava flow, trapping and heating air. This trapped air, and possibly also degassing of the lava, resulted in the oxidation. Fumarolic alteration cuts the air-fall and overlying lag breccia showing that the entire climactic eruption took place before the Cleetwood flow cooled completely (Bacon, 1983, Fig. 8). About 180 m N of the road is the first of a series of small rift valleys transverse to the direction of flow, formed as the brittle upper surface of the flow broke up while its fluid interior flowed N (presumably in response to seismicity associated with caldera collapse) and S into the caldera. Faults bounding the valleys cut climactic pumice fall and thin Wineglass Welded Tuff (stop 7). Lithic breccia is banked against the walls of these valleys showing that caldera collapse began before the ring-vent phase ended (Bacon, 1983, Fig. 12). 1.5

49.0 **STOP 7, Wineglass Welded Tuff, Palisade flow, and view of caldera walls.** Park at the W end of the paved turnout on R and walk to the caldera rim. You are standing on the Wineglass Welded Tuff, the top of which has been scoured away by the ring-vent-phase pyroclastic flows. Note how the Wineglass thins to the W of here as it rises to higher elevation and is thickest in the topographic depression between here and the prominent cliff to the E (Roundtop); the Wineglass is not present on the summit of Roundtop and other high points. The Wineglass consists of up to four flow units of rhyodacitic ignimbrite that form a single cooling unit. It was deposited by heat-conserving, valley-hugging ash flows that emanated from the collapsed eruption column of the single-vent phase (Bacon, 1983, p. 92-93, Fig. 9). Here the top of the ignimbrite shows gash fractures parallel to the caldera rim, indicating that, where sufficiently thick, the tuff slumped toward the caldera while still hot and plastic. This is compelling evidence for the rapidity of events during the climactic eruption. The large pumice slope on the wall S of the flat bench that forms Skell Head SSE of here shows a fine section of air falls of the Llao Rock, Cleetwood, and climactic eruptions overlain by the Wineglass Welded Tuff capped by lag breccia (Druitt and Bacon, 1986, Fig. 5). There the Wineglass shows four distinct flow units, the top one being a vitrophyre. The lag breccia appears crudely bedded into five units. The lower two contain rhyodacite pumice, the next two hornblende andesite pumice, and the top one mafic scoria, showing that the breccia mimics the stratigraphic order of the zoned valley-filling ignimbrite.

The craggy top of the Palisade flow (dacite) that forms the cliffs just to the W of stop 7 has been glaciated, and the flow occupies an old glacial valley carved against the older silicic andesite of Roundtop to the E. This is also a good place to see features on the E and S caldera walls. The cliff-forming flows below the rim N from Skell Head are inclusion-bearing silicic andesites, the younger of which is the subject of stop 8. The Redcloud Cliff rhyodacite flow (70-71% SiO_2) forms the prominent cliff on the E wall just S of Skell Head, filling its vent crater and spilling to the N over earlier dacite (**Figure 24**). It is compositionally similar to the rhyodacites of Grouse Hill and Steel Bay, and is late Pleistocene in age (ca. 20-30 ka). Immediately S and stratigraphically below Redcloud Cliff are Pumice Castle and related dacite flows described at stop 10. Andesites below the dacites fall into at least two groups between about 220 and 340 ka. Between Pumice Castle and Kerr Notch (the lower of two beheaded glacial valleys) is Sentinel Rock (best seen from the W in afternoon light) where thick intra-canyon low-silica dacite flows (about 300 ka) lie on glaciated older andesite (about 340 ka). Between Kerr and Sun Notches is Dutton Cliff. The oldest rocks exposed on the caldera wall, about 400 ka, are at water level below Dutton Cliff; they comprise the Phantom Cone (Williams, 1942). Stripes on the wall with moderate initial dips highlight pervasively altered rubbly tops of agglutinated andesite flows; their dense interiors, having been less permeable, show little effect of hydrothermal alteration from this distance. Phantom Ship is the small island and is partly composed of dikes related to the adjacent Phantom Cone. From Dutton Cliff to water level beneath Sun Notch three sets of andesite flows are exposed. The top is fresh olivine-bearing pyroxene andesite, the middle altered andesite, and the lowest altered andesite with abundant magmatic inclusions. Applegate Peak and Garfield Peak form the summits of the S wall, W of Sun Notch. All are pyroxene andesite and low-silica dacite flows except the top two flows of Garfield Peak, which are hornblende andesite (about 220 ka). The altered flows near water level below the big talus slopes between Applegate and Garfield Peaks comprise Chaski slide, a block of caldera wall that failed to slide completely into the caldera, presumably during the main collapse event. W of Garfield Peak is the head of Munson Valley, site of Crater Lake Lodge and Rim Village, which are built on the dacitic debris-flow and lithic-pyroclastic-flow deposits of stop 2. **1.0**

50.0 Wineglass. Small unpaved turnout on R where Rim Drive meets the caldera rim after traversing the andesite flow of Roundtop provides limited access to Wineglass. Wineglass refers to the shape of a scree chute below here. Excellent exposures of the Wineglass Welded Tuff lying on climactic pumice fall can be seen from the SE side of the valley. The uppermost 1 m of the fall deposit is oxidized because emplacement of the Wineglass on warm air fall trapped and heated air. Lag breccia lies on the Wineglass. This is the locality where hot fiamme oozed over the surface of an open fracture (a larger version of those seen from above at stop 7) before the tuff completely cooled; the block on the caldera side of the fracture has since fallen away revealing tongues of devitrified glass hanging over the inclined fracture surface. Because the caldera must have been present at the time of fracturing, caldera collapse took place shortly after emplacement of the Wineglass, i.e. during the ring-vent phase of the climactic eruption (Bacon, 1983, p. 102). Beneath the scree below, climactic pumice fall lies on Cleetwood air-fall pumice, which rests on Llao Rock pumice. The 7015±45 yr BP radiocarbon age of the Llao Rock eruption was determined on charcoal at the base of the Llao Rock pumice here. Below the air-fall units is till containing glassy andesite derived from the flow underneath. This lava flow displays spectacular radiating columnar joints that grade into shattered glassy blocks upward into the till. The flow evidently encountered ice. Similar features can be seen in many other less obvious examples of chilled lava around the caldera. **1.0**

51.1 **STOP 8, Inclusions in andesite.** Park in small picnic area on R. Walk 100 m back through the woods and then along shoulder of Rim Drive to a cut in lava on the SW side of the road. Observe abundant magmatic inclusions of andesite (60% SiO_2) in silicic andesite (62% SiO_2). The host is the younger of two compositionally identical flows that are about 70 ka (Bacon, 1985, Fig. 6; 1986, Figs. 9-15, Table 4). Inclusions larger than about 12 cm have exceptionally porous (diktytaxitic) cores with denser rinds a few cm thick. Because of a relatively small initial thermal contrast between porphyritic host and nearly aphyric inclusion magmas, incomplete undercooled crystallization of inclusions resulted in a large fraction of vapor-saturated residual liquid (75% SiO_2). Most of this liquid was expelled from cores of inclusions, by then rigid crystal meshes, by gas filter-pressing owing to differential vapor pressure between host lava and cores of large inclusions. Consequently, cores are more mafic (56% SiO_2) than the original inclusion bulk composition. Although the rind texture is not common, andesitic inclusions are abundant in many silicic andesite and dacite flows of Mt. Mazama, particularly in the relatively old rocks of Mt. Scott and the E and S caldera walls. The cut at this stop is high in the holocrystalline interior of the flow; the upper glassy zone of the flows is preserved locally near their distal ends, where glass-lined segregation vesicles can be seen in inclusions. Inclusions are most abundant near the tops of flows, apparently because tops of flows were erupted later and advanced more rapidly than the underlying lava. Inclusions may have been incorporated into the host from a deeper level in the magma reservoir by forced convection during flow of magma up the eruption conduit.

Returning to vehicles, on the R side 25 m beyond the last andesite lava is an exposure of climactic pumice fall lying on Cleetwood pumice separated by about 2 cm of fine ash which marks the beginning of the climactic eruption.

Between picnic area and Mt. Scott trail, roadcuts expose rhyodacite of Redcloud Cliff for about 1.5 mi, then dacite of Cloudcap and Scott Bluffs. **2.6**

53.6 Parking area on L for Mt. Scott trail and view of Mt. Scott. Mt. Scott consists of the oldest dated lavas of Mt. Mazama, about 420 ka. This cone is made up of sheets of agglutinated low-silica dacite. Andesitic inclusions are abundant. Glaciation has exposed the core of the volcano.

Hydrothermal alteration variably affects these rocks, and these features can be seen readily on the trail to the summit. The summit ridge is capped by climactic pumice fall and lag breccia. **0.2**

53.8 Turn R on Cloudcap Road. Cuts are in dacites as on Rim Drive below. **0.8**

54.6 **STOP 9, rhyodacite lava and pumice, welded dacite air-fall pumice, and view of NE flank of Mt. Mazama.** Pull off on paved shoulder to R. Cut on L exposes five air-fall pumice deposits and a steeply dipping rib of lava. From E to W, in stratigraphic order, the units are: (1) buff to orange dacite pumice with blocky plagioclase and pyroxene, similar to the tephra of Pumice Castle; (2) white hornblende rhyodacite pumice with abundant coarse lithic blocks, the proximal air fall for the Redcloud Cliff vent; (3) rhyodacite lava grading from pumiceous blocks up through obsidian into lithophysal and spherulitic rhyodacite, believed to be a glaciated "bathtub ring" of lava left after most of an oversteepened dome slid N to form the Redcloud Cliff flow; (4) Llao Rock and (5) Cleetwood air fall poorly exposed beneath scree; (6) climactic pumice fall supporting a small cliff; and (7) lag breccia of the ring-vent phase of the climactic eruption. Walk around the W end of this exposure and up through the trees over the lava to a brick-red 2-m cliff E of the cut. This is welded air-fall pumice similar to that of Pumice Castle. From here one can see many features. In the distance are Mt. Thielsen and tree-covered lateral moraines extending to the E, the Gibraltar-like W-facing scarp of Walker Mtn. 50 km NE near the town of Chemult (approximate limit of ring-vent-phase ignimbrite), the toad-back silhouette of Newberry Volcano 110 km NE, and the shield-like form of Yamsay Mtn. 55 km E. In the middle distance are the late Pleistocene andesitic shield and cone of Timber Crater to the N and ignimbrite-floored flats of Klamath Marsh to the E. Closer features are the higher domes of the Sharp Peak group of at least 12 small rhyodacite domes thought to have been erupted from the climactic chamber (Bacon, 1985, Fig. 8; Bacon and Druitt, 1988), several pre-Mazama rhyodacite domes surmounting thick flows (Bear Butte, Lookout Butte, Scout Hill, approximately 400 to 700 ka; climactic lag breccia occurs on the summit of Lookout Butte), cliffs to the N in the silicic andesite flow of stop 8, cliffs in the Redcloud Cliff flow, and the dacite flows of Scott Bluffs. **0.3**

54.9 Cloudcap overlook. Return to Rim Drive, bearing to R at Y junction. **1.1**

56.0 Stop sign at Rim Drive. Next cuts are in dacite similar to flows of Scott Bluffs and nearby cuts on Rim Drive and Cloudcap Road. **1.1**

57.1 **STOP 10, Pumice Castle overlook.** Large paved turnout where Rim Drive bends L. Best view is from a few feet W of the stone wall. The view of the NE wall of the caldera includes the Cleetwood flow, Palisade flow, Roundtop flow, and Wineglass. Redcloud Cliff dominates the E wall in the foreground. Cliff at rim S of Redcloud Cliff is in Pleistocene dacite flow lying on tephra of Pumice Castle. The latter consists of dacite pumice that is nonwelded at the S end of the exposure but becomes progressively more densely welded to the N. Ages of lava flows above and below the tephra of Pumice Castle suggest that this dacite erupted about 70 ka. Just S of Redcloud Cliff this deposit includes several vitrophyric layers and a stubby lava flow. The top of the air fall has been "fused" by the overlying lava flow. Pumice Castle is the prominent brick-red-orange set of towers with resistant welded layers. Below the pumice are sheets of basaltic andesite (about 220 ka) lying on altered andesite lavas (as old as 340 ka) and debris-avalanche deposits. Numerous dikes cut the pre-Pumice Castle rocks. In the trees just below the caldera rim immediately E of the turnout is an exposure of partly welded Wineglass Welded Tuff, the southeasternmost such outcrop at the rim. Confinement of the Wineglass to depressions from just S of Llao Rock clockwise to here indicates the vent for the single-vent phase of the climactic eruption was NE of the summit(s) of Mt. Mazama. Large lithic blocks in the climactic pumice fall above Pumice Castle are consistent with proximity to the vent, which may have been located between here and Cleetwood Cove. Here, as elsewhere, lag breccia lies on the Wineglass. Thick lag breccia also mantles air fall above Redcloud Cliff and the dacite flow above Pumice Castle. The next two turnouts provide good views of the S caldera walls. **2.5**

59.6 Road on L to Pinnacles, which has spectacular exposures of rather atypical ring-vent-phase ignimbrite. Note lag breccia on R. Cuts S of here are in silicic andesite and low-silica dacite flows. **1.9**

61.5 View to S of Klamath Graben, Upper Klamath Lake (48 km), and Medicine Lake Volcano (160 km), the broad shield-like form beyond upper Klamath Lake. Snow-capped Mt. Shasta (175 km) also may be visible. **1.0**

62.5 Turnout on L with view NW to Applegate Peak. Silicic andesite and low-silica dacite flows grade from agglutinated lava upward into bomb beds in each "flow" at the caldera rim, and S down the flank into streaky lava flows with rubbly tops. Lavas in cuts here on Rim Drive are strongly flow-banded olivine-bearing silicic andesites of Dutton Cliff (ca. 210 ka) except the last cut before Sun Notch, which exposes altered, older andesite (≈320 ka). **1.0**

63.5 Parking area for trail to Sun Notch. A 1/4 mi walk to the caldera rim yields a fine view of Phantom Ship and Dutton Cliff. A gully about 30 m up the road from the E end of the parking area exposes lag breccia resting on Llao Rock pumice fall. The breccia contains included blocks of climactic air fall and friable altered andesite. **1.4**

64.9 Vidae Falls on R. Visible from Rim Drive S of here are gray cliffs of distant canyon walls to SE in pre-Mazama rhyodacite flows. **0.6**

65.5 Crater Peak trailhead. Columnar base and platy interior of silicic andesite flow well exposed in cut where Rim Drive leaves valley of Sun Creek and crosses Vidae Ridge immediately E of trailhead. **1.5**

67.0 Roadcuts on R expose moraine on E wall of Munson Valley. **0.9**

67.9 Stop sign across from Park Headquarters. Turn L (S). **4.0**

71.9 Intersection of Oregon Highway 62. Turn R toward Medford. Just W of intersection, the road rises along and over scarp of E-facing normal fault cutting three Pleistocene olivine andesite flows. **6.0**

77.9 Columnar-jointed medial Mazama ignimbrite can be seen on R in the canyon of Castle Creek for 0.9 mi W from this point. **7.5**

85.4 Long (0.3 mi) cut on R exposes nonwelded distal silicic Mazama ignimbrite. At the far (W) end, a flow-unit contact is visible at the reversely graded base of the flow unit that makes up most of the vertical exposure; the tops of both flow units consist of coarse pumice blocks. Flow-unit breaks are rare in the ignimbrite. **2.5**

87.9 Junction Highways 62 and 230.

WESTERN CASCADES, SOUTHERN OREGON AND NORTHERN CALIFORNIA

by

James G. Smith

This section of the field trip passes through the volcanic rocks of the Western Cascade Range (**Figure 25**). This sequence of gently deformed and slightly metamorphosed volcanogenic rocks and associated sedimentary rocks is a more deeply eroded mid-Tertiary analogue of the younger rocks of the High Cascades.

In southern Oregon and northern California, the volcanic rocks of the Western Cascade Range were deposited 35 to 15 Ma and form a simple homocline dipping E to NE. Folds are lacking, and the sequence is cut by only a few faults. The present outcrop belt marks the approximate location of the mid-Tertiary arc. However, volcanic edifices of the High Cascades unconformably cover the eastern part of the arc, and the exact location of the axis of the Western Cascade Range is uncertain. In southern Oregon and northern California, there is a hiatus between the Western Cascade rocks (35 to 15 Ma) and the High Cascade rocks (<7 Ma).

Only a few geochemically oriented studies have been done in the volcanic rocks of the Western Cascade Range in southern Oregon and northern California. They suggest that there is neither a difference in chemistry between the Western and High Cascade sequences nor a systematic change in composition with time.

The Western Cascade sequence contains only a narrow range of lithologies, is stratigraphically complex, and is poorly exposed. The lack of lithologic diversity and the stratigraphic complexity result from the way in which volcanic and volcaniclastic rocks were formed, deposited, and reworked in this subaerial arc environment. Hundreds of small overlapping and intertonguing volcanogenic and sedimentary units comprise the range.

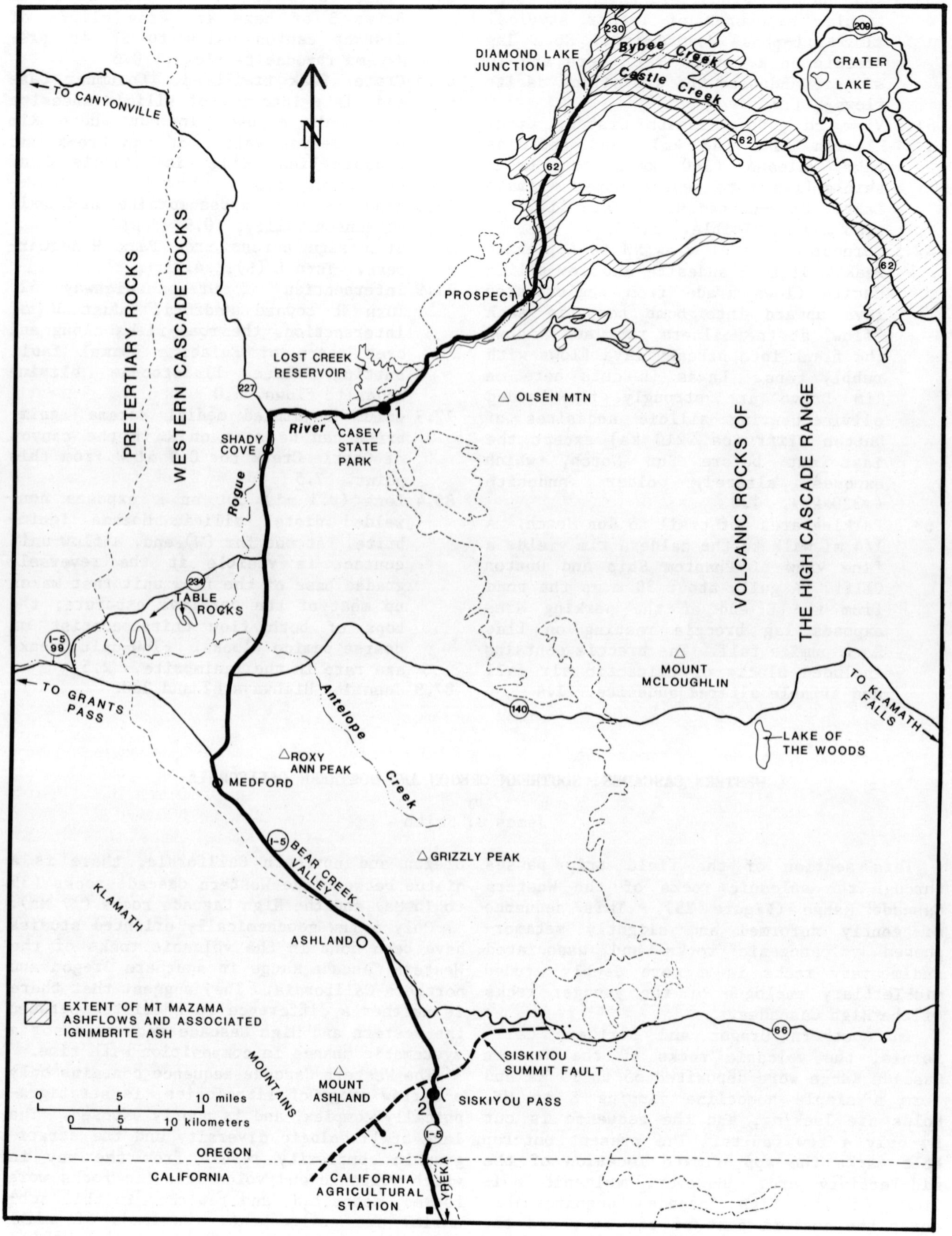

FIGURE 25 Route map for Western Cascade Range of southern Oregon and northern California.

Individual lithostratigraphic units are discontinuous and commonly interbedded in an intricate fashion. Widespread marker units are lacking throughout much of the sequence. Lithologic correlation, even of similar stratigraphic sequences, is unreliable without corroborating radiometric ages or detailed mapping. These factors have discouraged studies, and geologic knowledge of the Western Cascade Range lags behind that of the High Cascades.

Mileage

0.0 Diamond Lake Junction (intersection of Oregon Highways 62 and 230). Proceed S on Oregon Highway 62. From Bybee Creek (1.5 km N of Diamond Lake Junction) to the head of the Cascade Gorge (32 km S) the canyon of the Rogue River contains a succession of thin flows of primitive, low-K tholeiitic basalt that apparently issued from vents in the drainage of Castle Creek. These basalt flows transformed the 200-m-deep V-shaped ancestral valley of the Rogue River into a nearly flat valley floor 1.6 to 4.5 km wide that slopes about 12 m/km. Overlying the intra canyon basalt is up to 10 m of Mazama ash-flow tuff. The ash flows sped down the major drainages off the W and S flanks of Mt. Mazama, made a 90° turn at the Rogue River, and continued S for many more km. Locally, ash flows ran up the W side of the Rogue Valley as much as 30 m. **1.0**

1.0 Rogue River Gorge viewpoint. Good view of intracanyon basalt and Mazama ignimbrite. **1.5**

2.5 Natural Bridge campground. Rogue River flows through a lava tube in intracanyon basalt. **9.5**

12.0 Prospect Ranger Station on L. **1.9**

13.9 For the next several miles, Olsen Mtn., a Pliocene shield volcano with a prominent parasol drainage pattern, is visible at 10:00 to 12:00. The hills to the R (NW) are underlain by interstratified basalt and andesite flows, ash-flow tuff, tuff breccia, lapilli tuff, and volcanic sediment of the Western Cascade Range. In this area the rocks are between 25 and 15 Ma. Between here and Medford, Highway 62 proceeds generally down-section through this NNW-striking NE-dipping homoclinal sequence of volcanic and volcanic sedimentary rocks, which here total more than 5.5 km. **4.5**

18.4 Gorge Restaurant on L. Leave Pleistocene intracanyon basalt and start descent into Rogue River Canyon. Roadcuts on the R for the next 1.5 miles expose typical olivine basalt flows of the Western Cascade Range. The flows are approximately 20 Ma and fill a broad paleovalley cut into volcanic sediment and ash-flow tuff. These are exceptionally good exposures in which to observe flows and flow-contact features; elsewhere in the Western Cascade Range, distinguishing flows from sills is a common problem. **0.3**

18.7 Large roadcut at wide shoulder on R. Paleovalley in Western Cascade basalt is filled by lower-Pleistocene intracanyon flow of fresh pahoehoe basalt with local palagonitic hyaloclastite channel-fill at its base. **1.3**

20.0 N end of bridge over Lost Creek Reservoir in the valley of the Rogue River. For the next 0.6 mile, roadcuts expose intracanyon Pleistocene basalt flows. Just beyond a curve, the highway ascends to the top of these intracanyon flows and continues across them for a mile. To the L (S) is the eroded edge of the Olsen Mtn. shield volcano, of Pliocene age. To the R (N) across the reservoir are good views of the ENE-dipping volcanic rocks of the Western Cascade Range. **1.3**

21.3 To R is entrance to camping area at Stewart State Park. **0.3**

21.6 On L (S), roadcut of mafic hyaloclastite from Olsen Mtn. **0.6**

22.2 Entrance to day use area, Stewart State Park. **0.5**

22.7 Roadcut on L through massive nonvesicular olivine-bearing basalt or basaltic andesite of Olsen Mtn., part of a km-wide flow that moved N down a radial valley eroded into the shield and continued on to the valley of the Pliocene Rogue River. **0.9**

23.6 For the next 0.7 mile the roadcuts through another series of intracanyon olivine-bearing basalt or basaltic andesite flows from Olsen Mtn. Columnar joints spaced 10 to 20 cm apart are typical of the lower part of these flows. **0.6**

24.2 Roadcuts expose columnar-jointed, black, glassy, two-pyroxene, olivine-bearing andesite of Table Rocks (late Miocene age; 7 Ma). These distinctive flows are the oldest flows assigned to the volcanic rocks of the High Cascades in southern Oregon and northern California. Flat-topped erosional remnants crop out along the Rogue River's valley walls for 30 mi from 1 mi E of Prospect to the Table Rocks 10 mi N of Medford. Near Prospect the flows are 100 m thick, decreasing only slightly downstream to Table Rocks where they are still 75 m thick. Their volume was at least 6.3 km^3. Insignificant weathering of flow tops suggests that all the flows erupted within a short period. **0.5**

24.7 View at 2:00 of Lost Creek dam and part of Lost Creek Reservoir. **0.1**

24.8 **STOP 1, Basalt flows of the Western Cascade Range.** Make a sharp horseshoe bend R onto a nearly hidden, unmarked, dirt road. Exposures are in roadcuts on both sides of highway. **Even though the shoulder is wide, be extremely cautious about traffic, particularly log trucks.** Roadcuts expose 28-25 Ma basalt flows. Closely spaced jointing, recrystallization (indicated by dull mottled groundmass textures), and zeolites in vesicles and fractures are typical features of mafic and intermediate flows of the Western Cascade Range. Incised and channelized flows, irregular flow contacts, and cross-cutting dikes suggest a nearby source. Both abutments of Lost Creek dam are in this series of flows. **0.7**

25.5 Contact between basalt flows and altered ash-flow tuff of the Western Cascade Range. [Similar ash-flow tuffs can be viewed safely on the north side of Lost Creek reservoir along Takelma Drive, to the R just beyond the bridge over the Rogue River 0.8 mile ahead.] **1.0**

26.5 Casey State Park on L. For the next 7 mi, Highway 62 passes through poorly exposed interbedded andesite and basaltic andesite flows, tuff breccia, lapilli tuff, and small intrusives. **1.4**

27.9 Pasture land on both sides of the road in this area is probably underlain by reworked Mazama ash. **1.7**

29.6 Entrance to Rogue Elk County Park on L. On the R is a deposit of reworked Mazama ash that marks the farthest known extent of Mt. Mazama ash, 45 mi along the Rogue River from the Rim of Crater Lake. **3.3**

32.9 Junction with Oregon Highway 227 to Canyonville. [Those wishing to see a well exposed section of Western Cascade tuff breccia, lapilli tuff, and finer grained volcaniclastic beds typical of the the alluvial facies of Smedes and Prostka (1972) will find it along Oregon 227 and in the adjacent creek bed. The Western Cascade Range contains many thick sections like this one. Massive to poorly sorted, matrix-supported tuff breccia and lapilli tuff form beds 0.5 to 3 m thick with planar tops and bottoms. Both ash-flow tuff and debris flow deposits are present. Locally, diagenesis, weathering, and poor exposure limit the ability to distinguish between these two different types of deposits. Continental volcanic sandstone and siltstone sequences are interbedded with tuff breccia and lapilli tuff. Sandstone and siltstone beds range from a few cm to a few 10's of cm thick with sharp planar contacts. Locally, broad shallow channels with low-angle bed forms are present. This sequence is interpreted as having been deposited on aggrading volcanic dispersal aprons at some distance from volcanic vents. Volcanically induced high-energy depositional packages (e.g., debris flow deposits; ash-flow tuff) alternate with lower energy, fluvial, traction-current and overbank deposits.]

Continue on Highway 62, which turns L and heads S for the next 15 mi into the Medford Valley. The highway is parallel to the strike of the volcanic rocks of the Western Cascade Range, and this part of the trip crosses very little section. In many places rocks are poorly exposed and highly weathered. **0.2**

33.1 Roadcut in hypersthene andesite of the Western Cascade Range. **2.2**

35.3 Leave Shady Cove and cross Rogue River. Small quarry on L cuts through small mafic intrusion in the Western Cascade rocks. **2.6**

37.9 To the R are gently dipping slopes of Western Cascade lavas and sedimentary rocks. **2.3**

40.2 Hamell Road. The marshy pasture land to the L (E) is typical of low areas underlain by partly welded recrystallized ash-flow tuff of the Western Cascade Range. **1.6**

41.8 Junction with Oregon Highway 234 to Sams Valley, Gold Hill, and Grants Pass. Mt. McLoughlin is visible from time to time for the next several miles at about 9:30. **5.1**

46.9 Cross Antelope Creek. Road ascends onto a mid-Pleistocene fan terrace. A sequence of six stepped geomorphic surfaces was recognized by Parsons and Herriman (1976) in the Medford area. This surface, the Roxy Ann surface, is the next-to-oldest and is 30-45 m above the present flood plain of the

Rogue River. **2.5**

49.4 Junction with Oregon Highway 140 to Lake of the Woods and Klamath Falls. Mesas at 3:00 are capped by 7 Ma andesite flows of Table Rock. **2.4**

51.8 Central Point junction. Low hills at 10:30 are underlain by small Tertiary intrusives that were emplaced into upper Eocene continental sediment derived from the Klamath Mtns. to the W (R). **3.2**

55.0 Turn L onto Interstate 5 S. For the next 15 mi I-5 ascends Bear Creek valley. To the R (W) are the Klamath Mtns., underlain by several tectonostratigraphic terranes composed of complexly faulted and highly deformed late Paleozoic, Triassic, and Jurassic metamorphic and intrusive rocks. More than 1200 m of Upper Cretaceous unmetamorphosed dominantly marine sandstone, siltstone, and conglomerate (the Hornbrook Formation) was deposited unconformably as an overlap assemblage across the older terranes. A sequence of upper Eocene and lower Oligocene(?) fluvial conglomerate and sandstone derived from Mesozoic and Paleozoic rocks in the Klamath Mtns. unconformably overlies the Hornbrook Fm. This Tertiary sequence is as thick as 2150 m but thins abruptly to the S. Both the Hornbrook Fm. and the Tertiary rocks dip gently E and extend an unknown distance under the volcanic rocks of the Western Cascade Range. Volcaniclastic sediment and tuff coming from the nascent High Cascade volcanic arc are interbedded in the youngest part of the continental sequence. For a detailed description of the Hornbrook Fm. and lower Tertiary rocks between Medford and Yreka see the field trip guide of Nilsen et al. (1984).

The rounded tree-covered hill in the middle distance at 9:00 is Roxy Ann Peak. The change in slope at the shoulders of the peak marks the approximate contact between upper Eocene continental sandstone and conglomerate and the unconformably overlying volcanic rocks of the Western Cascade Range. A K-Ar date on the lowermost flows on Roxy Ann Peak gave an age of 30.8±2 Ma. **4.0**

59.0 On L (E), turbidites of the Hornbrook Fm. are exposed in roadcut, with upper Eocene fluvial sandstone in the upper part of the hill. **1.2**

60.2 Upper Eocene and lower Oligocene(?) continental sandstone and conglomerate make up the lower half of the mountainside rising to the L. The prominent cliff part way up the slope is a conglomerate lens. Basaltic andesite flows and interlayered tuff breccia of the Western Cascade Range dated at 25-30 Ma crop out along the upper half of the slope. **1.5**

61.7 On L are roadcuts exposing upper Eocene fluvial conglomerate. **1.5**

63.1 Just beyond the rest area, the highway crosses the contact between upper Eocene rocks and the Hornbrook Fm. **4.0**

67.1 Weigh station. The smooth slopes to the L (NE) behind the weigh station are the front of a large landslide. Most of the section visible from here is Western Cascade basaltic andesite and andesite flows and associated tuff breccia of the Western Cascade Range. The section is nearly 3.6 km thick. Rocks at the base of the section are about 30 Ma, and rocks at the top of Grizzly Peak (at 10:00) are about 25 Ma for an accumulation rate of 400 m/m.y. The flows and tuff breccia accumulated on coalescing dispersal aprons around volcanic centers. The lateral continuity of beds and uniformity of attitude suggest that the accumulation area was some distance from the source volcanoes. **4.1**

71.2 Second Ashland exit at Oregon Highway 66. **1.2**

72.4 At 10:00 in the middle of the valley is a small Tertiary plug intruded along the contact of the Hornbrook Fm. and the upper Eocene sedimentary rocks. **2.4**

74.8 Roadcut on R exposes sandstone and pebble conglomerate of the Hornbrook Fm. **0.6**

75.4 Contact between Hornbrook Fm. and small Tertiary intrusion. Both the intrusion and the Hornbrook Fm. are hydrothermally altered near the contact.

For the next 4 1/2 miles, roadcuts are in the Mt. Ashland pluton (Late Jurassic age) and screens of metamorphic rock of Paleozoic or Mesozoic age; intrusion of the Mt. Ashland pluton hornfelsed the regionally metamorphosed schist and gneiss. The unconformity with the overlying Hornbrook Fm. dips steeply to the E (L). **0.6**

76.0 Exposures of Mt. Ashland pluton. **1.0**

77.0 Exposures of roof pendants of metasedimentary biotite schist and quartz-biotite gneiss. **2.3**

79.3 Roadcuts expose the Mt. Ashland pluton (capped by a grus layer) and the overlying Hornbrook Fm. **0.6**

79.9 Offramp to Mt. Ashland. Highway crosses the poorly exposed Siskiyou Summit fault, a NNE-trending normal scissors fault down to the SE. W of I-5, the fault juxtaposes Hornbrook Fm. and Oligocene volcaniclastic sediments against the Mt. Ashland pluton. E of I-5, the fault puts Hornbrook Fm. and upper Eocene and lower Oligocene(?) continental sedimentary rocks against Oligocene volcanic rocks of the Western Cascade Range. The Siskiyou Summit fault is one of a series of widely spaced faults radial to Condrey Mtn., 25 km SW, and is thought to have formed as a result of some 7 km of vertical doming centered on Condrey Mtn. between about 14 and 5 Ma (Mortimer and Coleman, 1984). Displacement on the Siskiyou Summit fault SW of I-5 may be as much as 5.5 km, but only 5 miles E of I-5 no offset is detectable. **1.1**

81.0 **STOP 2, Recrystallized volcaniclastic sediments and ash-flow tuffs of the Western Cascade Range.** Truck brake check area at summit of Siskiyou Pass. Pull off on R. Volcaniclastic sediments are more prevalent in the N part of the roadcut. Note: (1) broad shallow channels, and (2) the wide variety of rock types, including roundstone conglomerate channel fill, fine overbank sand, lacustrine silt, and poorly sorted volcanic diamicton (interpreted as mudflows). The green color on fresh surfaces is typical of unweathered Western Cascade volcaniclastic rocks. The recrystallization assemblage includes zeolites (primarily heulandite with some clinoptilolite), chlorite (probably an iron-rich variety), calcite, and chalcedony (Bestland, 1985). A gray, white-weathering, partly welded ash-flow tuff is exposed, starting about one-third of the way through the cut. The ash-flow tuff fills a shallow channel which is best seen in the roadcut on the E side of the highway (**do not attempt to cross this dangerous interstate highway!**). Although the rock appears fresh, Bestland (1985) described very fine-grained clinoptilolite replacing the vitric matrix. At the S end of the roadcut, ash-flow tuffs 2-3 m thick are separated by volcaniclastic sandstone, conglomerate, and siltstone. **0.8**

81.8 The dark cliffs above the highway to the L are a 30 Ma intracanyon basalt flow. **0.5**

82.3 Yellow and buff volcaniclastic sandstone and siltstone are exposed in cuts on both sides of road. **0.4**

82.7 Light-colored outcrops on L are pumice lapilli ash-flow tuff dated at 29.9 ± 1.5 Ma by K-Ar. The section is repeated farther along in the roadcut by a small fault. A basaltic dike also intrudes the section here. **2.8**

85.5 California state line. On R, the roadcut exposes thin-bedded, rhythmically interbedded, basin-plain turbidites of the Hornbrook Fm. **1.6**

87.1 Prominent rim to the L (E) is formed by a Tertiary basaltic sill intruded into the Hornbrook Fm. **0.7**

87.8 Railroad overcrossing. The large rounded hill at 11:30 is Black Mtn., a large andesitic intrusion with many landslides on its sides, emplaced along the contact between the Hornbrook Fm. and the volcanic rocks of the Western Cascade Range. The railroad grade to the L is cut into Hornbrook Fm. Higher up the hill, volcanic rocks of the Western Cascade Range lie directly on the Hornbrook Fm. without any intervening upper Eocene sedimentary rocks. The cliff in the Western Cascade rocks is an ash-flow tuff. **3.6**

91.3 Agricultural inspection station.

MOUNT SHASTA AND VICINITY

by

Robert L. Christiansen and C. Dan Miller

Mt. Shasta is the largest stratovolcano of the Cascade chain. At nearly 500 km^3, it is comparable in volume to such stratocones as Fuji-san and Cotopaxi. The peak rises to an elevation of 4317 m, more than 3200 m above its base, and dominates the landscape of northern California.

The route of this trip (**Figure 26**) approaches Mt. Shasta from the N through Shasta Valley, largely floored with the debris of an enormous volcanic sector avalanche that virtually destroyed an ancestral Mt. Shasta about 350 ka (Crandell et al., 1984; Crandell, in press). Only a small remnant of this older volcanic edifice remains, but the volume of the Shasta Valley

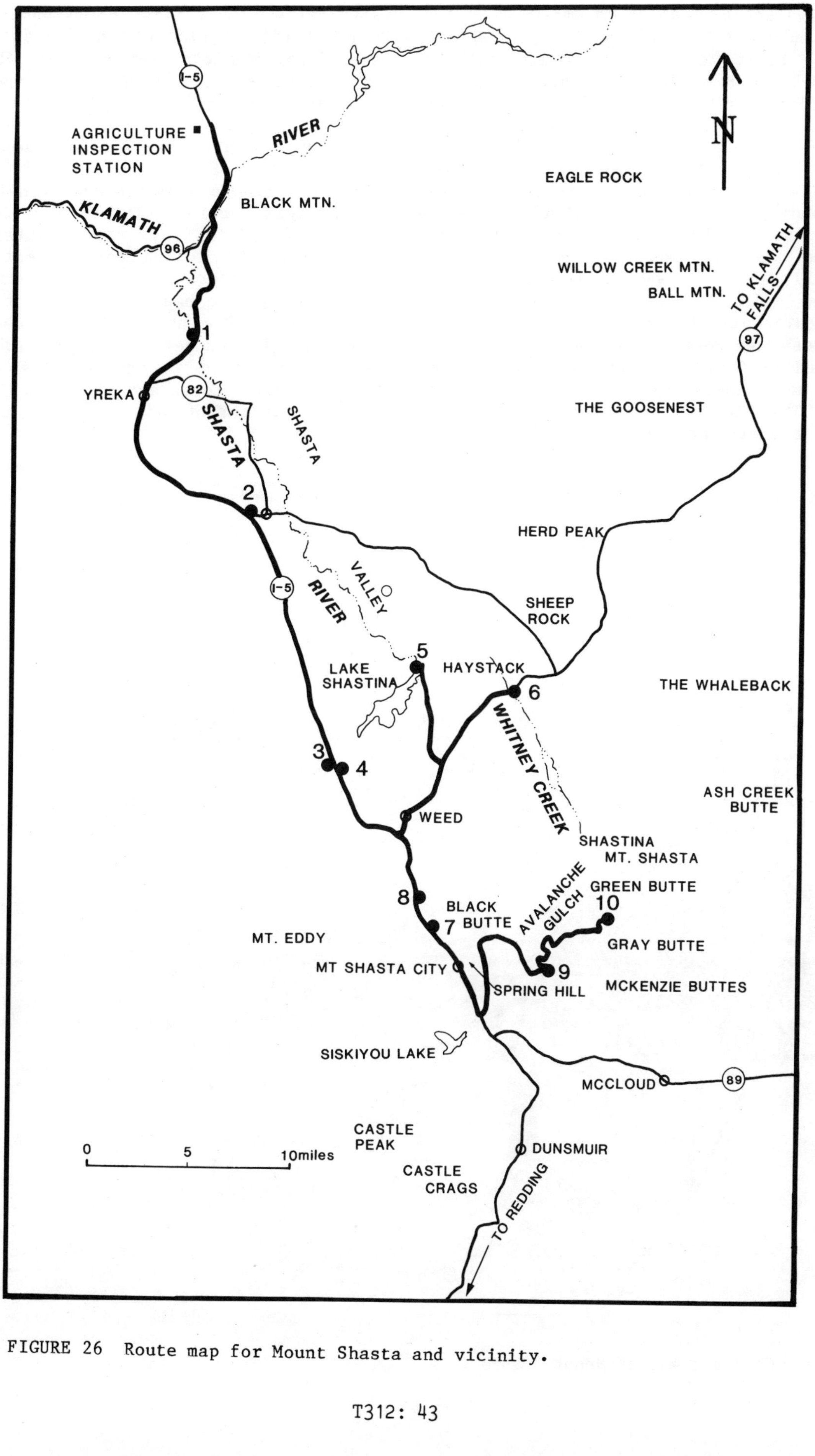

FIGURE 26 Route map for Mount Shasta and vicinity.

debris avalanche is about 45 km^3.

The bulk of Mt. Shasta is a composite feature built over the stump of the destroyed older volcano; it comprises four major cones built around discrete eruptive centers (**Figure 27**; Christiansen and Miller, 1976; Christiansen et al., 1977). Each of these cones appears to have grown mainly in a short episode, perhaps only a few hundred or a few thousand years (Christiansen, 1982; 1985), that produced numerous lavas from a central vent. Virtually all the lavas of the major cone-building episodes are 2-pyroxene silicic andesites. Following each of these episodes, the cones underwent significant erosion and less frequent eruptions from the central vent and, in most instances, from flank vents as well. The latest eruptions within the central craters produced dacitic domes and pyroclastic flows; flank eruptions typically produced either cinder cones and flows of basalt to basaltic andesite or domes and pyroclastic flows of dacite to rhyodacite. Pyroclastic flows are particularly conspicuous on the W flank of Shastina, the prominent cone on Mt. Shasta's W side (Miller, 1978).

The Mt. Shasta magmatic system has evolved more or less continuously since at least 590 ka. The Sargents Ridge cone, oldest of the four cones that formed the present volcano after major sector collapse, is younger than about 250 ka but has experienced two major glaciations. The next younger, the Misery Hill cone, is younger than about 130 ka and has been sculpted in one major glaciation. The two younger cones are Holocene: Shastina, W of the cluster of other central vents, was formed mainly between 9.7 and 9.4 ka; the

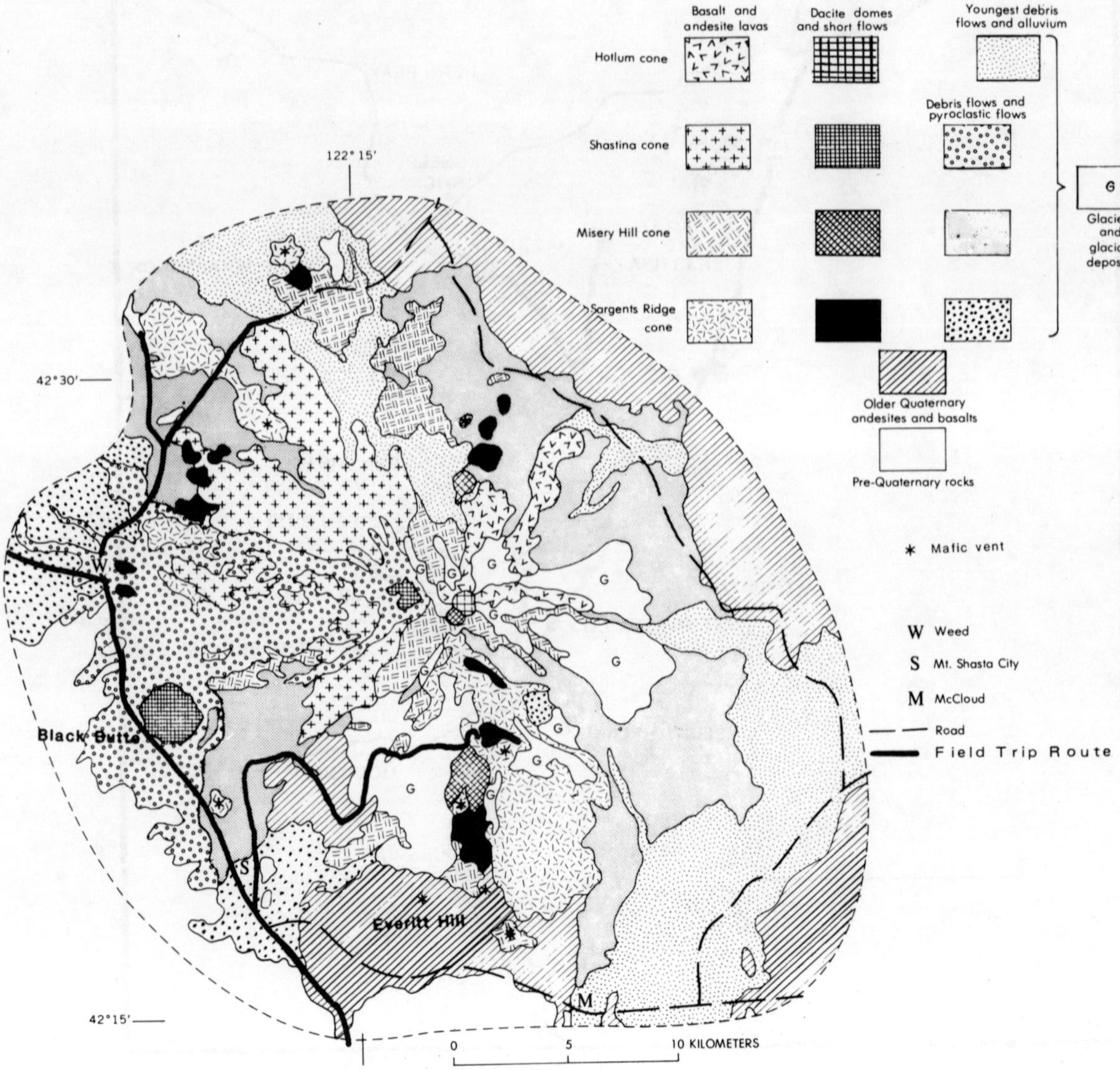

FIGURE 27 Geologic map of Mount Shasta.

Hotlum cone may overlap Shastina in age but appears to be mainly younger. Mt. Shasta has continued to erupt at least once every 600-800 yr since 10 ka (Miller, 1980). Its most recent eruption probably was in 1786 (Finch, 1930).

Mileage

0.0 Agricultural Inspection Station on US I-5 S. **1.6**

1.6 Offramp to Henley-Hornbrook. The low hills along the road are underlain by Upper Cretaceous Hornbrook Formation. At 11:00 is Black Mtn., a resistant knob of intrusive andesite of the Western Cascades. Sharper ridges to the R and in the distance are pre-Tertiary metamorphic rocks of the Klamath Mtns., mainly representing oceanic crust and marine metasedimentary and metavolcanic rocks of Paleozoic and early Mesozoic age. **1.5**

3.1 Large roadcut in sandstone of the Hornbrook Formation. The basal contact lies in the next gully to W. **0.5**

3.6 Large roadcut in greenstone of the Klamath Mtns. sequence. Most roadcuts beyond here for about the next 10 miles expose similar greenstone and some interstratified chert. **1.3**

4.9 Cross Klamath River. The river flows W to the ocean through a deep canyon. Large roadcuts along the road for the next several miles are all in metamorphic rocks of the Klamath Mtns. **6.2**

11.1 **STOP 1, View across the volcanic debris avalanche of Shasta Valley to Mt. Shasta.** Exit R at offramp for vista point. The conspicuously hummocky topography of Shasta Valley is the expression of a volcanic debris-avalanche deposit that is exposed S to Black Butte, the prominent steep-sided dome at the W foot of Mt. Shasta, 50 km away. The deposit resulted from an enormous sector collapse of an ancestral Mt. Shasta. The flat valley floor between the hummocky megablocks is a matrix facies of debris flows (lahars) that resulted from dewatering of the avalanche. These debris flows were forced through the narrow valley of the Shasta River just W of here (R) and ponded in the wider valley upstream to an average thickness of 75 m (Crandell, in press). Downstream (farther N), the debris flows must have entered the Klamath River. A few of the larger hills surrounded by the hummocky debris avalanche are eroded remnants of Tertiary Western Cascades volcanic rocks that once extended across Shasta Valley.

On the skyline to the E (L) are three young basaltic-andesite shields of the High Cascades built on eroded volcanic rocks of the Western Cascades (L to R): The Goosenest (late Pleistocene), Herd Peak (Pliocene), and The Whaleback (late Pleistocene). Outcrops NE across the highway are intrusive andesites of the Western Cascades.

The higher terrane to the W and SW is entirely underlain by metamorphic rocks of the Klamath Mtns.; prominent at the S end of the view is Mt. Eddy, underlain by the Trinity ophiolitic complex of early Paleozoic age.

Return to Highway I-5 S. **0.9**

12.0 Cross Shasta River. **5.4**

17.4 Offramp to California Highway 3 to Ft. Jones and Weaverville; continue on I-5 S. Note that the Shasta Valley volcanic debris avalanche did not enter the valley of Yreka, separated from the main Shasta Valley by a low bedrock ridge to the L. Hills on both sides are underlain by metasedimentary rocks of Ordovician and Silurian age. **7.3**

24.7 Exit R at offramp for Gazelle; at stop sign, turn L across I-5. **0.3**

25.0 **STOP 2, View across the Shasta Valley volcanic debris avalanche to Mount Shasta.** Park along the wide roadside while still high on the overpass ramp but just past the bridge over I-5. Low rounded hills across the road to NE are megablocks of the avalanche deposit; the flat surface in the immediate foreground is underlain by debris flows of the matrix facies. The steep-sided dome of Black Butte is prominent at the W foot of Mt. Shasta. On the skyline to the E is a group of late Pliocene to early Holocene basaltic andesite shield volcanoes along the main Cascades axis, including (L to R) Eagle Rock, Willow Creek Mtn., Ball Mtn., The Goosenest, Herd Peak, The Whaleback, and Ash Creek Butte. The Goosenest has a prominent agglutinate cone at its central vent, probably typical of many of these shields but not always well preserved on the older ones; Ash Creek Butte is carved by prominent glacial cirques that removed much of a similar agglutinate cone.

Low hills to the N and across the highway to the SW are phyllites of

Ordovician age.

Continue ahead. **0.4**

25.4 Gas station and restaurant at Grenada. Turn around and return across overpass to the W side of I-5. **0.5**

25.9 Turn L onto onramp to I-5 S. **1.5**

27.4 Cross railway overpass. I-5 near here traverses the flat surface of matrix-facies debris flows of the Shasta Valley debris avalanche. **2.1**

29.5 From here S for several miles, I-5 passes through hills that are megablocks in the volcanic debris avalanche; the flat surface between the hills is underlain by debris flows of the matrix facies. **3.3**

32.8 Offramp to Louie Road on R; continue on I-5 S. Beyond here I-5 climbs through megablocks of the Shasta Valley debris avalanche. **1.0**

33.8 To the R is the first of two large closed depressions in the surface of the debris avalanche. **3.8**

37.6 I-5 leaves the megablock terrain and crosses a flat-surfaced terrace developed on debris flows of the debris-avalanche matrix facies. The surface on this terrace is marked by prominent stone polygons and other patterned-ground features, produced by periglacial processes during the late Pleistocene glaciations. **0.7**

38.3 **STOP 3**, Exit R at offramp to rest stop, toilet facilities, and vista point. Return to vehicles and continue along frontage road. **0.4**

38.7 Turn R on road to Weed Airport. At intersection, turn L and continue along the road under the highway. **0.4**

39.1 **STOP 4, Matrix facies of the Shasta Valley debris avalanche.** Park at turnout on R, opposite junction with road to Weed airport. Walk a few meters S to cross a barbed-wire fence and descend to low cuts along the railroad. The variety of volcanic rock types in the deposits is representative of the ancestral Mt. Shasta, consisting mainly of silicic andesites and subordinate dacites. A few basaltic andesites are also present. Note the rounding of the fragments and the indurated fine-grained matrix. The deposit dates from about 350 ka and has a fairly well developed soil, mainly expressed as a deep zone of oxidation. From the banks above the railroad cuts is a good view S to Mt. Shasta and to Black Butte, a dacitic dome complex at its W foot.

Turn around and return by the same roads to the frontage road for I-5 S. **0.6**

39.7 Re-enter I-5 S. The highway continues along the patterned-ground surface on a matrix-facies terrace of the Shasta Valley debris avalanche. **2.3**

42.0 Offramp to Stewart Springs Road and Edgewood; continue on I-5 S. I-5 S climbs through large ridges that are megablocks in the Shasta Valley volcanic debris avalanche--virtually intact fragments of the ancestral edifice of Mt. Shasta. The roadcuts provide excellent exposures of these megablocks, which consist predominantly of unconsolidated volcanic diamictons that are locally disturbed and cut by numerous small faults but remain in upright, subhorizontal attitudes and have largely coherent stratigraphy. **2.7**

44.7 Exit R at offramp to N Weed and US Highway 97; at stop sign, turn L through underpass beneath I-5, then R at the next stop sign, following signs to Weed and Klamath Falls. **0.9**

45.6 Stop at intersection with flashing red light and then turn L onto US-97; just beyond, cross railroad on overpass. **0.7**

46.3 First of several roadcuts on both sides in megablocks of Shasta Valley debris avalanche. **1.2**

47.5 Apple Valley Road. Roadcut on R and a sand and gravel pit farther to R expose ash-cloud deposit from pyroclastic flows erupted at the summit of Shastina. Beyond here, the road drops onto a surface underlain by older pyroclastic flows of early Holocene age. Hills to L are megablocks in the debris avalanche that underlies the pyroclastic flows. Hills to R are andesitic lava flows from the Shastina central vent; they overlie the pyroclastic flows. Also to R are several older dacitic domes. **2.5**

50.0 Turn L onto road to Lake Shastina and Big Springs. The road crosses pyroclastic flows from Mt. Shasta dated at about 9.7 ka and bearing a poorly developed soil--about 70 cm of yellowish oxidation. In the distance to the L, beyond the debris-avalanche megablocks, are ridges of the Klamath Mtns. underlain by metamorphic and plutonic rocks. On the skyline to R are basaltic-andesite shields along the main Cascades axis. **2.5**

52.5 Irregular ridge on R and ahead is an andesite flow from the oldest part of modern Mt. Shasta, the Sargents Ridge cone. It is deeply weathered, overlies megablocks of the Shasta Valley landslide, and is overlain by early Holocene pyroclastic flows from Mt. Shasta. **0.3**

52.8 Entrance to golf resort and restaurant on L; continue on main road. **0.6**

53.4 Roadcut through a megablock of the Shasta Valley debris avalanche. The younger Sargents Ridge andesitic lava flow overlaps this megablock on the R. The flat surface along which most of the road passes is late Holocene pyroclastic flow. **0.7**

54.1 Entrance to Lake Shastina resort development on L; continue on main road. **0.9**

55.0 The road descends from the pyroclastic-flow surface onto the debris-flow fan of Whitney Creek, on which debris flows continue to be deposited this far from Mt. Shasta during years of heavy snowmelt and high runoff. On the L, the road passes a megablock of the Shasta Valley volcanic debris avalanche. **0.2**

55.2 **STOP 5, Megablock of the Shasta Valley volcanic debris avalanche and view of Mt. Shasta.** At a gate that closes off a paved road on the L, turn around and park. Cross the gate on foot and walk up the paved road straight ahead. About 200 m up the road, just past the crest of a hill, turn R onto a dirt road blocked by a cable and walk about another 250 m across the dam of Lake Shastina to a quarry on the other side. This quarry provides an excellent exposure into one of the megablocks of the Shasta Valley volcanic debris avalanche. The materials of the block are principally unconsolidated pyroclastic flows, volcanic debris flows, and sediments, including several successive buried soils. Numerous normal faults of varied orientations and displacements cut the block, but the section remains subhorizontal and upright, and its stratigraphy remains coherent.

To the SE is Mt. Shasta. The summit and N and NE flanks are the young Hotlum cone. Shastina is in front and to R but is not clearly outlined from this perspective. The small ridge between these cones is part of the breached crater rim of the Misery Hill cone. Below Mt. Shasta in a line toward us are andesitic lava flows erupted from the summit vent of Shastina. One lava flow from the older Sargents Ridge cone emerges from beneath the visible edifice of Mt. Shasta in a line toward Lake Shastina, abutting megablocks of the Shasta Valley debris flow. Black Butte, to the S, is a young dacitic flank dome of the Shastina cone; between it and Shastina but closer to us are several older dacitic flank domes that are overlapped by Shastina lavas.

Along the skyline to the E (R to L, N from Mt. Shasta) are The Whaleback, Sheep Rock (eroded andesitic breccias of the older Western Cascades), Herd Peak, and The Goosenest. The sparsely tree-covered area of Shasta Valley beyond the grassy surface of the Whitney Creek debris-flow fan is underlain by basalt of 300 ka that overlaps and helps to date the Shasta Valley debris avalanche.

Return to vehicle and retrace route S to US-97. **5.2**

60.4 Turn L onto US Highway 97. After the turn, the hill on the L is a megablock in the Shasta Valley debris avalanche. To the R, the early Holocene pyroclastic-flow surface is overlain by the snout of a lava flow from the central vent of Shastina. **0.6**

61.0 Roadcut on R exposes an early Holocene pyroclastic flow from Mt. Shasta that forms the surface in this area and shows a weak soil profile, mainly expressed as about 70 cm of yellowish oxidation. Near here this pyroclastic flow is dated at 9.7 ka. **0.3**

61.3 Turnout just past low crest. About 4.5 km at 2:00 is Cinder Cone, an old basaltic flank vent just beyond an even older flow (with the prominent railroad cut). To the L is a good view down Shasta Valley of the debris avalanche. **1.5**

62.8 Area to the L of highway displays outcrops of an andesitic lava flow from the Sargents Ridge cone. The lava postdates the Shasta Valley debris avalanche but underlies early Holocene pyroclastic flows. **1.6**

64.4 Turnoff on R, opposite the end of a dirt road with stop sign on L. A few meters S (R) of the road is the steep flow-front scarp, about 20 m high, of the Lava Park flow, an andesitic block lava that erupted from a flank vent of Shastina at about 9.5 ka. This is the only Shastina andesite not erupted from the central crater. Although not a stop on the regular field trip, a climb from here to the top of this flow provides a view superior to that of stop 6. (If this climb is made, however, be cautious of **poison oak,** which grows on the scarp and in the gully at its base and can cause a severe skin rash). **0.3**

64.7 Dirt road on R; continue on highway. View at 10:00 of Haystack, an old dacitic flank dome of Mt. Shasta, and a composite cinder cone farther L. The large shield volcano ahead on the skyline is The Whaleback. **0.9**

65.6 **STOP 6, View of the younger cones of Mt. Shasta.** Park at large turnout on R by green fence just before bridge across Whitney Creek. Walk through the gate (marked by a red post) near the bridge and proceed a short distance upstream along the bank of Whitney Creek. The creek is fed from daily meltwater runoff from the Whitney glacier and is often dry in the morning but runs later in the day; runoff may vary from a trickle to a torrent. Note the debris-flow levees along both banks. To the S, the Hotlum and Shastina cones of Mt. Shasta are prominent, with a remnant of the breached crater rim of the Misery Hill cone between them at the head of the Whitney glacier; the edge of the summit dome of Misery Hill is also visible to the L of the rim.

Low hills to the E are lava flows of the Misery Hill cone that emerge from beneath the younger Hotlum cone on the N side of the compound stratovolcano. The top of Haystack, an older dacitic flank dome, is just visible beyond them to the NE. The Lava Park flow, an early Holocene block lava from a flank vent of Shastina, is prominent to the SW (R).

Turn vehicles around and return on US-97 toward Weed. **5.2**

70.8 Lake Shastina Road on R; continue on US-97. **4.5**

75.3 Stop at flashing red light. Turn L onto Weed Blvd. Pass through Weed and under I-5. **0.5**

75.8 Turn L onto onramp to I-5 S. Hill to the L of the highway is an old dacitic flank dome of Mt. Shasta. **1.3**

77.1 Offramp to S Weed Boulevard; continue on I-5 S. The valley bottom here is a constructional surface of pyroclastic flows from the summit vent of Shastina. Ahead on the R is a large megablock of the Shasta Valley volcanic debris avalanche--the nearest to the ancestral cone that is still exposed. Roadcut ahead is in Shastina pyroclastic flows. On the L ahead is Black Butte, a dacitic flank dome of Shastina and its youngest eruptive product. At least 4 and perhaps 5 emplacement units of the dome are expressed geomorphically. **2.0**

79.1 Offramp to Summit Dr. and Truck Village Dr.; continue on I-5 S. Hummocky terrain is the surface of hot-avalanche deposits and pyroclastic flows from Black Butte. **2.3**

81.4 Exit R on offramp to Abrams Lake Road. At stop sign; turn L across highway. **0.4**

81.8 Turn L onto onramp to I-5 N. **1.2**

83.0 **STOP 7, Black Butte hornblende dacite.** Park at large turnout on R, 0.1 mile beyond sign for Black Butte summit. Talus from the Black Butte dome comes down to the roadside and offers opportunities to examine the lithology of this hornblende dacite. The fresh lithology from the interior of the dome is gray with dark hornblende phenocrysts, generally with hollow centers and plagioclase cores; a pink lithology with oxyhornblende represents the surficial carapace of the dome. **0.8**

83.8 Exit R on offramp for Truck Village Dr. At stop sign, turn R and pass Truck Village on R. **0.3**

84.1 At T intersection, turn L. **0.3**

84.4 Roadcut in a large, internally shattered block of dacite in a hot-avalanche deposit from the emplacement of the Black Butte dome.

Continue ahead on main paved road. **0.7**

85.1 **STOP 8, Shastina and Black Butte pyroclastic flows.** Turn around and park in open area past end of pavement. Walk W down into railroad cut. The major exposures are flows from the Shastina summit vent, distinguished by dacite blocks that contain phenocrysts of hornblende, hypersthene, and augite. A pyroclastic flow from Black Butte, recognized by large cored hornblende phenocrysts but lacking pyroxene, overlies the Shastina flow in pockets, especially at the S end of the cut. Black Butte rises prominently to the S; several emplacement units are recognizable topographically. To the E, Shastina rises conspicuously on the W side of Mt. Shasta, with a fan of pyroclastic flows mantling its W flank.

Return along paved road toward Truck Village. **1.0**

86.1 At intersection, turn R and continue beneath I-5. **0.2**

86.3 Turn L onto onramp to I-5 S and return to the highway. **1.9**

88.2 Offramp to Abrams Lake Road; continue

on I-5 S, which passes through hummocky topography of Black Butte hot-avalanche deposits and pyroclastic flows. **0.5**

88.7 Sand-and-gravel pit on R, where pyroclastic flows of Black Butte are quarried for aggregate. **0.6**

89.3 Exit R on offramp to Mt. Shasta City. On the L is Spring Hill, an old andesitic lava cone on the W flank of Mt. Shasta. **0.9**

90.2 Mt. Shasta City Park on R; continue ahead on main road. **0.4**

90.6 Cross railroad tracks, a spur from the main line to the R. **0.2**

90.8 Bear R at Y intersection. **0.3**

91.1 Traffic light at major intersection in Mt. Shasta City. Turn L onto E. Alma St. Proceed through the intersection with the flashing yellow light and follow the road with the double yellow line as it curves around a school and baseball field. **0.6**

91.7 Stop at flashing red light; turn L onto Everitt Memorial Highway. **0.4**

92.1 Cross railroad tracks. Beyond, the road climbs gradually on a fan of upper Pleistocene pyroclastic flows from Mt. Shasta, postdating the building of the Misery Hill cone. **1.0**

93.1 Gentle curves in road. On L ahead is Black Butte. High peak on the W skyline is Mt. Eddy, exposing the Trinity ophiolitic complex. **0.7**

93.8 Road on L to Black Butte Trail; continue on highway, which continues to climb the surface of upper Pleistocene pyroclastic flows. **1.7**

95.5 Road on L to McBride Springs Campground, at which a spring issues from the base of an andesitic lava flow from Shastina where it lies on the older pyroclastic flows. Continue on Everitt Memorial Highway. **0.2**

95.7 Wide turnout on R at orange sign "Elevation 4790 feet, 4 miles." Shastina on the L. Ahead, post-Misery Hill pyroclastic flows onlap steep tree-covered bluffs that rise to the L of the road, a remnant of ancestral Mt. Shasta. **0.6**

96.3 Road crosses gully onto a steeper slope. Roadcuts on the L ahead expose 590 ka andesitic lava flows, overlain by colluvial and pyroclastic deposits of varied thickness and weathered to a deep, brightly colored soil. The lavas are part of the only remnant of ancestral Mt. Shasta that existed before the 350-ka sector collapse. **2.4**

98.7 Orange sign "7 miles, elevation 5600 feet." At a sharp curve L and a small side road to the R, about 25 m past the sign, is good exposure of porphyritic andesite of ancestral Mt. Shasta. The prominent curved platy jointing is common in many Cascades andesitic lavas. **1.0**

99.7 Orange sign "Elevation 5900 feet, 8 miles." On L just beyond the sharp curve ahead is a good example of the deep weathering profile developed on the lavas and surficial deposits of ancestral Mt. Shasta. **0.5**

100.2 Turnoff on R for John Everitt Vista Point. To the R and behind is a view S to Castle Crags and Castle Peak, erosional features of a Jurassic granitic pluton. To the SW are Siskiyou Lake and Mt. Shasta City at the foot of debris-flow and pyroclastic-flow fans from Mt. Shasta. To the W is the ophiolite complex on Mt. Eddy. To the NW is the top of Black Butte. **0.3**

100.5 **STOP 9, View of the older cones of Mt. Shasta.** Park at turnout on R. On the L is the early Holocene cone of Shastina, overlapping lavas from the apparent summit cone, the Misery Hill cone. The true summit, on the Hotlum cone, is blocked from view by Misery Hill, a dacite dome that rises within the crater of the cone named for it. To the R of Misery Hill is a prominent ledge, Red Banks, formed by agglutinated fallout pumice erupted from the summit crater of the Misery Hill cone at about 9.7 ka. The prominent glaciated valley leading directly toward here from Red Banks is Avalanche Gulch. Just to the R of Red Banks is Sargents Ridge, which exposes lavas that dip N, back under the Misery Hill cone, and define the N flank of the Sargents Ridge cone; S dips on the skyline define the opposite flank of the Sargents Ridge cone.

A N-S line of flank vents marks the skyline to the E, on the S flank of the Sargents Ridge cone. The highest prominent vent feature is Gray Butte, a glaciated dacite dome; next below is an andesitic cinder cone; next are two more dacite domes (upper and lower McKenzie Buttes, with ski runs cut into the vegetation of the upper); less conspicuous at the foot of the mountain at the S end of the line is the top of Signal Butte, an andesitic lava cone. Farther R, to the S from here, is Everitt Hill, a 450 ka basaltic-andesite shield volcano.

The roadcut on the L side of the

road exposes the deep soil developed on ancestral Mt. Shasta and part of an andesitic lava flow. Spoil from the cut lies at the roadside in the turnout, providing relatively fresh samples of the andesitic lava. Note the abundant corroded plagioclase and moderately abundant hypersthene phenocrysts. **0.2**

100.7 Sign "Elevation 6260 feet, 9 miles." **0.6**

101.3 Red Fir Flat at the top of the remnant of ancestral Mt. Shasta. **0.1**

101.4 On L is the S end of the loop road to Sand Flat and Horse Camp; continue on paved road. **0.4**

101.8 Orange sign "Elevation 6650 feet, 10 miles." Large roadcut on the L cuts through the first of 3 morainal ridges, the right-lateral moraines of the youngest major Pleistocene glacier of Avalanche Gulch. **0.2**

102.0 On L is the N end of the loop road to Sand Flat and Horse Camp; continue on paved road. **0.3**

102.3 On L is large exposure of the right-lateral moraine of Avalanche Gulch. Ahead is Mt. Shasta and the glaciated valley of Avalanche Gulch, with Sargents Ridge on the R. To the R is Gray Butte, a glaciated dacitic dome at the head of the N-S vent alignment on the S flank of the Sargents Ridge cone. In the far distance to the S, Lassen Peak can be seen on a clear day. **0.4**

102.7 Sign "Elevation 6860, 11 Miles." Just beyond on L is Bunny Flat Trail Head. **0.2**

102.9 Roadcut on L in glaciated andesite flow of Misery Hill cone. Similar roadcuts ahead expose Misery Hill andesite flows, overlying till of the latest major glaciation, and pockets of the 9.7-ka pumice of Red Banks. Along the road beyond, note the widespread sandy mantle that covers all other units; this material is colluvially reworked lithic ash from Holocene vulcanian eruptions at the summit of the Hotlum cone. **1.1**

104.0 Low roadcuts on L in glacial deposits overlying Misery Hill andesitic lava. **0.4**

104.4 Panther Meadows, a thick accumulation of colluvial sand-sized lithic ash from vulcanian eruptions of the Hotlum cone. **0.4**

104.8 Sign "Elevation 7550 feet, 13 miles." **0.5**

105.3 **STOP 10, Sargents Ridge and Misery Hill cones.** Park in large parking area at Ski Bowl Trail Head. The view from the parking area or, better, from the old ski lodge just above, shows clearly the eroded edifice of the Sargents Ridge cone, with quaquaversal dips from a center just E of the sharp pinnacles of Sargents Ridge. Two glaciations have carved major valleys and left narrow arêtes between them, including the prominent ridge just E of here. Behind Sargents Ridge rises the massive bulk of the Misery Hill cone, also glaciated but in only one glaciation and thus less deeply eroded. The arête to the E exposes the interior of a silicic andesite flow of the Sargents Ridge cone with sparse but conspicuously large hornblende phenocrysts. The prominent butte on the W side of the glaciated Ski Bowl valley is Green Butte (although it is predominantly gray and reddish), a glaciated pile of basaltic-andesite agglutinates and spatter-fed lavas that had a component of highly magnesian basaltic magma; at an elevation of more than 2800 m, this may be the highest mafic vent in the Cascades (Anderson, 1974). Below the Ski Bowl, on the E side of the valley, is the glaciated dacite dome of Gray Butte (with a microwave antenna on one of its summits). The floor of the Ski Bowl valley is mantled by the irregular surface of a rock-glacier deposit of early Holocene age. Mantling it is a mixed colluvium of lithic ash from younger vulcanian eruptions near the summit.

To the SW and W can be seen several ranges of the Klamath Mtns., underlain by metamorphic and plutonic rocks of Paleozoic and Mesozoic age. To the S is the Jurassic granitic pluton of Castle Crags. The higher range to the SW is the Trinity Alps. To the W, partly hidden from view, is Mt. Eddy. Draining S from Mt. Shasta is the Sacramento River, which has carved a deep canyon through the Klamath Mtns. Everitt Hill, a shield volcano of about 450 ka on the S flank of Shasta, fed long flows of basaltic andesite down the canyon before it had reached its present depth.

Depending on time, some short walks can be taken from here. Particularly interesting are the outcrops of agglutinates and spatter-fed lavas of Green Butte, a round-trip walk of 30-45 minutes.

Return to Mt. Shasta City by Everitt Memorial Highway. **13.6**

118.9 Flashing red light just past Mt. Shasta High School; turn R and follow the road with the double yellow line as it curves around a baseball field and school. Proceed one block past the intersection with the flashing yellow light. **0.5**

119.4 Traffic light at intersection; turn L from East Alma St. onto Mt. Shasta Blvd. **0.2**

119.6 Traffic light at intersection; turn R from Mt. Shasta Blvd. onto West Lake Street toward Highway I-5.

The following section contains only a brief outline of features seen along I-5 S from Mt. Shasta City to Redding. It is intended only as a sketch of some features of interest, lists no stops, and is not an authoritative guide to the geology of this part of the eastern Klamath Mtns.

Mileage

0.0 Traffic light at intersection of Mt. Shasta Blvd. and West Lake St. in Mt. Shasta City. Proceed W on West Lake St. **0.1**

0.1 Cross railroad tracks. **0.4**

0.5 Cross Interstate Highway 5 on overpass, then curve R onto onramp to I-5 S. **1.9**

2.4 Offramp for California Highway 89; continue on I-5 S. Roadcuts for next 1.5 miles provide poor exposures of some volcanic debris flows from Mt. Shasta that date from the time of the Sargents Ridge cone. **1.5**

3.9 Weigh station on R. **1.1**

5.0 Offramp for Mott Road and Dunsmuir Ave.; continue on I-5 S. **0.5**

5.5 In a roadcut on L, across the other lanes of the freeway, basaltic-andesite lava from the 450-ka Everitt Hill shield S of Mt. Shasta is exposed beneath volcanic debris flows from the Sargents Ridge cone. **1.6**

7.1 Offramp for Dunsmuir Ave. and Siskiyou Ave.; here the highway crosses from a constructional surface of debris flows and underlying basaltic-andesite lava flow onto eroded metamorphic rocks of the eastern Klamath Mtns. For about the next 25 miles through the canyon of the Sacramento River, roadcuts along the highway expose mainly serpentinized ultramafic rocks and greenstone of lower Paleozoic age with well developed reddish soils. **1.4**

8.5 Cross the Sacramento River. **2.8**

11.3 Offramp for Crag View Dr. and Railroad Road; view on R to Castle Crags, a spectacularly exposed Jurassic granitic pluton. **3.2**

14.5 Offramp for Castella and Castle Crags State Park. **0.2**

14.7 View back toward the R up Castle Creek to Castle Crags. **4.1**

18.8 Flume Creek Road; continue on I-5 S. **0.8**

19.6 Roadcuts on both sides expose a columnar lava flow of basaltic andesite from Everitt Hill, the 450-ka shield volcano at the S foot of Mt. Shasta. This lava appears to have flowed down the canyon of the Sacramento River for at least 30 km, although it has not yet been studied in detail. **1.1**

20.7 Offramp for Sims Road; continue on I-5 S. **1.5**

22.2 For the next 1/2 mile, roadcuts on both sides expose the columnar lava flow of basaltic andesite from Everitt Hill overlying old gravels deposited by the Sacramento River. **Note that this section of I-5 is being re-routed in the summer of 1988**; the log follows the old highway and thus will give only a general idea of the geology along the new route. **0.4**

22.6 Shotgun Creek Road and Shotgun Cafe to R; continue on I-5 S. **0.5**

23.1 Columnar lava flow of basaltic andesite from Everitt Hill overlies gravels at top of high roadcuts on both sides of the highway. **2.4**

25.5 Columnar basaltic andesite flow overlies gravel in roadcut on R. **2.2**

27.7 Pollard Flat cafe and store on L. Most outcrops in roadcuts below about here are fine-grained and thin-bedded low-grade metasedimentary rocks of the Mississippian (Lower Carboniferous) Bragdon Formation. **1.1**

28.8 Columnar basaltic andesite flow overlies gravel in roadcuts on both sides. **0.4**

29.2 La Moine Road on R; continue on I-5 S. **2.6**

31.8 Red soil of Tertiary age developed in the Bragdon Formation and colluvium is particularly conspicuous near here. **0.4**

32.2 Dog Creek Road on R and Delta Road on L; continue on I-5 S. **0.4**

32.6 Cross Dog Creek; in roadcuts beyond are more exposures of columnar basaltic andesite lava flow. **5.2**

37.8 Beginning of bridge across Lake Shasta. **6.6**

44.4 Offramp for O'Brien and Shasta Caverns Roads; continue on I-5 S. **1.2**

45.6 View E (L) across Shasta Lake to prominent outcrops along a ridge of the Permian McCloud Limestone, which bears a well preserved Tethyan fauna. **2.2**

47.8 Beginning of bridge across Shasta Lake. View back NE up the McCloud River Arm to prominent outcrops of McCloud Limestone; view E up the Pitt River Arm to the Magee volcano, a glaciated basaltic andesite composite cone along the Cascades crest. **1.8**

49.6 Conspicuous outcrop of McCloud Limestone ahead, about 1 km to the E. **0.8**

50.4 Offramp for Fawndale Road and Wonderland Blvd.; continue on I-5 S past sign "Elevation 1000 feet." **1.5**

51.9 Pass from deeply weathered and eroded metamorphic rocks of the eastern Klamath Mtns. onto a surface of Pliocene and Quaternary gravels in the N part of the Sacramento Valley. **2.4**

54.3 Offramp for Central Valley and Shasta Dam; continue on I-5 S. Red soil in Upper Cretaceous sedimentary rocks poorly exposed near the highway. **1.2**

55.5 Offramp for Pine Grove Ave.; continue on I-5 S. Upper Cretaceous sedimentary rocks poorly exposed for about 2 miles along the highway. **2.4**

57.9 Offramp for Redding and Market St. Beyond here, the low roadcuts expose only eroded gravels of the northern Sacramento Valley.

REFERENCES

Anderson, A.T. Jr., Evidence for a picritic volatile-rich magma beneath Mt. Shasta, California, J. Petrol., 15, 243-267, 1974.

Anderson, C.A., The Tuscan formation of northern California, with a discussion concerning the origin of volcanic breccias, Univ. California, Dept. Geol. Sci. Bull., 23:7, 215-176, 1933a.

Anderson, C.A., Volcanic history of Glass Mountain, northern California, Am. J. Sci., 26, 485-506, 1933b.

Anderson, C.A., Hat Creek lava flow, Am. J. Sci. 38, 277-492, 1940.

Anderson, C.A., Volcanoes of the Medicine Lake Highland, California, Univ. California, Dept. Geol. Sci. Bull., 25:7, 347-422, 1941.

Bacon, C.R., Eruptive history of Mount Mazama and Crater Lake caldera, Cascade Range, U.S.A., J. Volcanol. Geotherm. Res., 18, 57-118, 1983.

Bacon, C.R., Implications of silicic vent patterns for the presence of large crustal magma chambers, J. Geophys. Res., 90, 11243-11252, 1985.

Bacon, C.R., Magmatic inclusions in silicic and intermediate volcanic rocks, J. Geophys. Res., 91, 6091-6112, 1986.

Bacon, C.R., Mount Mazama and Crater Lake caldera, Oregon, Geol. Soc. Am. Centennial Field Guide--Cordilleran Section, 1, 301-306, 1987.

Bacon, C.R., and T.H. Druitt, Compositional evolution of the zoned calcalkaline magma chamber of Mount Mazama, Crater Lake, Oregon, Cont. Mineral. Petrol., 98, 224-256, 1988.

Berge, P.A., and M.E. Monfort, Teleseismic residual study of the Lassen Volcanic National Park region in Calif., US Geol. Surv. Open-File Rep. 86-252, 71 p., 1986.

Berge, P.A., and D.A. Stauber, Seismic refraction study of upper crustal structure in the Lassen Peak area, northern Calif., J. Geophys. Res. 92:B10, 10571-10579, 1987.

Bestland, E.A., Stratigraphy and sedimentology of the Oligocene Colestin Formation, Siskiyou Pass area, Southern Oregon, M.S. thesis, 150 p., Univ. Oregon, Eugene, 1985.

Christiansen, R.L., Volcanic hazard potential in the California Cascades, Calif. Div. Mines Geol. Spec. Pub. 63, 41-59, 1982.

Christiansen, R.L., The Mount Shasta magmatic system, in Guffanti, M. and L.J.P. Muffler, Proc. Workshop on Geotherm. Resour. Cascade Range, US Geol. Surv. Open-File Rep. 85-521, 31-33, 1985.

Christiansen, R.L., and M.A. Clynne, The climactic eruptions of Lassen Peak, California, in May 1915 (abs.), Eos Trans. AGU, 67, 1247, 1986.

Christiansen, R.L., F.J. Kleinhampl, R.J. Blakely, E.T. Tuchek, F.L. Johnson, and M.D. Conyac, Resource appraisal of the Mt. Shasta wilderness study area, Siskiyou County, California, US Geol. Surv. Open-File Rep. 77-250, 53 p., 1977.

Christiansen, R.L., and C.D. Miller, Volcanic evolution of Mount Shasta, California (abs.), Geol. Soc. Am. Abstr. Programs, 8:3, 360-361, 1976.

Clynne, M.A., Stratigraphy and major-element geochemistry of the Lassen Volcanic Center, California, US Geol. Surv. Open-File Rep. 84-224, 168 p., 1984.

Crandell, D.S., Glaciation near Lassen Peak, northern California, US Geol. Surv. Prof. Pap. 800-C, C179-C188, 1972.

Crandell, D.R., Gigantic debris avalanche of Pleistocene age from ancestral Mount Shasta volcano, CA, and debris-avalanche hazard zonation, US Geol. Surv. Bull. [in press].

Crandell, D.R., C.D. Miller, H.X. Glicken, R.L. Christiansen, and C.G. Newhall, Catastrophic debris avalanche from ancestral Mount Shasta volcano, California, Geology, 12, 143-146, 1984.

Crandell, D.S., D.R. Mullineaux, R.S. Sigafoos, and Meyer Rubin, Chaos Crags eruptions and rockfall-avalanches, Lassen Volcanic National Park, California, US Geol. Surv. J. Res., 2:1, 49-59, 1974.

Diller, J.S., and H.B. Patton, The geology and petrography of Crater Lake National Park, US Geol. Surv. Prof. Pap. 3, 167 p., 1902.

Donnelly-Nolan, J.M., Medicine Lake Volcano and Lava Beds National Monument, California, Geol. Soc. Am. Centennial Field Guide --Cordilleran Section, 289-294, 1987.

Donnelly-Nolan, J.M., A magmatic model of Medicine Lake volcano, California, J. Geophys. Res., 93:B5, 4412-4420, 1988.

Donnelly-Nolan, J.M., and D.E. Champion, Geologic map of Lava Beds National Monument, northern California, US Geol. Surv. Misc. Invest. Series Map I-1804, scale 1:24,000, 1987.

Donnelly-Nolan, J.M., E.V. Ciancanelli, J.C. Eichelberger, J.H. Fink, and Grant Heiken, Roadlog for field trip to Medicine Lake Highland, in Johnston, D.A., and J.M. Donnelly-Nolan (eds.), Guides to some volcanic terranes in Washington, Idaho, Oregon, and northern California, US Geol. Surv. Circ. 838, 141-149, 1981.

Donnelly-Nolan, J.M., and K.M. Nolan, Catastrophic flooding and eruption of ash-flow tuff at Medicine lake volcano, California, Geology, 14, 875-878, 1986.

Druitt, T.H., and C.R. Bacon, Lithic breccia and ignimbrite erupted during the collapse of Crater Lake caldera, Or., J. Volcanol. Geotherm. Res., 29, 1-32, 1986.

Eichelberger, J.C., Origin of andesite and dacite: Evidence of mixing at Glass Mountain in California and at other circum-Pacific volcanoes, Geol. Soc. Am. Bull., 86, 1381-1391, 1975.

Eichelberger, J.C., Mechanism of magma mixing at Glass Mountain, Medicine Lake Highland Volcano, California, in Johnston, D.A., and J.M. Donnelly-Nolan (eds.), Guides to some volcanic terranes in Washington, Idaho, Oregon, and northern California, US Geol. Surv. Circ. 838, 183-189, 1981.

Eppler, D.B., Jonathan Fink, and Raymond Fletcher, Rheological properties and kinematics of emplacement of the Chaos Jumbles rockfall avalanche, Lassen Volcanic National Park, California, J. Geophys. Res., 92:B5, 3623-3633, 1987.

Finch, R.H., Activity of a California volcano in 1786, Volcano Lett. 308, 3, 1930.

Fink, J.H., Structure and emplacement of a rhyolitic obsidian flow, Little Glass Mtn., Medicine Lake Highland, California, Geol. Soc. Am. Bull., 94, 362-380, 1983.

Fink, J.H., and D.D. Pollard, Structural evidence for dikes beneath silicic domes, Medicine Lake Highland, California, Geology, 11, 458-461, 1983.

Greeley, Ronald, and Roger Baer, Hambone, California and its magnificent lava tubes --Preliminary report (abs.), Geol. Soc. Am. Abstr. Programs, 3:2, 128, 1971.

Grove, T.L., and M.B. Baker, Phase equilibrium controls on the tholeiitic versus calc-alkaline differentiation trends, J. Geophys. Res., 89, 3253-3274, 1984.

Grove, T.L., R.J. Kinzler, M.B. Baker, J.M. Donnelly-Nolan, and C.E. Lesher, Assimilation of granite by basaltic magma at Burnt Lava flow, Medicine Lake volcano: Decoupling of heat and mass transfer, Contrib. Mineral. Petrol., 99, 320-343, 1988.

Grove, T.L., and J.M. Donnelly-Nolan, The evolution of silicic lavas at Medicine Lake Volcano, California: Implications for the origin of compositional gaps in calc-alkaline series lavas, Contrib. Mineral. Petrol., 92, 284-302, 1986.

Guffanti, Marianne, and C.S. Weaver, Distribution of late Cenozoic volcanic vents in the Cascade Range: Volcanic arc segmentation and regional tectonic considerations, J. Geophys. Res., 93:B6, 6513-6529, 1988.

Hart, S.R., K, Rb, Cs, Sr and Ba contents and Sr isotope ratios of ocean floor basalts, Philos. Trans. R. Soc. London, Ser. A, 268, 573-587, 1971.

Hart, W.K., J.L. Aronson, and S.A. Mertzman, Areal distribution and age of low-K, high-alumina olivine tholeiite magmatism in the northwestern Great Basin, Geol. Soc. Am. Bull., 95, 186-195, 1984.

Heiken, Grant, Plinian-type eruptions in the Medicine Lake Highland, California, and the nature of the underlying magma, J. Volcanol. Geoth. Res., 4, 375-402, 1978.

Ingebritsen, S.E., and M.L. Sorey, A quantitative analysis of the Lassen hydrothermal system, north central California, Water Resour. Res., 21:6, 853-868, 1985.

Kane, P.S., Pleistocene glaciation, Lassen Volcanic National Park, California, Geology, 35:5, 95-105, 1982.

Kinzler, R.J., A field, petrologic, and geochemical study of the Callahan lava flow, a basaltic andesite from Medicine Lake shield volcano, California, M.S. thesis, 100 p., Mass. Inst. Tech., 1985.

Luedke, R.G., and R.L. Smith, Map showing distribution, composition, and age of late Cenozoic volcanic centers in California and Nevada, US Geol. Surv. Misc. Invest. Map I-1091-C, scale 1:1,000,000, 1981.

Luedke, R.G., and R.L. Smith, Map showing distribution, composition, and age of late Cenozoic volcanic centers in Oregon and Washington, US Geol. Surv. Misc. Invest. Map I-1091-D, scale 1:1,000,000, 1982.

Lydon, P.A., Geology and lahars of the Tuscan Formation, northern California, Geol. Soc. Am. Memoir 116, 441-475, 1968.

Maynard, L.C., Geology of Mount McLoughlin, M.S. thesis, 139 p., Univ. Oregon, Eugene, 1974.

McBirney, A.R., Volcanic evolution of the Cascade Range, Ann. Rev. Earth Planet. Sci., 6, 437-456, 1978.

Mertzman, S.A., Jr., The petrology and geochemistry of the Medicine Lake Volcano, California, Contr. Mineral. Petrol., 62, 221-247, 1977a.

Mertzman, S.A., Jr., Recent volcanism at Schonchin and Cinder Buttes, northern California, Contr. Mineral. Petrol., 61, 231-243, 1977b.

Mertzman, S.A., and R.J. Williams, Genesis of Recent silicic magmatism in the Medicine Lake Highland, California: Evidence from cognate inclusions found at Little Glass Mountain, Geochim. Cosmochim. Acta, 45, 1463-1478, 1981.

Miller, C.D., Holocene pyroclastic-flow deposits from Shastina and Black Butte, west of Mount Shasta, California, US Geol. Surv. J. Res., 6:3, 611-624, 1978.

Miller, C.D., Potential hazards from future eruptions in the vicinity of Mount Shasta volcano, northern California, US Geol. Surv. Bull. 1503, 43 p., 1980.

Mortimer, N., and R.G. Coleman, A Neogene structural dome in the Klamath Mountains, California and Oregon, in Nilsen, T.H. (ed.), Geology of the Upper Cretaceous Hornbrook Formation, Oregon and California, Soc. Econ. Paleontol. Mineral., Pacific Section, Book 42, 179-186, 1984.

Muffler, L.J.P., N.L. Nehring, A.H. Truesdell, C.J. Janik, M.A. Clynne, and J.M. Thompson, The Lassen geothermal system, US Geol. Surv. Open-File Rep. 82-926, 1982.

Muffler, L.J.P., Raymond Jordan, and A.L. Cook, Thermal features and topography of Bumpass Hell and Devils Kitchen, Lassen Volcanic National Park, California, US Geol. Surv. Misc. Field Studies Map 1484, scale 1:2,000, 1983.

Nilsen, T.H., M.A. Elliott, and W.B. Purdom, Description of field trip stops and road-log, Upper Cretaceous Hornbrook Formation, southern Oregon and northern California, in Nilsen, T.H. (ed.), Geology of the Upper Cretaceous Hornbrook Formation, Oregon and California, Soc. Econ. Paleontol. Mineral., Pacific Section, Book 42, p. 9-41, 1984.

Nockolds, S.R. and R. Allen, The geochemistry of some igneous rock series, Geochim. Cosmochim. Acta, 4, 105-142, 1953.

Parsons, R. B., and R.C. Herriman, Geomorphic surfaces and soil development in the upper Rogue River Valley, Oregon, J. Soil Sci. Am., 40, 933-938, 1976.

Peterson, N.V., and J.R. McIntyre, The reconaissance geology and mineral resources of eastern Klamath County and western Lake County, Oregon, Oregon Dept. Geology and Mineral Industries Bull. 66, 70 p., 1970.

Philpotts, J.A., Wayne Martin, and C.C. Schnetzler, Geochemical aspects of some Japanese lavas, Earth Planet. Sci. Lett., 12, 89-96, 1971.

Riddihough, R.P., Recent movements of the Juan de Fuca plate system, J. Geophys. Res., 89, 6980-6994, 1984.

Sammel, E.A., and D.L. Peterson, Hydrologic reconnaissance of the geothermal area near Klamath Falls, Oregon, with a section on preliminary interpretation of geophysical data, US Geol. Surv. Water Resourc. Invest. 76-127, 129 p., 1976.

Smedes, H.W., and J.J. Prostka, Stratigraphic framework of the Absaroka Supergroup in the Yellowstone National Park region, US Geol. Surv. Prof. Paper 727-C, 33 p., 1972.

Smith, R.L., and R.G. Luedke, Potentially active volcanic lineaments and loci in western conterminous United States, in Explosive Volcanism: Inception, Evolution, and Hazards, p. 47-66, Nat. Acad. Press, Washington, DC, 1984.

Sparks, R.S.J., S. Self, and G.P.L. Walker, Products of ignimbrite eruptions, Geology, 1, 115-118, 1973.

Veen, C.A., Gravity anomalies and their structural implications for the southern Oregon Cascade Mountains and adjoining Basin and Range province, M.S. thesis, 86 p., Oregon State Univ., Corvallis, 1982.

Williams, H., Geology of the Lassen Volcanic National Park, Calif, Univ. Calif., Dept. Geol. Sci. Bull. 21:8, 195-385, 1932a.

Williams, Howel, Mount Shasta, a Cascade volcano, J. Geol. 40:5, 417-429, 1932b.

Williams, Howel, The geology of Crater Lake National Park, Oregon, Carnegie Inst. Wash. Pub. 540, 162 p., 1942.

Yoder, H.S., Jr., and C.E. Tilley, Origin of basalt magmas: An experimental study of natural and synthetic rock systems, J. Petrol., 3, pt. 3, 342-532, 1962.

Cenozoic Volcanism in the Cascade Range and Columbia Plateau, Southern Washington and Northernmost Oregon

Seattle, Washington to Portland, Oregon
July 3–8, 1989

Field Trip Guidebook T106

Leader: *Donald A. Swanson*

Associate Leaders:
Joseph A. Vance *Geoffrey A. Clayton* *Russell C. Evarts*

American Geophysical Union, Washington D.C.

Published 1989 by American Geophysical Union

2000 Florida Ave., N.W., Washington, D.C. 20009

ISBN: 0-87590-604-4

Printed in the United States of America

COVER Mount St. Helens from Harry's Ridge, showing 1980 crater, dome, north flank, and pumice plain. Dome is 750 m wide. Photo by Lyn Topinka on May 20, 1988. Optional hiking trip to crater starts on pumice plain at left side of photo, traverses west, climbs to crater along west side of deep canyon, and leaves east part of crater before traversing east to vehicles.

Leader:

Donald A. Swanson
U.S. Geological Survey
Cascades Volcano Observatory
5400 MacArthur Blvd.
Vancouver, WA 98661

Associate Leaders:

Joseph A. Vance and Geoffrey A. Clayton
Department of Geological Sciences, AJ-20
University of Washington
Seattle, WA 98195

Russell C. Evarts
U.S. Geological Survey, MS 901
345 Middlefield Road
Menlo Park, CA 94025

IGC FIELD TRIP T106: CENOZOIC VOLCANISM IN THE CASCADE RANGE AND COLUMBIA PLATEAU, SOUTHERN WASHINGTON AND NORTHERNMOST OREGON

Donald A. Swanson[1], Kenneth A. Cameron[1], Russell C. Evarts[2], Patrick T. Pringle[1], and Joseph A. Vance[3]

INTRODUCTION

This guidebook is for a six-day excursion between Issaquah, Washington (east of Seattle), and Portland, Oregon, that emphasizes the Tertiary and Quaternary volcanic geology of the western Columbia Plateau and the Cascade Range of southern Washington and northern Oregon (Figures 1 and 2). The guidebook summarizes the geology of selected areas along the route and provides a brief introduction to the general volcanic history of the Columbia River Basalt Group and the southern Washington Cascades. An extensive but not exhaustive bibliography accompanies the guidebook. The road logs are designed to be self-guiding; as such, they are more complete than necessary for guided bus excursions.

TECTONIC AND GEOPHYSICAL SETTING OF THE CASCADE RANGE IN SOUTHERN WASHINGTON

The Cascade Range has been an active arc for about 36 m.y. as a result of plate convergence [*Atwater*, 1970]. Volcanic rocks between 55 and 42 Ma occur in the Cascades [*Vance et al*, 1987] but are probably related to a rather diffuse volcanic episode that created the Challis arc extending southeastward from northern Washington to northwest Wyoming [*Armstrong*, 1978; *Vance*, 1979]. Convergence between the North American and Juan de Fuca plates continues at about 4 cm/yr in the direction of N50°E, a slowing of 2-3 cm/yr since 7 Ma [*Riddihough*, 1984]. According to most interpretations, volcanism in the Cascades has been discontinuous in time and space [*McBirney*, 1978; *Luedke and Smith*, 1982; *Guffanti and Weaver*, 1988], with the most recent episode of activity beginning about 5 Ma and resulting in more than 3000 vents [*Guffanti and Weaver*, 1988]. In Oregon, the young terrane is commonly called the High Cascades and the old terrane the Western Cascades, terms that reflect present physiography and geography. The terms are not useful in Washington, where young vents are scattered across the dominantly middle Miocene and older terrane.

Oligocene and Miocene rocks in the southern Washington Cascades and adjacent western Washington and Oregon are tectonically rotated in a clockwise sense according to paleomagnetic data [*Simpson and Cox*, 1977; *Beck and Burr*, 1979; *Bates et al.*, 1981; *Wells and Coe*, 1985]. The amount of rotation in general increases with increasing age. Rocks probably about 30-35 Ma in the southern Washington Cascades are rotated about 34° [*Bates et al.*, 1981], whereas the 12-Ma Pomona Member of the Saddle Mountains Basalt is rotated only 8-10° at the western end of the Columbia Gorge [*Wells and England*, in press; *Wells et al.*, in press]. Middle Eocene rocks in the Oregon Coast Range are rotated from 20° to as much as 65° [*Wells and Coe*, 1985], and Eocene rocks in central Oregon (Clarno Formation), more than 300 km inboard of the continental margin, an average of 16° [*Grommé et al.*, 1986]. In some models, the rotation pattern implies that the Paleogene trend of the Cascades was northwestward rather than northward as at present [*Magill et al.*, 1981]. In other models, the rotation is confined to domains tens of meters to as much as 30 km across and is related to shear along dextral northwest-trending faults or Riedel shears rather than to wholesale regional block rotation of microplates [*Beck*, 1980; *Wells and Coe*, 1985]. The continuum model of *Wells and England* [in press] suggests that the average rate of rotational shear is about 1.3° per million years near the southwest margin of the southern Washington Cascades. The origin of the rotations remains unresolved, but appropriate strike-slip faults in the region make the dextral-shear model attractive. No evidence for significant rotation younger than 12 Ma has been recognized, owing in large measure to the paucity of upper Miocene rocks in southern Washington and northern Oregon.

In Washington and Oregon, a striking contrast has existed for the past 5 m.y. in the style of volcanism in the Cascades relative to geography [*Weaver and Malone*, 1987; *Guffanti and Weaver*, 1988]. North of Mount Rainier, young volcanism is concentrated in only a few isolated andesitic and dacitic composite cones (notably Glacier Peak, Mount Baker, and the volcanoes of the Garibaldi belt in British Columbia), whereas south of Mount Hood moderate-sized andesitic and dacitic composite cones are relatively unimportant features of a landscape dominated by small andesite and basalt vents (Figure 1) [*Guffanti and Weaver*, 1988]. The area between Mounts Rainier and Hood is transitional; large andesite and dacite composite cones (Rainier, Adams, St.

[1]U.S. Geological Survey, Cascades Volcano Observatory, Vancouver, Washington.

[2]U.S. Geological Survey, Menlo Park, California.

[3]Department of Geological Sciences, University of Washington, Seattle, Washington.

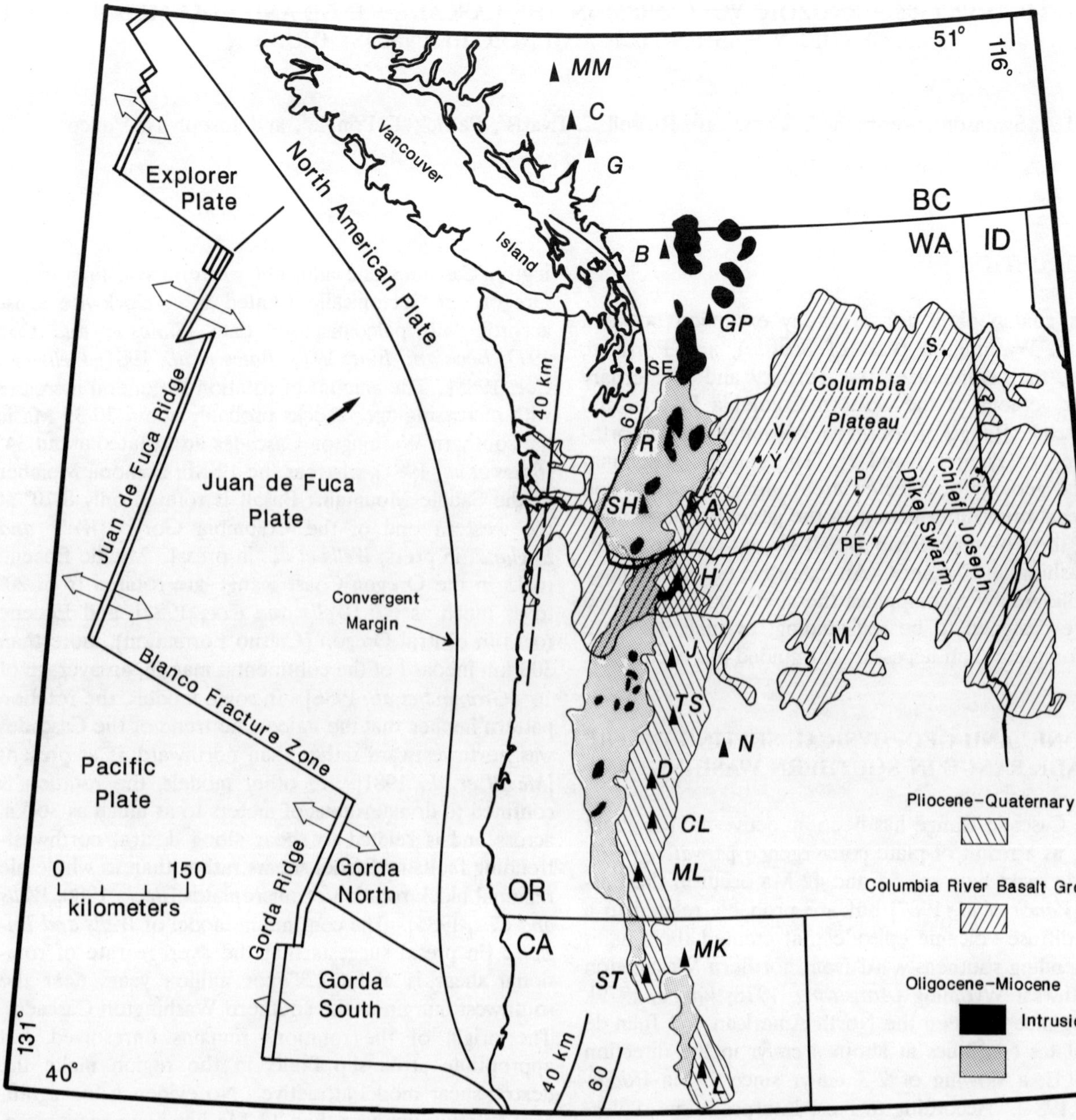

FIGURE 1. Features of Juan de Fuca-North American subduction system [*Riddihough*, 1984] relative to Cascade Range and Columbia Plateau. Open arrows, ridge-spreading directions; solid arrow, convergence direction. The 40- and 60-km contours show depth of seismicity in upper part of downgoing slab [*Weaver and Baker*, 1988]. Generalized subdivision of Cascade Range from *Hammond* [1979]. Distribution of Columbia River Basalt Group from *Tolan et al.* [in press]. Solid triangles, major Quaternary volcanoes: MM, Meager Mountain; C, Mount Cayley; G, Mount Garibaldi; B, Mount Baker; GP, Glacier Peak; R, Mount Rainier; SH, Mount St. Helens; A, Mount Adams, H, Mount Hood; J, Mount Jefferson; TS, Three Sisters; N, Newberry; D, Diamond Peak; CL, Crater Lake (Mount Mazama); ML, Mount McLoughlin; MK, Medicine Lake; ST, Mount Shasta; L, Lassen Peak. M, Monument dike swarm; SE, Seattle; V, Vantage; Y, Yakima; P, Pasco; PE, Pendleton.

FIGURE 2 (facing page). Geologic map (faults not shown) for southern Washington Cascades and adjacent Columbia Plateau, simplified from *Walsh et al.* [1987]. Units in order of descending age: Qa, basin fill, alluvium, drift, and landslide deposits; Ql, lahars, including 1980 debris avalanche at Mount St. Helens; Qv, Quaternary volcanic rocks; QPv, Quaternary and Pliocene volcanic rocks; Ms, Miocene sedimentary rocks, mostly younger than bulk of Yakima Basalt Subgroup; Mcr, Yakima Basalt Subgroup; Mv, lower and middle Miocene volcanic rocks; MOv, Miocene and Oligocene volcanic rocks; Ov, Oligocene volcanic rocks; OEv, Oligocene and Eocene volcanic rocks; Ev, Eocene volcanic rocks; Es, Eocene sedimentary rocks; pT, pre-Tertiary rocks; solid, intrusions, mostly of early and middle Miocene age. Dashed line, route of field trip, with selected stops numbered.

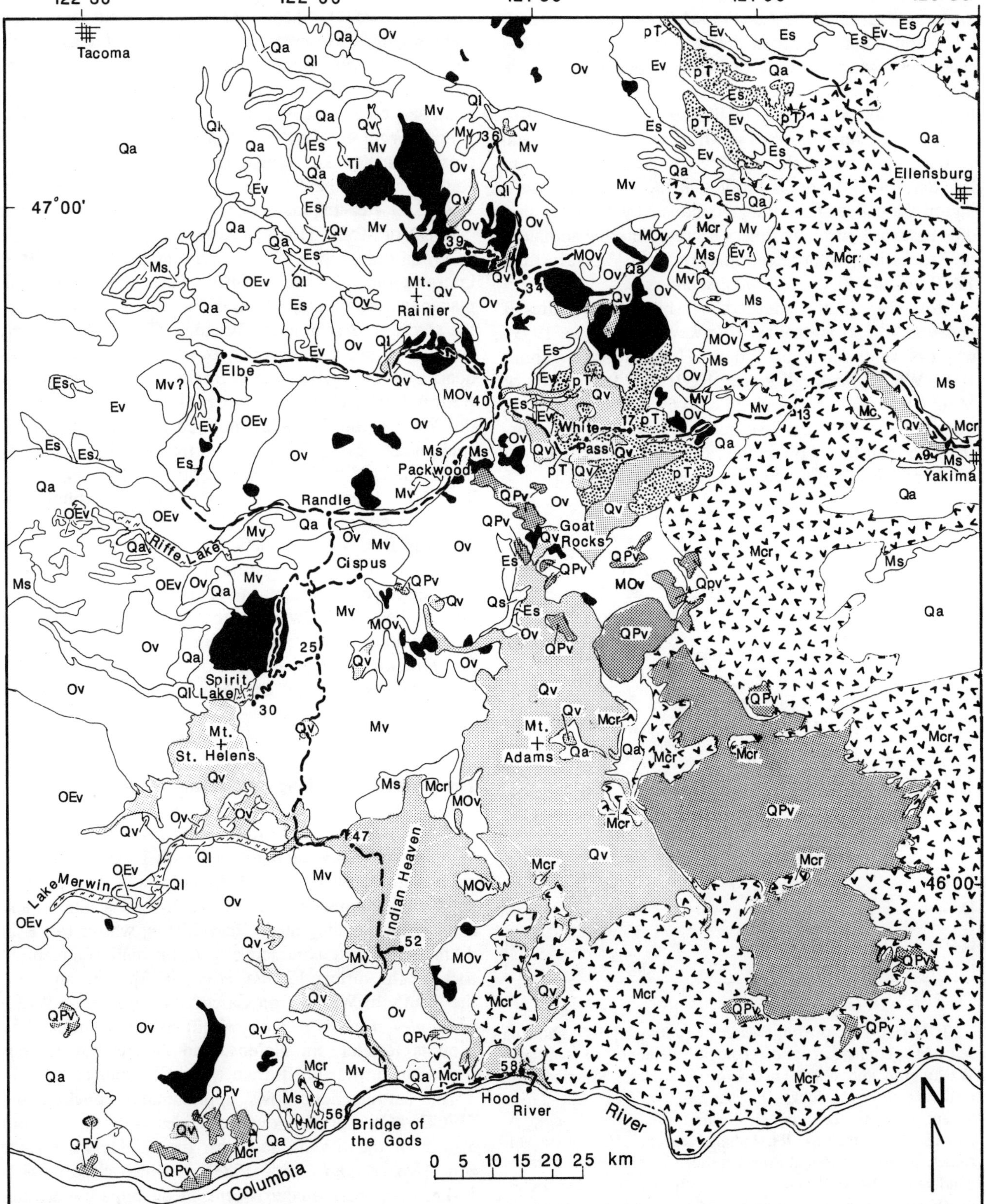

Helens, Hood, and the extinct Goat Rocks volcano) occur together with fields and scattered vents of olivine basalt (Indian Heaven, Simcoe Mountains, and the King Mountain fissure zone south of Mount Adams).

Weaver and Malone [1987] point out that the southern Washington Cascades are also transitional geophysically (Figure 3). An east-west saddle in an otherwise west to east Bouguer gravity gradient occurs along the

Columbia River between Oregon and Washington [*Thiruvathukal et al.*, 1970; *Couch and Baker*, 1979). North of Mount Adams a broad gravity low follows the Cascade crest, whereas south of Mount Hood the Cascade gravity low merges eastward into a low related to extension of southeast Oregon [*Weaver and Malone*, 1987].

Magnetotelluric profiles south of Mount Hood reveal a flat-lying, relatively simple conductivity structure [*Stanley*, 1984], whereas profiles in the southern Washington Cascades show a high-conductivity anomaly that dips eastward and may reflect accreted marine sedimentary rocks in a compressed forearc basin [*Stanley et al.*, 1987] (Figure 3; see following section).

Heat flow in the southern Washington Cascades is between 70 and 90 mWm^{-2} [*Blackwell and Steele*, 1983], whereas that south of Mount Hood averages about 100 mWm^{-2} [*Blackwell et al.*, 1982] and that north of Mount Rainier a relatively low 40-60 mWm^{-2}. A rapid transition occurs from low heat flow west of the Cascades to high heat flow in the range in both Oregon and southern Washington [*Blackwell and Steele*, 1983].

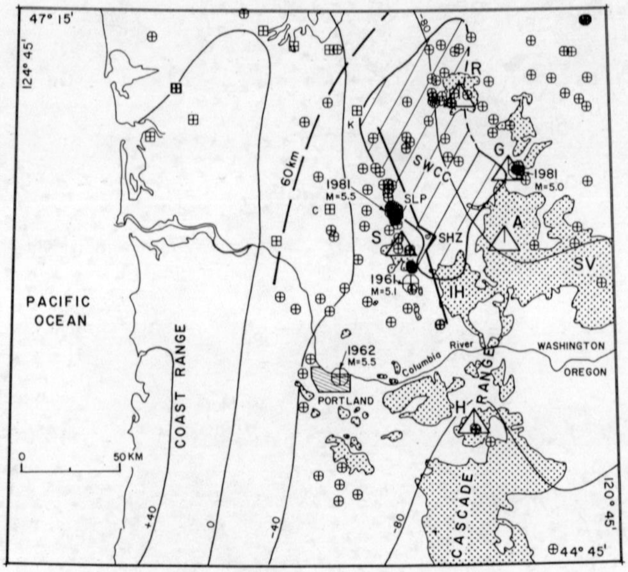

FIGURE 3. Map from *Weaver and Malone* [1987] showing selected geophysical and geologic elements of southern Washington and northern Oregon Cascades. Stippled, late Cenozoic volcanic rocks [*Luedke and Smith*, 1982]. Large letters, volcanic centers: R, Rainier; G, Goat Rocks; A, Adams; S, St. Helens; IH, Indian Heaven; SV, Simcoe volcanics; H, Hood. Ruled area, southern Washington Cascades conductor (SWCC) [*Stanley et al.*, 1987]. Earthquakes: circles, depth <30 km; squares, depth >30 km; date and magnitude given for magnitudes >5; small symbols for magnitudes between 2.5 and 4.9. Bold bracket (SHZ), St. Helens seismic zone. Bold dashed line, top of Juan de Fuca plate at 60 km depth. Light lines, Bouguer gravity contours in milligals, filtered to pass wavelengths >100 km at reduction density of 2.43 g/cm^3. SLP, Spirit Lake pluton.

The southern Washington Cascades are seismically active (Figure 3). Most earthquakes occur along the 100-km-long, north-northwest trending St. Helens seismic zone [*Weaver and Smith*, 1983], where most focal mechanisms show dextral slip parallel to the trend of the zone and consistent with the direction of plate convergence [*Grant et al.*, 1984; *Weaver et al.*, 1987]. Other crustal earthquakes concentrate just west of Mount Rainier and in the Portland area [*Weaver and Malone*, 1987]. Few earthquakes occur north of Mount Rainier or south of Mount Hood [*Weaver and Michaelson*, 1985].

From tomography, *Rasmussen and Humphreys* [1988] interpret the subducted Juan de Fuca plate as a quasi-planar feature dipping about 65° to about 300 km under the southern Washington Cascades. The plate is poorly defined seismically, however, owing to a lack of earthquakes within it [*Weaver and Baker*, 1988]. *Guffanti and Weaver* [1988] show that the present volcanic front of the Washington Cascades, defined by the westernmost young vents, parallels the curved trend of the subducting plate reflected by the 60 km-depth contour (Figure 1). The front trends northwest in northern Washington—where Glacier Peak, Mount Baker, and the volcanoes of southern British Columbia occur along a virtually straight line—and northeast in southern Washington. A 90-km gap free of young volcanoes between Mount Rainier and Glacier Peak is landward of that part of the subducting plate with the least average dip to a depth of 60 km [*Guffanti and Weaver*, 1988]. South of Portland, the volcanic front is offset 50 km eastward and extends southward into California, probably still parallel to the trend of the convergent margin [*Guffanti and Weaver*, 1988].

The Southern Washington Cascades Conductor

A recent magnetotelluric survey of the southern Washington Cascades, combined with geomagnetic variation data [*Law et al.*, 1980], defines a broad, high-conductivity anomaly with a resistivity of 1-4 ohm m and a thickness possibly more than 15 km, whose top is 2-8 km below less conductive volcanic and sedimentary rocks at the surface [*Stanley et al.*, 1987]. This anomaly, the "southern Washington Cascades conductor" (SWCC) of *Stanley et al.*, is roughly within the area bounded by Mounts Rainier, St. Helens, and Adams but extends somewhat farther north than Mount Rainier (Figure 3). Three aeromagnetic lows approximately enclose the SWCC. One of the lows coincides with the western margin of the SWCC and with the St. Helens seismic zone [*Weaver and Smith*, 1983], in which clusters of 5-15-km deep earthquakes occur in an apparent dextral shear zone. Another aeromagnetic low occurs within the SWCC west and south of Mount Rainier and correlates with a zone of folds mapped by *Hammond* [1980] and *Fiske et al.* [1963]; the north end of the low coincides with an elongate belt of seismicity just west of Mount Rainier. The third low coincides with the southern end of the SWCC but extends farther south into Oregon.

Stanley et al. [1987] interpret the SWCC as caused by "marine sedimentary rocks of early Jurassic to Eocene age, containing hypersaline fluids and posssibly dominated by shale facies" that form an accreted seamount complex in a compressed forearc basin. Mount Rainier lies on the eastern edge of the SWCC [*Stanley et al.*, 1987], and Mount St. Helens is on the western edge as well as on the St. Helens seismic zone (Figure 3). Tectonic controls on the locations of these volcanoes may reflect the accretionary history of the region.

TRAVERSE ACROSS THE CASCADES ON INTERSTATE 90

The rocks along and south of this route are largely Cenozoic, but Mesozoic basement is locally exposed in inliers and fault blocks. The route follows a transition from dominantly Tertiary volcanic rocks south of I-90 to the largely pre-Tertiary North Cascades. Four major groupings of rocks are present:

Mesozoic—Jurassic ophiolite and island-arc rocks and their associated Jurassic and Cretaceous sedimentary cover; Early Cretaceous high P-T metamorphic rocks, largely blueschist, greenschist, and phyllite; and Cretaceous plutonic rocks.

Eocene strata—Thick sections of nonmarine arkose and interbedded volcanic rocks.

Rocks of the Cascade magmatic arc—Lava flows and volcaniclastic rocks of the Oligocene Ohanapecosh Formation and the overlying upper Oligocene and lower Miocene Fifes Peak Formation, and the shallow granitoid plutonic rocks of the Miocene Snoqualmie batholith.

Columbia River Basalt Group—Miocene flood basalt erupted from vents far east of the Cascade Range.

Smith [1904] and *Smith and Calkins* [1906] worked out much the basic geology of this area. *Porter* [1976] described the Quaternary deposits along I-90.

Interstate 90 crosses the southern end of the Straight Creek fault, a major Tertiary structure. This fault, more than 500 km long, extends southward down the Fraser River in British Columbia into the Washington Cascades to I-90, where it splits into a series of southeast-trending splays that disappear beneath the Columbia River Basalt Group. About 90-110 km of dextral strike slip occurred on the fault [*Vance*, 1985; *Kleinspehn*, 1985]. Marked lithologic differences distinguish the Mesozoic and Eocene units on either side of the Straight Creek fault.

Mesozoic Rocks

A wide variety of units, mostly of Jurassic or Cretaceous age, represent several exotic terranes that accreted to western North America certainly before the Eocene and probably before the end of the Mesozoic.

The Mesozoic section west of the Straight Creek fault consists of graywacke and argillite with local chert and limestone. These sedimentary rocks are associated with greenstone, gabbro, and minor tonalite; all rocks are highly folded, faulted, and sheared. Contacts between units are commonly tectonic. *Frizzell et al.* [1987] mapped the suite as two tectonic-melange belts distinguished by differences in relative abundance of sedimentary and igneous rocks. The western melange belt, consisting largely of Jurassic marine turbidites, is widely exposed along the western front of the Cascades north of North Bend. The eastern melange belt, distinguished by more abundant greenstone, chert, and limestone, occurs in scattered outcrops north of Snoqualmie Pass. U-Pb ages for tonalite and gabbro in the melanges are Jurassic, and fossils from the chert and clastic sedimentary rocks have mostly Late Jurassic and Early Cretaceous ages. The lithology and age of the melanges led *Frizzell et al.* [1987] to conclude that the igneous rocks are an oceanic suite (root of an island arc?) that were intruded into the sedimentary rocks prior to the deformation that formed the melange.

Several Mesozoic units lie just east of the Straight Creek fault and between its southeasterly splays. The most widespread is the high P-T Easton Metamorphic Suite consisting of blueschist, greenschist, and phyllite. Farther north, lithologically equivalent rocks of the Shuksan Metamorphic Suite [*Misch*, 1966] occur extensively west of the Straight Creek fault and have been offset about 90 km from the Easton by dextral slip [*Vance*, 1985]. Most workers interpret the Easton and related rocks as metamorphosed oceanic basalt of MORB affinity and its sedimentary cover [*Brown*, 1986; *Dungan et al.*, 1983]. Radiometric dating suggests a Late Jurassic age for the Easton protolith and an Early Cretaceous age for its metamorphism [*Brown et al.*, 1982]. Other Mesozoic units between the Straight Creek splays include medium-grade pelitic schist [*Stout*, 1964] and gneissic tonalite of Jurassic age [*Frizzell et al.*, 1984].

Two other Mesozoic units are visible from I-90 in the Stuart Range 30 km north of Cle Elum. The older, the upper Jurassic and lower Cretaceous Ingalls Tectonic Complex, consists of a faulted sequence of ultramafic rocks, gabbro, diabase, pillow basalt, and associated sedimentary rocks generally interpreted as an ophiolite [*Southwick*, 1974; *Tabor et al.*, 1982b, 1987; *Miller*, 1985a]. The younger unit (95-88 Ma) consists of tonalite, diorite, and granodiorite of the upper Cretaceous Mount Stuart batholith, which intrudes the Ingalls.

Eocene Sedimentary and Volcanic Rocks

Eocene strata dominate the bedrock along I-90 between Issaquah and Cle Elum. They include volcanic and fluvial arkosic rocks, in places interbedded. The sedimentary units are quartzose to arkosic micaceous sandstone, shale, and conglomerate derived from a plutonic and metamorphic terrane farther east or northeast. They are locally marine but mostly fluvial and deltaic. Major coal production came from the Puget Group and the Roslyn Formation. The Eocene units are thick (2-4 km) and were deposited in rapidly subsiding, fault-bounded extensional basins [*Tabor et al.*, 1984; *Johnson*, 1985]. Many of the Eocene volcanic assemblages are bimodal, composed of basalt (or andesite) and high-silica rhyolite. Radiometric ages of the volcanic rocks are 51-42 Ma, mostly middle Eocene [*Tabor et al.*, 1984; *Turner et al.*, 1983; *Frizzell et al.*, 1984]. Invertebrate faunas in the rare marine sedimentary units are middle to late Eocene. The volcanic rocks do *not* belong to the Cascade arc but instead to the older and much broader field Challis arc [*Armstrong*, 1978], which encompasses large areas of Eocene volcanic and shallow plutonic rocks in the Pacific Northwest. *Ewing* [1980] and *Heller et al.* [1987] discussed the regional tectonics and paleogeography during the Eocene.

Figure 4 is a simplified correlation chart for some of the Eocene units along the field-trip route. These units pose several stratigraphic and tectonic problems. Obvious contrasts exist between Eocene rocks east and west of the Straight Creek fault. The extent to which the fault merely separated adjacent basins with different histories or juxtaposed the basinal sequences by major dextral movement is unresolved. Stratigraphic relations are not fully understood for the Eocene units separated by the southeastern splays of the Straight Creek fault.

Some Eocene volcanic units exhibit clear structural control. For example, eruption of basalt in the Teanaway Formation was localized in the southeastern bend of the Straight Creek fault. However, no structural control has been identified for other thick volcanic units, such as the Tukwila Formation interbedded in the Puget Group between North Bend and Seattle.

The Puget Group is widely exposed in the eastern Puget Sound basin and in the western Cascade foothills, where it is at least 2.7 km thick [*Buckovic*, 1979]. East of Seattle the Puget is underlain by the Raging River Formation and is subdivided into three formations [*Vine*, 1962, 1969]. From the base upward the four units are:

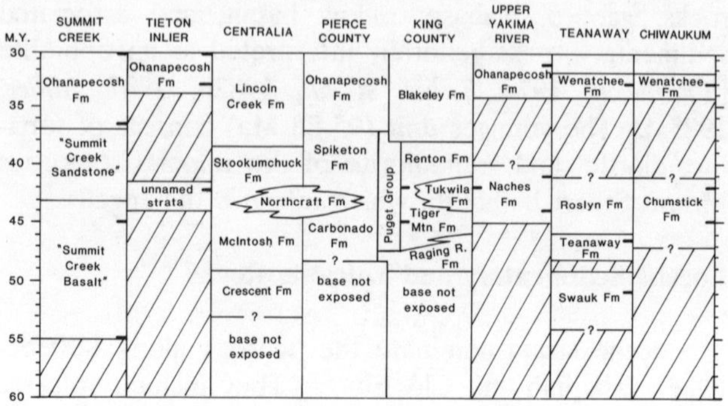

FIGURE 4. Correlation of Eocene rocks of central Washington Cascades and vicinity. Diagram and sources of data from *Vance et al.* [1987].

1) Raging River Formation, fossiliferous and tuffaceous marine sedimentary rocks of middle Eocene age;

2) Tiger Mountain Formation, nonmarine arkosic sedimentary rocks;

3) Tukwila Formation, largely coarse andesitic pyroclastic and volcaniclastic rocks with arkosic interbeds;

4) Renton Formation, nonmarine arkose.

Abrupt facies changes characterize the Puget Group. The volcanic rocks of Mount Persis [*Tabor et al.*, 1982a], probably correlative with the Tukwila Formation, make up the entire Eocene section in the western Cascade foothills of southern Snohomish County but are absent in the thick Puget section of southern King County.

Interbedded volcanic and arkosic sandstones of the Naches Formation occur along I-90 between Snoqualmie Pass and Easton west of the Straight Creek fault. The Naches is as thick as 3 km and is mostly coeval with and lithologically similar to the Puget Group. Silicic lava flows and ash-flow tuffs are abundant in the Naches; one prominent silicic welded tuff, the Mount Catherine Rhyolite Member [*Foster*, 1960; *Hammond*, 1977; *Tabor et al.*, 1984], has been mapped separately.

The Puget over most of its extent is deformed into tight plunging folds. The Oligocene marine Blakeley formation (informal), which conformably overlies the Puget, is likewise folded, but farther north the volcanic rocks of Mount Persis, coeval with the Blakeley, are not.

The Naches Formation is modestly to strongly deformed, particularly near the Straight Creek fault, where tight overturned folds occur. The Naches was folded in at least two episodes. The first predates the Oligocene Ohanapecosh Formation, which overlies the Naches with marked angular unconformity near the Straight Creek fault. This unconformity is absent above the Puget farther west. The second, weaker eposide of folding warped both the Ohanapecosh and Naches.

The Eocene section near the bend of the Straight Creek fault (Figure 5) was described and named by *Smith* [1904] and *Smith and Calkins* [1906] and further studied by *Foster* [1960], *Stout* [1964], and *Tabor et al.* [1982b, 1984]. Field-trip guides [*Hammond*, 1977; *Gresens et al.*, 1977] provide additional data. The Eocene section in this area is quite different from that west of the Straight Creek fault. The oldest unit, the Swauk Formation, is a thick nonmarine arkosic unit that unconformably overlies the Ingalls Tectonic Complex. Silicic volcanic rocks (Silver Pass Volcanic Member) interbedded in the Swauk Formation are dated at about 50 Ma. Consequently the Swauk is older than the Eocene rocks west of the Straight Creek. The Swauk was moderately folded about 48 Ma. Basalt flows of the Teanaway Formation (about 47 Ma) and overlying sandstone of the Roslyn Formation were deposited unconformably on the eroded Swauk. The Swauk-Teanaway unconformity is the earliest clear evidence of movement on the Straight Creek fault. A swarm of north-northeast-trending feeder dikes for the Teanaway reflects continued dextral shear on the Straight Creek fault during Teanaway time. The Teanaway and Roslyn were then deformed into a broad syncline, probably in response to late movement on the Straight Creek fault.

Cascade Arc

The Cascade magmatic arc in Washington is defined by a narrow north-south belt of Oligocene and Miocene shallow plutons and coeval volcanic rocks. This belt coincides closely with the zone of large Quaternary stratovolcanoes that dominate the present landscape. In Washington, the Cascade arc is superposed on the older, unrelated Eocene rocks of the much broader and more diffuse Challis magmatic field [*Vance*, 1982]. Inception of the Cascade arc occurred about 36 Ma and is marked by the outbreak of volcanism (Ohanapecosh Formation) and emplacement of the earliest Cascade plutons in the Index and Chilliwack batholiths [*Vance et al.*, 1986].

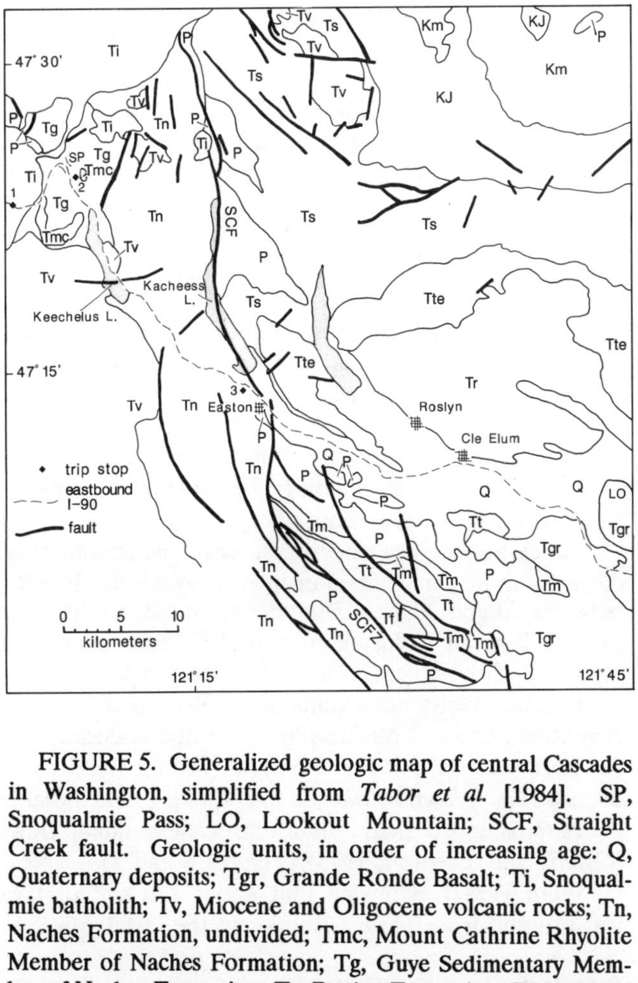

FIGURE 5. Generalized geologic map of central Cascades in Washington, simplified from *Tabor et al.* [1984]. SP, Snoqualmie Pass; LO, Lookout Mountain; SCF, Straight Creek fault. Geologic units, in order of increasing age: Q, Quaternary deposits; Tgr, Grande Ronde Basalt; Ti, Snoqualmie batholith; Tv, Miocene and Oligocene volcanic rocks; Tn, Naches Formation, undivided; Tmc, Mount Cathrine Rhyolite Member of Naches Formation; Tg, Guye Sedimentary Member of Naches Formation; Tr, Roslyn Formation; Tf, basalt of Frost Mountain; Tte, Teanaway Formation; Tt, Taneum Formation; Tm, Manastash Formation; Ts, Swauk Formation; P, pre-Tertiary rocks, undivided. Stops 1-3 labeled.

Rocks of the Cascade arc are widely exposed south of I-90. The two major units are the Ohanapecosh and the uppermost Oligocene and lower Miocene Fifes Peak Formations [*Fiske et al.*, 1963; *Vance et al.*, 1987]. The Ohanapecosh includes thick sequences of well-bedded volcaniclastic rocks of intermediate composition and debatable depositional enivronment [*Fiske*, 1963; *Vance et al.*, 1987], as well as andesitic and local silicic lava flows and silicic ash-flow tuffs. *Hammond* [1977] described thick, near-vent rhyodacitic ash-flow tuff along the shore of Lake Keechelus that correlates with the Ohanaceposh on the basis of field relations and radiometric ages [*Tabor et al.*, 1984]. The volcaniclastic rocks of Wildcat Creek [*Swanson*, 1978] in the Tieton River area on the east flank of the Cascades are a distal facies of the Ohanapecosh [*Vance et al.*, 1987]. Oligocene forearc-basin deposits west of the Cascades (Blakeley formation of the Puget Sound basin and Lincoln Creek Formation of southwest Washington) consist largely of volcaniclastic debris reworked from the Ohanapecosh.

The Ohanapecosh Formation is overlain, commonly unconformably, by the Fifes Peak Formation, which consists mostly of andesitic to basaltic lava flows, pyroclastic rocks, and lahars, as well as several thick silicic ash-flow tuffs. Vents for the Fifes Peak are known in the Fifes Peaks themselves [*Carkin*, 1985] and in the Tieton River area [*Swanson*, 1978]. *Fiske et al.* [1963] defined the Stevens Ridge Formation to include silicic ash-flow tuffs between the Ohanapecosh and Fifes Peak Formations in Mount Rainier National Park. Recent work shows that such tuffs are common throughout the section and are not useful for regional stratigraphic subdivision. Therefore, *Vance et al.* [1987] referred the Stevens Ridge to member status within the Fifes Peak Formation.

Volcanic rocks in the range 18-5 Ma are sparsely represented in the Cascade arc in Washington. However, scattered plutons in this interval are present, and recent sedimentological studies and radiometric dating [*Smith et al.*, 1988] confirm that the dacitic volcaniclastic detritus in the Ellensburg Formation, which overlies and interleaves with the Columbia River Basalt Group, is the distal volcanic debris from eruptions of this time.

The Snoqualmie batholith, a typical shallow plutonic complex of the Cascade arc, is exposed in many places along I-90 for about 15 km west of Snoqualmie Pass. This composite intrusive body ranges from gabbro to granite but is dominantly hornblende-biotite tonalite and granodiorite [*Erikson*, 1969; *Tabor et al.*, 1982a; *Frizzell et al.*, 1984]. Radiometric dating indicates that the batholith consists of two major phases, each with a wide compositional range. A northeastern phase is about 25 Ma and correlates with the Grotto batholith and related smaller plutons farther north. This phase intruded and postdates movement on the Straight Creek fault. The southern part of the batholith, including the exposures along I-90, is in the range 20-17 Ma.

MOUNT RAINIER

Mount Rainier, highest (4392 m) and third-most voluminous volcano in the Cascades after Mounts Shasta and Adams, dominates the Seattle-Tacoma area, where more than 1.5 million know it fondly as *The Mountain*. The Mountain is, however, the most dangerous volcano in the range, owing to the large population and to the huge area and volume (92×10^6 m^2 and 4.4×10^9 m^3, respectively [*Driedger and Kennard*, 1986]) of ice and snow on its flanks that could theoretically melt to generate debris flows during cataclysmic eruptions. In addition, sector collapses of clay-rich, hydrothermally altered debris have generated at least three huge ($>2 \times 10^8$ m^3) debris flows in the last 5000 years. Yet surprisingly little is known of Mount Rainier's eruptive history, composition, or age. For example, probably fewer than two dozen chemical analyses of Rainier's products have been published. Probably the dominantly nonexplosive nature of past eruptions and the challenging logistics of studying the cone contribute to the relatively limited knowledge. Outstanding work, however, has been completed on its fragmental deposits, and most of what is known about the volcano derives from this work.

Underpinnings

Mount Rainier is underlain by middle Tertiary volcanic rocks of the Ohanapecosh, Stevens Ridge, and Fifes Peaks Formations [*Fiske et al.*, 1963; *Vance et al.*, 1987], described elsewhere in this guide. These rocks were gently warped along a northwest-trending system of folds and intruded by the Tatoosh pluton, chiefly granodiorite and quartz monzonite. The main body of the Tatoosh is 17.5-14.1 Ma, but dikes, sills, and various volcanic deposits interpreted as forerunners to the emplacement of the pluton to its final level are as old as 26 Ma, judging from U-Pb dating of zircons [*Mattinson*, 1977]. *Fiske et al.* [1963] interpreted the Tatoosh to have "broken through to the surface" several times, giving rise to eruptions such as that which created The Palisades, a 250-m-high cliff of 25-Ma silicic welded tuff 5 km northeast of Yakima Park [*Mattinson*, 1977].

Forerunner to Mount Rainier

No evidence has been found for magmatism near Mount Rainier between about 14 and 3 Ma. The Lily Creek Formation, a thick sequence of debris flows and related volcanic deposits [*Crandell*, 1963a], crops out just west of the volcano and was hypothesized by *Fiske et al.* [1963] to have been erupted from the Tatoosh. However, *Mattinson* [1977] dated the Lily Creek as no older than 2.9 Ma and therefore agreed with *Crandell* [1963a] that it likely was formed during the earliest activity of Mount Rainier or from a center that just preceded the cone. Stratigraphic relations with glacial deposits suggest that Lily Creek volcanism began before 0.84 Ma [*Easterbrook et al.*, 1981; *Smith*, 1987]. Hornblende occurs in juvenile clasts of the Lily Creek but not in rocks from Mount Rainier [*Fiske et al.*, 1963]. No chemical analyses have been published for the Lily Creek.

The Main Cone

Mount Rainier was built on a rugged surface with more than 700 m of relief eroded mainly into the Tatoosh pluton and the Stevens Ridge Formation [*Fiske et al.*, 1963]. Early andesite flows from the volcano, undated but presumably several hundred thousand years old, were channelled along deep canyons, some of which were oblique to the present radial drainage pattern. The eruptions were apparently frequent, because in places one flow rests on the undissected surface of its predecessor [*Fiske et al.*, 1963]. Laharic deposits and till locally occur between the early flows. Eventually the flows stacked up to form a mound near the main vent that became the foundation of the present cone.

Most of Mount Rainier's cone was built by hundreds of thin lava flows interbedded with breccia and minor tephra. The flows are rarely more than 15 m thick high on the cone, where they drained down the steep slopes. They thicken near the base, and flows more than 60 m thick occur on the apron around the cone [*Fiske et al.*, 1963]. Some flows entered canyons radial to the volcano. Much breccia on the cone was probably derived from moving flows, but some probably reflects explosions and lahars. Radial dikes are prominent in places [*Smith*, 1897; *Coombs*, 1936; *Fiske et al.*, 1963]; possibly they fed some of the flows. The flows and dikes are petrographically uniform two-pyroxene andesite; the few chemical analyses available (Figures 6 and 7) are medium-K silicic andesite, with three analyses marginally dacitic. The flows and breccia eventually built a cone standing 2100-2400 m above its surroundings before the end of the latest major glaciation about 10 ka.

A thick pumice layer northeast, east, and southeast of the volcano may have been erupted from Mount Rainier between 70 and 30 ka. Estimates from limited outcrops suggest it is an order of magnitude more voluminous than any of the volcano's Holocene tephra layers.

Two late Pleistocene vents, Echo Rock and Observation Rock, erupted olivine-phyric basaltic andesite near the northwest base of the cone after Mount Rainier was almost fully grown [*Fiske et al.*, 1963]. The basaltic andesite is more mafic than the cone-building flows [*Condie and Swenson*, 1973] (Figures 6 and 7).

Smith [1987] estimates that about 270 km^3 of lava was erupted from Mount Rainier in the last 1 Ma.

Postglacial Eruptive History

Eleven tephra layers record evidence of Holocene explosive volcanism at Mount Rainier [*Mullineaux*, 1974].

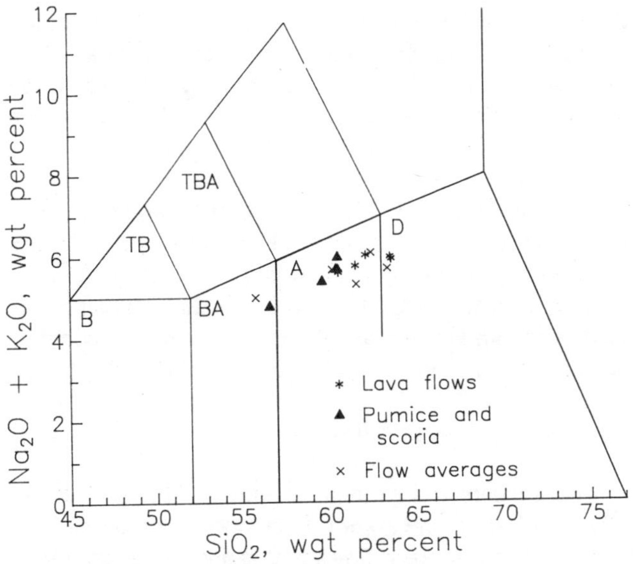

FIGURE 6. Total alkali-silica diagram for Mount Rainier. Analyses from *Coombs* [1936] (lava flows), *Condie and Swenson* [1973] (flow averages), and *Mullineaux* [1974] (pumice and scoria), calculated water-free to 100%. See Figure 17 for field names.

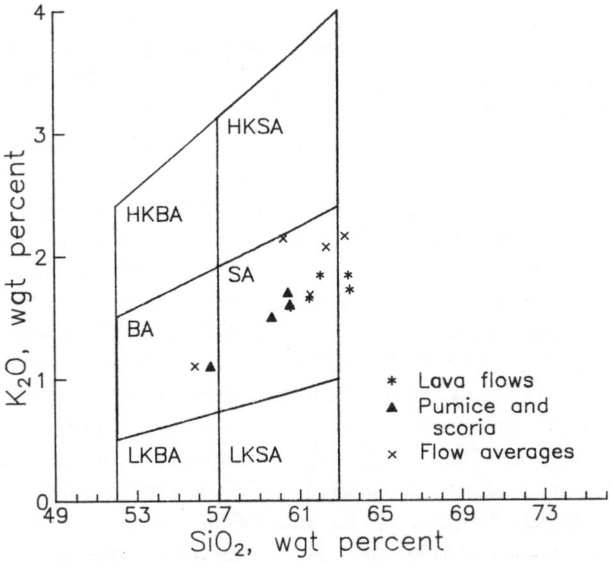

FIGURE 7. Andesite classification for Mount Rainier, from *Gill* [1981] except that basalt-andesite division is 52%. Data sources in Figure 6. HKBA, high-K basaltic andesite; BA, basaltic andesite, LKBA, low-K basaltic andesite; HKSA, high-K silicic andesite; SA, silicic andesite; LKSA, low-K silicic andesite.

Eight of the tephras fell between 6500 and 4000 ^{14}C years B.P. (Table 1). Only one tephra-producing eruption, between about 1820 and 1854 A.D., is known from the last 2200 years [*Mullineaux*, 1974]; it was very small and left a scanty deposit that could easily be overlooked if it were older. Chemical analyses indicate that layer D is basaltic andesite and layers L and C are silicic andesite (Figure 7) [*Mullineaux*, 1974].

The tephra layers rich in lithic fragments are probably products of phreatic or phreatomagmatic eruptions. Layer F contains 5 to 25% clay-sized material, as much as 80 percent of which locally consists of clay minerals, chiefly montmorillonite, that was derived from altered rocks within Mount Rainier [*Mullineaux*, 1974]. Layer F and the Osceola debris flow [*Crandell*, 1971] have similar clay contents and ages, and are likely correlative. However, layer F has not been found on the Osceola and so could be slightly older.

Layer C, the most widespread and voluminous of the Mount Rainier tephras, covers the east half of the National Park with 2-30 cm of lapilli, blocks, and bombs. Overall it is the coarsest of the tephras, with 25-30-cm bombs 8 km from the summit. A block-and-ash flow in the South Puyallup valley west of the volcano contains blocks emplaced above the Curie isotherm and charcoal dated at 2350 ± 250 ^{14}C years [*Crandell*, 1971]. Its age and lithologic similarity suggest correlation with layer C.

Isopachs and isopleths for layer C indicate an origin at the summit of Mount Rainier, yet the layer does not occur on snow-free parts of Columbia Crest cone, a young andesite cone standing 250 m above the former summit of the volcano [*Mullineaux*, 1974]. Columbia Crest cone is therefore younger than about 2200 years.

TABLE 1. Holocene tephras from Mount Rainier

Layer	Age	Predominant materials	Volume
X	±150	Pumice	1
C	2200	Pumice, scoria, lithic frags.	300
B	>4000	Scoria, lithic fragments	5
H	>5000	Pumice, lithic fragments	1
F	5000	Lithic frags, pumice, crystals, clay	25
S	5200	Lithic fragments	20
N	5500	Lithic fragments, pumice	2
D	6000	Scoria, lithic fragments	75
L	6400	Pumice	50
A	6500	Pumice, lithic fragments	5
R	>8750	Pumice, lithic fragments	25

From *Mullineaux* [1974]. Age of X from tree rings; all others in radiocarbon years B.P. Volume in millions of m^3.

Crandell [1971] found numerous lahars and flood deposits in valleys surrounding the mountain that postdate layer C but predate layer Wn (1480 A.D.) from Mount St. Helens; some of the flowage deposits have ^{14}C ages older than 1000 years B.P. The eruptions that formed Columbia Crest cone likely produced some of those deposits. If so, the layer-C explosions might have initiated activity that formed Columbia Crest, and the cone would be about 2000 years old [*Mullineaux*, 1974].

Crandell [1971] identified more than 55 lahars and

debris flows of Holocene age from Mount Rainier. At least some were probably associated with eruptions, most notably the Paradise lahar [*Crandell*, 1963b, 1971] (possibly associated with tephra layers A, L, or D) and the Osceola debris flow (layer F). The Osceola is described in the road log for Mount Rainier. In general, *Crandell* [1971] associates lahars that lack much clay at Mount Rainier with magmatic eruptions, and those that contain much clay with phreatic or phreatomagmatic eruptions or with collapses of the hydrothermally altered edifice. Glacier-outburst floods from Little Tahoma Glacier, typically in late afternoon of warm days or after heavy rain, repeatedly scoured Tahoma Creek in the late 1960s and the middle and late 1980s. Outburst floods frequently modified many other drainages, most notably Kautz Creek and Nisqually River, during historical time.

About 20 small earthquakes occur yearly at Mount Rainier, more than at other composite cones in the Cascades except Mount St. Helens [*Malone and Swanson*, 1986]. Trilateration and tilt networks established in 1982 indicate no definite deformation. Seven significant thermal areas above 3350 m on the volcano, including one at the summit, reflect "a narrow, central hydrothermal system...forming steam-heated snowmelt at the summit craters and localized leakage of steam-heated fluids within 2 km of the summit" [*Frank*, 1985].

SUMMARY OF THE GEOLOGY OF THE MOUNT ST. HELENS AREA

Mount St. Helens rests on the deeply-dissected remains of the mid-Tertiary Cascade magmatic arc, represented by a diverse assemblage of variably altered subaerial volcanic and plutonic rocks (Figure 8). Near the volcano, 5-7 km of Tertiary rocks crop out in the east-dipping limb of a broad, south-trending and plunging anticline. Exceptional exposures occur near Spirit Lake, where vegetation and surficial deposits were removed by the May 1980 eruption and subsequent erosion.

Many age determinations show that volcanic and plutonic activity that constructed the Tertiary arc in southern Washington occurred largely between 36 and approximately 17 Ma [*Frizzell et al.*, 1984; *Phillips et al.*, 1986; *Evarts et al.*, 1987; *Vance et al.*, 1987]. Volcanism was more or less continuous during this period. Although unconformities are widespread in the section in southern Washington [*Fiske et al.*, 1963; *Evarts et al.*, 1987; *Vance et al.*, 1987], none has regional extent. K-Ar and ^{40}Ar-^{39}Ar ages from just north of Mount St. Helens indicate a deposition rate of about 360 m/m.y, similar to those compiled by *Smith* [in press a] for other well-mapped and dated mid-Tertiary sequences in the Washington Cascades. Radiometric ages suggest that the rate of volcanism in southern Washington declined between 20 and 17 Ma, and that a subsequent further downturn coincided with eruption of the Grande Ronde Basalt on the Columbia Plateau. Beds of Cascade-derived tephra between flows of Grande Ronde Basalt indicate that explosive volcanism did not stop entirely, but reliable age determinations from the Cascades in the 17-5-Ma range are rare [*Smith et al.*, 1988; *Smith*, in press a].

The rate of volcanism in southern Washington increased in the early Pliocene, after regional folding, uplift, and extensive erosion. The volcanism continues but is localized and sporadic relative to that of the mid-Tertiary. Products of this period include basalt fields such as Indian Heaven as well as the prominent composite cones of Mounts Rainier, Adams, and St. Helens.

Mid-Tertiary Volcanic Rocks

The Tertiary section near Mount St. Helens contains rocks typical of near-vent depositional environments, such as abundant lava flows and domes, coarsely pumiceous pyroclastic rocks, coarse lithic tuff-breccia and lapilli tuff, fine-to medium-grained intrusive rocks, and widespread, locally intense, hydrothermal alteration. The units are mostly discontinuous and lenticular. Well-bedded, finer-grained volcaniclastic rocks of more distal character, such as those in the Ohanapecosh Formation near Mount Rainier [*Fiske et al.*, 1963], are common only near Cougar south of Mount St. Helens (Figure 8).

The eruptive products range from olivine-phyric basalt to pyroxene rhyodacite. The distribution of these chem-

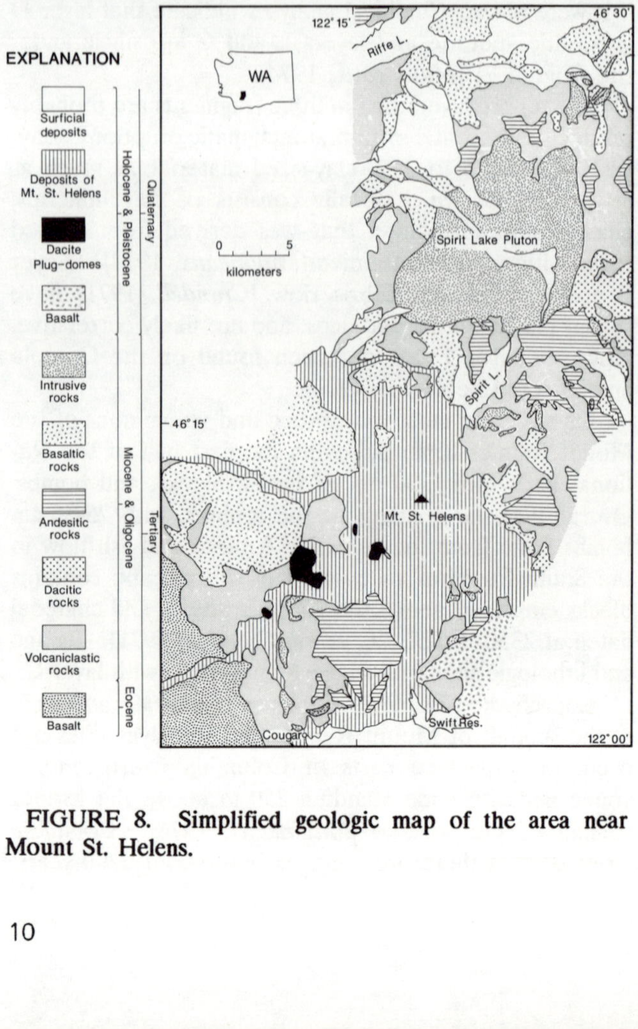

FIGURE 8. Simplified geologic map of the area near Mount St. Helens.

ically diverse rocks is complex but not random. The upper Eocene and lower Oligocene section west of Mount St. Helens is almost entirely basalt and basaltic andesite, whereas lower Miocene rocks farther east are mostly pyroxene andesite and dacite. The relative volume of pyroclastic rocks increases upsection. Vent areas of several kinds have been recognized, including a swarm of mafic dikes west of Spirit Lake, a silicic dome field on the east flank of Strawberry Mountain north of Bear Meadow, and an inferred caldera along Quartz Creek.

Most rocks are phyric, but sparsely phyric to aphyric andesite and dacite are common. Phenocrysts in basalt are plagioclase, olivine, and in some samples augite; andesite and dacite typically contain plagioclase, augite, and hypersthene. Rhyodacite and silicic tuff rarely contain quartz phenocrysts. Notably absent from the Tertiary volcanic rocks are hornblende and biotite, although hornblende occurs rarely in a few subvolcanic intrusions.

Chemical analyses of the least altered samples have subalkaline tholeiitic to calc-alkaline compositions typical of volcanic arcs [*Ewart*, 1982]. The ratio of calc-alkaline to tholeiitic rocks increased with time. The most notable chemical feature is relatively low K_2O (Figure 9); in this respect the rocks resemble those of evolved ensimatic arcs such as the Aleutians more than those of the Quaternary Cascades [*Evarts et al.*, 1987].

Plutonic Rocks

Intrusive rocks are widespread near Mount St. Helens and elsewhere in southern Washington [*Hammond*, 1980]. They range from shallow subvolcanic dikes and plugs to medium- to coarse-grained epizonal granitic bodies typified by the Spirit Lake pluton [*Evarts et al.*, 1987]. Compositions of the intrusive and volcanic rocks span the same range, from olivine gabbro to granite, and many probably fill synvolcanic conduits. Most of the smaller bodies are undated. A composite sill complex along Windy Ridge east of Spirit Lake has a K-Ar age (plagioclase) of about 24 Ma (roughly 2 Ma younger than its host rocks), and several K-Ar and fission-track ages on the Spirit Lake pluton are about 21 Ma. Most of the intrusive bodies are probably only slightly younger than their volcanic country rock, late Oligocene to early Miocene. Intrusions elsewhere in the Cascades generally have similar ages [*Tabor and Crowder*, 1969; *Engels et al.*, 1976; *Mattinson*, 1977; *Power et al.*, 1981; *Tabor et al.*, 1982a; *Frizzell et al.*, 1984; *Vance et al.*, 1987].

The Spirit Lake pluton consists of three major phases, each containing many smaller units. The earliest phase comprises phyric quartz diorite and granodiorite that occur as relatively fine-grained dikes with hornfels screens in the northwest part of the pluton and as isolated bodies in the main phase. Most of these rocks display intense deuteric alteration and are intruded by rocks resembling those of the main phase. The pluton consists mainly of the main and granite phases, which are coarser and more massive than the quartz diorite. The pluton has a configuration of a flat-roofed cylinder tilted eastward with its country rock. Its contact sharply truncates wall rock and is generally simple, but complex to gradational contacts are common internally. Stratigraphic evidence suggests a emplacement less than 3 km deep.

Chemical and petrographic attributes of the Spirit Lake pluton are those of I-type granitic rocks found in most arc environments [*Brown et al.*, 1982]. Plagioclase, augite, and hypersthene generally form euhedral to subhedral crystals in a matrix of coarse anhedral to micrographic quartz and K-feldspar. Minor late hornblende and biotite occur sparingly, mainly in the deeper western half of the pluton. The magma was probably relatively dry. SiO_2 contents vary from 56 to 75%.

Metamorphism and Hydrothermal Alteration

The Tertiary rocks were affected by very low-grade, zeolite-facies burial metamorphism. Some massive flows and densely welded ash-flow tuffs still contain glass (albeit hydrated), but generally the glass was replaced by iron-rich smectitic clay minerals that provide the green colors characteristic of the Tertiary units. More intense recrystallization took place around plutonic bodies. The Spirit Lake pluton produced a well-developed contact metamorphic aureole consisting of an inner zone of fine-grained black amphibole hornfels and an outer zone of green albite-epidote hornfels. Broad zones of epidote-bearing hornfels commonly surround even small phaneritic intrusive bodies, which may therefore be cupolas extending from a much larger pluton at shallow depth.

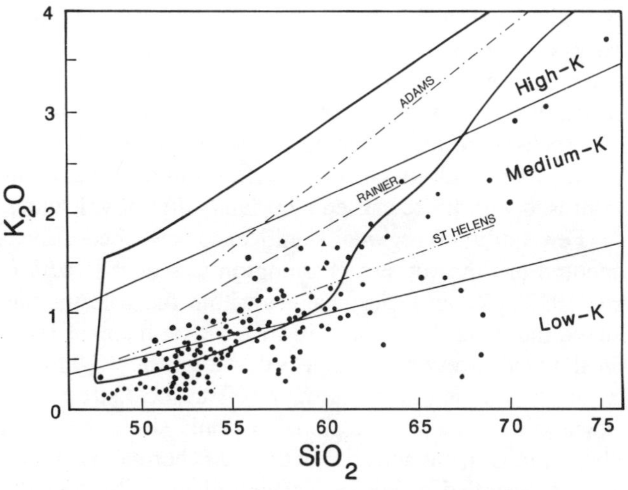

FIGURE 9. K_2O-SiO_2 variation for Tertiary volcanic rocks near Mount St. Helens. Heavy line encircles field of Quaternary volcanic rocks in southern Washington Cascades [*Hammond and Korosec*, 1983]. Generalized trends for Quaternary rocks at Mounts St. Helens, Rainier, and Adams from *Greeley and Hyde* [1972], *Condie and Swenson* [1973], *Hoblitt et al.* [1980], and *Hildreth and Fierstein* [1985]. Modified from *Evarts et al.* [1987].

Intense metasomatic alteration of several types has been recognized in the area [*Evarts et al.*, 1987]. Widespread localized low-temperature alteration to carbonate and clay is particularly common near abundant dikes and along minor faults, as on Johnston Ridge. Higher temperature argillic to advanced argillic assemblages occur in three areas near the north end of the Spirit Lake pluton. A few kilometers farther south is the Earl porphyry copper-molybdenum deposit, the largest of many such occurrences in the Washington Cascades. The Earl is within the Spirit Lake pluton, yet multiple age determinations indicate that it is about 4 m.y. younger and thus not genetically related to the pluton.

Quaternary Volcanism Exclusive of Mount St. Helens

Quaternary rocks near Mount St. Helens are restricted to isolated vents south and southwest of the volcano. Basalt and basaltic andesite dominate, but a few silicic domes and plugs follow an east-northeast trend southwest of the volcano [*Evarts et al.*, 1987].

In contrast to the Tertiary rocks, many of the younger intermediate to silicic rocks contain hornblende and biotite phenocrysts. They are typically more phyric (especially the dacite) and richer in K_2O at equivalent silica contents than the Tertiary rocks [*Evarts et al.*, 1987].

Structural Elements

Broad open folds with north- to northwest-trending axes dominate the structure of the southern Washington Cascades [*Hammond*, 1980; *Walsh et al.*, 1987]. The age of folding is poorly constrained, but much probably occurred in the middle to late Miocene, for all Tertiary rocks are affected whereas Pliocene rocks apparently are not. A few low-angle unconformities within the mid-Tertiary section suggest that minor folding took place as the rocks were accumulating [*Evarts et al.*, 1987; *Vance et al.*, 1987], although such unconformities are easily confused with those caused by primary dips at volcanoes.

Few faults of regional significance have been documented in the southern Washington Cascades [*Walsh et al.*, 1987]. Rapid facies changes within the volcanic pile make faults hard to find. However, even mapped faults in the well-exposed area north of Mount St. Helens are clearly minor features that may reflect local stress near volcanoes rather than regional tectonic stress. Many of these faults are associated with hydrothermal alteration or are occupied or cut by Tertiary dikes. The only demonstrably young faulting was related to emplacement of the Goat Mountain dacite plug west of Mount St. Helens. No surface offset along the north-northwest-trending St. Helens seismic zone [*Weaver and Smith*, 1983] is apparent; this zone connects several clusters of shallow earthquakes, including those beneath Mount St. Helens, whose preferred focal mechanisms suggest dextral displacement parallel to the trace of the zone.

Minor but pervasive north- and northwest-trending dextral shear zones, and northeast-trending sinistral shear zones, occur in the area east of Mount St. Helens. Subhorizontal slickensides on stepped fault surfaces allow the sense of displacement to be determined, but offsets are less than a few centimeters. These shear zones are part of a regional system of similar structures, chiefly dextral, recognized on the western Columbia Plateau in Washington and northern Oregon; in places they are simply shear zones with little displacement and in other places faults with offsets of hundreds of meters or more [*Bentley and Anderson*, 1980; *Anderson et al.*, 1987].

Evarts et al. [1987] and *Weaver et al.* [1987] discuss linear trends of geologic features that may reflect deep crustal structures in the Mount St. Helens area (see also Stop 45). Mount St. Helens, Mount Rainier, and Glacier Peak define a north-northeast-trending segment of the modern arc. This trend coincides with that of the major epizonal Miocene intrusions of the region, such as the Silver Star (east of Vancouver [*Felts*, 1939]), Snoqualmie, Tatoosh, and Spirit Lake bodies, which define the core of the Tertiary arc. Thus the position of the volcanic front in southern Washington has been more or less fixed for 25 Ma. The position of Mount St. Helens may be determined by the intersection of this broad trend with younger crustal flaws manifested in a line of Quaternary silicic vents and the St. Helens seismic zone.

Geophysics

Williams et al. [1987] inferred a large, shallow "intrusive complex" under Mount St. Helens on the basis of gravity data. The inferred complex is interpreted to be 5-6 km thick and to intrude the putative compressed forearc marine basin in the southern Washington Cascades conductor [*Stanley et al.*, 1987]. Conceivably this complex could be the source of gabbroic inclusions [*Heliker*, 1984] in the dacite of Mount St. Helens, although its size (10 by 20 km to as large as 18 by 22 km) is much larger than the inferred current magma reservoir.

Finn and Williams [1987] found a residual magnetic high over Mount St. Helens after correcting pre- and post-1980 aeromagnetic data for terrane effects. The anomaly is shallow, probably near or at the base of the volcano. They suggest that the high, which is not reflected by the gravity data, results from terrain that predates Mount St. Helens, such as a buried andesitic or basaltic cone or a valley filled with lava flows. These two possibilities cannot be distinguished, although the presence of an eroded and buried cone, perhaps fed by the inferred intrusive complex, is appealing. Such a cone would probably be older than 50 ka, the approximate age of the oldest dacite from Mount St. Helens.

MOUNT ST. HELENS

Mount St. Helens is young. Its oldest known deposits were erupted about 50-40 ka, and the cone that partly collapsed in 1980 is only 2200 years old. Since its birth it has produced more than 60 tephra layers [*Mullineaux*, 1986], several tens of volcanically-induced debris flows (at least six of which entered the Columbia River 100 km downstream [*Scott*, 1988]), and the equivalent of 60 km^3 of dacitic lava [*Smith*, 1987]. It has been the most active volcano in the Cascades during the Holocene, and for that reason its eruption in 1980 came as no surprise. The volcano has been studied intensively, and its eruptive history is known with greater clarity than that of any other Cascade volcano.

Eruptive History

Crandell [1987] divides the history of the volcano into "four extended stages of intermittent activity, each lasting two thousand years or longer. The volcano is now in such a stage that began about 4,000 radiocarbon years ago" (Table 2). The stages are separated by dormant intervals of thousands of years. Each stage contains eruptive periods with durations of decades to centuries; those periods for the current stage are named, but those for past stages are not. Many ^{14}C and dendrochronologic ages [*Crandell et al.*, 1981; *Mullineaux*, 1986; *Yamaguchi*, 1983, 1985] provide a well-constrained context within which to interpret the volcanic history.

Mullineaux [1986] recognized at least one tephra set, comprising several layers of tephra, in each eruptive stage and period except the Sugar Bowl and Goat Rocks periods, each of which has only one layer. The tephra sets and some single tephra layers form distinctive markers, recognizable by the assemblage and relative proportions of ferromagnesian phenocrysts (hypersthene, hornblende, biotite, cummingtonite, olivine, and augite). The tephra sets and some layers can be correctly identified in the field with a mortar, pestal, and binocular microscope using methods developed by D. R. Mullineaux.

The tephras are the most useful means of placing deposits of Mount St. Helens in proper stratigraphic context. They are also used to date events far from the volcano. For example, set S, about 13 ka, is interbedded with deposits of the Missoula floods on the Columbia Plateau and provides one of the best ages for the floods [*Mullineaux et al.*, 1978; *Waitt*, 1985]. Tephra layer Cs, about 37 ka, spread southward into the Lake Lahontan basin of Nevada, where it is probably equivalent to the Marble Bluff Bed [*Davis*, 1978; *Mullineaux*, 1986]. Layer Yn, about 4 ka, occurs near Entwhistle, Alberta, 900 km north-northeast of Mount St. Helens [*Westgate et al.*, 1970; *Mullineaux*, 1986]. Layer Ye occurs in northeast Oregon, 300 km from the volcano [*Borchardt et al.*, 1973]. Layers Wn (1480 A.D.) and We (1482 A.D.) [*Yamaguchi*, 1983, 1985] are known 400 km northeast and east of the volcano, respectively [*Smith et al.*, 1977]. Layer T (1800 A.D.) occurs in northwestern Montana, 375 km away [*Okazaki et al.*, 1972].

Pyroclastic flows and lahars formed in each eruptive stage and period [*Mullineaux and Crandell*, 1981; *Crandell*, 1987]. Many pyroclastic flows carry lithic fragments probably derived from coeval domes. Lava flows are known with certainty from only the Cougar stage and Castle Creek and Kalama periods, although the "floating island" lava flow [*Lawrence*, 1941; *Hoblitt et al.*, 1980] probably erupted in the 19th century [*Crandell*, 1987].

All chemical analyses reported by *Smith and Leeman* [1987; see also *Smith*, 1984] and *Crandell* [1987] from pre-Castle Creek tephras are silicic andesite or dacite (Figure 10), mostly with water-free SiO_2 contents of 61-68% but rarely as low as 57%.

The lack of mafic tephra suggests that no basaltic andesite or basalt was erupted before Castle Creek time. Proximal exposures are needed to test this suggestion, however, because such flows can be erupted with little tephra. Large dacite or silicic andesite domes of Pine Creek age, but no mafic flows, occur in the lower half

TABLE 2. Eruptive History of Mount St. Helens

Eruptive Stage or Period[1]	Age[2]	Tephra Set	Nature of Volcanism[3]
Present	1980-	1980	Phreatic explosions, cryptodome, landslide, blast; dacite t., p.f., and dome.
Goat Rocks	180-123[4]	T[6]	Dacite t. and dome; andesite flow
Kalama	500-350[5]	X W	Dacite t., p.f., and dome; andesite t.
Sugar Bowl	1150[5]	D[6]	Dacite dome, blast, t., and p.f.
Castle Creek	2200-1700[5]	B	Andesite, dacite, and basalt t.; andesite and basalt flows; andesite and dacite p.f.
Pine Creek	3000-2500[5]	P	Dacite t., dome, p.f.
Smith Creek	4000-3300[5,7]	Y	Dacite t., dome, p.f.
SWIFT CREEK	13-10 ka[7]	J S	Dacite t., dome, p.f.
COUGAR	21-18? ka[7]	K M	Dacite t., dome, p.f. and lava flows.
APE CANYON	<50-36? ka	C	Dacite t. and p.f.

Simplified from *Crandell* [1987]. [1]Stages are capitalized; periods are in Spirit Lake eruptive stage. [2]Approximate ^{14}C age in years before 1950. [3]t., tephra; p.f., pyroclastic flow; lahars formed in each stage. [4]Years before 1980, from tree-ring ages and historical records. [5]Years before 1980, from tree-ring and ^{14}C ages. [6]One tephra layer only. [7]Probably includes dormant intervals of at least a few centuries.

of the present crater wall, so it is unlikely that mafic flows were produced during Pine Creek time. Three dacite or silicic andesite domes of possible Smith Creek age or older crop out in the northern part of the crater; their age assignment is based on the presence of cummingtonite, which is absent in younger deposits. However, domes and tephras of a given period need not have the same phenocryst assemblage. Aside from these questionably older domes, no pre-Pine Creek deposits occur in the crater. Near-vent mafic flows, perhaps in a small cone [*Finn and Williams*, 1987], *could* underlie the Pine Creek domes. No direct evidence exists for such flows, but gabbro inclusions resembling those in the present dome [*Heliker*, 1983, 1984] occur in Cougar and Pine Creek deposits (J.S. Pallister, written commun., 1988) and suggest that mafic magma was in the reservoir system during much of Mount St. Helens history, possibly even before 50 ka [*Williams et al.*, 1987]. Nonetheless, the volcano has surely been *dominated* by silicic domes throughout most of its history.

A major compositional change occurred at the onset of Castle Creek time about 2200 years B.P. The period began, was dominated by, and ended with basalt, basaltic andesite, and lesser andesite, although dacite tephras, pyroclastic flows, and possibly a dome also formed (Figure 10) [*Mullineaux and Crandell*, 1981; *Smith and Leeman*, 1987; Crandell, 1987]. Numerous mafic lava flows, mostly borderline trachybasalt and trachybasaltic andesite [*Le Bas et al.*, 1986] (Figure 10) although more often called olivine basalt and two-pyroxene basaltic andesite, were erupted on all sides of the volcano, especially the southwest and north. The Cave Basalt (the one basalt plotted in Figure 10) issued from a vent probably near the southwest base of the cone about 1700 ^{14}C years B.P.; it contains 3.4-km-long Ape Cave, the longest known uncollapsed segment of a lava-tube, as part of its 8.3-km-long tube system [*Greeley and Hyde*, 1972].

Sugar Bowl dome on the north flank of the cone formed about 1200 ^{14}C years B.P. Two lateral blasts, the largest of which threw lithic lapilli 10 km northeast of the vent, probably accompanied dome growth [*Crandell and Hoblitt*, 1986]. East dome, at the east foot of the volcano, chemically resembles Sugar Bowl; it is undated, but bracketing ages allow it to be of Sugar Bowl age. The rhyodacitic compositions of both domes (Figure 10) are the most silicic (69-70%) yet found at Mount St. Helens [*Smith and Leeman*, 1987].

The Kalama eruptive period began in winter or early spring of 1479-1480, as deduced from dendrochronologic dating of the dacitic Wn tephra [*Yamaguchi*, 1983, 1985], the most voluminous tephra from Mount St. Helens since the Y tephras about 4000 years B.P. Another widespread tephra, We, fell in winter or early spring of 1481-1482. Episodic activity thereafter produced voluminous silicic andesite lava flows. The symmetric pre-1980 cone, known as North America's Fuji, was built by these and the older Castle Creek

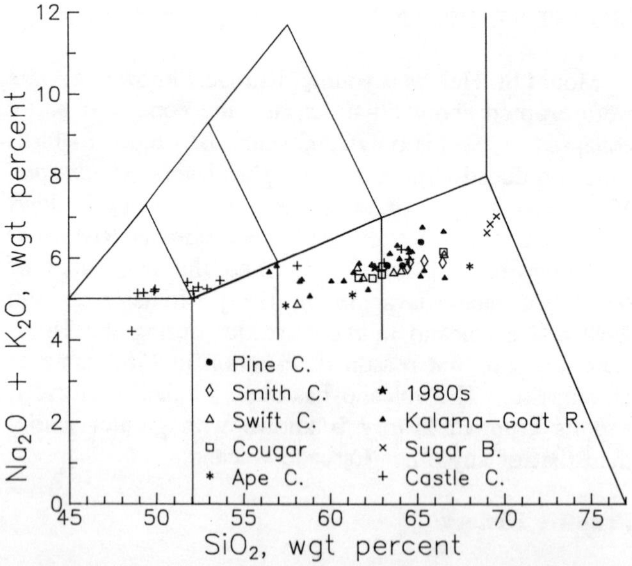

FIGURE 10. Total alkali-silica diagram for Mount St. Helens. Analyses from *Smith* [1984], *Smith and Leeman* [1987], and *Crandell* [1987], calculated water-free to 100%. Symbols indicate age (see text). Figure 17 gives field names.

flows. A dacite dome formed at the summit during Kalama time, and lahars and pyroclastic flows were frequently produced.

The Goat Rocks eruptive period began in 1800 A.D. with the dacitic tephra layer T and ended in 1857 [*Crandell*, 1987]. The "floating island" silicic andesite flow was erupted before 1838, and an explosion sent lithic ash 100 km downwind in 1842. The Goat Rocks dome was extruded on the northwest flank of the volcano 600-700 m below the summit within several years after the 1842 explosion, possibly during or before 1847, when Paul Kane, a Canadian artist, painted a famous canvas of its growth. A large fan of debris spread downslope from the Goat Rocks dome; prismatic jointing is common in blocks of the fan, and paleomagnetic measurements indicate that some of the blocks were deposited above the Curie point and others below (R. P. Hoblitt in *Crandell* [1987]). Contemporary accounts indicate activity several times during the 1840s and 1850s but are non-specific and even contradictory. The last significant activity before 1980 was "dense smoke and fire" in 1857, although minor, unconfirmed eruptions were reported in 1898, 1903, and 1921 [*Majors*, 1980].

Volcanic Activity, 1980-1988

This activity has been thoroughly documented and is familiar to most volcanologists. See especially papers in *Lipman and Mullineaux* [1981], a series of nine papers in Science (v. 221, no. 4618, 1983), and *Swanson et al.* [1987] for general summaries. Only a synopsis is given here; other specifics are mentioned in the road log.

Seismicity began several days before March 20, 1980,

when an earthquake (M=4.2) centered under the volcano commanded wide attention. The first of a series of small phreatic explosions occurred on March 27, accompanying the opening of a crater within a horseshoe-shaped graben concave northward at the summit of the cone. Strong seismicity continued, at times with bursts of volcanic tremor [*Endo et al.*, 1981; *Qamar et al.*, 1983]; deep tremor was felt state-wide in early April but died away without returning. By mid-April a bulge was obvious on the north flank of the volcano; geodetic measurements began shortly thereafter and documented horizontal growth of the bulge at a steady maximum rate of >1.5 m/day [*Lipman et al.*, 1981a]. The bulge was surface evidence of a cryptodome intruding the volcano. Seismicity continued into May, with fewer but larger earthquakes, and phreatic activity was intermittent. No magmatic gas was detected, although new fumaroles appeared in the crater and at the head of the bulge.

At 0832 on May 18, a complex earthquake (M=5.1) shook the volcano, probably causing (but possibly caused by) a huge, 2.7-km^3-landslide that in three different blocks successively removed the bulge and upper 400 m of the volcano [*Voight et al.*, 1981, 1983], leaving a 600-m-deep crater 2 km wide rim-to-rim. The landslide quickly developed into a debris avalanche that sped at 110-240 km/hr for 24 km down the North Fork Toutle River; arms of the avalanche entered Spirit Lake, 8 km from the summit, and overtopped 300-380-m high Johnston Ridge north of the Toutle. The avalanche buried the Toutle valley to a depth of nearly 50 m. Its hummocky deposit is distinctive; similar morphology at other volcanoes has been reinterpreted in light of its observed origin [*Siebert et al.*, 1987].

The landslide removed confining pressure on the cryptodome and its surrounding hydrothermal system. Juvenile gas was rapidly released from the cryptodome and superheated groundwater flashed to steam, causing a blast that exploded laterally from the collapsing north flank. The blast, called a "stone wind" by local journalists, knocked down most trees (the equivalent of about 150,000 houses) in a 600-km^2 area. Its maximum velocity may have been supersonic [*Kieffer*, 1981b]. The degree to which juvenile gas or flashed groundwater drove the blast is debated [*Eichelberger and Hayes*, 1982; *Kieffer*, 1981a, b; see also *Brugman*, 1988], as is the question of whether the blast was, in volcanologic terms, a low-aspect-ratio ignimbrite [*Walker and McBroome*, 1983] or a surge [*Hoblitt and Miller*, 1984; *Waitt*, 1984b].

Soon after the blast, a lahar rushed down the South Fork Toutle River and several streams draining the south and east flanks of the volcano. Whether water for these lahars came from snowmelt or from groundwater ejected by the eruption is hotly argued. The largest lahar, down the North Fork Toutle, did not start until early afternoon; it was fed as the debris avalanche dewatered [*Janda et al.*, 1981]. In addition to causing havoc along the rivers themselves, the lahars fed so much debris into the Columbia River that 31 ships were stranded in upstream ports until the 4-m-deep channel was dredged to its pre-eruption depth of 12 m—the first in a series of similar dredgings to maintain Portland as a seaport.

Juvenile dacite pumice and ash mixed with lithic debris began erupting soon after the blast [*Criswell*, 1987], perhaps from the shallow root of the cryptodome. The flux increased about noon, apparently with arrival of pumice from a 7-10-km-deep reservoir [*Rutherford et al.*, 1985; *Carey and Sigurdsson*, 1985; *Scandone and Malone*, 1985]. Experimental work suggests that just before eruption this reservoir was at a pressure of 220 ± 30 MPO, P_{water} was 0.5-0.7 P_{total}, and the temperature was 930° ± 10°C [*Rutherford et al.*, 1985]. Pyroclastic flows fed by the eruption column covered the debris avalanche in the upper North Fork Toutle basin, forming the pumice plain. Hydroexplosions created phreatic pits on the pumice plain, possibly as the pyroclastic flows covered the dewatering debris avalanche [*Moyer and Swanson*, 1987].

Plinian to subplinian explosions took place on May 25, June 12, July 22, August 7, and October 16-18, 1980. Products of the explosions decreased in SiO_2 content with time from 65 to 63% or less, possibly because they tapped a magma body zoned chemically [*Lipman et al.*, 1981b], mineralogically [*Kuntz et al.*, 1981], and in gas content [*Melson*, 1983; *Scandone and Malone*, 1985].

Small domes grew in June and August but were destroyed by the next explosion. The current dome began growing after the last major explosion on October 18. As of December 1988, 17 episodes of dome growth have taken place, each lasting several days to 1 year (1983-1984); the latest was in October 1986. The dome stands 267 m above its vent and 350 m above its north base, is 860 m by 1060 m across, and contains 74 x 10^6 m^3. The dome is highly phyric (30-35% plagioclase, 5% hypersthene, 1-2% hornblende, 1-2% Fe-Ti oxides, and <0.5% clinopyroxene [*Cashman*, 1988]), about 50% crystalline, and has about 63% SiO_2. Geodetic and seismic precursors enabled prediction of growth events [*Swanson et al.*, 1983, 1985; *Chadwick et al.*, 1988]. The geometry and volumetric rate of growth of the dome followed consistent patterns with time [*Swanson and Holcomb*, in press]. Several hundred small explosions occurred from the dome between 1980 and 1986, and a few large rockfalls spawned minor lithic pyroclastic flows and surges, none of which had sufficient volume to leave the crater [*Mellors et al.*, 1988]. Several small lahars formed when snow melted during explosions and rockfalls [*Waitt et al.*, 1983; *Waitt and MacLeod*, 1987].

Petrologic interpretation

Smith and Leeman [1987] found that the St. Helens dacite has similar or even lower contents of many incompatible elements than to basalt and andesite from the volcano but is relatively enriched in Ba, Rb, K, Cs, and Sr. The unusual depleted nature of the dacite, and low bulk distribution coefficients for numerous trace elements, preclude an origin by fractionation of the basalt or andesite. *Smith and Leeman* [1987] favor an interpretation involving melting of metabasaltic crustal rocks enriched in Ba, Rb, Cs, and Sr owing to interbedded sedimentary rocks or metasomatic enrichment of the source region. They consider mantle-derived basaltic magma to be the heat source for crustal melting.

WHITE PASS-GOAT ROCKS NEOGENE VOLCANISM

The White Pass-Goat Rocks area (Figure 2) has been host to volcanism since 4 Ma, producing at least 25 volcanoes and >175 km^3 of material [*Clayton*, 1983]. Geoff Clayton has been working in this area for many years, and the following is excerpted from his M.S. thesis [*Clayton*, 1983] and a USGS open-file report on the Goat Rocks Wilderness [*Swanson and Clayton*, 1983].

Pliocene volcanism (4-2.8 Ma) resulted in domes, tuff, and shallow intrusions of hornblende dacite north of White Pass. South of the pass, high-silica rhyolite tuff, many layers of which are welded, form a 650-m-thick section on the east flank of the present Goat Rocks area. Flow-layered rhyolite and autobreccia probably belong to one or more domes and have a zircon fission-track age of 3.2 Ma. *Smith* [in press b] mapped part of a caldera margin enclosing the rhyolitic deposits.

Silicic volcanism ended about 3 Ma, and undated olivine basalt was locally erupted onto the rhyolitic rocks. Soon thereafter lava flows of high-K_2O andesite, dominantly pyroxene-phyric but including some flows with abundant hornblende, began to form the large composite Goat Rocks volcano. Eruptions probably took place between about 2.5 Ma and 0.6 Ma, as judged from zircon fission-track ages and magnetic polarity. Some large-volume lava flows entered paleovalleys and advanced many kilometers from their source. The most notable of these flows is 2-km^3-Tieton Andesite, which moved about 80 km eastward down the valleys of the ancestral Tieton and Naches Rivers. Thick sections of lava flows with radial dips of 10-20° surround the hydrothermally altered core of the volcano; dikes cut the flows and define several sectors of a radial swarm. The Cispus Pass pluton occupies the southern part of the altered core and may be as young as 1 Ma (based on one fission-track age).

After a period of erosion, hornblende andesite erupted from vents near the core of Goat Rocks volcano. The highest point in the Goat Rocks area, Gilbert Peak (2476 m), is capped with hornblende andesite. Old Snowy Mountain along the Cascade crest erupted lava flows of hornblende andesite that poured westward into the glaciated Cispus River valley. The hornblende andesite is glaciated and at least as old as late Pleistocene. Whether it represents rejuvenation of the Goat Rocks volcano or independent volcanism is unclear.

More than 200 olivine basalt and basaltic andesite lava flows formed the 700-m-high Hogback Mountain shield volcano just south of White Pass during the late Pliocene and early Pleistocene. The flows intertongue with those from Goat Rocks. A magnetic reversal high in the shield may mark the base of the Olduvai event (1.7-1.9 Ma [*Mankinen and Dalrymple*, 1979]).

As volcanism at Goat Rocks and Hogback Mountain waned in the early Pleistocene, activity shifted to the Tumac Plateau north of White Pass. Most of the plateau is composed of hornblende andesite and olivine-bearing basaltic andesite. At least five volcanoes on the plateau are of late Pleistocene age. The youngest, Tumac Mountain, is a basaltic shield that last erupted 20-30 ka according to tephra studies [*Clayton*, 1983].

Many Pliocene and Quaternary lava flows of olivine basalt erupted in a zone extending 25 km south from the Goat Rocks to Mount Adams [*Hildreth and Fierstein*, 1985]. One vent, the subglacial Walupt Lake volcano, may date from the Evans Creek maximum *ca.* 18-15 ka [*Hammond*, 1980; *Swanson and Clayton*, 1983].

INDIAN HEAVEN VOLCANIC FIELD

The Indian Heaven volcanic field (Figure 2), midway between Mount St. Helens and Mount Adams, is a Quaternary center, chiefly of basalt. Paul Hammond has worked here for many years but has published few detailed results [*Hammond et al.*, 1976; *Hammond and Korosec*, 1983; *Hammond*, 1984, 1987]. He graciously allowed some of his work to be summarized, mainly from an informal field guide he wrote in 1985.

About 60 eruptive centers lie on the 30-km-long, N10°E-trending, Indian Heaven fissure zone [*Hammond*, 1984; *Hammond et al.*, 1976]. The 600-km^2 field has a volume of about 100 km^3 and forms the western part of a 2000-km^2 Quaternary basalt field in the southern Washington Cascades, including the King Mountain fissure zone along which Mount Adams was built [*Hammond et al.*, 1976; *Hildreth et al.*, 1983].

All lava flows at Indian Heaven have normal magnetic polarity [*Hammond et al.*, 1976] and so are assumed to be younger than about 0.73 Ma, consistent with their morphology and geomorphic relations. Two K-Ar ages [*Hammond and Korosec*, 1983; *Hammond*, 1987] suggesting ages between 3 and 4 Ma are probably too old [*Korosec*, 1987a, b]; they are based on trace contents of radiogenic argon, and no evidence exists for a long hiatus or marked erosion to explain the lack of reversely magnetized flows in the section. The youngest eruption produced Big Lava Bed about 8200 ^{14}C years B.P.

The field is dominated by small shield volcanoes surmounted by cinder and spatter cones. Subglacial vents occur on the northwest flank of the field [*Hammond*, 1987]. The overall shape of the field is that of a north-south elongate shield, with basal diameters of 30 km and 10-15 km, whose center rises about 1 km above its base. *Hammond* [1984] noted that most of the flows erupted from the center of the field. Hence Indian Heaven can be interpreted as a large, complex shield volcano, fed by a central reservoir, that supports numerous flank vents.

Pahoehoe and a'a typify the basalt flows; some of the andesite is block lava. Tephra production was minimal, although the basaltic tephra accompanying extrusion of the Big Lava Bed flow has a volume of about 10^5 m^3. Big Lava Bed is known for its tubes and for the remarkable microtopography on its surface, which mostly results from inflation of the flow when its feeder tubes clogged.

The field is dominantly basaltic (Figure 11) [*Smith*, 1984; *Hammond*, 1984], but basaltic andesite and andesite also occur. Mann Butte, a rhyolite dome or plug east of the field, was once considered to be Pleistocene [*Hammond et al.*, 1976] but now is interpreted as Tertiary (P.E. Hammond, oral commun., 1986). About half of the basaltic andesite and andesite was erupted on the flank of the field, possibly from fractionated stranded magma bodies, and the rest in the central part of the field [*Hammond*, 1984]. Both low-K, olivine-normative tholeiitic basalt and high-Al, hypersthene-normative calc-alkaline basalt occur [*Smith*, 1984; *Hammond*, 1984], as well as transitional types. In this regard the basalt at Indian Heaven is representative of basalt elsewhere in the southern Washington Quaternary field [*Smith*, 1984].

The field occupies a 100-mgal gravity low [*Hammond et al.*, 1976; *Williams et al.*, 1988] that may reflect light Tertiary rocks below the field, a hydrothermally altered or fractured zone, or a shallow magma reservoir as yet undetected by other means.

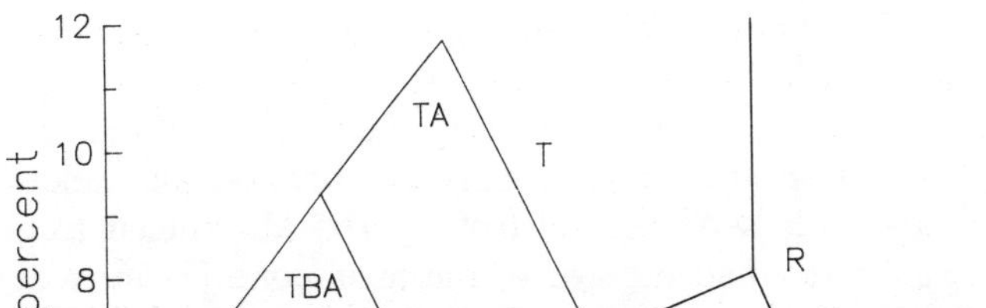

FIGURE 11. Total alkali-silica diagram for Indian Heaven. Unpublished analyses courtesy of P. E. Hammond; published analyses from *Smith* [1984]. Field names as in Figure 17; R, rhyolite; T, trachyte.

The elongate field follows the direction of regional extension and faulting as well as overall volcanism in the Cascades. A few north-trending normal faults are mapped just northeast and southwest of the field [*Walsh et al.*, 1987], chiefly on the basis of relations within the Tertiary section. Little is known of their ages, not surprising in view of the overall lack of detailed knowledge of the Tertiary rocks in the southern Washington Cascades.

MOUNT HOOD

Mount Hood (Figure 12), 3428 m high, is the fourth highest peak in the Cascades and the highest in Oregon. It was named after a British admiral and first described in 1792 by William Broughton, member of an expedition under command of Captain George Vancouver [*Broughton*, 1929]. The first geologic reconnaissance primarily described the existing glaciers [*Hague*, 1871]. A complete geologic study of Mount Hood is still wanting; no detailed geologic map exists, and few ages have been obtained on lava flows that form most of the cone.

Early History

The Mount Hood area has hosted volcanic activity since at least the middle Miocene. More than 400 m of locally derived intermediate and silicic volcaniclastic rocks (14-11-Ma Rhododendron Formation) interfinger with and overlie the Wanapum Basalt of the Columbia River Basalt Group [*Wise*, 1968, 1969; *Priest et al.*, 1982; *Keith et al.*, 1982, 1985]. The Rhododendron underlies flows of the 10.5-Ma Last Chance andesite of *Priest et al.* [1982]. Both units are cut by the Laurel Hill and Still Creek quartz diorite plutons and related offshoots, with K-Ar ages of *ca.* 9.3-8.5 Ma [*Bikerman*, 1970; *Priest et al.*, 1982; *Keith et al.*, 1985]. In the Pliocene local vents erupted thick flows of andesite and minor basalt that cap many ridges surrounding Mount Hood (Zigzag Mountain, Tom Dick and Harry Mountain) [*Wise*, 1969].

The late Pliocene Sandy Glacier volcano (Figure 12) produced basalt, basaltic andesite, and minor andesite near the site of Mount Hood [*Wise*, 1968, 1969]. More than 900 m of this composite cone are exposed on the west flank of Mount Hood under Sandy and Reid Glaciers and contain the most mafic rocks known on Mount Hood (Figures 13 and 14). The youngest exposed flows of the Sandy Glacier volcano have a K-Ar whole-rock age of about 1.3 Ma [*Keith et al.*, 1985]. Reversely magnetized andesite flows of approximately the same age [*Keith et al.*, 1985] occur in inverted topography on Vista Ridge and apparently came from a vent now buried under the north flank of Mount Hood. Aeromagnetic data indicate a reversely magnetized body, perhaps the source of the flows, beneath the northern flank [*Flanagan and Williams*, 1982; *Williams and Keith*, 1982].

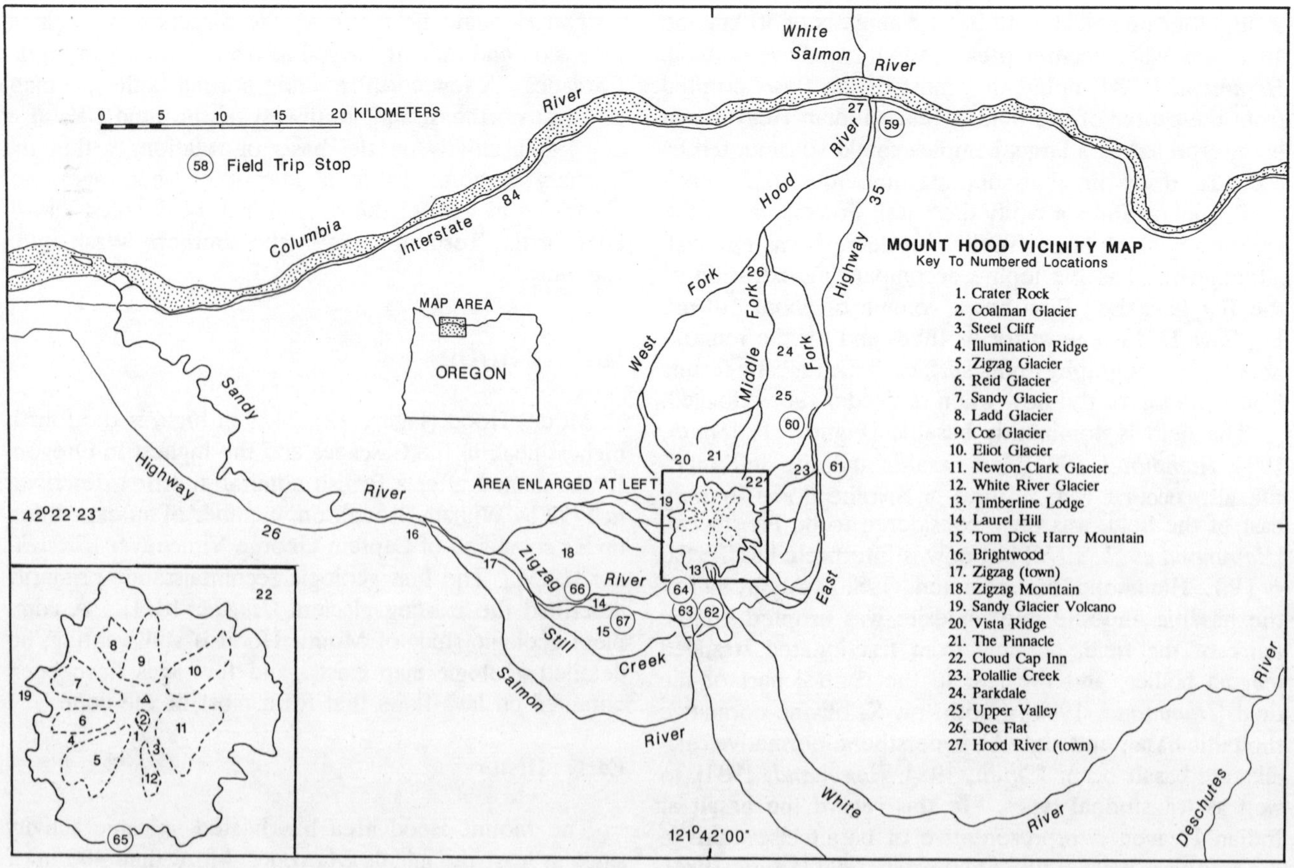

FIGURE 12. Map of Mount Hood area, showing localities mentioned in text and locations of stops.

Construction of the Cone

The age of the main edifice of Mount Hood is poorly known. All tested lava flows have normal magnetic polarity, so the cone is probably younger than 0.73 Ma. Whole-rock K-Ar ages for two flows and one dike of the main cone range from 0.6 to 0.3 Ma [*Keith et al.*, 1985]; the dated flows are from stratigraphically low parts of the section in the upper Zigzag River canyon.

The volcano comprises approximately 70% andesite flows and 30 percent clastic material (mostly concentrated high on the cone) [*Wise*, 1969]. Flows near the summit dip away from a source higher than, and north of, the present summit. Most of the flows are less than 3 m thick, but notable exceptions occur in Steele Cliff and Illumination Ridge, where lava apparently ponded to depths of over 100 m. Eruptions were relatively nonexplosive, and significant tephra deposits were limited to the flanks and a small area east of the mountain, where they rarely total more than 1 m thick.

Most of the cone-building flows are medium-K silicic andesites (Figure 14); a few others are mafic dacite. The cone-building flows include no basaltic andesite or basalt, in contrast to the older Sandy Glacier volcano. The cone-building flows are phyric, chiefly two-pyroxene andesite with lesser olivine andesite; hornblende is a disequilibrium phase in about half of the flows. Neither *Wise* [1969] nor *White* [1980] recognized an overall change with time in major-or trace-element compositions.

After construction of most of the cone, relatively small flows erupted from satellite vents on Vista Ridge and The Pinnacle (Figure 12). A flow from The Pinnacle has a K-Ar whole-rock age of 0.15 ± 0.02 Ma [*Keith et al.*, 1985]. *Wise* [1969 interpreted flows erupted near Cloud Cap Inn to be young, but one sample yielded whole-rock K-Ar ages of 0.49 to 0.65 Ma, similar to or even older than the age of the main cone [*Keith et al.*, 1985]. The satellite flows including Cloud Cap are more mafic than the andesite forming the main cone [*Wise*, 1969; *White*, 1980], comprising medium-K mafic andesites or basaltic andesites (Figures 13 and 14).

Between about 0.05 and 0.1 Ma (based on profiles of soil development), a sector collapse on the north or northeast side of the volcano formed a debris avalanche that traveled the length of Hood River, crossed the Columbia River, and moved 5 km up the White Salmon River–a distance of at least 40 km (Figure 12) [*Vallance*, 1986]. The avalanche deposits locally are more than 40 m thick; much of Hood River town is built on them. No source for the avalanche is evident, and no avalanche deposits crop out within 10 km of the volcano, because of a blanket of glacial outwash. This leaves as a mystery the exact point of origin of what is probably the largest single event to occur on Mount Hood.

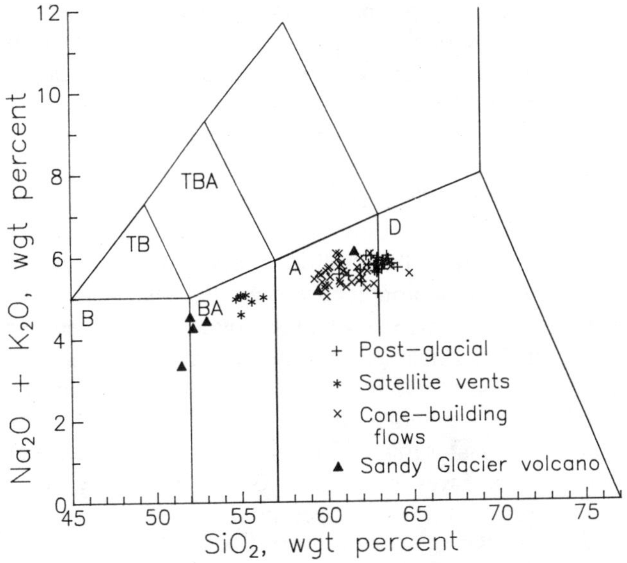

FIGURE 13. Total alkali-silica diagram for Mount Hood. Representative analyses from *Wise* [1969] and *White* [1980], recalculated water-free to 100%. Figure 17 gives field names.

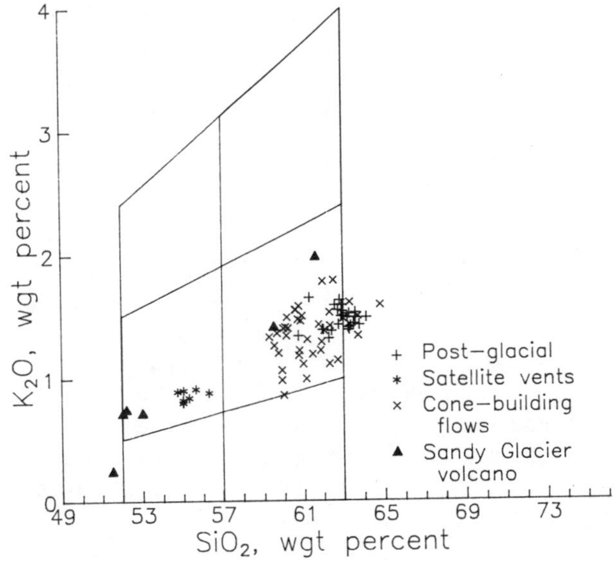

FIGURE 14. Andesite classification for Mount Hood, from *Gill* [1981] except that basalt-andesite division is 52%. Data sources as in Figure 13; fields as in Figure 7.

The 5-km-long Parkdale flow erupted in Upper Hood River Valley about 6 ka. It chemically resembles the basaltic andesite from The Pinnacles [*Wise*, 1969].

The latest addition to the cone was a composite hornblende dacite dome, Crater Rock, just south of the summit. *Wise* [1969] considered the dome as coeval with the main stage of cone building, but *Crandell* [1980] and *Cameron and Pringle* [1986] interpreted it to have formed 200-300 years ago.

The summit of Mount Hood is about 1 km south of the apex of a gravity high of at least 8 mGals. *Williams and Keith* [1982] interpret the high to reflect a dense intrusive body that fed Mount Hood and its forerunners.

Glaciation

Twelve glaciers and named snowfields cover approximately 80% of the cone above the 2100-m level and contain about 0.35 km^3 of ice [*Driedger and Kennard*, 1986]. Most of the glaciers have remained roughly constant in size over the last few decades, after retreating from a neo-glacial maximum early in the 18th century [*Lawrence*, 1948].

Modern glacier termini are at about 2100 m, but in the last major alpine glaciation (Fraser, about 29-10 ka) glaciers reached the 700-800 m level. During this time, ice spread 15 km from the summit area [*Crandell*, 1980].

Lacustrine siltstone from near-terminus periglacial lakes plaster valley walls just upstream from the mouth of Polallie Creek on the east side of the mountain. Highway 35 crosses White River near the maximum extent of Fraser ice, and the left-lateral moraine is prominent just upstream from the bridge. The full extent of the Fraser-age glaciers has not been accurately mapped.

Glacier retreat released large volumes of outwash, some of which filled the ancestral Hood River Valley near Parkdale, forming the flat surfaces of Upper Hood River Valley and Dee Flat. Outwash also formed a debris fan in the upper East Fork Hood River.

Evidence of older glaciation is seen in roadcuts on the southeast side of the volcano and in rolling morainal landscape near Brightwood west of the volcano. The deposits are not dated but may be coeval with the Hayden Creek Drift near Mount Rainier [*Crandell*, 1980], probably about 0.14 Ma [*Colman and Pierce*, 1981].

Postglacial Eruptive History

Mount Hood has had at least four eruptive periods in and after late Fraser time, in order of decreasing age: 1) Polallie (15-12 ka [*Crandell*, 1980]), 2) Timberline (1400 to 1800 years B.P. [*Crandell*, 1980; *Cameron and Pringle*, 1986]), 3) Zigzag (400 to 600 years B.P. [*Cameron and Pringle*, 1986]), and 4) Old Maid (170 to 220 years B.P. [*Cameron and Pringle*, 1987]). Work in progress will determine the details of these episodes in order to assess hazards of a future eruption.

The Polallie eruptive period occurred during the final stage of the Fraser Glaciation [*Crandell*, 1980]. Lahars, thin tephras, and pyroclastic flows intertongue with late Fraser-age outwash in Upper Hood River Valley. Elsewhere, Polallie deposits mantle ridge crests and valley walls but not valley floors. Probably glacial ice still occupied valley floors at the time of the eruptions. No radiometric ages have been obtained for the Polallie.

The Timberline eruptive period broke the apparent 10-ka-long post-Polallie quiescence. The vent shifted from its summit location during Polallie time to the high southwest flank. Erupted material except airfall tephra was consequently confined to the Sandy, Salmon, and

Zigzag River drainages, where it formed the broad, gently-sloping debris fan that dominates the southwest flank of the volcano. Pyroclastic flows dated at 1440 ± 155 years B.P. [*Cameron and Pringle*, 1986] moved at least 8 km down the Zigzag River, and lahars reached the mouth of the Sandy River more than 80 km from the volcano. Small debris fans formed in the canyons of the upper Salmon and Sandy Rivers. An upper age for the Timberline of 1830 ± 50 years B.P. was obtained from a pyroclastic-flow deposit near Zigzag (K. A. Cameron and P. T. Pringle, unpub. data).

The Zigzag eruptive period was apparently minor, feeding several lahars and related floods into the Zigzag River and one pyroclastic flow into the Sandy River. The pyroclastic flow has an age of 455 ± 135 years B.P., and trees buried by the first lahars are 550 ± 130 years old [*Cameron and Pringle*, 1986].

The Old Maid eruptive period apparently began with emplacement of the Crater Rock hornblende dacite dome high on the north flank of the cone. Reconnaissance work shows that the dome comprises three lobes of markedly differing internal structure [*Cameron and Pringle*, 1986] but does not clarify how each lobe relates to the others. Numerous lahars probably fed by avalanches from the dome and accompanying snowmelt entered the Sandy, Zigzag, Salmon, and White Rivers; a pyroclastic flow traveled from the Crater Rock area at least 9 km along the White River. One lahar extends 65 km along the White River, and the sandy run-out deposit from another is identifiable 80 km from the mountain. At least 60 m of lahar and fluvial deposits partly fill the upper White River canyon near Timberline Lodge. A terrace made of a lahar overlain by reworked eruptive debris is more than 13 m thick on the lower Sandy River, 60 km from the mountain. Dendrochronologic dating of some of these events [*Cameron and Pringle*, 1987] indicates that the Sandy River lahar occurred in the mid-1790s, the pyroclastic flow in the upper White River about 1800, and the lahar that traveled 80 km down the White River between 1800 and 1810.

The post-glacial products are dominantly mafic dacite and silicic medium-K andesite [*Wise*, 1969; *White*, 1980; *Crandell*, 1980], generally more silicic than the cone-building flows. *White* [1980, p. 5] claims that, within the post-glacial sequence, "a general trend can be seen in which rocks from the younger units are slightly richer in SiO_2 and poorer in MgO, CaO, and Fe_2O_3."

Historical Activity

No major eruptive events have occurred at Mount Hood since systematic records began in the 1820s. Reports of steam and tephra emissions accompanied by red glows or "flames" from the area of the post-glacial vent are known from 1859, 1865 (twice) and 1903 A.D. No tephra deposits have been correlated with these events, though the summit area is capped by stratified tephra and scattered pumice blocks and breadcrust bombs.

Present thermal activity is in fumarole fields near Crater Rock, at the apex of a semi-circular zone of fumaroles and hydrothermally-altered, heated ground. In summer 1987, maximum ground temperatures were near 85°C and maximum fumarole temperatures were about 92°C [*Cameron*, 1988], slightly above the boiling point of water at 3100 m. Many of the fumaroles are actively precipitating crystalline sulfur. Comparison of modern and historical photographs shows that the amount of perpetually snow-free ground surrounding the fumarole fields has been increasing since last century. Until the 1980 eruption of Mount St. Helens, the only volcanically related human fatality in the Cascades occurred in the thermal area at Mount Hood in 1934, when a climber exploring ice caves in Coalman Glacier melted by fumaroles suffocated in the oxygen-poor gas.

Jökulhlaups (glacial-outburst floods) have been recorded from the Zigzag, Ladd, Coe, and White River Glaciers. In 1922, a dark debris flow issued from a crevasse high on Zigzag Glacier and moved 650 m over the ice before entering another crevasse; this event initiated a scare that Mount Hood was erupting [*Conway*, 1921]. The Ladd Glacier jökulhlaup in 1961 destroyed sections of the road around the west side of the mountain and partly undermined a tower of a major powerline [*Birch*, 1961]. The Coe Glacier outburst occurred around 1963, causing a section of trail to be abandoned and the "round-the-mountain" trail to be rerouted farther from the glacier. Jökulhlaups from White River Glacier were reported in 1926, 1931, 1946, 1949, 1959, and 1968; the Highway 35 bridge over the White River was destroyed during each episode. The more frequent outbursts from White River Glacier may be due in part to an increase in size of the fumarole field at the head of the glacier at Crater Rock [*Cameron*, 1988].

A rainfall-induced debris flow on Polallie Creek on Christmas 1980 killed one and destroyed an 8-km section of Highway 35. The flow started as a moderate slope failure of only 3800 m^3 but rapidly bulked up and deposited over 76,000 m^3 of debris at the mouth of Polallie Creek [*Gallino and Pierson*, 1984]. The debris dammed the East Fork Hood River, creating a temporary lake; the dam breached, and flooding destroyed the highway.

Felt earthquakes occur on Mount Hood occur every 2 years on average. Seismic monitoring, in effect since 1977 [*Weaver et al.*, 1982], indicates a generalized concentration of earthquakes just south of the summit area and 2-7 km below sea level. A seismic swarm in July 1980, during which nearly 60 earthquakes (mostly 5-6 km deep with a maximum bodywave magnitude of 2.8) recorded in a 5-day period [*Rite and Iyer*, 1981], prompted development of an emergency response plan to coordinate local authorities in the event of future eruption.

Geodetic surveillance of the volcano was initiated in 1980, and 30 EDM lines and several tilt stations were resurveyed in 1983 and 1984 [*Chadwick et al.*, 1985;

Cascades Volcano Observatory, unpub. data]. Observed changes are within the range of expected error.

COLUMBIA RIVER BASALT GROUP

The Columbia River Basalt Group (CRBG) is the youngest and most studied flood basalt. The province underlain by the basalt is loosely termed the Columbia Plateau (Figure 1). Such an overall designation is a misnomer, however, for the basalt has been sharply folded and broadly warped, so that its top varies in elevation from slightly below sea level in the Pasco Basin to more than 2.5 km above sea level in the Wallowa Mountains of northeast Oregon. On this trip we see a small part of the western section of the plateau, as well as part of the route taken by flows as they followed the ancestral Columbia River across the Cascades.

Stratigraphy and age

The group is formally divided into five formations (Figure 15), which in turn are broken into formal and informal members [*Swanson and others*, 1979a; *Camp*, 1981; *Beeson et al.*, 1985; *Reidel et al.*, in press; *Bailey*, in press]. On this trip we will visit flows in the Grande Ronde and Wanapum Basalts only.

The group has a volume of about 174,000 km^3 and covers about 164,000 km^2 [*Tolan et al.*, in press]. These figures have been revised downward from previous estimates. It was erupted between 17.5 and 6 Ma, as measured by K-Ar and ^{40}Ar-^{39}Ar ages [*Long and Duncan*, 1983; *McKee et al.*, 1977, 1981; *Swanson et al.*, 1979a]. Early eruptions (17.5-17 Ma) fed the Imnaha Basalt, which is confined to the southeast part of the province [*Hooper et al.*, 1984]. Most of the group was formed during a 1.5-m.y. period between about 17 and 15.5 Ma (Figure 16), resulting in the Grande Ronde Basalt [*Mangan et al.*, 1986; *Reidel et al.*, in press] and the greatly subordinate and geographically limited Picture Gorge Basalt [*Waters*, 1961; *Bailey*, 1986]. Later eruptions formed the Wanapum Basalt (about 15.5-14.5 Ma) and the Saddle Mountains Basalt (about 14-6 Ma) [*Swanson et al.*, 1979a; *Camp*, 1981; *Beeson et al.*, 1985]. Relatively little erosion took place between flows, owing to the rapid rate of accumulation, except during Saddle Mountains time. However, a regionally extensive saprolite (fossil soil) or a sedimentary interbed separates the Grande Ronde and Wanapum in most places; flows just below and above the contact typically are normally magnetized, so that the time represented by the break is probably less than a few hundred thousand years, most likely less than 100,000 years. In Saddle Mountains time, however, interflow erosion was significant, and most contacts are erosional unconformities.

series	group	sub-group	formation	member	mag polarity
Miocene	Columbia River Basalt Group	Yakima Basalt Subgroup	Saddle Mountains Basalt	Lower Monumental Member	N
				Ice Harbor Member	N,R
				Buford Member	R
				Elephant Mountain Member	R,T
				Pomona Member	R
				Esquatzel Member	N
				Weissenfels Ridge Member	N
				Asotin Member	N
				Wilbur Creek Member	N
				Umatilla Member	N
			Wanapum Basalt	Priest Rapids Member	R3
				Roza Member	T,R
				Frenchman Springs Member	N2
				Eckler Mountain Member	N2
			Grande Ronde Basalt		N2
					R2
			Picture Gorge Basalt		N1
					R1
			Imnaha Basalt		R1
					T
					N0
					R0

FIGURE 15. Stratigraphic subdivision of Columbia River Basalt Group. N, normal magnetic polarity; R, reversed polarity; T, transitional polarity. Subscripts denote magnetostratigraphic units.

Chemical composition

All major flows within the Yakima Basalt Subgroup (focus of this trip) are tholeiitic, with MgO contents from 2.75 to 8.25% (Table 3; Figure 17). Specific chemical compositions characterize each formation (*Wright et al.*, in press]. Flows of the Grande Ronde Basalt are aphyric and have low MgO (3-6%), high SiO_2 (52-58%), relatively low TiO_2 (1.5-2.5%) and FeO (9-13%), and a

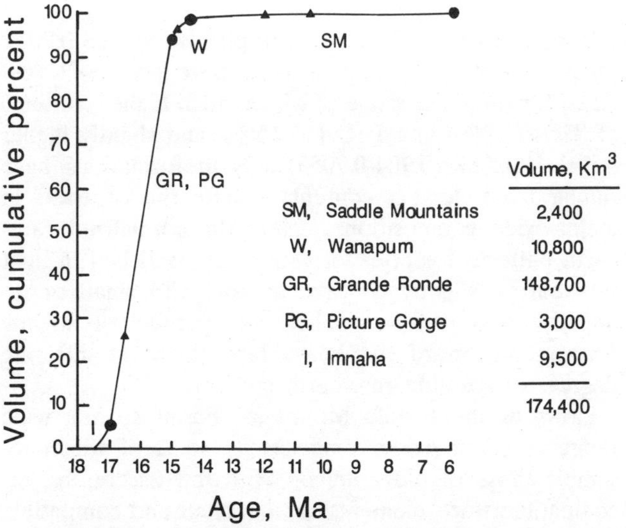

FIGURE 16. Cumulative volume erupted vs. age for units of Columbia River Basalt Group. Dots separate formations; solid triangles are selected radiometric dates. Data from *Tolan et al.* [in press].

TABLE 3. Representative Major-Oxide (%) and Trace-Element (ppm) Analyses of Grande Ronde, Wanapum, and Saddle Mountains Basalts [*Wright et al.*, in press]

	1	2	3	4	5	6	7	8	9	10	11	12	13	14	15	16
SiO_2	53.84	55.69	49.90	52.37	51.70	50.94	49.95	50.12	54.76	54.34	50.87	54.26	52.00	51.27	47.51	50.20
Al_2O_3	14.37	13.78	16.82	15.47	13.48	14.27	13.44	14.17	14.24	14.49	16.05	13.85	15.04	13.28	13.79	14.17
"FeO"	11.37	11.85	10.16	10.78	14.35	13.50	14.78	13.93	12.29	11.18	9.54	12.65	10.45	14.79	15.03	14.21
MgO	5.25	3.51	8.21	5.91	4.44	4.57	4.50	5.16	2.75	4.51	8.25	3.98	7.19	4.18	6.05	4.89
CaO	8.97	7.11	10.83	9.90	8.08	8.56	8.67	8.82	6.21	8.42	10.79	7.40	10.39	8.35	9.75	8.63
Na_2O	2.92	3.20	2.43	2.91	2.74	2.85	2.80	2.59	3.39	2.65	2.29	2.58	2.23	2.44	2.28	2.69
K_2O	1.10	2.01	0.37	0.68	1.36	1.25	1.21	1.05	2.56	1.76	0.39	1.71	0.65	1.25	0.78	1.44
TiO_2	1.75	2.23	0.97	1.46	3.02	3.12	3.60	3.23	2.79	1.90	1.48	2.92	1.62	3.61	3.70	2.93
P_2O_5	0.23	0.43	0.17	0.36	0.61	0.68	0.84	0.73	0.82	0.51	0.19	0.44	0.24	0.60	0.87	0.64
MnO	0.19	0.20	0.14	0.15	0.22	0.25	0.20	0.19	0.19	0.23	0.14	0.22	0.18	0.23	0.24	0.20
Ba	472	730	175	348	549	556	600	539	3313	876	303	592	243	485	611	541
Ce	38.0	54.2	16.3	32.3	51.0	54.2	62.7	57.4	88.2	80.5	31.5	73.8	35.3	68.7	85.6	67.8
Cr	54.9	12.5	151.4	159.7	39.5	44.3	13.6	82.0	3.6	36.6	283.7	18.9	109.0	18.5	148.8	24.0
La	19.5	27.6	7.6	15.2	25.4	27.0	31.1	28.0	46.7	43.7	15.7	38.2	17.0	33.7	42.9	35.3
Nb	11.5	13.0	6.0	8.7	13.0	17.0	17.0	15.8	21.3	16.7	13.0	21.8	13.0	24.5	23.7	25.7
Ni	15.5	--	117.0	42.7	39.3	33.6	27.0	35.0	2.7	47.0	203.3	16.0	51.3	30.7	42.0	28.5
Rb	30.5	45.4	--	15.8	35.4	31.4	34.6	26.6	47.0	38.2	10.0	47.3	10.0	27.6	19.8	18.0
Sm	5.6	7.3	2.7	4.2	6.1	6.9	7.6	7.9	9.8	7.8	4.3	8.4	4.7	9.4	10.7	7.1
Sr	335	359	353	397	317	313	286	302	276	277	252	257	227	213	227	350
Th	3.4	6.38	0.37	1.16	3.88	3.98	4.04	3.56	7.25	6.65	2.05	8.67	2.55	5.91	2.19	4.98
Yb	3.25	3.13	2.10	3.08	3.52	3.89	4.62	3.92	4.52	4.23	2.37	3.65	2.73	4.58	5.64	3.13
Zr	135	178	96	136	196	223	238	223	520	282	--	240	--	316	293	350

Grande Ronde Basalt:
1, High MgO
2, Low MgO
Wanapum Basalt:
3, Eckler Mountain Member, Robinette Mountain flow
4, Eckler Mountain Member, Dodge flow
5, Frenchman Springs Member
6, Roza Member
7, Priest Rapids Member, Rosalia type
8, Priest Rapids Member, Lolo type

Saddle Mountains Basalt:
9, Umatilla Member
10, Wilbur Creek Member
11, Asotin Member
12, Esquatzel Member
13, Pomona Member
14, Elephant Mountain Member
15, Ice Harbor Member, Basin City flow
16, Lower Monumental Member

relatively low initial $^{87}Sr/^{86}Sr$ isotopic ratio (0.703-0.705). Most flows of the Wanapum Basalt have lower SiO_2 (49-52%) for the same range of MgO, much higher contents of TiO_2 (>3%) and FeO (13-15%), and slightly higher initial $^{87}Sr/^{86}Sr$ (0.704-0.7055). Both formations have similar trace-element contents despite the contrasts in major-oxide composition. Chondrite-normalized rare-earth patterns for both formations are enriched in light rare earths, slightly concave upward with small or no negative europium anomalies, and parallel or slightly convergent toward the heavy rare earths at differing degrees of absolute rare-earth content.

Flows of the Saddle Mountains Basalt span a wide range of major oxide compositions, some of which resemble those of older units. With few exceptions, incompatible trace elements are enriched and compatible elements (particularly Sr and Sc) depleted relative to older units, and $^{87}Sr/^{86}Sr$ ratios are higher (>0.707). Chondrite-normalized rare-earth patterns for many units are steeper, with a more marked negative europium anomaly, than are those of the older formations. The Umatilla is the only member in the Saddle Mountains with a large positive europium anomaly.

Vent Systems

Linear vent systems occur only in the eastern half of the province, except for feeders of the Picture Gorge (the Monument dike swarm) near the southern limit of the province (Figure 1) [*Waters*, 1961; *Swanson et al.*, 1975, 1979a; *Tolan et al.*, in press]. Some vent systems are longer than 150 km, and all trend within a few degrees of due north, mostly north-northwest. The systems are correlated with specific stratigraphic units chiefly by the presence of dikes of appropriate chemical composition, petrography, magnetic polarity, and stratigraphic position. Hundreds of dikes have been identified, although many probably represent *en echelon* segments of one vent system [*Waters*, 1961; *Taubeneck*, 1970]. Most dikes are known from Chief Joseph dike swarm in the tri-state area of Washington, Oregon and Idaho [*Taubeneck*, 1970]. Distribution patterns for some flows of the Grande Ronde Basalt suggest that their feeder dikes are hidden beneath younger flows north of the Chief Joseph swarm [*Reidel et al.*, in press]. The dikes typically are a few meters wide, but some are wider than 20 m. Composite dikes, consisting of basalt of two or more different compositions and hence ages, have not been found, but multiple dikes, with internal jointing patterns suggestive of two or more pulses during the same intrusive event, are common. Remnants of spatter cones and other near-vent features, similar to those of modern Kilauea, occur locally [*Swanson et al.*, 1975]. The degree of vesicularity and breakage of pyroclasts in these deposits resembles that of modern basaltic tephra, so it is unlikely that the magma was unusually rich in gas.

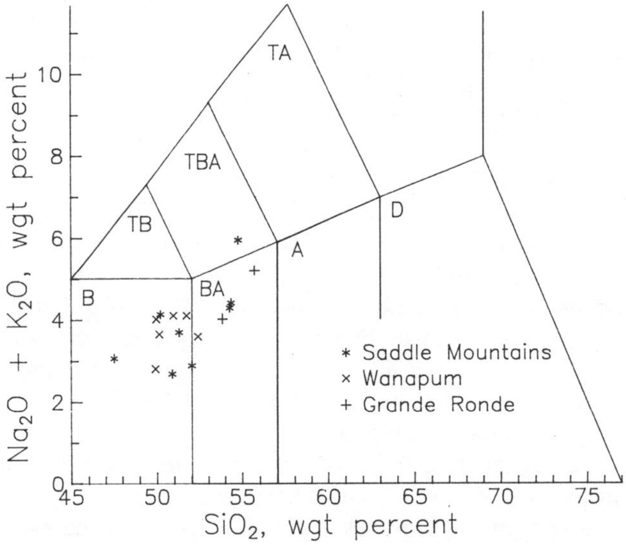

FIGURE 17. Total alkali-silica diagram for Yakima Basalt Subgroup. Selected analyses from *Wright et al.* [in press]. Rock types and fields from *Le Bas et al.* [1986]. B, basalt; BA, basaltic andesite; A, andesite; D, dacite; TB, trachybasalt; TBA, trachybasaltic andesite; TA, trachyandesite.

Volumes and Eruption Rates

Flood-basalt provinces by definition contain flows of huge volume, on the order of 5-10 km^3 or more; this is the major distinction between flood-basalt and plains-basalt provinces [*Greeley*, 1977], such as the Snake River Plain and Iceland, in which flows are generally much less than 1 km^3. Recent work by *Tolan et al.* [1987, in press] suggests that about 300 *great* flows were erupted on the Columbia Plateau, with an average volume per flow of about 580 km^3. Numerous low-volume flows are probably present near vents, but the great flows form most of the province. Flows with volumes more than 100 km^3 occur in all of the formations, but most such flows were erupted during Grande Ronde time, when at least 110 major flows with volumes of 90 km^3 to more than 5000 km^3 were produced [*Reidel et al.*, in press; *Tolan et al.*, 1987, in press]. Such a huge eruption, if from a shallow crustal source, would probably have resulted in recognizable subsidence along the trace of its fissure system. No such evidence has been found, so the inference on geologic grounds alone is that the eruptions tapped a deep reservoir. Viewed regionally, the overall saucer-like subsidence of the province might reflect withdrawal of magma, but its amplitude and diameter would demand a deep source. Petrologic evidence [*Helz*, 1978; *Hooper*, 1984; *Wright et al.*, in press] supports a deep storage reservoir, perhaps at the base of the lithosphere.

Model calculations [*Shaw and Swanson*, 1970], based on evidence that little cooling occurred during flowage of hundreds of kilometers, suggest that eruption and emplacement spanned only a few days for these huge outpourings. The flows evidently moved 5-15 km/hour or faster [*Shaw and Swanson*, 1970] and advanced as sheet floods; no lava tubes have been found. The evidence for little cooling during transport is that flows quenched to glass when they entered water after traveling several hundred kilometers; the crystal content of the glass is no higher than that of chilled margins of feeder dikes. The effective viscosity of the lava was *not* unusually low; in fact, calculations of the viscosity based on the chemical composition for a range of reasonable water contents indicate that the lava was somewhat more viscous than modern Kilauea lava. According to *Shaw and Swanson* [1970], the high rate of eruption (about 1 km^3/day/linear kilometer of fissure or higher, 3-4 orders of magnitude faster than rates of Hawaiian and Iceland eruptions [*Swanson et al.*, 1975]) combined with the huge volume of available magma enabled the flows to travel so far, a relation that *Walker* [1973] found on empirical grounds (Figure 18). Regional mapping [*Swanson et al.*, 1979b, 1980, 1981] indicates that the flows ponded against opposed slopes, including locally recognizable natural levees, and formed low-aspect-ratio (0.0002-0.0001) lava lakes generally 30-40 m thick and 200-400 km in diame-

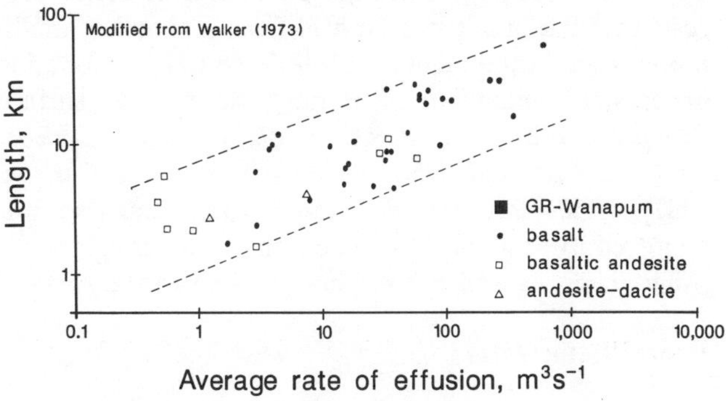

FIGURE 18. Relation of length and rate of effusion for representative lava flows of Grande Ronde and Wanapum Basalts and for historical flows throughout world. Modified from *Walker* [1973]. Rates were observed except those for Grande Ronde and Wanapum, which were calculated using model of *Shaw and Swanson* [1970].

ter. The lakes cooled to ambient temperatures within a few years to a few tens of years [*Long and Wood*, 1986].

Average Magma Supply Rate

About 5.5% of the group was erupted during the first 0.5 m.y. of volcanism to form the Imnaha Basalt [*Tolan et al.*, in press]. Activity peaked during 1.5-m.y. Grande Ronde time, when 87% of the group was erupted, 85% as the Grande Ronde Basalt and 2% as the Picture Gorge Basalt and other units. Activity waned thereafter, accounting for 6% of the group during Wanapum time, which lasted about 1 m.y., and 1.5% during prolonged Saddle Mountains time from about 14 to 6 Ma.

During peak activity the average interval between major eruptions was about 13,500 years, the average volume per major flow was about 1350 km^3, and the average magma supply rate was 0.1 km^3/year. The average supply rate is identical to that calculated for historical time at Kilauea [*Swanson*, 1972; *Dzurisin et al.*, 1984]. On this basis, a larger heat source for Grande Ronde Basalt than for modern Kilauea is unnecessary. An important distinction between the two provinces is that magma "leaks" from Kilauea almost continuously, whereas magma for the CRBG, if produced at the Kilauea rate, must have been stored for thousands of years before eruption in order to account for the huge volumes of single flows. (*Baksi* [1988] believes the maximum rate of supply, in late Grande Ronde time, was 2-3 times that at Kilauea; this interpretation is based on his revision of the geomagnetic time scale and seems overly dependent on his model for the time scale.)

The difference in eruption style between the CRBG and Kilauea may result from the 40-km-thick continental crust [*Catchings and Mooney*, 1988; *Michaelson and Weaver*, 1986] above the melting zone for the CRBG compared to the 12-km-thick oceanic crust above the melting zone for Kilauea. The continental crust may have provided a density trap keeping magma from rising until the trap was broken tectonically. Petrologic evidence suggests that the basalt was derived from a complex, multisource zone in the mantle with little crystal fractionation or crustal contamination during rise to the surface [*Hooper*, 1982, 1984; *Wright et al.*, in press; *Carlson*, 1984; *Carlson and Hart*, in press; *Hart and Carlson*, 1987]. This interpretation implies that the rise was rapid, consistent with the idea that fracturing, not a rising diapir, caused the magma to move to the surface.

Coeval Deformation

The basalt was erupted into a subsiding basin centered north of Pasco, Washington, in which sediment was accumulating [*Reidel*, 1984; *Reidel et al.*, in press]. How much of the subsidence owes itself to isostatic response caused by magma withdrawal, to weighting of the crust, and to tectonism is unclear [*Waitt and Swanson*, in press; *Catchings and Mooney*, 1988]. Seismic-refraction data, in concert with other geophysical data, suggest that rifting beneath the central part of the province occurred in the Eocene and possibly later [*Catchings and Mooney*, 1988]. If part of the Miocene subsidence was caused by magma withdrawal, significant lateral transport of magma from beneath the center of the province to its eastern flank—the locus of venting—is required.

Tectonic deformation was clearly taking place as the basalt accumulated, because single flows change thickness in anticlines and synclines [*Reidel*, 1984; *Reidel et al.*, in press]. The fold pattern is complex, and no fully satisfactory model for its development exists [*Laubscher*, 1981; *Davis*, 1981; *Barrash et al.*, 1983; *Hart and Carlson*, 1987]. Most folds trend within 20° of east, but some of the largest structures, such as the Pasco Basin and the Naneum Ridge anticline-Hog Ranch uplift, trend nearly north, as does the Cascade Range, whose uplift raised the basalt as much as 2 km. Most of the folds are confined to the western half of the province, but the longest structure, the Blue Mountains uplift, bounds the Columbia Plateau along most of its southern margin.

Ongoing folding, coupled with Cascade uplift and broad subsidence centered on the Pasco Basin, markedly influenced drainage patterns, particularly during Saddle Mountains time when quiet periods were long and canyons had time to form [*Swanson and Wright*, 1979; *Fecht et al.*, 1987; *Waitt and Swanson*, 1987, in press]. This topography in turn controlled the distribution of many flows of the Saddle Mountains Basalt. The most notable example is the Pomona Member, which erupted east of Lewiston, Idaho, flowed down the ancestral Clearwater and lower Snake Rivers, spread across the sagging central plateau, and left the plateau via structurally controlled channels that led through the Cascade Range and finally to the Pacific Ocean [*Swanson et al.*, 1979a; *Anderson*, 1980; *Camp*, 1981; *Tolan and Beeson*, 1984c; *Anderson and Vogt*, 1987; *Wells and Simpson*, 1987].

Invasive Flows and Sedimentary Interbeds

Sediment deposited between eruptions was commonly invaded by the next basalt flow that entered the area [*Schmincke*, 1967c; *Byerly and Swanson*, 1978, 1987; *Swanson et al.*, 1979a; *Niem and Niem*, 1985; *Wells and Niem*, 1987], producing pépérites and sill-like bodies called *invasive flows* that closely resemble classic intrusions. The only way to distinguish invasive flows from intrusions in such a setting is to know the stratigraphy and map the relations between subaerial and invasive components of the flow.

The sediment between the basalt flows has four general sources: erosion of highlands along the margin of the province, erosion of basalt within the province, deposition of diatomite in lakes within the province, and explosive volcanism in the adjacent Cascade Range.

Evidence for coeval Cascade volcanism is clear. Primary pyroclasts, generally mixed with epiclastic detritus but forming pure tuff in places, occur in interbeds in the section of basalt [*Schmincke*, 1964, 1967d; *Swanson*, 1967; *Priest et al.*, 1982]. However, no lava flows from the Cascades have been shown to be interbedded with the Grande Ronde or Wanapum Basalt; drilling in these formations just west of Mount Hood, Oregon, encountered andesite once thought to be flows [*Priest et al.*, 1982] but now interpreted by some as intrusions [D. R. Sherrod and T. L. Tolan, oral commun., 1988]. Apparently active vents in the Cascades were beyond the limit of the basalt flows, so that only tephra and laharic debris reached them. Cascade-derived debris in interbeds of Grande Ronde and Wanapum age is so sparse relative to that in beds of Saddle Mountains age and younger [*Schmincke*, 1964, 1967d; *Smith*, 1988; *Smith et al.*, 1988] that some workers [e.g. *Smith*, in press a; *Sherrod and Smith*, in press] suggest a downturn in Cascade activity followed by an upsurge during Saddle Mountains time. This idea is reasonable, although the short span of Grande Ronde and Wanapum time would by itself have resulted in minimal interbedding, all else being equal.

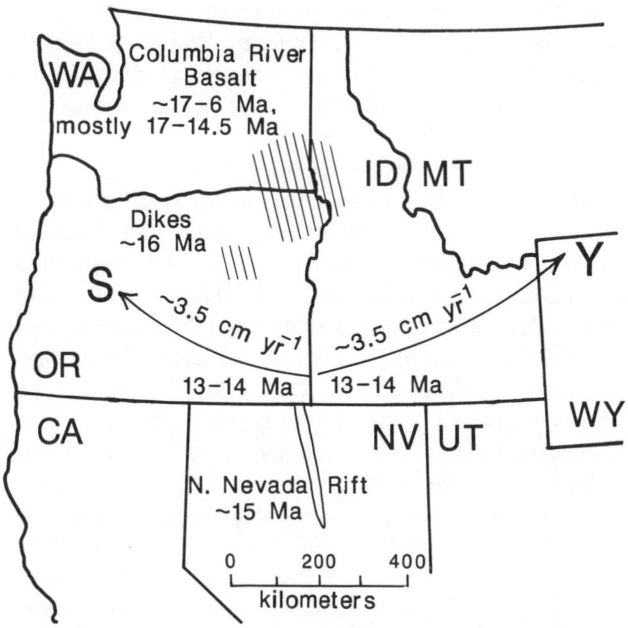

FIGURE 19. Map showing spatial and temporal relations of dikes of Columbia River Basalt Group, Northern Nevada Rift, and direction, onset time, and rates of migrating silicic volcanism toward Quaternary vents at South Sister, Oregon (S) and Yellowstone (Y). See text for discussion.

Speculations on Origin

Rampino and Stothers [1988] and *Alt et al.* [1988] recently suggested that flood-basalt provinces are generated by boloid impacts. The CRBG was probably not so generated. Geophysical work and deep drilling show no evidence of such an impact; in fact, the rift-like topography indicated by seismic refraction [*Catchings and Mooney*, 1988] beneath the center of the plateau would be nearly impossible to duplicate by impact. Moreover, calc-alkaline volcanism in the Cascades continued during flood-basalt volcanism but did not upsurge when basaltic eruptions began. One might expect that activity in the nearby Cascades would have followed suit if impact had triggered the basalt. Apparently the flood-basalt volcanism was independent of calc-alkaline volcanism, and neither was caused by impact.

Morgan [1981, 1988] suggested that new hot spots (mantle plumes) give rise to flood basalts, and *Duncan* [1982] argued that the CRBG is the product of the Yellowstone hot spot. At least two major problems exist with this idea. First, no fossil hot-spot track is preserved in the province, despite 11 m.y. of volcanism (17-6 Ma). Instead, feeder dikes are distributed randomly across the eastern half of the province; for example, the 8.5-Ma feeders for the Ice Harbor Member are on the east flank of the Pasco Basin [*Swanson et al.*, 1979a], the 12-Ma feeder for the Pomona Member is more than 210 km *east* of there (in the opposite direction from that expected of a hot-spot trace) [*Camp*, 1981], and the other dikes lie between. Second, the feeder system is only the northernmost part of a complex zone of basaltic dikes extending onto the Columbia Plateau from the [*Zoback and Thompson*, 1978; *Christiansen and McKee*, 1978]. This zone is not only a locus of basaltic magmatism but also a "plane of symmetry" separating traces of silicic volcanism migrating west-northwest across eastern Oregon and east-northeast across Idaho [*Armstrong et al.*, 1975; *MacLeod et al.*, 1975; *Eaton et al.*, 1978]. The basaltic activity along the zone began about 17 Ma, and the initiation of the migrating silicic volcanism about 13-14 Ma. The Yellowstone hot spot cannot easily explain these observations.

Lachenbruch and Sass [1977] and *Christiansen and McKee* [1978] pointed out that the high heat flow of the Basin and Range province requires the presence of a huge volume of Neogene basaltic magma, possibly equivalent to that produced on the Columbia Plateau, in the form of cooling intrusions in or underplated [*Furlong and Fountain*, 1986] on the base of the crust. In this context, *Christiansen and McKee* [1978] (R.L. Christiansen, numerous oral communications) suggested that stress relief at the base of the lithosphere, in response to crustal extension resulting from plate interactions, generated the voluminous basaltic magma. In this hypothesis, extension was greatest in the Basin and Range, so that magma could be stored in crustal reservoirs that developed in response to the extension, and least on the Columbia Plateau, so that crustal reservoirs were not formed and most of the magma that intruded the crust continued through it to the surface and fed the flood-basalt eruptions. This idea does not explain the bilaterally symmetrical nature of migrating silicic volcanism (no published hypothesis does, although the traction concept of *Eaton et al.* [1978] is stimulating) but does link the middle Miocene basaltic magmatism throughout the zone to plate interaction and as such is more appealing than either the hot spot or boloid-impact models.

Wright et al. [in press], building on work of *Carlson* [1984], conclude that the source for the basalt is a mixed oceanic-continental mantle that includes depleted oceanic mantle, subducted oceanic crust, original continental mantle, and old continental crust. They postulate that large bodies of magma generated from this material were underplated as sill-like bodies at the base of the continental crust and then tapped during deep continental extension or rifting. All of the magmas are inferred to have equilibrated with clinopyroxene during either melting or storage, but little evidence exists for fractionation of olivine, augite, and plagioclase as the petrogenetic models of *Cox* [1980] and *Hooper* [1984] demand. *Wright et al.* [in press] interpret the Grande Ronde as derived from relatively young oceanic mantle contaminated with subducted oceanic sediment, possibly with minor interaction with lower crustal rocks prior to or during ascent. They interpret the Wanapum to have much the same origin, except from a part of the oceanic mantle that had undergone Fe-Ti-P metasomatism. They interpret the Saddle Mountains to have a source related in some way to those for the Grande Ronde and Wanapum, owing to the similarity in major-element chemistry. However, the different trace-element abundances and ratios require additional complexity such as a second metasomatic event or melting of 2.5-Ga continental mantle with relatively high incompatible element contents.

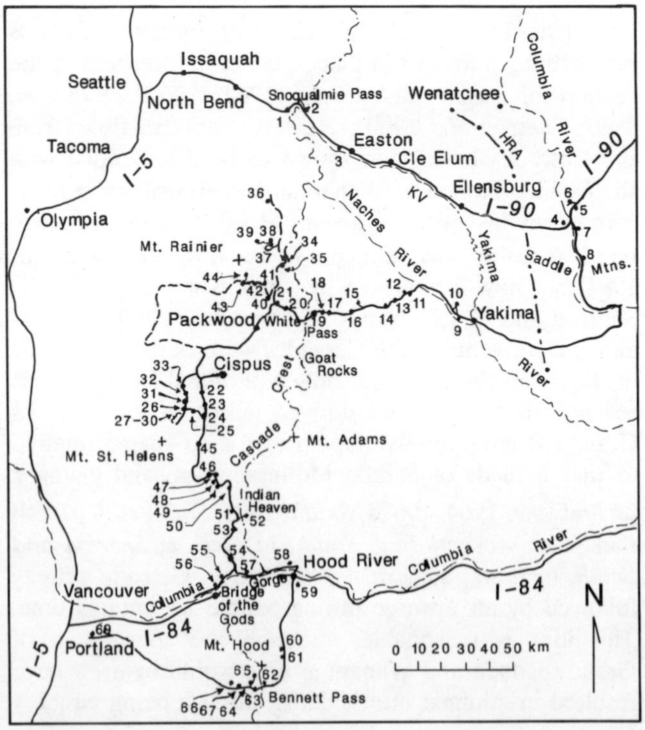

FIGURE 20. Map showing route of field trip between Issaquah, Washington, and Portland, Oregon. Numbers are stops. KV, Kittitas Valley; HRA, Hog Ranch Axis.

ROAD LOGS

These logs are planned for use with a large bus, but we designed optional stops and routes for use by vans and automobiles. We list too many stops (Figure 20) for a bus load each day, but vans and passenger vehicles should have little trouble in visiting all of them.

Each day has a specific emphasis. The first stresses the Columbia River Basalt Group but also introduces the middle Tertiary geology of the Cascades. This day involves considerable driving on high-speed roads and ends in Yakima. The second day crosses a section of Oligocene to Quaternary volcanic rocks on the east flank of the range west of Yakima, ending in Packwood. Two options exist for the third day, each starting in Packwood and ending at Cispus Center. The first takes in the Tertiary and Quaternary volcanic geology of Mount Rainier National Park. Snow is a likely problem through July 4 for this trip. If so, the third day will involve a trip to the Mount St. Helens area, stressing the Tertiary geology as well as the May 18, 1980, eruption. The fourth day has two options dependent on the third day, both again ending at Cispus Center. If we visit Mount Rainier on the third day, the fourth day will be the trip to the Mount St. Helens area. Otherwise the fourth day will be devoted to hiking trips around Mount St. Helens, including an arduous trip into the crater and a less strenuous excursion onto the Pumice Plain. No guide has been written for this trip, because the area is not open to the public and we do not want to encourage self-guided tours, but we will supply handouts. The fifth day involves a long drive on back-country roads to see Quaternary basalt of the Indian Heaven area, including subglacial basalt, and some of the geology of the Columbia Gorge, ending in Hood River, Oregon. The sixth day emphasizes the Quaternary volcanic geology of Mount Hood and ends in Portland.

Cumulative mileage is listed for each day of the trip, and interval mileages are given in bold face at the end of each entry. Numerous mileage check points are available. I indicates Interstate Highway, as I-90.

Log for Day 1, Issaquah to Yakima via Vantage

Drive east on Interstate 90 from Seattle to Issaquah.

0.0 Continue on I-90 at Exit 17 to Issaquah.

0.7 Climb onto kame terrace formed about 12 ka by the Puget Lobe during the Vashon stade of the Fraser glaciation. Large gravel pit visible to left just before start of log is in this terrace. **0.7**

1.7 Hills are underlain by the Puget Group. **1.0**

3.1 High Point off-ramp. Tukwila Formation, andesitic

mudflow breccias of middle Eocene age, in cuts to left. The Tukwila is probably correlative with the Naches Formation, about 44-40 Ma [*Tabor et al.*, 1984]. **1.4**

4.6 Tukwila Formation in roadcut to left. **1.5**

9.5 Exposures of steeply dipping arkose and volcaniclastic rocks of the Eocene Puget Group. The Puget is moderately folded in most places. It crops out discontinuously from here to North Bend. **4.9**

11.6 Craggy peak to left is Mount Si, underlain by Jurassic gabbroic and diabasic greenstone, possibly a dismembered ophiolite, and its sedimentary cover in the western melange belt of *Frizzell et al.* [1984]. The belt contains mostly Jurassic and Lower Cretaceous rocks and forms the west flank of the Cascades north of here. **2.1**

16.6 Rugged ridge crest at 11 o'clock is underlain by the Snoqualmie batholith. **5.0**

17.6 Low ridge ahead is a kame terrace formed by the Puget Lobe about 12 ka. Such features were built as deltas formed in lakes when the Puget Lobe blocked drainage from the Cascades. **1.0**

19.6 Range front of the Cascades, here mostly underlain by the volcanic rocks of Huckleberry Mountain [*Frizzell et al.*, 1984], correlative with the Ohanapecosh Formation seen later in the trip. **2.0**

23.1 Gabbro related to Snoqualmie batholith in roadcut on left. From here road stays in the batholith for 9-10 miles and enters the area of Figure 5 near Stop 1. **3.5**

31.8 STOP 1. Pull off on shoulder of freeway. If this is deemed too dangerous, vehicle can continue for 0.4 miles to a small road where parking is available, and geologists can walk back to stop. In roadcut view granodiorite of the Snoqualmie batholith, cut by aplite veins and containing mafic enclaves. The Snoqualmie is a large composite intrusive body and one of a series of shallow mid-Tertiary plutons in a narrow north-south belt in the Cascades. Radiometric ages are consistently between 20 Ma and 17 Ma [*Frizzell et al.*, 1984], mostly about 19 Ma, in this area, which is the main outcrop area of the batholith. The eastern part of the batholith, where individual plutons such as the Grotto have been delineated, is older, about 25 Ma. The batholith consists mainly of hornblende-biotite granodiorite and tonalite but includes gabbro, diorite, monzonite, and granite.

Sandstone float from the Naches Formation occurs along the road. **8.7**

32.8 Bare Snoqualmie Mountain in distance, 1890 m elevation. Contact with hornfelsed basalt of the Naches Formation in roadcut nearby. **1.0**

34.9 Roadcuts in argillite and sandstone of the Guye Sedimentary Member of the Naches Formation [*Tabor et al.*, 1984], metamorphosed by the batholith. **2.1**

35.5 Summit of Snoqualmie Pass, 921 m, the lowest pass across the Washington Cascades owing to the presence of non-resistent argillite in the Guye. **0.6**

37.6 Leave freeway at Hyak-Rocky Run exit. **2.1**

37.8 Turn left at stop sign and go under freeway. **0.2**

38.1 Turn right on pavement toward Rocky Run. **0.3**

38.5 STOP 2. Bus can stop on road, which has little traffic. View the Mount Catherine Rhyolite Member of the Naches Formation [*Tabor et al.*, 1984], a welded ash-flow tuff. Two fission-track ages of about 35 Ma for the Mount Catherine have probably been reset by the Snoqualmie batholith from true ages of about 44-40 Ma [*Tabor et al.*, 1984]. Note steeply dipping (about 65° SE), vague foliation in the welded tuff. **0.4**

38.7 Bus turnaround. High country to left up Gold Creek is underlain chiefly by basalt and feldspathic sandstone of the Naches Formation. **0.2**

Return to freeway and turn east toward Spokane.

39.9 Lake Keechelus on right is a natural lake further dammed for irrigation use. Roadcuts expose sandstone of the Naches Formation to mileage 42.6. **1.2**

42.6 Start of series of cuts in massive dacitic ash-flow tuff of Lake Keechelus [*Frizzell et al.*, 1984]. Zircon fission-track ages are 32-30 Ma, equivalent to the age of the Ohanapecosh Formation farther south, where similar tuff is interbedded with the Ohanapecosh. The tuff is quartz phyric and contains numerous large pumice lapilli and lithic inclusions. Columnar jointed in places, and commonly thoroughly welded and lithophysal. The tuff has a relatively gentle dip and rests on the eroded surface of the steeply dipping Naches Formation. Forms both north and south shores of Lake Keechelus. **2.7**

44.4 Last roadcut in the tuff of Lake Keechelus. **1.8**

45.9 Roadcuts in an ash-flow tuff in the Naches Formation dipping about 45 degrees. The freeway stays in the Naches Formation until it crosses the Straight Creek fault. The Naches consists chiefly of interbedded sandstone and a variety of volcanic rocks, chiefly basalt, basaltic andesite, and rhyolite, forming a more or less bimodal suite. It is about 4 km thick, comparable to the Puget Group farther west, and was apparently deposited in a subsiding extensional basin. The Naches is highly deformed; near-vertical dips are common in this area. It comprises most of the Cabin Creek block of *Tabor et al.* [1984] and *Frizzell et al.* [1984], a structural block west of the Straight Creek fault. The freeway follows the southeast strike of the formation. **1.5**

54.7 Basalt flows of the 48 to 47- Ma Teanaway Formation [*Tabor et al.*, 1984] occur at 10 o'clock on a ridge in the Teanaway River structural block east of the Straight Creek fault [*Frizzell et al.*, 1984]. **8.8**

55.2 Leave I-90 at Exit 70. At stop sign follow road toward Easton. **0.5**

55.6 STOP 3. Look up Lake Kachess valley, which is the trace of the Straight Creek Fault (Figure 5), to see the Teanaway overlying the steeply dipping Silver Pass Volcanic Member of the Swauk Formation [*Tabor et al.*, 1984] in the Teanaway River block. The Silver Pass, chiefly andesitic to dacitic flows and pyroclastic rocks, is about 1.8 km thick and is interbedded with fluvial subquartzose sandstone of the Swauk. The Silver Pass is 52-50 Ma on the basis of zircon fission-track ages [*Vance and Naeser*, 1977; *Tabor et al.*, 1984].

The Straight Creek fault is a major dextral strike-slip structure extending from central Washington into Canada [*Misch*, 1977; *Davis et al.*, 1978]. The northern segment of the fault in Washington separates low-grade metamorphosed rocks on the west from plutonic and high-grade metamorphic rocks of the North Cascades on the east. The southern segment separates lower Eocene sedimentary and volcanic rocks on the east (Teanaway River block) from upper Eocene to Oligocene(?) rocks on the west (Cabin Creek block) [*Tabor et al.*, 1984]. Movement of the southern part of the fault began at or before 48-47 Ma, as reflected by the angular unconformity separating the Silver Pass from the overlying Teanaway. Movement continued until at least 42 Ma, because the Naches Formation is cut by fault splays. The 25-Ma eastern part of the Snoqualmie batholith is not offset by the fault [*Vance and Miller*, 1981]. **0.4**

Bus can continue along road, turn around, and come back to the site. Return to freeway.

56.0 Reenter east-bound freeway. Just to south the Straight Creek fault bends southeast and splits into splays covered by the Grande Ronde Basalt. **0.4**

61.4 Tenaway Formation underlies ridge to left. The light gray exposures on the ridge behind Cle Elum are sandstones of the Roslyn Formation. **5.4**

68.7 At 12 o'clock is Lookout Mountain (Figure 5), underlain by N_2 and R_2 flows of the Grande Ronde Basalt near the western limit of the area underlain by the Columbia River Basalt Group. The basalt dips about 5° SE and defines the plunging trough of the Kittitas Valley syncline. Grande Ronde Basalt also forms most of Cle Elum Ridge south of the valley, where it rests on the Teanaway Formation [*Tabor et al.*, 1982b]. **7.3**

As we drive, observe the Stuart Range northwest of here. The brown rocks are ophiolite of the upper Jurassic and lower Cretaceous Ingalls Tectonic Complex [*Miller*, 1985a; *Tabor et al.*, 1987]. The gray peaks are tonalite of the upper Cretaceous Mount Stuart batholith.

77.8 Elk Heights pass. From here the route is within the Kittitas Valley syncline, developed principally in the Grande Ronde Basalt. The basalt is overlain in most places by the Pliocene Thorp Gravel capped by modern stream alluvium and glacial drift. **9.1**

83.2 White bluffs to north are in volcaniclastic rocks of the Ellensburg Formation interbedded with flows of the Grande Ronde Basalt. The Ellensburg here is pumiceous diamict and sandstone that indicates contemporaneous dacitic volcanic activity in the Cascades as the Grande Ronde Basalt was being erupted. **5.4**

85.0 Dark exposures north of road are type locality of Thorp Gravel, deposited by the ancestral Yakima River and its tributaries. The gravel contains two facies: mainstream alluvium dominated by durable silicic volcanic rocks derived from the Naches and Ohanapecosh Formations, and sidestream alluvium eroded from the Grande Ronde Basalt in the limbs of the syncline [*Waitt*, 1979; *Tabor et al.*, 1982b]. *Waitt* [1979] recognized three east-trending faults cutting the Thorp, which is about 4 Ma on the basis of fission-track ages on interbedded tuff and is the youngest deformed unit in the area. **1.8**

87.3 To south is Manastash Ridge, a complex thrusted anticline. Manastash is one of the most northern uplifts in the Yakima fold belt [*Bentley*, 1977]. **2.3**

94.7 Continue past Exit 109 to Canyon Road, which follows Yakima Canyon across several well-exposed, ESE-trending anticlines and synclines. The river follows structural sags in the anticlines and thus is considered consequent [*Waitt and Swanson*, 1987]. Continued folding caused the river to incise the anticlines within the broad N-S sag between the Hog Ranch and Cascade uplifts [*Schmincke*, 1964; *Waitt and Swanson*, 1987]. **7.4**

95.5 Continue past junction with I-82. Cuts along I-82 show deformed Wanapum Basalt and interbedded Ellensburg Formation [*Bentley*, 1977] but are hard to visit owing to laws against walking along the freeway. **0.8**

98.1 Kittitas Valley gradually ends to the east, where a low south-trending ridge system, the Hog Ranch uplift [*Mackin*, 1961], cuts off the southeast-trending Kittitas Valley syncline. The uplift, largely defined by the culminations of transverse anticlines, is a southward extension of the Naneum Ridge anticline [*Tabor et al.*, 1982b]. Major east-trending folds cross the uplift. **2.6**

105.5 Cuts in coulee left of freeway show pillowed flow in Frenchman Springs Member, which in this area is separated from the Grande Ronde Basalt by tuffaceous sandstone in the Vantage Member of the Ellensburg Formation. Exposures of the Frenchman Springs (in places pillowed), the Vantage, and the upper flow of the Grande Ronde occur for many miles as the highway crosses the Hog Ranch uplift. **7.4**

110.9 Ryegrass Crest, 773 m elevation. From here highway descends from Hog Ranch uplift to Columbia River. White tuffaceous sandstone of the Vantage crops out in places between the Frenchman Springs and underlying Grande Ronde Basalt. **5.4**

121.2 Exit 136. Leave I-90 toward Vantage. **10.3**

121.5 Stop, turn left, and drive through Vantage. **0.3**

122.2 Turn right to Ginkgo museum. **0.7**

122.7 STOP 4. Visit the museum, which displays petrified wood from the Vantage Member. The trees and logs were engulfed by and mixed with pillows and hyaloclastic debris when the basal flow of the Frenchman Springs crossed a swamp. The wood, chiefly cypress and ginkgo, indicates a warm, humid to wet setting, quite different from today's aridity.

Across the Columbia River, I-90 follows a prominent bench eroded along the Vantage Member. The bench rises northward toward the crest of the Frenchman Hills anticline. Flows in the Grande Ronde below the Vantage have been given informal names. The two best known are the Museum, the thin upper flow just below the east bridge abutment and the museum, and the underlying Rocky Coulee, a thick, complex flow that forms most of the cliff near the east end of the bridge.

Flows of Wanapum Basalt above the Vantage likewise have only informal names but belong to formal members (Figure 15). The lowest, named the Ginkgo flow by *Mackin* [1961], is mostly covered by talus above the east end of the bridge; it is the flow beneath which the petrified forest occurs. The overlying flow, the Sand Hollow, has a "double-barrelled" jointing aspect [*Mackin*, 1961] that suggests a compound cooling unit. The highest flow is the Sentinel Gap, here consisting of foreset-bedded pillows and hyaloclastite near the flow terminus. All three flows belong to the Frenchman Springs Member. On the skyline is the Roza Member, and still farther back from the river is the Priest Rapids Member. These three members can be traced across much of the Columbia Plateau and form marker units whose distributions, vent areas, and chemical compositions are among the best known of the province. **0.5**

Return to eastbound I-90.

125.1 Straight ahead note foreset-bedded pillows in Sentinel Gap flow on top of cliff. North dip of foresets indicates direction of flow. **2.4**

126.9 West of river, note museum and bench formed on top of Grande Ronde by stripping of the Vantage. **1.8**

131.4 Exit 143. Turn right on Champs deBrionne Winery road. **4.5**

131.7 STOP 5 just before intersection with the winery road. Plagioclase-phyric Roza Member invades diatomite to form a pépérite, which is overlain by silicic tuff capped by very sparsely plagioclase-phyric Priest Rapids Member (Figure 21). Note the stringers and dikelets of Roza mixed with diatomite. Note the sharp contact between the pépérite and the younger tuff.

Observe the sparsely phyric Priest Rapids and contrast it with the highly plagioclase-phyric Roza. Note that the vesicular top of the Roza is relatively less phyric than the rest of the flow. **0.3**

131.8 Turn left on Champs deBrionne Winery road. **0.1**

FIGURE 21. Person points to Roza Member of Wanapum Basalt invading and mixed with diatomite to form pépérite at Stop 5. Priest Rapids Member (top of photo) overlies thin silicic tuff (massive unit above lithic-rich pépérite near upper right edge of photo) deposited on pépérite.

132.5 Turn left at stop sign. **0.7**

132.7 Roza Member in roadcut. **0.2**

133.1 Bench on top of Frenchman Springs Member. Farther along, roadcuts show the vesicular top on this flow, the Sand Hollow. **0.4**

133.7 STOP 6. Spiracle at base of Roza Member. Spiracles presumably form as steam blasts through a flow from a wet substrate. A natural "Stonehenge" of platy columns has been eroded from the Roza Member just beyond the stop. Platy jointing is common in relatively coarse flows; the joints do not cross column boundaries and probably form during cooling. Frenchman Coulee below is a flood-carved coulee eroded as waters poured from the Quincy basin to the northeast and descended to the Columbia. The coulee is dry except for excess irrigation water. White mounds on the Roza north of the coulee are spoil from diatomite mines in the Squaw Creek Member of the Ellensburg, the deposit invaded by the Roza. Note the varied jointing patterns of the flows in the coulee wall, which illustrate the problems in using jointing habits for flow correlation. **0.6**

133.8 Good bus turnaround. Return to I-90. **0.1**

135.7 Turn right onto I-90. **1.9**

136.9 At 1200 is good view of high sharp peak in distance, Whiskey Dick Mountain. Shell Oil drilled an exploration hole near Whiskey Dick, as well as in several other areas in the western plateau, during the 1980s. The Whiskey Dick hole penetrated about 1.5 km of basalt, and a hole in the Saddle Mountains south of here went through more than 3.5 km of basalt, about 3.2 km of which was Grande Ronde only. No oil was found in the target beneath the basalt, presumably Eocene sandstone. Few data are publicly available for these holes. Drilling was continuing in summer 1988. **1.2**

140.7 Roadcut in Museum and underlying hackly jointed Rocky Coulee flows of the Grande Ronde Basalt. **3.8**

141.4 Take Highway 26 at Exit 137. **0.7**

142.1 Good exposure in cliff at 9 o'clock of "double-barrelled" Sand Hollow flow. **0.7**

142.8 Keep left at junction with Highway 243. **0.7**

143.0 STOP 7. Mouth of Sand Hollow. Walk 100 m to next roadcut. Pillows and palagonite in Ginko flow of Frenchman Springs Member. Note the dipping foresets in a lava delta that indicates local movement toward the west or northwest. **0.2**

145.2 Bus turnaround at road junction. **2.2**

147.6 Return to Highway 243 and turn left. **2.4**

147.9 Start up grade across Frenchman Springs and overlying Roza Members. Type locality for the Wanapum Basalt, named from nearby Wanapum Dam. **0.3**

152.2 View at 12 o'clock of Sentinel Gap, eroded across the Saddle Mountains anticline [*Reidel*, 1984]. **4.3**

154.9 White layer high on west wall of Sentinel Gap is sandstone of Vantage Member. **2.7**

157.8 STOP 8 at turnout along right side of highway in Sentinel Gap. The Sentinel Bluff section here has been closely studied by geologists formerly of Rockwell Hanford Operations, in particular Steve Reidel [*Reidel*, 1984] and Phil Long [*Long and Wood*, 1986; *McMillan et al.*, 1987], during investigation for a potential nuclear-waste repository within the basalt on the Hanford Reservation. The four oldest exposed flows are in the Grande Ronde Basalt, have a relatively low-MgO composition [*Myers and Price*, 1979], and are assigned to the Schwana unit. The oldest Schwana flow, not exposed here but cropping out just to north, has reversed magnetic polarity. The youngest Schwana flow, the Umtanum, is approximately 60 m thick divided equally between a lower colonnade and an upper entablature. The Umtanum was the main candidate flow for the nuclear-waste repository at Hanford. Overlying the low-MgO flows are eight relatively high-MgO flows in the Sentinel Bluffs unit of the Grande Ronde [*Myers and Price*, 1979], the uppermost of which is the Museum flow. A prominent bench, eroded back along the Vantage, separates the Sentinel Bluffs from the overlying Frenchman Springs. Work based on jointing in this section and elsewhere led *Long and Wood* [1986] and *DeGraff and Aydin* [1987] to conclude that entablatures are produced by rapid cooling as water percolates downward into opening joints.

A prominent tongue of basalt southwest of the gap (the Huntzinger flow of *Mackin* [1961]) belongs to the Asotin Member of the Saddle Mountains Basalt; it partly fills a paleovalley cut into the Priest Rapids along the course of the ancestral Columbia River about 13 Ma.

The Saddle Mountains anticline is asymmetric with a faulted north limb. The fold axis on the east side of the river is 2.5 km south of that on the west side, perhaps because of a tear fault [*Reidel*, 1984]. The Saddle Mountains reverse fault is best exposed across the river on the north limb of the fold. Some flows thin and even pinch out against the uplift, as does the Vantage Member [*Reidel*, 1984]. Apparently the fold was growing during deposition of the basalt and interbeds.

Ash erupted about 7 ka from Mount Mazama (Crater Lake, Oregon) 475 km to the south occurs in colluvium. Evidence of the Missoula floods are obvious. **2.9**

159.0 High on ridge at 8 o'clock is an abandoned quarry exposing two layers of vitric rhyolite tuff between the Wanapum and Saddle Mountains Basalts. The lower tuff, as thick as 15 m, contains prominent layers rich in accretionary lapilli [*Schmincke*, 1967a, 1967c]. The Pomona Member of the Saddle Mountains Basalt invades the tuff just above the lapilli layers. This duo–Pomona lying above accretionary lapilli in a vitric tuff (commonly welded) [*Schmincke*, 1967c]–occurs in many places on the Columbia Plateau as far east as Clarkston, Washington. The ash and far-flung accretionary lapilli testify to a powerful explosion in the Cascades just before eruption of the Pomona in northern Idaho about 12 Ma. **1.2**

164.2 Ahead is Umtanum Ridge, one of the larger and most studied anticlines on the Plateau [*Price*, 1982]. **5.2**

169.7 At 12 o'clock is view of rollover in Umtanum Ridge anticline. The north limb is locally overturned. The fold is cut by thrust faults [*Price*, 1982]. **5.5**

170.8 Steeply overturned flow at 3 o'clock. **1.1**

172.5 Road travels along the 82-km *Hanford Reach*, the only free-flowing stretch of the Columbia River from the Canadian border to Bonneville Dam near Portland. **1.7**

175.7 Stay right at junction to Vernita Bridge. **3.2**

178.2 Pomona Member overlies Umatilla Member of Saddle Mountains Basalt on hillslope along grade up from Vernita Bridge. Ridges in distance, Gable Mountain and Gable Butte, are local culminations of largely buried Umtanum Ridge anticline. **2.5**

182.0 Turn right on Highway 24 toward Yakima. Road goes up Cold Creek Syncline, whose lower, broader part was the favored site of a planned storage repository for high-level nuclear waste. The site was abandoned in 1988 in favor of one in Nevada, owing to a complex decision involving hydrologic questions and politics. **3.8**

186.9 Ridge at 10 o'clock is Rattlesnake Hills. Near its crest the first deep hole was drilled on the Columbia Plateau in 1957-58, a wildcat owned by Standard of California. The hole bottomed at 3248.6 m in Grande Ronde Basalt, as interpreted from the chemical compositions of the cuttings [*Reidel et al.*, 1982]. Drilling by Shell in the 1980s confirmed the great thickness of the basalt, particularly the Grande Ronde.

Scattered outcrops for the next several miles expose the Pomona and Elephant Mountain Members of the Saddle Mountains Basalt. **4.9**

189.2 Keep right at junction with road to Sunnyside, and follow the axis of the Black Rock syncline. **2.3**

191.7 Ridge to left is continuation of Rattlesnake Hills, which extends from 50 km west of Yakima to 23 km southeast of Pasco, a total length of about 175 km–the longest anticlinal structure on the Columbia Plateau. **2.5**

206.2 Rounded hill at 11 o'clock is Elephant Mountain, local high point on the Rattlesnake Ridge anticline and the type locality of the Elephant Mountain Member of the Saddle Mountains Basalt [*Waters*, 1955; *Schmincke*, 1967a]. The Elephant Mountain erupted in northeast Oregon about 10.5 Ma and flowed via a canyon system to the western plateau [*Swanson et al.*, 1979a]. **14.5**

215.3 At 9 o'clock is Union Gap, a water gap exposing Wanapum and Saddle Mountains Basalt cut by the Yakima River through the Rattlesnake Hills anticline. **9.1**

219.1 Turn north on I-82. **3.8**

222.2 Exit 31 to North First Street in Yakima. End of a long first day. **3.1**

Log for Day 2, Yakima to Packwood

0.0 Entrance to freeway bypass at north end of North First Street in Yakima.

0.1 Selah Gap in Yakima Ridge anticline exposes flows of Frenchman Springs and Priest Rapids Members. **0.1**

2.8 At 11 o'clock is bluff of Tieton Andesite. **2.7**

3.5 Turn left on Ackley Road. **0.7**

3.6 At stop sign turn left and park. **STOP 9.** Walk

100 m west and climb stairs to view the reversely magnetized, 1.0 Ma Tieton Andesite. This is an archaeological site (Indian paintings); **do not sample here** (a sampling opportunity may be possible at Stop 11). This is the most voluminous of several flows erupted from the Goat Rocks volcano, a Pliocene and Pleistocene andesitic cone on the crest of the Cascades 80 km west of here. The flow poured down the ancestral canyon of the Tieton River until debouching into the lower-gradient Naches Valley, where it spread out, developed pressure ridges and tumuli, and terminated at this site [*Warren*, 1941; *Becraft*, 1950; *Swanson*, 1964, 1978; *Clayton*, 1983]. The flow has about 60% SiO_2 and 2.8% K_2O and is a high-potassium andesite. It contains 30-35% phenocrysts of plagioclase (greatly dominant), clinopyroxene, hypersthene, and magnetite. These characteristics suggest rather low fluidity, yet the flow is the longest known andesite, longer than any in *Walker's* [1973] compilation. Apparently the flow advanced so far because of its large volume (conservatively estimated as 2 km^3), possibly high eruption rate, and confined course in the canyon. Note the complicated columnar joints, which probably record cooling against a vertical channel wall eroded into volcaniclastic rocks of the Ellensburg Formation (since removed). Note the glassy texture, typical of its entire length. **0.1**

3.7 Return to highway and turn left. **0.1**

4.5 Turn right on Old Naches Highway and stay left at Y-junction immediately after turn. **0.8**

5.3 Turn right on Mapleway Road. Bluffs ahead and to left are considered by many workers as the "type" locality of the Ellensburg Formation, although Ellensburg itself is more than 30 km farther north. **0.8**

5.9 Roadcut exposes tuffaceous sandstone and dacite-rich conglomerate representative of the volcaniclastic upper part of the Ellensburg Formation in this area. The dacite was produced by explosive activity in the Cascades mostly between 12 and 8 Ma [*Smith et al.*, 1988]. Specific sources have not been identified, but a dacite-porphyry intrusion near Bumping Lake 60 km west-northwest of here has a K-Ar age of 8.8 ± 0.2 Ma and could be an Ellensburg feeder [*Smith et al.*, 1988]. The Ellensburg here is overlain by dacite-poor conglomerate that contains reworked tephra with a K-Ar age of 7.4 ± 0.4 Ma [*Smith*, 1988]. *Smith* [1988] interprets the dacite-poor conglomerate, forming part of his Pleasant Hill conglomerate unit, as partly filling a paleovalley eroded into the volcaniclastic Ellensburg. **0.6**

6.2 STOP 10. Examine pumiceous debris flow with fine-grained base in Pleasant Hill conglomerate (Figure 22). The origin of the fine-grained base, a photo of which appears in *Schmincke* [1967b], is debatable. The debris flow contains scattered prismatically jointed blocks of dacite, rests on dacite-poor conglomerate, and underlies dacite-rich conglomerate. Apparently the debris flow formed during a period of renewed dacitic volcanism in the ancestral Naches River basin [*Smith et al.*, 1988]. Walk to bottom of grade from here to examine deposits described at mileage 5.9. Bus continues up hill. **0.3**

6.5 Bus turns around at top of hill, returns to bottom of hill, and waits for group. **0.3**

7.6 Stop at Old Naches Highway and continue straight on Mapleway Road. **1.1**

8.7 Turn left on McCormick Road and then right on Highway 12. **1.1**

8.9 At 9 o'clock is bluff of Tieton Andesite. The present flow margin probably approximates its original margin, a conclusion consistent with its overall jointing characteristics. If so, the Naches Valley has widened about 2 km to its present position in the last 1 m.y., presumably when high discharge during glacial periods eroded the relatively nonresistant deposits of the Ellensburg. Wells drilled on the bluff tap water from river gravel below the Tieton Andesite, good evidence that the flow advanced down the main channel of the ancestral Naches. Presumably the Naches then re-established itself along the margin of the flow and commenced widening its valley asymmetrically northward. **0.2**

11.4 At 3 o'clock the Pomona Member rests on white tuffaceous rocks of Ellensburg Formation. **2.5**

12.2 Roadcut in Frenchman Springs Member just before substation. This is the farthest west the Frenchman Springs and Pomona can be traced in this area. **0.8**

12.9 Bluffs to north show Ellensburg Formation stratigraphically higher than the Pomona Member. **0.7**

13.8 Ahead is Cleman Mountain, one of the highest anticlines on the Columbia Plateau. It stood high enough to trap most coarse Ellensburg detritus in the Nile basin just west of the mountain, keeping it from reaching the Naches Valley (R.S. Fiske, oral commun., 1962; [*Smith*, 1988]). The anticline continued to grow after Ellensburg time to its present elevation. **0.9**

16.3 Good views of Tieton Andesite left of road. Note the hummocky nature of its surface, caused by large pressure ridges. Cleman Mountain looms at 1 o'clock.

FIGURE 22. Fine-grained base of pumiceous debris flow containing rare "hot" dacite blocks. Stop 10.

The south limb of the anticline is locally overturned. The structure of the anticline is poorly known, owing to the absence of marker flows younger than Grande Ronde and to the complexity of the structure itself. **2.5**

19.8 Turn left toward White Pass on Highway 12. **3.5**

20.0 Cross Naches River just upstream from mouth of Tieton River. View to left of Tieton Andesite, which flowed down the Tieton River canyon. Note the angular discordance with the underlying northeast-dipping Grande Ronde Basalt. Remnants of Tieton Andesite hang on the walls of the canyon, carved in Grande Ronde Basalt and the Fifes Peak Formation, for the next 30 km into Tieton basin. Tieton Andesite can generally be recognized in the canyon by its unconformable contact and notable columnar jointing, commonly with an entablature above long subvertical columns. **0.2**

21.9 Entrance to Oak Creek Wildlife Recreation Area. Exposures at 2 o'clock show Tieton Andesite only a few meters above river level. Apparently base level did not change much during the last 1 m.y., so that no downcutting took place except through the lava flow. **1.9**

22.6 STOP 11. Dangerous curve. Pull bus to left of road if traffic is light. Tieton Andesite rests on imbricated boulder gravel of ancestral Tieton River (Figure 23). East of river, erosional unconformity between Grande Ronde and Tieton Andesite is well exposed. **0.7**

23.9 STOP 12. Micaceous sandstone and siltstone between two N_2 flows of Grande Ronde Basalt (Figure 24; locality 2 in table 10 of *Swanson* [1967]). The lower flow is invasive into wood-bearing subarkose, probably unconsolidated at the time of invasion; note the dense glassy top on the flow and the disruption and tilting of the sediment. Mica-bearing sediment and the accompanying metamorphic and plutonic heavy-mineral suite [*Swanson*, 1967] indicate a northern provenance and therefore imply a south-flowing river during Grande Ronde time, which is interpreted from regional evidence to be the ancestral Columbia or a major tributary. During Grande Ronde time, no ancestral Columbia is known from within the Columbia Plateau itself. A reasonable interpretation is that the river had been diverted to the margin of the province by the basalt flows [*Fecht et al.*, 1987; *Waitt and Swanson*, 1987]. Mixed with the metamorphic and plutonic suite is a pyroclastic suite, including euhedral oxyhornblende, glass shards, and glass-rimmed plagioclase and mafic minerals. This suite provides clear evidence of explosive activity from Cascade volcanoes during Grande Ronde N_2 time, 17-16 Ma. Pyroclastic debris in older interbeds shows that Cascade activity also occurred in N_1 and R_2 time. **1.3**

24.7 Nearly vertical contact between Tieton Andesite and Grande Ronde Basalt at 11:30 o'clock. **0.8**

26.2 Roadcut shows flow of Grande Ronde Basalt that invaded unconsolidated micaceous sand (locality 1, table 10 of *Swanson* [1967]). **1.5**

27.9 Windy Point measured section of Grande Ronde Basalt ahead. Lower 3 flows are in magnetostratigraphic unit R_2, and the upper 9 in N_2 [*Swanson*, 1978]. **1.7**

FIGURE 23. Base of columnar Tieton Andesite resting on gravel of ancestral Tieton River, Stop 11.

28.5 Polished fault plane cutting pillows of Grande Ronde. The fault cannot be traced from the cut. **0.6**

28.8 Pillowed Grande Ronde Basalt west of bridge. **0.3**

29.0 STOP 13. Local margin of Grande Ronde Basalt, where it lapped onto the eroded remnant of Tieton volcano [*Swanson*, 1966, 1978], a large stratocone in the Fifes Peak Formation (Figure 25). The contact is obscured by landsliding. Note the foreset-bedded lava delta in one of the lower R_2 flows of the Grande Ronde that indicates westward advance of the basalt.

The base of Tieton volcano consists of gently dipping two-pyroxene basaltic andesite flows that lap against the deformed flanks of an older cone and form a shield about 200 m thick (Figure 26). More than 1500 m of bedded tuff, lithic-lapilli tuff, and breccia overlie the shield (Figure 27); the steep radial dips of the fragmental rocks define a volcaniclastic cone. The reconstructed volcano had a basal diameter of at least 1.5 km and probably rose 2400-3000 m above its base before erosion and partial inundation by the Grande Ronde Basalt.

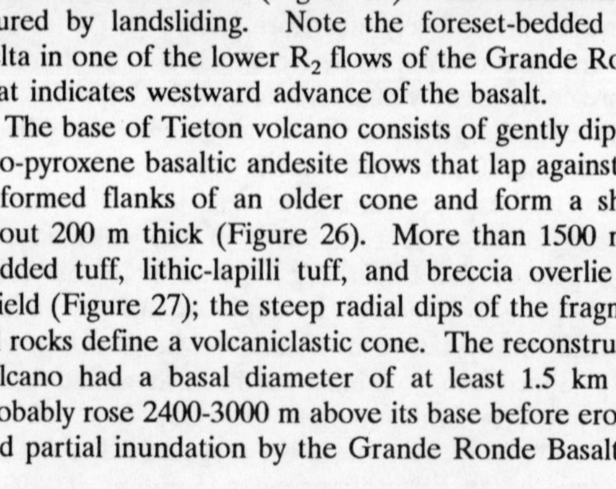

FIGURE 24. Micaceous sandstone and siltstone between flows of Grande Ronde Basalt at Stop 12. Lower flow invades sandstone. Hammer at contact.

Holocene	Alluvium and colluvium
Pleistocene	Intracanyon flows of hypersthene-augite Tieton Andesite (1 Ma) and olivine basalt erupted from vents near Goat Rocks and the White Pass area. Overlain and underlain by till.
Pleistocene and Pliocene	Small plugs of olivine basalt and andesite.
	UPLIFT
Pliocene(?)	Gravel on top of Grande Ronde Basalt
	EROSIONAL AND ANGULAR UNCONFORMITY
Middle Miocene	Grande Ronde Basalt (approx. 17-15 Ma). As much as 520 m of gently eastward-tilted flows. Local discontinuous interbeds of clastic sediment, tuff, and palagonite.
	EROSIONAL AND POSSIBLE ANGULAR UNCONFORMITY
Early Miocene	Fifes Peak Formation (approx. 23 Ma in this area). More than 1 km-thick section of andesite and basaltic andesite lava flows and breccia, and dacite-rhyolite pyroclastic flows, erupted from at least two large composite cones. Prominent radial dike swarm. Cut by andesitic and microdioritic intrusions in western part of area.
	ANGULAR UNCONFORMITY
Oligocene	Volcaniclastic rocks of Wildcat Creek, probable correlatives with Ohanapecosh Formation (about 32-33 Ma in this area). More than 300 m of volcaniclastic rocks, many of which are graded, have continuous bedding across outcrops, generally lack current structures, and have poorly developed sole marks. Vertebrate fossils. Zeolitized. Probably deposited either by subqueous turbidity currents or subaerial debris flows.
	EROSIONAL AND POSSIBLE ANGULAR UNCONFORMITY
Eocene	Welded tuff of Spencer Creek (about 42 Ma) and associated epiclastic sandstone and conglomerate.
	ANGULAR UNCONFORMITY
Cretaceous and Jurassic	Russell Ranch Formation, graywacke and argillite with lesser pillow basalt (greenstone) and chert. Also includes Indian Creek unit, mainly sheared orthogneiss of tonalitic composition but varying from gabbro to trondhjemite. Together comprise the "Tieton inlier" of pre-Tertiary rocks.

FIGURE 25. **Stratigraphy of Tieton River area, simplified from *Swanson* [1967]. Ages for rocks older than Grande Ronde Basalt from *Vance et al.* [1987]. Data for Indian Creek unit from *Miller* [1985].**

On Bethel Ridge, 6-10 km west of the center of the volcano, at least 300 m of gently dipping debris-flow deposits (mainly lahars), lava flows, and low-silica rhyolitic pyroclastic flows can be traced nearly continuously into the cone. Some breccias thin and pinch out to the west, and one andesite flow ends with westward-inclined foreset lenses of breccia and lava at its snout. The Bethel Ridge rocks probably formed part of a lowland apron at the foot of Tieton volcano.

The volcano erupted block-lava flows, coarse- to fine-grained bedded tuff, and chaotic breccia of multiple origins. Intruding them are more than 200 dikes that form the southern half of a radial swarm centered on the volcano. Single dikes are 1.5-6 m wide and dip 70-90°. They decrease in abundance upward in the volcano and were injected subvertically, not laterally, as indicated by lineations on the dike margins. The dikes and lava flows are medium-K mafic and slightly silicic andesite (or basaltic andesite and lesser andesite.

A lava flow in the apron of Tieton volcano yielded a fission-track age of 23.3 ± 2.0 Ma [*Vance et al.*, 1987, sample JV 140]. Other isotopic ages for the Fifes Peak Formation range from 20.0 ± 2.0 to 27.3 ± 2.9 Ma [*Vance et al.*, 1987]. Centers similar to those in the Tieton River area occur at the type locality in the Fifes Peaks themselves, 40 km northwest of Tieton volcano.

Tieton volcano cannot be seen well from the highway, because the canyon is so narrow, but it is splendidly displayed from a high ridge road, inaccessible to buses, that turns off at mileage 28.3. Many of the radial dikes are visible from the highway, however. Keep track of their trends as we drive through the canyon to convince yourself that they radiate from a common center. **0.2**

29.2 Dike in roadcut, with strike of about 295°. Wall-rock is andesite flow and breccia of Tieton volcano. **0.2**

30.0 Slope above road shows rocks of older Fifes Peak volcano cut by radial dikes of Tieton volcano. **0.8**

30.6 STOP 14. Bus stops here. Group walks 100 m ahead to examine roadcut exposing at least 11 basaltic andesite dikes belonging to the radial dike swarm cutting rocks of the older volcano. **0.6**

31.8 Prominent knob to left is Sentinel Rock, capped by Tieton Andesite. **1.2**

34.0 Radial dikes at 3 o'clock. Remnant of Tieton Andesite high on ridge. **2.2**

35.7 Breccia and platy basaltic andesite of the older Fifes Peak volcano. The units resemble those of Tieton volcano but can be distinguished by map relations. **1.7**

36.2 Enter Tieton basin, chiefly underlain by landslide debris. **0.5**

37.0 Goose Egg Mountain microdiorite ahead. **0.8**

37.9 Start series of roadcuts in bedded volcaniclastic rocks of Wildcat Creek, age and possibly stratigraphic correlatives of the Ohanapecosh Formation. A fission-track age of 32.2 ± 3.3 Ma was obtained on a pumiceous ash-flow tuff near the west end of these cuts [*Vance et al.*, 1987, sample JV 139]. **0.9**

38.7 Straight ahead is Westfall Rocks, a microdiorite twin of Goose Egg Mountain. **0.8**

39.0 Turn right on Wildcat Road 1306. **0.3**

39.6 Cattle guard. **0.6**

39.8 STOP 15. Climb steep slope to examine deposits typical of the upper part of Wildcat Creek unit (Figure 28), which consists dominantly of laterally extensive, graded beds 1-6 m thick separated by thinly bedded tuff. The lower several centimeters of each thick bed is typically framework-supported, zeolite-cemented lapillistone, which locally has sole marks on its base (Figure 29). The lapillistone grades abruptly upward into poorly sorted lapilli tuff, which grades gradually upward into well-sorted tuff. The thin beds are commonly graded, and some beds are crudely laminated; small-scale cross

FIGURE 26. Nearly horizontal andesite flows in basal shield of Tieton volcano, overlain by volcaniclastic cone. Base of shield (dashed line) overlies tilted flows and breccia of older volcano. Between Stops 13 and 14 on north side of canyon.

FIGURE 27. Bedded volcaniclastic rocks forming cone of Tieton volcano on north side of canyon between Stops 13 and 14. Note concave-upward attitude.

FIGURE 29. Lapillistone base of graded lapilli tuff at Stop 15. Note possible load casts at base of bed.

FIGURE 28. Continuous bedding in the volcaniclastic rocks of Wildcat Creek, Stop 15.

bedding and channelling occur in some thin beds.

Clasts are freshly erupted crystals (commonly exhibiting bubble-wall texture), dense lithic fragments (chiefly andesite and dacite), pumice (mostly of the long-tube type), and ash. Accretionary lapilli, both broken and whole, occur in the lapilli tuff and tuff, and fragmentary vertebrate fossils have been found [*Grant*, 1941; *Swanson*, 1964; *Lander*, 1977]. All glass is entirely altered to, in order of formation, opal or chalcedony, clay and celadonite, heulandite, mordenite, and calcite.

At the base of the outcrop is a poorly exposed pumice-lapilli tuff, probably an ash flow; it is better exposed in roadcuts along Highway 12 and is the one dated by *Vance et al.* [1987] (see mileage 37.9). Farther down section, obvious fluvial deposits occur; current directions indicate southeast-directed transport.

The origin of these beds is debatable. *Swanson* [1964; 1978] interpreted the thick graded units as products of subaqueous turbidity currents carrying a mixture of freshly erupted and older detritus. Recent work on lahars and hyperconcentrated streamflows [*Scott*, 1988; *Pierson and Scott*, 1985; *Pierson*, 1985; *Smith*, 1986] raises the possibility of such subaerial origins for these rocks. The volcaniclastic rocks of

Wildcat Creek are interpreted as a distal facies of the Ohanapecosh Formation in Mount Rainier National Park, where lava flows and coarser, thicker graded beds occur [*Vance et al.*, 1987]. **0.2**

40.2 Bus turnaround. Return to Highway 12. **0.4**

41.3 Junction with Highway 12. Turn right. **1.1**

41.5 Cross Wildcat Creek. At 5 o'clock is remnant of a Pleistocene olivine basalt flow that elsewhere fills a channel eroded into Tieton Andesite. **0.2**

41.8 Westfall Rocks microdiorite unit. **0.3**

42.9 STOP 16. Margin of Westfall Rocks microdiorite unit, with dark hornfelsed Wildcat Creek lapilli tuff hanging on margin (Figure 30). This intrusion, as well as Goose Egg Mountain and two other dikelike bodies of similar lithology, Kloochman Rock and Chimney Peaks, are of mid-Tertiary age but have not been dated. Whether they are feeders for andesite in the Fifes Peak Formation is problematic. The Westfall Rocks and Goose Egg intrusions are probably connected at depth but have joint patterns that suggest they are separate bodies at the present level of exposure. Both bodies were intruded along the contact between the volcaniclastic rocks of Wildcat Creek and pre-Tertiary rocks of the Rimrock Lake inlier [*Miller*, 1985b] that underlie Rimrock Lake. The nature of this contact is not clear, but *Miller* [1985b] believes that contacts of the Rimrock Lake inlier and the Cenozoic rocks "are generally pronounced angular unconformities that only locally show minor modification by faulting."

Cliffs on the southern skyline expose the Divide Ridge section of Grande Ronde Basalt, about 500 m of basalt and interbedded micaceous sandstone [*Swanson*, 1967]. Twelve to 15 flows are present; the lower 3 or 4 are in the N_1 magnetostratigraphic unit, the middle 6 or 7 in the R_2 unit, and the upper 3 or 4 in the N_2 unit. At least 3 sill-like invasive flows occur in the section.

The Grande Ronde dips gently eastward on the flank of the Cascade uplift. The uppermost flow, which correlates with the upper part of the Sentinel Bluffs sequence at Sentinel Gap [*Mangan et al.*, 1986], is at a maximum elevation of about 2100 m at the west end of the ridge [*Swanson*, 1967, 1978]. A similar stratigraphic position below the museum at Vantage is at an elevation of less than 150 m and in the Pasco Basin is as low as 600 m below sea level [*Myers and Price*, 1979]. Structural relief of more than 2700 m is therefore indicated. East of Pasco Basin, a general elevation for the top of the Grande Ronde is about 730 m near the northern and northeastern periphery of the province, where the basalt has not been markedly deformed [*Swanson et al.*, 1979c]. If this reflects a "stable" datum, then the Divide Ridge area has been uplifted about 1370 m absolutely.

Whether the uplift is truly absolute is uncertain, but the calculation emphasizes the general observation that the Cascade Range at this latitude has been significantly uplifted in the last 15 m.y. About 75% of this uplift is confined to the western 16 km of the present Columbia

FIGURE 30. Hornfelsed lapilli tuff along margin of Westfall Rocks microdiorite unit at Stop 16.

Plateau, for the top of the Grande Ronde is lower than 580 m near Tampico, about 19 km southeast of here, whereas it is higher than 2100 m only 16 km farther west [*Swanson*, 1967]. West of Divide Ridge the basalt projects into a younger volcanic center (Goat Rocks volcano) built on an uplifted platform of lower Tertiary rocks [*Swanson and Clayton*, 1983], and the basalt does not occur farther west at this latitude on the western flank of the Cascades.

The gently sloping, flat-topped ridge south of Rimrock Lake is Pinegrass Ridge, underlain by Tieton Andesite and several other Pleistocene lava flows. This marks the pathway by which the long flow of Tieton Andesite moved from Goat Rocks volcano into Tieton basin. **1.1**

43.5 Sheared trondhjemitic phase of the Jurassic Indian Creek unit [*Miller*, 1985b; *Swanson*, 1964]. **0.6**

43.6 Fault contact of trondhjemite with the "eastern greenstone belt" [*Miller*, 1985b] of the Russell Ranch Formation. **0.1**

44.8 Roadcuts of Jurassic and Cretaceous Russell Ranch Formation [*Miller*, 1985b] to Indian Creek. This assemblage (dominantly graywacke and argillite, with lesser pillow basalt, radiolarian chert, green tuffs, and red shale) dominates the pre-Tertiary Rimrock Lake inlier, which is the southernmost occurrence of pre-Tertiary rocks in the Washington Cascades. *Miller* [1985b] believes that "the various rock types originally formed at a considerable distance from each other" and were juxtaposed by faulting and major lateral transport. **1.2**

48.8 Cut along highway at junction is olivine basalt of Tumac Mountain, a late Pleistocene cone on the Cascade crest 10 km to the northwest [*Clayton*, 1983]. **4.0**

50.8 STOP 17. Sheared Indian Creek unit, a weakly foliated orthogneiss that *Hopson and Mattinson* [1973] suggested was part of a dismembered ophiolite. **2.0**

50.9 Upper Pleistocene hornblende andesite in roadcut. Cuts for next 3 km expose this andesite. **0.1**

51.5 STOP 18. Park just **before** "No Parking" sign. Platy upper Pleistocene hornblende andesite flow erupted from Tumac Plateau north of road. Its vent is not

known, but similar flows issued from vents now filled by small domes [*Clayton*, 1983]. Chemical analyses from farther west in the roadcuts indicate silicic andesite (about 62.8% silica) [*Clayton*, 1983]. **0.6**
53.0 Spiral Butte at 12 o'clock is a hornblende dacite (64-66% silica) dome that merges outward into a 4-5 km long flow [*Clayton*, 1983]. *Clayton* [1983] suggests that Spiral Butte is the youngest hornblende andesite or dacite vent in the area and is no more than 45 ka, judged from sedimentation rates on the butte below Mount St. Helens tephra layer C (about 36 ka). **1.5**
53.8 STOP 19. Clear Creek Falls Overlook. Restrooms. The hornblende silicic andesite of Deer Lake Mountain [*Clayton*, 1983] forms the falls. It was erupted from a vent about 5 km farther west (now occupied by a 250-m-high dome 250) and advanced into the glaciated North Fork Clear Creek valley. The flow is more than 100 m thick, overlies a basalt dated at 0.65 ± 0.08 Ma, and is normally magnetized [*Clayton*, 1983]. **0.8**
54.0 Outlet for Dog Lake, with Spiral Butte towering above. Growth of Spiral Butte dome impounded Dog Lake, according to *Clayton* [1983]. **0.2**
54.9 Roadcuts in Deer Lake Mountain flow. **0.9**
56.2 Summit of White Pass (1362 m), with Deer Lake Mountain flow to north dipping from its dome-like vent. The pass is eroded into relatively nonresistant graywacke and argillite of the Russell Ranch Formation. Chair lift crosses gentle slopes of Russell Ranch capped by more resistant olivine basalt and basaltic andesite erupted in the late Pliocene or early Pleistocene from Hogback Mountain 5 km south of White Pass. **1.3**
57.3 Russell Ranch in roadcuts here and following. **1.1**
58.5 Ridge ahead of Ohanapecosh Formation, in places overlain by quartz-phyric ash-flow tuff of the Stevens Ridge Formation (*Swanson and Clayton*, 1983). **1.2**
60.1 Roadcuts still in Russell Ranch Formation. **1.6**
61.0 Bench to left across canyon is on top of Clear Fork andesite flow (see Stop 20) [*Ellingson*, 1972; *Clayton*, 1983; *Swanson and Clayton*, 1983]. **0.9**
61.4 Cortright Creek [*Hammond*, 1980] fault zone between Russell Ranch and Ohanapecosh Formations. *Miller* [1985b] found that the "straightness of the map pattern for much of this contact is compatible with a steep fault" but saw "little evidence to indicate that there has been anything more than minor movement along this contact." Extensive landslides obscure the contact south of here [*Swanson and Clayton*, 1983]. **0.4**
61.7 Cuts in Ohanapecosh Formation, riddled by fine-grained plagioclase-phyric andesitic sills. The sills resemble those in nearby Mount Rainier National Park that *Fiske et al.* [1963] and *Mattinson* [1977] ascribe to an early hypabyssal stage of the Tatoosh pluton, a Miocene granodiorite and quartz monzonite body. **0.3**
63.7 Start of cuts in the Pliocene or Pleistocene olivine basalt of Hogback Mountain. **2.0**
65.0 Prominent columnar-jointed sill cutting Ohanapecosh Formation in cut. Near base of measured section

FIGURE 31. **Columnar-jointed hornblende andesite of Clear Fork at Stop 20. Trees on flow about 25 m high.**

of the Ohanapecosh [*Fiske et al.*, 1963]. Lower part of formation is exposed farther north, where it intertongues with arkosic sandstone [*Vance et al.*, 1987]. **1.3**
66.2 Turn left into Palisades viewpoint. **1.2**
66.3 STOP 20. Restrooms. View the remarkably columnar hornblende andesite of Clear Fork (Figure 31) [*Ellingson*, 1972; *Hammond*, 1980; *Clayton*, 1983]. The flow has 59-62% SiO_2 [*Clayton*, 1983], slightly lower than other hornblende andesites in the White Pass area. The flow was erupted near Coyote Lake, 9 km to the southeast [*Swanson and Clayton*, 1983], and moved down the canyon in the late Pleistocene. It partly fills a glaciated canyon but is overlain by Evans Creek Drift (*ca.* 18-15 ka). *Hammond* [1980] suggests, because of the abrupt termination and vertical columns indicative of ponding, that the flow was blocked by the Cowlitz glacier during the Hayden Creek stade (*ca.* 140 ka). **0.1**
66.4 Leave viewpoint and turn left on highway. **0.1**
66.9 Ohanapecosh Formation along road. **0.5**
68.8 Junction of Highways 12 and 123. Turn right and then immediately left into parking area. **1.9**
68.9 STOP 21. Restrooms. Roadcut in Ohanapecosh Formation shows wood-bearing, well-bedded volcanic siltstone, lapillistone, and lithic-lapilli tuff above an ash-flow tuff. Thin sill cuts the ash-flow tuff. **0.1**
69.0 Return to Highway 12 heading west. **0.1**
70.6 Roadcut in bedded volcaniclastic rocks of the Ohanapecosh Formation. **1.6**
70.8 Quartz-phyric ash-flow tuff above bedded volcaniclastic rocks. This tuff, formerly considered the basal unit of the Stevens Ridge Formation, has been fission-track dated as 36.4 ± 3.6 Ma [*Vance et al.*, 1987, sample JV 126]. This age is consistent with assignment to the lower part of the Ohanapecosh [*Vance et al.*, 1987]. Several roadcuts from here to Packwood expose ash-flow tuff and volcaniclastic rocks of the Ohanapecosh. **0.2**
75.9 Quarry to right in volcanic siltstone under sill. **5.1**
76.2 Forest Service station on left in Packwood, opposite junction with Forest Road 52. End of log. **0.3**

Log for Day 3 or Day 4, Packwood to Cispus Center via Windy Ridge

0.0 Leave Forest Service station in Packwood.
1.0 Hills to left and ahead have not yet been mapped in detail. Reconnaissance suggests a middle Tertiary terrane intruded by dikes and sills. **1.0**
8.4 Cowlitz River, which heads on Cowlitz Glacier at Mount Rainier. The river flows across a terraced flood plain on Evans Creek Drift (about 18-15 ka) of the Fraser glaciation [*Crandell and Miller*, 1974]. **7.4**
16.2 Turn left at light in Randle onto Cispus Road. **7.8**
17.1 Keep right on Road 25 at junction. **0.9**
17.5 At 2 o'clock is basalt and basaltic-andesite center of Huffaker Mountain, about 23.5 Ma [*Evarts et al.*, 1987], with east dip of 20° toward synclinal trough. **0.4**
24.8 Turn left on Road 25 at junction just after bridge across Cispus River. **7.3**
25.0 Bedded volcaniclastic rocks, chiefly sandstone. **0.2**
25.9 Continue on Road 25 at junction. Landslide. **0.9**
30.2 Active landslide disrupts road. Volcaniclastic rocks dip east parallel to the slope, facilitating sliding. **4.3**
30.3 Lithic-lapilli tuff in cuts near Benham Creek, about 23 Ma as judged by field relations between here and junction of Roads 25 and 99. **0.1**
30.5 STOP 22. Warped volcanic sandstone and siltstone, probably deformed during landsliding or soft-sediment deformation. Similar small outcrops of disturbed fine-grained volcaniclastic rocks occur throughout the area, with no consistency of attitudes. **0.2**
31.4 Lapilli tuff and sandstone at Fourmile Creek. **0.9**
33.5 STOP 23. Basaltic andesite with phenocrysts of plagioclase and two pyroxenes, representative of the middle Tertiary section in this area. Not hornfelsed, as are other flows seen later today. **2.1**
35.6 Basaltic andesite in roadcut. This flow caps the dominantly volcaniclastic section since last stop. **2.1**
36.0 Turn right on Road 99 toward Windy Ridge. **0.4**
36.1 STOP 24. Vitrophyre at base of dacitic ash-flow tuff. Contact with underlying fine-grained tuffaceous siltstone exposed. Columnar, with columns indicating east dip. Vitrophyre grades up into lithic-lapilli tuff. Dated as 23.3 ± 1.8 Ma by Ar-Ar on plagioclase (R. C. Evarts and J. G. Smith, unpub. data, 1987]. **0.1**
37.0 Road traverses through ash-flow tuff, with andesite flows visible on hillslope to north (right). **0.9**
39.4 Thin andesite in a mainly tuffaceous section. **2.4**
40.1 Altered, fine-grained, phyric granodiorite in cut. Sill-like body underlies hill south of Bear Meadow. **0.7**
40.5 STOP 25. Bear Meadow parking area. Restrooms. This is the site from which the two most famous series of photographs of the May 18, 1980, landslide and blast from Mount St. Helens were taken by Gary Rosenquist and Keith Ronholm. The following description, abstracted from *Rosenbaum and Waitt* [1981], is from accounts of eyewitnesses at Bear Meadow:

One observer watching with binoculars saw the north side of the volcano get "fuzzy, like there was dust being thrown down the side." Several seconds later the north face began to slide. The "bulge was moving...the whole north side was sliding down." The first cloud appeared to form at the base of a "cirque-like" wall from which the bulge had moved. In about 20 sec the landslide was out of view behind a ridge. At this time, two distinct clouds seemed to issue from separate vents. An extremely dark cloud grew vertically from the summit. A lighter cloud (the blast cloud), which seemed to come from the area vacated by the landslide, expanded uniformly except for a large "arm" that shot out to the north in the direction of the avalanche. As the blast cloud grew, what appeared to be a shock wave similar to that associated with a nuclear explosion moved ahead of it. About 1-1/2 min after the blast, a noise resembling a thunder clap was heard at Bear Meadow, accompanying a sudden pressure change. This noise was followed by continuous rumbling "like a freight train." The blast cloud moved toward Bear Meadow, reaching the ridge nearest the mountain 25-30 sec after the start of the landslide. When the cloud hit the ridge, it rose and "boiled" upward. Just before the top of the mountain became obscured, observers saw the south side of the summit crumble into the hole formed by the landslide. Seven or eight minutes after the eruption's start, rocks began falling on one of the witnesses 3-5 km northeast of Bear Meadow. He collected two "golfball-sized" rocks (determined later to be cryptodome dacite) from the fall, which continued for roughly 30 sec. As he drove, this material was replaced by "mud drops" that flattened on impact. The largest observed flattened mud drop was 19 mm across. Gradually mudfall abated, but fine ash fell more heavily until the witness could no longer see to drive. Ashfall then abated slowly.

Note that the blast stopped little more than 3 km from Bear Meadow. An anecdote illustrates the serendipity of the remarkable Rosenquist photographs. His camera was mounted on a tripod and centered on the volcano. When the eruption began, Rosenquist was looking away from the volcano. As he quickly turned, he hit the tripod, rotating the camera slightly clockwise so that the mountain was no longer centered but instead was far to the left side of the image. This accident improved coverage of the north flank of the volcano, thereby enabling the sequence of events to be captured in the series of photographs that was later made into a kind of time-lapse video.

Examine the altered, flow-layered dacite, typical of the Strawberry Mountain dacite dome and flow field, in roadcuts 100 m beyond Bear Meadow. North of here, the dacite generally dips eastward and interfingers with the section of ash-flow tuff that we traversed beginning at Stop 24. Thus the dome and flow field may be the

source area for the tuffs. **0.4**
40.7 Volcaniclastic rocks beneath the dacite flow. **0.2**
40.8 Flow-layered dacite; breccia just beyond. **0.1**
41.1 Andesite flow in roadcut. **0.3**
41.7 Epidote-bearing andesite in the outer part of the hornfels zone around the Spirit Lake pluton. From this point, the road roughly parallels the margin of the pluton, and most of the rocks are recrystallized to assemblages of the albite-epidote hornfels facies. **0.6**
42.0 Margin of 1980 devastated zone just above to right. Standing dead trees typify the outer few hundred meters of the devastated area. The *standing dead* were killed by the heat of the blast cloud, which had lost so much lithic debris at its margins that it became lighter than air and lifted off the ground. **0.3**
42.5 Enter the devastated area. Still in andesite. **0.5**
43.9 Lithic-lapilli tuff interbedded in the dominantly andesitic section. **1.4**
44.9 STOP 26. Junction with Road 26. Park just beyond junction and walk back to roadcut exposing Mount St. Helens tephra deposits: 1980 pumice and blast, layer T (early 1800) [*Yamaguchi*, 1983], set X (16th century?), layers Wn and We (early 1480 A.D. and early 1482 A.D., respectively) [*Yamaguchi*, 1985], and sets B (2.2-1.6 ka), P (2.5 ka), Y (3.5-3.0 ka), and J (11 ka) [*Mullineaux*, 1986]. All the tephras are dacitic except for T (andesitic) and part of B (basaltic and andesitic). Yamaguchi's tree-ring ages recognize that the 1800, 1480, and 1482 eruptions could have occurred in the last 3-4 months of the preceding year—remarkable precision for geologic events of prehistorical vintage! **1.0**

Continue straight on Road 99.
45.0 Miner's car, destroyed in blast and preserved here as a much-vandalized relic. Interbedded andesite flows and volcaniclastic rocks on skyline to west are in albite-epidote hornfelsed zone. **0.1**
45.1 Trailhead to Meta lake, which was covered by ice on May 18, 1980, and so retained abundant aquatic life, including trout. Young evergreens around the lake survived the blast because of snow cover. **0.1**
45.6 Nice example of blowdown on hill to right. **0.5**
46.1 Pyroxene diorite dikes, nearly parallel to road, cut hornfelsed volcaniclastic rocks in roadcut. Minor intrusions ranging from pyroxene diorite to monzogranite are common in the Tertiary rocks of the area. Few have been dated directly, but most are believed to be of early Miocene age (25-17 Ma). **0.5**
46.4 STOP 27. Cascade Peaks Viewpoint and Crater House Restaurant and Gift House. Restrooms. Views of Mounts St. Helens, Adams, and Hood (in Oregon). Roadcut opposite restaurant exposes Mount St. Helens tephra deposits, including layer D (1200 B.P.), a thin dacitic blast deposit formed during emplacement of the Sugar Bowl dome [*Crandell and Hoblitt*, 1986], between sets W and B. The tephra overlies hornfelsed pumice-lapilli tuff, probably ash-flow tuff. **0.3**
46.7 Dacite plug visible in hill west of road. **0.3**
47.3 Up creek to right is relatively fresh hornblende andesite dike cutting hornfelsed dacite breccia. Hornblende dated as 19.9 ± 0.7 Ma [*Evarts et al.*, 1987]. **0.6**
47.4 Diorite in roadcut. **0.1**
47.9 STOP 28. Trail to Independence Pass. Hike 200 m to small bench on crest of low hill. The blast deposit is easily recognized by its gray color, poor sorting, and content of shredded wood. Note the tack driven in the sawn end of log at switchback; it marks a period of stressed growth (narrow tree rings) in 1801-1805 A.D. caused by fall of the dacitic T tephra in 1800. From the bench are good views of Spirit Lake, tree blowdown, and the northwest corner of the dacite dome in the crater. Standing tree trunks on and behind ridge to north indicate that the blast lofted but snapped off their crowns. Note zone of tree removal on Harry's Ridge west of the lake; trees were removed as the debris avalanche scoured the lower part of the ridge. Below is Harmony Falls basin, through which a trail leads to the shore of Spirit Lake, whose high level drowns most of the falls. Hill between here and Mount St. Helens is underlain by an andesite flow (cliffs along Harmony Falls trail) capped by the dacite dome seen at Stop 29.

The Tertiary rocks visible across the lake are east-dipping andesite flows and volcaniclastic rocks similar to those already seen. North of the lake, the rocks are progressively more thoroughly recrystallized as the southern contact of the Spirit Lake pluton is approached. The ridge on the skyline is made of hard flinty hornblende hornfels adjacent to the contact, which is just north of the ridgeline. **0.5**
48.0 Coarse lithic-lapilli tuff, containing clasts of dacite seen at next stop, cut by a northwest-trending (335°) shear with horizontal slickensides and steps indicating dextral slip. The southern Washington Cascades and adjacent Columbia Plateau are host to hundreds of such northwest-trending dextral shears, in most places with little demonstrated displacement. **0.1**
48.6 On inside of curve, brecciated andesite flow fills a vertically walled channel eroded into wood-bearing lithic-pumice lapilli tuff cut by zeolite veins. **0.6**
49.2 Trail to Harmony Falls. Enjoyable walk to shore of Spirit Lake, with well-exposed volcaniclastic rocks near water's edge. Harmony basin was site of floods fed by water displaced from Spirit Lake by the debris avalanche [*Waitt*, 1984a; *Doukas and Swanson*, 1987]. **0.6**
49.5 STOP 29. Cedar Creek Viewpoint. Excellent view of Spirit Lake. Walk back along road 150 m to quarry in dacite flow. The dacite, probably a dome or small intrusion, is very fine grained, essentially aphyric, and has numerous limonitic stains along flow layers that probably represent oxidized pyrite. **0.3**
50.4 Donnybrook Viewpoint. Complex of granodiorite sills at 2 o'clock. Good exposure of Mount St. Helens tephra deposits in roadcut opposite viewpoint. **0.9**
50.9 Smith Creek Viewpoint. Good views into the Smith Creek drainage, devastated by the blast and later

lahars [*Brantley and Waitt*, 1988], and, farther east, the shield-like Indian Heaven Wilderness, made of Quaternary olivine basalt. The huge andesitic cone of Mount Adams, the second-most voluminous Cascade volcano after Mount Shasta, dominates the eastern skyline. Tertiary rocks underlying Mount St. Helens below Plains of Abraham include many small intrusions, like those seen along this road, cutting andesite flows and volcaniclastic rocks. Much of the host rock is hornfelsed even between the small intrusions, perhaps because a master intrusion is at shallow depth. **0.5**

51.0 Roadcuts in multiphase granodiorite intrusive complex of Windy Ridge. Some phases contain hornblende, a sample of which was K-Ar dated at 24.3 ± 1.3 Ma [*Evarts et al.*, 1987]. **0.1**

51.9 STOP 30. Windy Ridge Viewpoint. Restrooms. Ascend steps for best views of Spirit Lake, Harry's Ridge, hummocky debris avalanche that dams Spirit Lake, the Pumice Plain, and lahar fan directly north of crater. The Pumice Plain was at one time entirely covered by 1980 pyroclastic flows, but lahars generated by eruption-induced snow melt in the crater have swept away or covered most of the flows due north of the crater. Part of the Pumice Plain west of the lahar fan still has an intact, though deeply gullied, primary surface.

The debris avalanche dammed and partly filled Spirit Lake, displacing water so that the surface of the new lake was about 61.5 m higher than before (1036.3 m *vs.* 974.8 m). Harry Truman, who stubbornly refused to vacate his lodge on the shore of Spirit Lake, was buried beneath about 45 m of debris and 25 m of water in the southwest corner of the lake. The avalanche formed an unstable debris dam that threatened to fail catastrophically when lake level rose to the approximate height of the low point on the dam. The Army Corps of Engineers began pumping the lake in November 1982 to stabilize water level at an acceptable elevation (1055.2 m). This costly venture was ended in April 1985 when a 2.6-km, 14-million-dollar, tunnel was opened through Harry's Ridge and began to serve as a gravity-fed controlled outlet for the lake [*Sager and Chambers*, 1986]. The tunnel is 3.4 m in diameter, has a grade of 1%, and was constructed with a tunnel-boring machine advancing at an average rate of 20 m per day for 5.5 months. The portal to the tunnel, which crosses the St. Helens seismic zone with a capability of a M=7 earthquake [*Weaver and Smith*, 1983], is visible about halfway along the western shore in basaltic andesite and breccia. Opening the tunnel caused lake level to drop about 6.7 m to the elevation of the portal, resulting in the prominent shoreline with stranded logs. Water from the tunnel eventually enters the Toutle River after flowing through South Coldwater Creek and Coldwater Lake.

The northwest corner of the 1980-86 dacite dome can be seen from the hilltop [*Swanson et al.*, 1987; *Swanson and Holcomb*, in press]. At the time of writing (December 1988) the dome had last grown in October 1986 and was about 267 m high, 1,060 m long, 860 m wide, and had a volume of 74.1 x 10^6 m^3. Its volumetric rate of growth and geometry were notably regular [*Swanson and Holcomb*, in press], and each period of growth was easily predicted on the basis of precursory ground deformation and seismicity [*Swanson et al.*, 1983; 1985].

A small shed on top of Harry's Ridge houses surveying and radio equipment used by volcanologists of the U.S. Geological Survey's Cascades Volcano Observatory (CVO) to monitor volcanic activity. Most monitoring is now conducted within the crater and on the dome. CVO notifies the U.S. Forest Service and other governmental agencies when monitoring indicates significant changes in activity. All decisions regarding access and land use in Mount St. Helens National Volcanic Monument are made by the Forest Service. **0.9**

59.1 Return to Road 26 and turn left (north). **7.2**

59.7 The family of three who owned the car on display were killed in a small cabin near the portal to their prospect high on the hillside at 9 o'clock. They had reopened an adit originally excavated during a period of mining in the southern Washington Cascades around the turn of the century. The adit was driven along a polymetallic sulfide (pyrite, chalcopyrite, sphalerite) vein, one of a set of northwest-trending veins peripheral to the Earl porphyry copper deposit. The only known production from the St. Helens district was a few hundred kilograms of copper from the Sweden Mine, now under 30 m of water at the northeast end of Spirit Lake. All the copper reportedly found its way into a statue of Sacajewea, the guide for the Lewis and Clark expedition, that stands in a park in Portland [*Moen*, 1977]. **0.6**

60.0 Trailhead to Norway Pass, which affords excellent views of Spirit Lake and the crater. From Norway Pass, trails connect to Independence Pass and continue into the high country north of Spirit Lake. **0.3**

60.4 Cross from hornfelsed wallrock into Spirit Lake pluton. No place to stop. The pluton underlies the Green River basin and tree-covered hillside ahead, beneath which most of the 17-Ma Earl porphyry-copper deposit resides. This large-tonnage, low-grade deposit would probably have been open-pit mined if it were in Arizona, but mining wasn't done here, owing to 1) the eruption and presumed hazards therefrom, 2) the steep decline in copper prices in the early 1980s, and 3) environmental concerns about acid waste. Molybdenum would have been a byproduct of the copper mining. **0.4**

61.9 Looking west down Green River at 9 o'clock, all exposures are of the Spirit Lake pluton. **1.5**

62.6 Note margin of tree-blowdown area. Much of this area had been logged before 1980, as seen by the sawn stumps, but the standing dead that fringe some clearcuts provide evidence of the waning blast. **0.7**

62.8 Turn left on Road 2612 beyond Ryan Lake. **0.2**

62.9 STOP 31. Granodiorite porphyry, representative of the main phase of the Spirit Lake pluton, occurs in a nearby roadcut. The pluton consists of three phases, each displaying substantial internal compositional and textural variation [*Evarts et al.*, 1987]. The main phase, constituting most of the intrusion, ranges from diorite to granodiorite and is well dated at 21 Ma. The porphyritic texture here is typical of the main phase in the eastern half of the pluton. To the west, coarser-grained, more equigranular textures predominate. This pattern suggests that the pluton and its host rocks have been tilted eastward and eroded more deeply in the west.

Ryan Lake was once nestled in old growth forest out of sight (and mind) of Mount St. Helens. The campground was in use on May 18, 1980. Imagine the terrifying sensation when the blast cloud appeared over the hill with little if any warning. A body buried by ash alongside the outhouse was found a month after the explosion, and others were killed in the Green River basin. One miner managed to walk from his work site farther west to his pickup parked here, to drink a bottle or two of beer to try to clear his throat of the choking ash, and then to walk (downed trees kept him from driving) several kilometers down Road 26 before collapsing. An autopsy showed his lungs coated with ash.

Bus can go 0.3 miles farther to Goat Mountain trailhead to turn around and return to stop. **0.1**

63.0 Return to Road 26 and continue north. **0.1**

63.7 Note standing and fallen dead trees on east side of Quartz Creek basin. The prominent scars on the hillside record pumiceous debris flows of the last several winters. Such flows were a problem before 1980 that has been exacerbated by the May 18 deforestation. **0.7**

65.8 Cross contact from main phase, seen at Ryan Lake, into granite phase of Spirit Lake pluton. The contact is hard to define here owing to an overprint of intense deuteric alteration, but it appears gradational. Elsewhere, sharp contacts locally marked by intrusion breccia indicate that the granite phase is younger. **2.1**

66.1 STOP 32. Examine the granite phase of the Spirit Lake pluton in roadcut. Rocks of the granite phase are distinctly lighter in color, finer-grained, and more felsic than those of the main phase. Outcrops of the granite phase are concentrated along the eastern margin (top) of the exposed part of the pluton. **0.3**

66.4 At 9 o'clock, note standing dead along road. **0.3**

66.7 Leave devastated area. Green trees again! **0.3**

67.2 Roadcut in main phase of pluton. **0.5**

67.8 Back in granite phase of pluton. **0.6**

68.6 Series of roadcuts in granite phase of pluton. **0.8**

69.7 STOP 33. Screens of hornfelsed host rock, originally lithic- and pumice-lapilli tuff, cut by several aplite dikes and at least one granite dike. Contact with Spirit Lake pluton is obscured by forest just up road from here. The wallrock outside of the hornfelsed zone is coarse breccia and dacite flows, whose attitudes suggest the site of a Miocene caldera [*Evarts et al.*, 1987]. **1.1**

70.1 Hornfelsed dacite, in places with relict flow layering, and dacite breccia. **0.4**

70.7 Clastic dacite and andesite, possibly caldera fill, cut by two andesite dikes. Entire assemblage is recrystallized to albite-epidote hornfels. Similar rocks occur in discontinuous roadcuts for next 3 miles. **0.6**

71.1 At 9 o'clock, dacite flow in quarry dips east. **0.4**

72.8 Bedded volcaniclastic rocks, chiefly sandstone and siltstone, with Cispus River to right. Some of the sandstone is normally graded. **1.7**

73.1 Turn right at junction with Road 25. **0.3**

73.3 Same unit as at mileage 72.8. **0.2**

74.2 Turn left onto road 76 to Cispus Center. **0.9**

74.3 Bridge across Iron Creek. Bedded sandstone in stream cut. **0.1**

77.1 Columnar ash-flow tuff over tuffaceous sandstone, siltstone, and lithic lapilli tuff. **2.8**

79.7 Keep right at intersection. At 2 o'clock is Tower Rock monolith, an intrusion whose face was greatly steepened by glaciation in Cispus valley. **2.6**

81.2 Turn left into Cispus Center. End of log. **1.5**

Log for Optional Day 3, Packwood to Cispus Center via Mount Rainier

0.0 Depart Forest Service station in Packwood and retrace route of Day 2 on Highway 12 to Highway 123.

7.2 Turn left on Highway 123. **7.2**

7.8 Bridge across Carlton Creek, on either side of which are roadcuts in bedded Ohanapecosh Formation on east limb of Unicorn Peak syncline. Cuts for next several miles are in Ohanapecosh cut by sills. **0.6**

12.5 Stevens Canyon Road. Stay right. **4.7**

15.4 View straight ahead of Ohanapecosh Formation on Double Peak. **2.9**

17.4 Roadcuts of lapilli tuff in the Ohanapecosh cut by sills emplaced during the early stage of the Tatoosh pluton [*Fiske et al.*, 1963; *Mattinson*, 1977]. **2.0**

21.9 Good roadcuts in Ohanapecosh Formation. **4.5**

23.4 Junction with Highway 410 at Cayuse Pass (1431 m). Turn right toward Chinook Pass. **1.5**

23.6 All roadcuts from here to Chinook Pass are in Ohanapecosh Formation cut by sills. **0.2**

23.7 Good exposure of thin-bedded Ohanapecosh. **0.1**

25.1 Roadcuts at Stop 35 after bus turns around. **1.4**

26.9 Summit of Chinook Pass, 1899 m. **1.8**

27.0 STOP 34. Turn right into parking area. Restrooms across highway at lower parking lot. Scenic view down American River. Retrace route to next stop. **0.1**

28.9 STOP 35. Bus park on right side of road at curve. View of Governor's Ridge toward west, underlain by bedded Ohanapecosh. Craggy higher ridge behind Governor's Ridge is Cowlitz Chimneys and Sarvent Glaciers area, both in an Ohanapecosh lava-flow complex interpreted as a vent area that fed pyroclastic detritus into the Ohanapecosh [*Fiske et al.*, 1963].

Walk down road to observe 27-m-thick bed of pumice-lithic-lapilli tuff (probably a pyroclastic flow) and underlying thin beds of volcaniclastic rocks figured by *Fiske* [1963, Plate 4]. Fiske interpreted the thick deposit as a subaqueous pyroclastic flow, but *Vance et al.* [1987] believe it was deposited subaerially. *Fiske* [1963] interpreted the thin beds as products of turbidity currents, whereas *Vance et al.* [1987] think they were deposited by airfall or surges on land. Many sedimentologists have argued at this outcrop with no clear concensus. **1.9**

Return to Cayuse Pass.

30.8 Cayuse Pass. Turn right on Highway 410. **1.9**

33.0 Road goes down canyon of Klickitat Creek. Good view of Mount Rainier and source (Fryingpan, Emmons, and Inter Glaciers) of White River to left. **2.2**

34.0 Hidden contact of Tatoosh pluton cutting Ohanapecosh Formation [*Fiske et al.*, 1963]. **1.0**

34.3 Turn left at junction with road to Sunrise. **0.3**

Optional Trip to See Osceola Debris Flow

(0.0) Continue straight on Highway 410 at Sunrise junction. All roadcuts for next 5 km are in Tatoosh pluton, but exposure is poor because of till.

(2.4) Approximate contact of Tatoosh with volcaniclastic rocks of the Ohanapecosh [*Fiske et al.*, 1963]. **2.4**

(11.8) Pass entrance to Dalles Campground. **9.4**

(12.8) Turn left on Road 73 and at all junctions. **1.0**

(13.4) Keep left on Road 120; follow power line. **0.6**

(13.9) STOP 36. Gravel pit. Observe a cross section of the Osceola debris flow (see Stop 39 for discussion). The Osceola has a high clay content, presumably from hydrothermally altered rocks on the volcano, but here 50-60% of the clasts were derived from stream gravel and till, not the volcano. Here the upper 1-3 m of the Osceola is yellow, and the lower 2-4 m pinkish gray; this reflects ground-water oxidation in the aereated zone and is **not** a contact. The Osceola here overlies iron-oxide-coated hardpan on outwash gravel of probable Evans Creek age (*ca.* 18-15 ka). The hummocky terrain 200 m northwest of the gravel pit was once thought to be a separate lahar (the Greenwater lahar [*Crandell*, 1971]) but is now considered a marginal facies of the Osceola (J. W. Vallance, oral commun., 1988). **0.5**

Retrace route to Sunrise junction.

End of optional trip. **13.9**

39.4 White River bridge. Road starts up ridge, underlain by Tatoosh pluton, zircon-dated here as 14.1 Ma [*Mattinson*, 1977]. Till mantles much of slope. **5.1**

41.6 Start across intracanyon lava flow. **2.2**

41.8 STOP 37. Bus pullover along right side of road. View highly plagioclase-clinopyroxene-hypersthene-phyric andesite flow from Mount Rainier that partly filled the ancestral canyon of White River [*Fiske et al.*, 1963]. This site is near the terminus of the flow, and its columns are nearly horizontal where the flow cooled against canyon walls. Walk back 100 m, staying outside rock wall, to view the cross-sectional shapes of the columns and the transition to steeply platy andesite (Figure 32). The flow is one of the earliest intracanyon flows from the volcano and extends from Burroughs Mountain through Yakima Park to here [*Fiske et al.*, 1963]. It is 240 m thick at Burroughs Mountain. Opposite Yakima Park, White River has entrenched 400 m since the andesite was erupted [*Fiske et al.*, 1963]. **0.2**

42.2 Switchback at margin of the intracanyon andesite flow. Till obscures the contact. Tatoosh pluton along road to top of ridge. **0.4**

46.3 Bench on hillside below road at 3 o'clock is developed along top of intracanyon andesite flow. **4.1**

46.8 STOP 38. Parking area for viewpoint at switchback on crest of Sunrise Ridge, 1860 m. View of ridge-like Sourdough Mountains to northwest, capped mostly by lapilli tuff of the Ohanapecosh Formation intruded by the Tatoosh. Good view to south of Cowlitz Chimneys and Sarvent lava complex in the Ohanapecosh.

Examine calc-alkalic granodiorite and quartz monzonite of the Tatoosh pluton in roadcut and stone wall. Granodiorite is the dominant lithology of the pluton, but quartz monzonite is common in the roof areas, and quartz diorite and granophyre are common locally. Isotopic zircon ages suggest that the plutonic complex "was emplaced in a series of intrusive and extrusive events over a span of about 12 m.y." [*Mattinson*, 1977]. According to *Mattinson* [1977], the initial activity was intrusion of a dike and sill swarm about 26 Ma. Eruptions from the plutonic complex formed thick welded tuffs about 25 Ma and 22 Ma. The core of the pluton rose and intruded its earlier volcanic and hypabyssal cover in two stages, 17.5 and 14.1 Ma. This locality is interpreted as in the roof zone of the main pluton [*Fiske et al.*, 1963], and *Mattinson* [1977] dated this rock as 24.1 Ma. The Palisades, visible north of here, is underlain by welded tuff thought to grade downward into the pluton and has been dated as 25.1 Ma [*Mattinson*, 1977]. **0.5**

49.0 Approximate contact between Tatoosh pluton and intracanyon flow we saw earlier. This flow underlies Yakima Park and the Sunrise Lodge area. **2.2**

FIGURE 32. Columnar intracanyon andesite flow from Mount Rainier at Stop 37.

49.6 STOP 39. Flag pole at Sunrise, 1950 m elevation. The 10-12 cm-thick layer of crunchy brown pumice belongs to tephra layer C, erupted from Mount Rainier 2200-2300 years B.P. [*Mullineaux*, 1974]. Walk to Emmons Nature Trail Overlook above White River for view of the source area of the Osceola debris flow (2 km^3) that poured down the White River and West Fork White River and spread across more than 260 km^2, mostly in the Puget Sound lowland 40-70 km from the volcano [*Crandell and Waldron*, 1956; *Crandell*, 1971]. Logs in the Osceola give a bristlecone-corrected ^{14}C age of about 5700 years B.P. [*Crandell*, 1969, 1971].

The Osceola everywhere has a high content of clay (6-12% [*Crandell*, 1971]), presumably derived from hydrothermally altered rocks high on the volcano. The source area is now hidden beneath the broad upper part of the Emmons and Winthrop Glaciers and possibly includes material removed from a former summit region above 4250 m [*Crandell*, 1963b; *Crandell*, 1971]. As *Crandell and Waldron* [1956] pointed out, "The very existence of this broad expanse of ice suggests the possibility that the ice occupies a depression that resulted from the explosive destruction of part of the volcano, or from eruptive action followed by partial collapse of this area." The Osceola occurs on ridgetops above Glacier Basin and on top of Steamboat Prow (2957 m), so clearly it originated even higher on the volcano [*Crandell and Waldron*, 1956; *Crandell*, 1971]. A remnant of the debris flow is visible from here as a terrace along Inter Creek just upstream from the White River. Conceivably the debris flow could have been generated by collapse of the altered flank of the cone without accompanying eruptive activity. Whether an eruption occurred is uncertain. Tephra layer F [*Mullineaux*, 1974] from Mount Rainier is clay-rich like the Osceola and has a corrected ^{14}C age of 5.7-5.8 ka; it has been found neither above nor below the Osceola, however, perhaps owing to unfavorable wind directions. A reasonable supposition is that layer F records phreatic activity preceding or accompanying generation of the avalanche and resulting debris flow. The Osceola is 2-7 m thick along its margins but presumably much thicker in its center. Remnants of the debris flow on the sides of the White River and West Fork White River valleys show that the thickness was locally at least as much as 150 m while the debris flow was moving. A three-dimensional model in the visitor center helps to visualize the geometry and enormity of the Osceola. A measure of its size is the profound effect it had on Puget Sound more than 100 km from the volcano; the shoreline was displaced seaward 27-50 km as more than 460 km^2 of new land was created! **0.6**

Return to Highway 410 and thence southward past Cayuse Pass to junction with Stevens Canyon road.

79.3 Stevens Canyon road. Turn right toward Paradise. Road traverses through volcaniclastic rocks of the Ohanapecosh for next 8.5 km. **29.7**

84.4 Start of roadcuts on Backbone Ridge in upper part of Ohanapecosh Formation, according to *Fiske et al.* [1963]. Note the abundance of wood and presence of what *Fiske et al.* called a saprolite (fossil soil). **5.1**

84.7 Basal zone of Stevens Ridge Formation above the Ohanapecosh. **0.3**

84.8 STOP 40. Bus can let group off at narrow paved area on right side of road and park at trailhead 0.1 mile farther. Walk back 180 m to view basal zone in quartz-plagioclase-phyric welded ash-flow tuff of Stevens Ridge Formation. The ash flow plowed into loose "soil" atop the Ohanapecosh and incorporated blocks of the soil into its base (Figure 33) [*Fiske et al.*, 1963]. Note the intimate mixture of soil with the ash-flow tuff, as well as larger discrete pieces of soil within the tuff. Same site as that photographed in *Fiske et al.* [1963, figure 22]. Rock hammers can be used here; we are just outside the National Park! The ash-flow tuff here has been dated at 27.3 ± 2.9 Ma by *Vance et al.* [1987, sample JV 32].

Quartz-phyric ash-flow tuffs are abundant in the Stevens Ridge Formation, but according to *Vance et al.* [1987] they also occur in the underlying Ohanapecosh and hence are not useful in discriminating the two units. The tuff of Lake Keechelus (Day 1) is an example. *Vance et al.* [1987] therefore propose to demote the Stevens Ridge as defined by *Fiske et al.* [1963] to member status and assign it to the Fifes Peak Formation. **0.1**

87.0 Base of Stevens Ridge Formation hidden in forest. Road re-enters Ohanapecosh. **2.2**

89.6 STOP 41. Restrooms. Box Canyon of the Cowlitz River, a narrow gorge cut into glacially smoothed and polished diorite. Roadcut exposes the diorite, a minor but ubiquitous component of the Tatoosh pluton. **2.6**

Roadcuts for next 1.2 miles show diorite and granodiorite of Tatoosh pluton.

91.5 Prominent roadcuts in welded ash-flow tuff of Stevens Ridge Formation. **1.9**

91.8 Good view across Stevens Canyon to young intracanyon andesite flow just above creek. Note the complex columnar joints, which *Fiske et al.* [1963] suggest result from flowage along the side of a glacier. **0.3**

92.3 Granodiorite of the Tatoosh pluton in roadcuts. This unit continues for about 1 km. **0.5**

FIGURE 33. Dark pods of "soil" mixed in base of ash-flow tuff in Stevens Ridge Formation, Stop 40.

92.7 STOP 42. View granodiorite of the Tatoosh pluton here, if the Sunrise road is snowed in. See comments about the Tatoosh under description of Stop 38. The roadcuts are some of those used by *Dunn et al.* [1971] to document a magnetic field reversal (reversed to normal) during cooling of the pluton. **0.4**
93.5 Bridge across Stevens Creek. Cross a narrow intracanyon flow of andesite, then quickly return to granodiorite and in turn back into andesite. **0.8**
94.4 Intracanyon andesite visible high at 3 o'clock. **0.9**
94.7 Same andesite in roadcut at switchback. **0.3**
95.6 A narrow tongue of granodiorite in roadcut, with andesite around next curve. **0.9**
95.9 Louise Lake to right. Road from here to Reflection Lakes is along contact of Mount Rainier andesite flow and granodiorite. **0.3**
96.8 Reflection Lakes. Continue in granodiorite. **0.9**
97.8 STOP 43. Panoramic view of Mount Rainier from turnout on left: from east to west, top of Little Tahoma Peak, Gibraltar Rock, Nisqually Glacier, Wapowety Cleaver, and Success Cleaver. Note columnar Mount Rainier andesite above road. **1.0**
98.1 Keep left at junction. **0.3**
98.3 Turn right. Roadcut in granodiorite. **0.2**
98.9 Roadcut in andesite flow from Mount Rainier. Similar cuts continue sporadically to Paradise. **0.6**
99.7 Stay left at junction. **0.8**
100.2 STOP 44. Paradise visitor center. Restrooms, cafeteria, gift shop, and museum. General views, especially of the mountain and the Tatoosh Range.

Leave Paradise on Longmire road. **0.5**
102.1 Keep right toward Longmire at junction. Next roadcuts are in andesite. **1.9**
103.3 Platy andesite in roadcut. **1.2**
105.0 Continue right at junction. Road passes from andesite into the Tatoosh pluton as it descends into Nisqually River valley. Note snout of Nisqually Glacier covered by moraine. In early 20th century, the glacier extended nearly to the bridge. **1.7**
106.9 Roadcuts in granodiorite. **1.9**
108.2 View through trees across canyon of thick, complexly columnar intracanyon andesite flow. This is the flow that we were driving on near Paradise. **1.3**
109.0 View ahead of granodiorite on Cougar Rock. **0.8**
109.3 Roadcuts and natural exposures in volcaniclastic rocks of the Ohanapecosh Formation. **0.3**
114.2 Kautz Creek, scene of a series of lahars on October 2-3, 1947, induced by a cloudburst possibly enhanced by a glacier-outburst flood [*Crandell*, 1971]. **3.1**
116.4 Tahoma Creek, site of numerous debris flows in the middle and late 1980s fed by glacier-outburst floods from South Tahoma Glacier. The floods typically occurred in late afternoon of warm days. Water was photographed leaving the glacier well above its terminus. The debris flows repeatedly covered the Westside Road, which was closed to public travel in summer 1988. **2.2**
117.6 Leave Mount Rainier National Park. **1.2**

End of geologic log. To drive to Cispus Center, continue on Highway 706 from National Park to Elbe (10 miles), turn left on Highway 7 to Morton (16 miles), and then turn left on Highway 12 to Randle (17 miles). Thence follow log from Randle to Cispus Center.

Alternatively, turn left 3.3 miles outside National Park on Road 52 toward Packwood. After 1.8 miles turn left on Road 52 (here called Osborn Road) and follow it to the Forest Service Station in Packwood, 21.4 miles farther. (Most of this road was gravel in fall 1988 but was being readied for paving.) Thence follow log from Packwood to Cispus Center.

Total mileage to Cispus Center by either route is 177 miles (205 miles with the optional trip to see the Osceola debris flow.)

Log for Day 5, Cispus Center to Hood River via Indian Heaven

Retrace route back to Road 25 and then to junction with Road 99, where log starts.
0.0 Continue south on Road 25.
0.6 Andesite flow in roadcut. **0.6**
2.8 Pumice-lithic-lapilli tuff, probably ash-flow tuff. **2.2**
4.9 Basaltic andesite flow cut by narrow, petrographically similar dike trending 280°. This trend typifies that of a major dike swarm in this area. **2.1**
5.7 Basaltic andesite, with complex platy jointing. **0.8**
6.1 Lithic-pumice-lapilli tuff, probably an ash flow, that slides easily because of clay alteration. Underlies andesite flow seen in previous roadcut. **0.4**
6.8 At 2 o'clock is stack of thin near-vent basaltic andesite flows, slightly older than flows in roadcuts. **0.7**
6.9 Start cuts in mafic intrusion. Probably subvolcanic, but overall shape is unknown. **0.1**
8.8 Continue straight on Road 25 at junction. *Note: Major road work was underway when the trip was logged from here to Road 2573. Exposures and mileages may differ significantly from those given below.* **1.9**
9.0 Clearwater Overlook. Cut in finely hornblende-phyric fragmental dacite flow or ash-flow tuff. **0.2**
9.4 This unit cut by 5-m-thick columnar sill of plagioclase-phyric andesite. Similar exposures occur for next 0.4 miles. **0.4**
10.2 Andesite flow(?) in roadcut. **0.8**
10.4 STOP 45. Start of 0.6 miles of roadcuts in Quaternary olivine basalt of Paradise Falls [*Korosec*, 1987a], which erupted on the ridge crest just east of the road and poured into the valley of Clearwater Creek. It contains large phenocrysts of olivine (rimmed by iddingsite in oxidized samples) but none of plagioclase.

The vent for this flow lies midway along a N35°E line of Quaternary vents extending for about 50 km from Tumtum Mountain south of the Lewis River to Badger Ridge south of Cispus Center (Figure 34). Six vents lie along the Tumtum-Badger Ridge line, and only a few other Quaternary vents occur near it, except for a cluster around Mount St. Helens. The line can be extended with only slight bends to include a vent east of Cispus Center and the Battle Ground Lake maar [*Mundorff*, 1964] and adjacent vent [*Walsh et al.*, 1987], a total length of about 87 km (Figure 34). K-Ar whole-rock ages are available for the basalt of Paradise Falls (0.04 ± 0.03 Ma [*Hammond and Korosec*, 1983]), a flow on Marble Mountain (0.16 ± 0.01 Ma [*Hammond and Korosec*, 1983]), and the olivine basaltic andesite of Badger Ridge (0.65 ± 0.02 Ma; J. G. Smith, written commun., 1986). *Mundorff and Eggers* [1988] interpret Tumtum Mountain as slightly younger than the Amboy drift (a likely correlative [*Walsh et al.*, 1987] to the *ca.* 0.14-Ma Hayden Creek Drift). Battle Ground Lake maar is older than 20.2 ± 0.4 ka (the age of the oldest dated sediment in the lake [*Barnosky*, 1985]. The undated vents have normal magnetic polarity (except for that east of Cispus Center) and so are younger than about 0.73 Ma. Seven vents produced olivine basalt or basaltic andesite, but Tumtum Mountain erupted hornblende dacite (68% SiO_2) [*Mundorff and Eggers*, 1988]. Marble Mountain, the largest Quaternary vent area in southern Washington west of Indian Heaven (except for Mount St. Helens), grew where the line crosses the St. Helens seismic zone [*Weaver and Smith*, 1983].

Whether the Tumtum-Badger Ridge line reflects tectonic control is unclear. Relatively few Tertiary dikes in the area parallel the line, which is slightly oblique to the plate-convergence direction of N50°E [*Riddihough* [1984]; minor sinistral shears have northeast trends. The line does not extend northeast of the Cispus-Lewis fault zone mapped by *Hammond* [1980] or southwest of the Lacamas Lake fault and associated lineament *Walsh et al.* [1987]. The line is not directly reflected by the southern Washington Cascades conductor (SWCC) or other anomalies [*Stanley et al.*, 1987], although its projected trend merges into and parallels that of conductance contours in the south part of the SWCC. **0.2**

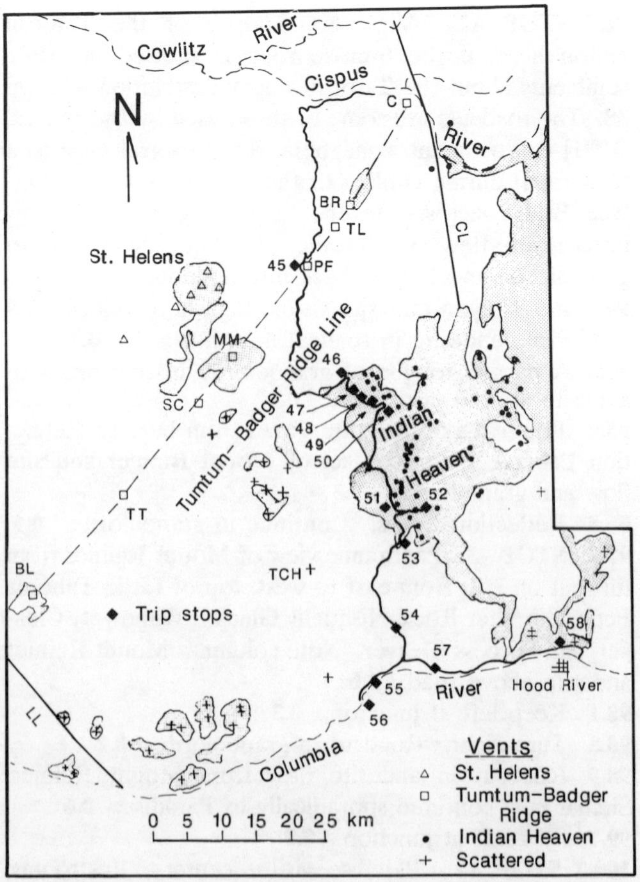

FIGURE 34. Map showing Pliocene and Quaternary vents (keyed to location) and major flow fields (shaded) near field-trip route between Cispus Center and Hood River. Tumtum-Badger Ridge line connects isolated vents. BL, Battle Ground Lake; TT, Tumtum Mountain; SC, vent south of Swift Dam; MM, Marble Mountain; PF, basalt of Paradise Falls; TL, The Loaf; BR, Badger Ridge; C, basalt east of Cispus Center; TCH, Trout Creek Hill; LL, Lacamas Lake lineament; CL, Cispus-Lewis fault zone. See text for discussion.

11.6 Plagioclase-phyric, oxidized, Tertiary andesite. Similar cuts occur sporadically for next 5 miles. **1.2**

17.0 Pumice-lapilli tuff, probably an ash flow. **5.4**

19.3 Section of Mount St. Helens tephra deposits in roadcut above oxidized Tertiary ash-flow tuff. **2.3**

20.0 Small roadcut of ash-flow tuff, with andesite flow just beyond. **0.7**

20.4 Alder-covered surface of lahar of May 18, 1980, which was generated on the southeast flank of Mount St. Helens by rapid melting of eroded snow incorporated into the blast surge and possibly enhanced by water ejected from the volcano [*Pierson*, 1985]. Mudlines 4-6 m above this surface (still evident on some trees in 1988) indicate maximum thickness of flowing lahar. The lahar reached a peak velocity of 30-40 m/s near the base of the volcano but had slowed to an approximately constant 3-6 m/s by the time it reached here [*Pierson*, 1985]. The lahar was generated soon after the start of eruption at 0832 and reached the site of the bridge (destroyed by the lahar) at about 0856. A smaller, pumiceous lahar, generated by a pyroclastic flow on the volcano's southeastern slope between 1530 and 1610, probably reached the site of the bridge at about 1625 [*Pierson*, 1985]. **0.4**

20.7 Bridge across Muddy River. From here to Pine Creek the road crosses pyroclastic-flow and laharic deposits older than 2200 years B.P. [*Phillips*, 1987]. **0.3**

24.3 Pine Creek. Another branch of the first May 18 lahar roared down Pine Creek at a peak discharge of about 29,000 m^3/s [*Pierson*, 1985]. Here erosion on one side of the channel was balanced by deposition on the other [*Pierson*, 1985]. The Muddy River and Pine Creek lahars destroyed 16 bridges, buried several kilometers of road, and raised the level of Swift Reservoir 0.8 m by depositing 13.4 million m^3 of slurry [*Cassidy et al.*, 1980 (quoted in *Pierson*, [1985]); *Schuster*, 1981]. **3.6**

25.1 Turn left at junction with Road 90. **0.8**

25.5 Cross Lewis River at east end of Swift Reservoir. Most logs along river were carried by lahars on May 18,

1980. Eagles Cliff to left is columnar welded ash-flow tuff, probably of early Miocene age [*Phillips*, 1987], overlain by Quaternary(?) basalt of Thomas Lake [*Hammond*, 1987; *Korosec*, 1987a; *Walsh et al.*, 1987]. **0.4**
26.1 Roadcuts on left are in lahar deposit that predates the Castle Creek eruptive period [*Phillips*, 1987]. **0.6**
26.4 Start long roadcut in highly plagioclase-phyric, densely welded ash-flow tuff of Eagles Cliff. The tuff underlies ridge left of road for next 2 km. **0.3**
28.0 Roadcuts of bedded lapillistone, tuff, and wood-bearing mudstone. Andesite dike 3 m wide cuts section. Similar roadcuts occur for next 600 m. **1.6**
30.2 Landslide blocks of the basalt of Thomas Lake [*Korosec*, 1987a]. **2.2**
30.6 Stay on Road 90 at junction with Road 51. **0.4**
31.5 Lithic-pumice lapilli tuff at Rush Creek. Cliff of basalt of Thomas Lake high ahead. **0.9**
32.3 Roadcuts of augite-plagioclase-olivine-phyric, vesicular Quaternary(?) basalt of Thomas Lake. A whole-rock K-Ar age of 3.8 ± 0.5 Ma was obtained for this flow from next series of cuts [*Hammond and Korosec*, 1983]; however, this age is analytically questionable (only 2.7% radiogenic argon), and the normally magnetized basalt is mapped as Quaternary on the basis of field relations and general appearance [*Korosec*, 1987a]. Nonetheless, the basalt is one of the oldest units of the Indian Heaven volcanic field [*Korosec*, 1987a; *Hammond*, 1987]. It was erupted just south of East Crater (15 km southeast of here) and flowed northwest into the ancestral Lewis River. The unit has a volume of about 4 km^3 [*Korosec*, 1987a] and is locally as thick as 150 m in the ancestral Lewis River valley [*Hammond*, 1987]. **0.8**
32.5 Basalt of Thomas Lake in discontinuous roadcuts for next mile. **0.2**
33.5 STOP 46. Possible bus parking at last chance to observe basalt of Thomas Lake. **1.0**
33.9 Turn right on obscure gravel road (3211) in middle of straight segment of Road 90. **0.4**
36.1 Intraglacial basalt of Burnt Peak in roadcut (Stop 14 of *Hammond* [1987]). Radially jointed olivine basalt is partly enclosed in pillow lava. **2.2**
36.6 Pillowed basalt of Burnt Peak, with minor palagonitized hyaloclastite. **0.5**
37.0 STOP 47. Stop 13 of *Hammond* [1987]. Pillows in the basalt of Burnt Peak dominate the outcrop. They invade thinly bedded lacustrine and fluvial or turbidite "black sand" hyaloclastite (Figure 35). The sediments terminate against a nearly vertical invasive or intrusive contact near the left side of the cut. The interstices between the pillows are filled by bedded sediment, which is locally distorted, has truncated bedding, and forms a pépérite with glassy basalt. A lower layer of pillows has similar interstitial fillings and clearly invades white laminated lacustrine(?) siltstone. The siltstone is overlain by bedded sand possibly deposited on top of the basalt; the bedded sand contains a few large blocks of glassy basalt that lack pillow forms and may have broken off a pillow complex. Overlying the pillow basalt and sediment is lithic-rich hyaloclastite in erosional contact with the bedded sand. The hyaloclastite is crudely layered and has west-dipping foreset bedding. A small fault offsets the diamict-basalt contact near the left end of the cut. **0.4**
37.2 Narrow basalt dike intrudes pillows and hyaloclastite. Similar exposures continue for 200 m. **0.2**
37.5 Broken pillow breccia and hyaloclastite. **0.3**
37.8 STOP 48. Stop 12 in *Hammond* [1987]. Well-bedded, reworked hyaloclastite, white because of opal cement, grades laterally into primary hyaloclastite cored by jointed and quenched basalt. Exposure extends about 100 m in roadcut. Bus park at start of cut. **0.3**
38.0 Top of plateau in Crazy Hills, so named by early surveyors who were puzzled by the morphology, which is dominated by erosionally modified moberg hills (hills consisting of palagonitized hyaloclastite and pillows) [*Kjartansson*, 1959]. The Crazy Hills moberg field covers about 21 km^2, has a volume of about 1.7 km^3, and contains hills and ridges 122 to 366 m high [*Hammond*, 1987]. According to Hammond, the hills are composed chiefly of palagonitized hyaloclastite in massive, thick to thin beds containing clasts of silt to boulder size. The basal parts of the hills are thought to consist of pillow lava, and the entire complex is interpreted to overlie subglacial and intraglacial lava flows and pillow lavas. The higher hills are held up by dikes, sills, and lava flows. View straight ahead of Lone Butte [*Hammond*, 1987], a tuya (volcano that erupts initially beneath a glacier, melts through the ice, and develops an upper, subaerial, commonly flat top [*Jones*, 1969]).

Lone Butte and the Crazy Hills apparently formed during the Hayden Creek Stade of the Salmon Springs glaciation, probably about 0.14 Ma [*Hammond*, 1987; *Crandell and Miller*, 1974; *Coleman and Pierce*, 1981]. The glacier filled the Lewis River valley, spilled onto the site of the Crazy Hills, and merged with a local glacier from the northern end of Indian Heaven. The hills and tuya have been modified by the later Evans Creek glaciation, which climaxed about 18-15 ka. **0.2**
38.1 Hyaloclastite in cut on curve. After rounding

FIGURE 35. Subglacial pillowed basalt invading lacustrine silt (light unit in floor of depression) overlain by sand at Stop 47. Crudely layered hyaloclastite overlies the sand. Height of exposure about 10 m.

curve, Burnt Peak is straight ahead. **0.1**
38.5 Keep right on Road 3211 at junction. **0.4**
38.9 Broken pillow hyaloclastite. **0.4**
39.1 Crudely bedded palagonitized hyaloclastite, possibly reworked. Stop 5 in *Hammond* [1987]. **0.2**
40.2 STOP 49. Stop 4 in *Hammond* [1987]. Small quarry shows pillowed basalt of Burnt Peak invasive into crudely bedded hyaloclastite. Pépérite halos surround sediment clasts between pillows. At west end of exposure, steeply dipping dikes of hyaloclastite cut bedded hyaloclastite, as if squeezed up from below by weight of pillow complex. In middle of cut above the pillow zone is a possible pipe of explosion breccia containing abundant angular lithic clasts of basalt with differing degrees of vesicularity. The pipe is cut off by the pillows. The explosion pipe dies out upward, but traces of it continue to top of second cut not visible from road. Other narrow sediment dikes cut the hyaloclastite in the second cut; walk above the exposure to see this. Small faults cut tilted hyaloclastite at east end of cut. **1.1**
40.4 Turn right at junction with Road 32. **0.2**
40.6 Turn left at junction with Road 30. **0.2**
40.7 View ahead of 430-m-high, 0.3-km^3 Lone Butte tuya [*Hammond*, 1987]. According to Hammond, the lower 230 m of the tuya is palagonitized hyaloclastite, overlain by 40 m of hyaloclastic surge deposits at the eastern end of the tuya and 40 m of foreset-bedded pillow breccia at the western end (exposed in the quarry visible from here). These subaqueous deposits are capped by a 145-m-thick remnant of a subaerial scoria cone, whose western flank is overlain by 65 m of subaerial basalt flows. Four dikes cut the hyaloclastite. At least 3 hours hiking time are needed to see these relations. **0.1**
41.9 At 10 o'clock is subaerial cap on Lone Butte. **1.2**
42.3 Turn right on Road 65. **0.4**
43.6 STOP 50. Bus can stop just before bridge over Rush Creek. Basalt of Outlaw Creek, a fine-grained, olivine-phyric diktytaxitic basalt with about 51.6% SiO_2 and 0.96% K_2O (P. E. Hammond, unpub. data), crops out along the creek. This flow is stratigraphically the oldest in the Indian Heaven field. **1.3**
47.9 Roadcut in basalt of Indian Heaven, diktytaxitic and highly plagioclase- and olivine-phyric [*Hammond*, 1987]. Flow was erupted from west vent of 120-m-high East Crater in the Indian Heaven Wilderness between 29 ka and Evans Creek time [*Hammond*, 1987]. The basalt has about 50.2% SiO_2 and 0.45% K_2O (P. E. Hammond, unpub. data, 1985). **4.3**
48.4 Roadcut in basalt of Indian Heaven. **0.5**
52.1 Continue straight at junction, crossing drift-covered andesite of Black Creek [*Korosec*, 1987b]. **3.7**
53.3 At 9 o'clock is The Wart, a 145-m-high cinder cone that erupted a small basaltic andesite flow just prior to Evans Creek time. Also visible is top of 1514-m-high Red Mountain (with lookout tower), a vent complex that erupted andesite, basaltic andesite, and olivine basalt several times, most recently between the Hayden Creek and Evans Creek stades (P. E. Hammond, written commun., 1983; [*Wise*, 1970]). **1.2**
53.8 STOP 51. Roadcut in basaltic andesite of The Wart (P. E. Hammond, written commun., 1983), included with andesite of Black Creek by *Korosec* [1987b]. Fine-grained and finely plagioclase-olivine-phyric. Contains about 53.8% SiO_2 and 0.8% K_2O (P. E. Hammond, unpub. data, 1985). **0.5**
54.0 Turn left toward Trout Lake on Road 60 and stay on this road to Stop 52. **0.2**
54.9 Small cut in basalt of Sheep Lakes, erupted from east base of Red Mountain just before eruption at The Wart (P. E. Hammond, written commun., 1983). **0.9**
55.3 Basalt of Big Lava Bed visible right of road. **0.4**
56.3 Cross west margin of Big Lava Bed tephra. **1.0**
56.7 STOP 52. Glaciated cinder cone covered by 3 m of Big Lava Bed tephra deposit along edge of Big Lava Bed flow. The cinder cone is oxidized, weathered, and clay-rich. It is locally overlain by 5 m of Evans Creek Drift (P. E. Hammond, written commun., 1983).

The Big Lava Bed tephra deposit was erupted from a 295-m-high cone about 3.5 km east of here. This cone is the source of the tube-fed basalt flow that forms Big Lava Bed itself. The tephra blanket is oval, extends northeastward toward Trout Lake, covers about 200 km^2, and contains about 10^5 m^3 (P. E. Hammond, written commun., 1985). The Big Lava Bed flow moved west from the cone before turning south, eventually reaching the Little White Salmon River nearly 17 km from its source [*Korosec*, 1987b]. It consists of many thin flow units fed by lava tubes. Pressure ridges and plateaus testify to inflation of the flow when lava poured into blocked tubes. A platform at the west base of the source cone was apparently uplifted about 60 m during the eruption (P. E. Hammond, written commun., 1985). The Big Lava Bed flow contains about 0.9 km^3, covers 59 km^2 [*Korosec*, 1987b], and is 8200 ± 100 years ^{14}C years B.P. [*Hammond and Korosec*, 1983]. **0.4**

Return to Road 65.

59.3 Turn left on Road 65. Andesite of Black Creek underlies Evans Creek Drift. A K-Ar whole-rock age of 3.3 ± 0.25 Ma was obtained on the andesite about 1 km south of here [*Hammond and Korosec*, 1983]; *Korosec* [1987b] considers this is too old, partly because no reversely magnetized flows overlie the andesite (or occur anywhere in the Indian Heaven area). **2.6**
62.2 Road winds into canyon of Panther Creek, which is eroded into middle Tertiary volcaniclastic rocks unconformably capped by the andesite of Black Creek [*Wise*, 1970; *Korosec*, 1987b]. **2.9**
62.6 Roadcuts in bedded volcaniclastic rocks. *Korosec* [1987b] considers these and other Tertiary rocks along Panther Creek to be of late Oligocene age. *Wise* [1970] tentatively correlated them with the Ohanapecosh Formation on the basis of lithology and reconnaissance tracing of the Ohanapecosh southward from Mount Rainier by R. S. Fiske and A. C. Waters in the early

1960s. A firm basis for this correlation is lacking. **0.4**
62.8 STOP 53. Quarry in columnar intracanyon flow of olivine-phyric andesite of Black Creek. This flow came down the valley of ancestral Panther Creek from an unknown vent near Red Mountain. A similar remnant occurs in 0.9 miles. **0.2**
64.1 Start series of cuts in middle Tertiary rocks. **1.3**
70.2 Turn right at junction with Old State Highway. **6.1**
70.3 Turn left at T-intersection. For next 8 km the road follows the Wind River valley on the surface of a basalt flow erupted from Trout Creek Hill 12 km northwest of here [*Wise*, 1970; *Korosec*, 1987b]. **0.1**
71.3 Skyline hill with tall antenna at 12 o'clock is Mount Defiance, Oregon, a middle or late Pleistocene shield volcano of basaltic andesite and andesite on the rim of the Columbia Gorge (D. R. Sherrod, written commun., 1988; [*Hammond*, 1980]). **1.0**
72.0 Small hill of Tertiary lapilli tuff surrounded by basalt of Trout Creek Hill. **0.7**
72.2 STOP 54. Bus parking just before roadcut in basalt of Trout Creek Hill. Coarsely olivine-phyric (clots to 5 mm or more) diktytaxitic basalt. Several flows were erupted from Trout Creek Hill about 0.34 ± 0.075 Ma (whole-rock K-Ar age) [*Hammond and Korosec*, 1983; *Korosec*, 1987b]. The cumulative thickness of flows in the Wind River valley is at least 70 m, measured below the bridge 500 m southeast of here [*Wise*, 1970; *Korosec*, 1987b]. The flows moved down valley to the Columbia River and built a hyaloclastite dam across the Columbia [*Waters*, 1973]. The Wind River formed a delta 45 m thick into the temporary lake [*Waters*, 1973] before the Columbia breached the dam and drained the lake. **0.2**
72.5 Bridge across Wind River, where the river crosses the 70-m-thick basalt fill just upstream from the mouth of Panther Creek. Before basalt flooded the ancestral drainage, Panther Creek and Wind River must have joined near the spot where we first met the Wind River Road. The flows diverted Panther Creek along the northeast margin of the valley fill and Wind River along the southwest margin. Enter Carson. **0.3**
72.9 Roadcut in basalt of Trout Creek Hill. **0.4**
74.6 Conical hill between 9 and 10 o'clock is Wind Mountain, a quartz diorite intrusion (see Stop 57). **1.7**
75.3 Roadcut in columnar andesite flow in the lower Miocene volcanic rocks of Stevenson Ridge [*Korosec*, 1987b]. Roadcuts from here to the outskirts of Stevenson are in this unit, which *Korosec* [1987b] estimates from K-Ar ages and geology to be 24-21 Ma. **0.7**
76.1 Turn right on Highway 14 and drive west along the Columbia River, ponded behind Bonneville Dam. **0.8**
76.5 Start prominent series of roadcuts in andesite flows and minor volcaniclastic rocks of Stevenson Ridge. **0.4**
78.9 Enter Stevenson, on volcaniclastic rocks of the 22-17-Ma Eagle Creek Formation [*Korosec*, 1987b]. **2.4**
79.2 Turn left on Russell Avenue at flashing caution light in front of Court House. **0.3**
79.3 Cross railroad tracks, turn right, and park opposite Waterfront Park. Walk to viewing area. **STOP 55.** (Stop 18 in *Tolan and Beeson* [1984b]). A thick red hyaloclastite of Pliocene or Pleistocene high-alumina basalt is exposed 500-600 m above the river. The hyaloclastite must have formed in water, yet the deposit is too low to have formed subglacially. *Tolan and Beeson* [1984b] reason that the most likely source of water was from the Columbia River and "think that the complex probably formed near present river level when the river was invaded by high-alumina basalt lava flows and was later elevated by Cascade upwarping while the river incised the Columbia River Gorge" (p. 112). They feel that the incision would have occurred in the last 2 m.y., the approximate age of the youngest alluvial deposits of the ancestral Columbia (the Troutdale Formation) that were incised during uplift [*Tolan and Beeson*, 1984a, c; *Lowry and Baldwin*, 1952; *Trimble*, 1963]. Their conclusion is rather controversial, but the main point is that Cascade uplift in this area is of late Neogene age, whatever the exact age might be.
Back bus up and return across tracks. **0.1**
79.6 Turn left on First Street, right at stop sign, and left on Highway 14. **0.3**
79.7 View ahead of part of the Columbia Gorge. On north side, dark, nearly shear cliffs of Grande Ronde Basalt overlie light-colored bedded volcaniclastic rocks of the Eagle Creek Formation that dip south, are rich in clay, and slide easily. The area south of the cliffs is the giant Bonneville landslide complex, parts of which are still active and on which the northern end of Bonneville Dam, the first dam constructed across the Columbia, is based. On the Oregon side of the gorge, the cliffs are mainly Grande Ronde Basalt capped in places by Pliocene and Pleistocene basalt and andesite flows. Locally, the Priest Rapids and Pomona Members partly fill channels carved in the Grande Ronde [*Tolan and Beeson*, 1984c]. From here to Bridge of the Gods, Highway 14 crosses three of the four units that form the Bonneville landslide complex: the Red Bluff, Mosley Lakes, and Bonneville slides [*Wise*, 1961; *Minor*, 1984]. **0.1**
82.1 Turn left at Historical Marker sign and stop in parking lot. **STOP 56.** View of Bridge of the Gods at the site of the original "Bridge of the Gods" of Native American legend, which formed in about 1100 A.D. [*Minor*, 1984] when the Bonneville landslide temporarily dammed the Columbia. The legend tells of crossing the river on dry land. The river was about 1.5 km wide and relatively straight before the slide displaced it at least 1 km southward. The resulting dam impounded a lake whose surface elevation reached 60-90 m above sea level and whose water drowned forests for 60 km upstream. After a few weeks the dam broke and generated a major flood [*Minor*, 1984]. The breaching resulted in the narrowest constriction in the Columbia Gorge and produced four main rapids with a total fall of 11 m [*Hodge*, 1938]. Native Americans established villages and seasonal camps, now important archaeologic sites, along the

rapids for fishing and trade control. **2.4**

Return to road to Carson (mileage 76.1).

87.9 Continue east at junction with road to Carson. **5.8**

88.4 Start series of cuts in flow and talus of basalt of Trout Creek Hill. **0.5**

90.5 Wind Mountain intrusion at 11 o'clock. **2.1**

91.8 STOP 57. Wind Mountain intrusion, a fine-grained plagioclase-phyric quartz diorite, in places with hypersthene and hornblende phenocrysts. The intrusion contains inclusions of Grande Ronde Basalt, and basalt on the west side of Dog Mountain, 3 km farther east, was hornfelsed by the intrusion. A similar quartz diorite occurs south of the river at Shellrock Mountain. These shallow bodies, and several others in the area, presumably were related to volcanic activity. K-Ar whole-rock ages of 4.9 ± 0.1 Ma [*Korosec*, 1987b] and 6.6 ± 0.7 Ma (T. L. Tolan in *Korosec* [1987b]) have been obtained for the Wind Mountain intrusion, and one of 5.7 ± 0.6 Ma (T. L. Tolan in *Korosec* [1987b]) exists for Shellrock Mountain. No clasts of these quartz diorite bodies have been identified in the coeval Troutdale Formation [*Tolan and Beeson*, 1984b]. If the intrusions fed volcanoes, where is the eruptive debris? **1.3**

93.8 Start of long series of roadcuts in tectonically fractured and faulted flows of Grande Ronde Basalt in the Dog Mountain section, the thickest (more than 1.3 km) unrepeated section of Grande Ronde Basalt (almost entirely the R_2 magnetostratigraphic unit) exposed west of the Columbia Plateau. **2.0**

95.9 At 12 o'clock is narrow dike-like fault breccia 10-20 m above road. **2.1**

97.3 Bridge across the Little White Salmon River. Thin flows of Quaternary olivine basalt from Underwood shield volcano, about 7 km northeast of here, cap tilted flows of Grande Ronde Basalt at 10 o'clock. Start series of five tunnels through tilted Grande Ronde. **1.4**

99.5 Good exposure of olivine basalt flows of Underwood volcano at 12 o'clock. One of these flows has a K-Ar age of 0.85 ± 0.02 Ma [*Korosec*, 1987b], although at least some flows from the volcano are normally magnetized, not reversely as would be expected for this age. From here to sawmill at Hood these flows cap Grande Ronde and are pillowed where they poured into the ancestral Columbia. **2.2**

103.5 Quartzite-bearing gravel and sand of the ancestral Columbia River. The deposit is younger than the 12-Ma Pomona Member of the Saddle Mountains Basalt and correlates with part of the Troutdale Formation farther west and the Snipes Mountain conglomerate on the Columbia Plateau [*Tolan and Beeson*, 1984b]. **4.0**

103.8 Turn left on Cook Underwood Road at mouth of White Salmon River. **0.3**

104.4 STOP 58. Deposit of a huge debris avalanche probably formed by collapse of the north or northeast flank of an ancestral Mount Hood 0.05-0.1 Ma. The avalanche traveled down Hood River, crossed the Columbia, and back-filled 5 km up the White Salmon River, a distance of 40 km [*Vallance*, 1986]. Here the deposit overlies gravel of the ancestral White Salmon River and underlies sand and gravel of the Late Pleistocene Missoula floods. The base of the avalanche deposit is 30 m above present lake level and 30 m thick. The deposit is about 80 m thick across the Columbia in Hood River, where its top is at about 105 m elevation and its base only slightly above river level (22 m). **0.6**

105.2 Turn left at fire-station sign. **0.8**

105.3 Turn around at Underwood Comunity Center and return to Highway 14. **0.1**

106.8 Turn left on Highway 14. **1.5**

107.0 Stratified hyaloclastite and pillowed olivine basalt erupted from White Salmon volcano about 2 km farther east. Its age is unknown, but *Korosec* [1987b] suggests it is coeval with the basalt of Underwood Mountain. The basal flows have normal magnetic polarity and are probably younger than 0.73 Ma. Similar exposures continue to White Salmon. **0.2**

108.6 Turn right across Hood River Toll Bridge. **1.6**

109.7 Intersection south of toll booth. End of log. **1.1**

Log for Day 6, Hood River to Portland via Mount Hood

0.0 First 4-way stop south of toll bridge in Hood River. Follow Highway 35 south over railroad tracks.

0.4 Stop and continue south on Highway 35, which climbs up Hood River Valley. Roadcuts in Frenchman Springs Member of Wanapum Basalt [*Swanson et al.*, 1981; *Korosec*, 1987b]. **0.4**

1.8 Roadcut in olivine basalt underlying much of Hood River Valley. Pliocene or Pleistocene in age [*Korosec*, 1987b]. **1.4**

2.1 Roadside quarry in this olivine basalt. **0.3**

2.2 Start of cuts in debris flow from Mount Hood (see Stop 58). **0.1**

2.4 Turn left at junction to Panorama Point. **0.2**

2.9 Turn left at T-intersection. **0.5**

3.5 Turn left into Panorama Park. **0.6**

3.7 STOP 59. Panorama Point. From right to left: Underwood Mountain volcano on west side of White Salmon River; Columbia Gorge, with Hood River Valley in foreground capped near shopping center by deposits of Missoula floods; Pleistocene Mount Defiance volcano across terraced Hood River Valley; main part of Hood River Valley; Middle Mountain separating Upper Valley from the main valley; Mount Hood; and Hood River fault zone bounding the east side of the valley. Several cones of olivine basalt occur in Hood River Valley south of here. The orchards grow on the Parkdale soils [*Harris*, 1973], probably ash-cloud deposits (perhaps rain-flushed as suggested by accretionary lapilli) associated with pyroclastic flows erupted from Mount Hood during late Polallie time about 12 ka [*Crandell and Rubin*, 1977; *Crandell*, 1980]. **0.2**

Retrace route toward Highway 35.

4.4 Turn right on Whiskey Creek Road and then left

on Highway 35. **0.7**
5.1 Cuts in olivine basalt. **0.7**
9.3 Parkdale soils on basalt opposite Odell road. **4.2**
9.9 Road gradually climbs to saddle east of Middle Mountain, underlain by faulted N_2 flows of the Grande Ronde Basalt and Frenchman Springs Member [*Swanson et al.*, 1981; *Korosec*, 1987b]. Pliocene or Pleistocene olivine basalt occurs in and east of saddle. **0.6**
10.5 Hills ahead are eroded vents for the Pliocene or Pleistocene olivine basalt. **0.6**
12.3 At 10 o'clock is ridge in R_2 and N_2 flows of Grande Ronde Basalt and Frenchman Springs Member in Hood River fault zone [*Swanson et al.*, 1981]. **1.8**
18.7 Roadcut in Frenchman Springs cut by fault of Hood River fault zone [*Swanson et al.*, 1981]. **6.4**
20.3 Cuts in faulted N_2 Grande Ronde and Frenchman Springs in Hood River fault zone. **1.6**
20.5 Overbank deposits from Christmas 1980 flood down Polallie Creek and East Fork Hood River [*Gallino and Pierson*, 1984]. **0.2**
21.6 STOP 60. Be alert for rockfalls! Cliff face exposes intracanyon flows of normally magnetized olivine andesite of Cloud Cap [*Keith et al.*, 1982; *Wise*, 1969], erupted from a vent on the northeast flank of Mount Hood 10 km southwest of here. Five flows or more totalling about 150 m thick were produced from this vent [*Wise*, 1969]. The eruptions supposedly occurred after the main cone-building stage of Mount Hood but before the Fraser glaciation about 29 ka [*Wise*, 1969; *Crandell*, 1980]. However, three K-Ar whole-rock ages from one sample near the Cloud Cap vent are about 0.49, 0.63, and 0.65 Ma [*Keith et al.*, 1985], coeval or even older than the main cone. *Wise* [1969] believes that the flows are not genetically related to those forming the Mount Hood edifice. Glaciofluvial deposits and Polallie-age lahars abut and overlie the andesite flows here. **1.1**
22.6 To left on ridge is white pumiceous ash-flow tuff interleaved with hornblende dacite flows partly filling old valley [*Wise*, 1969]. Erupted from dacite domes at Mill Creek Buttes, with a K-Ar (hornblende) age of 6.2 ± 1.3 Ma [*Fiebelkorn et al.*, 1983]. Rugged point at 12 o'clock is olivine andesite of Cloud Cap. **1.0**
23.2 Cut in upper Miocene pyroxene andesite [*Wise*, 1969; *Keith et al.*, 1982]. *Keith et al.* [1985] obtained a K-Ar whole-rock age of 8.18 ± 0.06 Ma near here. **0.6**
23.6 Fault zone in upper Miocene andesite flow. **0.4**
24.5 Deposits of Christmas 1980 Polallie Creek flood along river, in places reworked by man. **0.9**
24.9 Good exposures east of river of upper Miocene andesite K-Ar dated as 7.0 ± 0.8 Ma [*Wise*, 1969]. The andesite is overlain by Quaternary olivine basalt and basaltic andesite [*Wise*, 1969]. **0.4**
25.2 STOP 61. Mouth of Polallie Creek. Observe deposits and effects of the Christmas 1980 debris flow, whose volume is more than 20 times that of the initial slumped material owint to sediment-bulking during transport [*Gallino and Pierson*, 1984]. The debris flow, generated by intense rainfall, rushed down Polallie Creek, which cut into volcaniclastic deposits of Polallie age that contributed to the bulking. The debris flow dammed the river for 12-18 minutes; once the dam breached, a flood surged downstream, destroying 13 km of road and causing $13-million of damage. **0.3**
25.4 Upper Miocene andesite in roadcuts. **0.2**
26.6 Roadcut exposing interbedded lake and lahar deposits of Polallie age [*Crandell*, 1980]. Several cuts farther along show similar deposits. Polallie deposits, mainly pyroclastic flows near the volcano and lahars farther away, formed on all sides of the volcano. Here on the northeast side, the deposits form a broad triangular apron converging on Cooper Spur, 825 m below the present summit [*Crandell*, 1980]. The deposits flank both sides of Polallie Creek and form a fill (mostly eroded) in the valley of East Fork Hood River downstream from the mouth of the creek [*Crandell*, 1980]. The deposits created temporary dams and lakes upstream from Polallie Creek; the thinly bedded fine sand and silt in the cuts record these lakes [*Crandell*, 1980]. **1.2**
26.9 Good exposure of interbedded lake and lahar deposits in cut opposite Sherwood Campground. **0.3**
28.6 Smooth topography is caused by huge fan of Fraser-age outwash that fills valley and extends down from Newton Clark Glacier. **1.7**
34.3 Roadcut in Quaternary andesite flow of the cone-building stage [*Wise*, 1969] (called Main Stage by *White* [1980]) of Mount Hood, with outwash beyond. **5.7**
35.2 Bennett Pass. Roadcuts expose three tills, the youngest of which is of Fraser age [*Crandell*, 1980]. The ages of the older tills are not known. **0.9**
35.5 Next few cuts are in Miocene andesite flows. **0.3**
36.8 Cone-building andesite flow of Mount Hood. **1.3**
37.2 Cross White River. Terrace upstream is of Old Maid age [*Crandell*, 1980], about 1760-1810 A.D. [*Cameron and Pringle*, 1987]. Outburst floods from White River Glacier destroyed bridges here 5 times in this century, most recently in 1959 and 1968. **0.4**
37.3 Turn right into parking area and then right on gravel road. **0.1**
37.8 Turn right onto top of levee. If road is bad, continue straight and park wherever possible. **0.5**
38.0 STOP 62. Walk to stream cuts exposing dacitic lithic pyroclastic flow (as thick as 10 m) of Old Maid age capped locally by at least two thin lahars [*Crandell*, 1980; *Cameron and Pringle*, 1987]. The pyroclastic flow contains thoroughly charred wood and radially-jointed blocks that magnetic measurements indicate were above the Curie point when deposited [*Crandell*, 1980]. The charred wood has a ^{14}C age of 260 ± 150 years [*Crandell*, 1980]. The pyroclastic flow traveled down the easternmost drainage of the White River, probably from near Crater Rock, and entered the main drainage at a high angle, running about 75 m up the west valley wall; *Cameron and Pringle* [1987] calculated a minimum velocity of 135 km/hr from this relation. A Fraser-age lateral

moraine stands above the left bank of the river. **0.2**

Return to Highway 35 and turn right.

39.1 Next several roadcuts are in Miocene andesite. **1.1**

41.6 Roadcut in Miocene platy andesite. **2.5**

42.5 Keep right toward Government Camp, merging onto Highway 26. Cross Salmon River drainage. **0.9**

43.8 Cuts in platy andesite of cone-building stage. **1.3**

44.0 STOP 63. Quarry on right in platy, plagioclase-hypersthene andesite flow from cone-building stage of Mount Hood. Hypersthene is visible as light brown phenocrysts. Flow contains abundant rounded inclusions of fresh holocrystalline microdiorite. More than 90% of the cone consists of preglacial (older than 29 ka) lava flows, breccia, and pyroclastic rocks, most of which are andesitic and about 70 percent of which are lava flows [*Wise*, 1969; *White*, 1980. No rocks with reversed magnetic polarity have been found, so the cone is assumed to postdate 0.73 Ma [*White*, 1980; *Mankinen and Dalrymple*, 1979]. *Keith et al.* [1985] present K-Ar whole-rock ages ranging from 0.57 to 0.35 Ma for two cone-building lava flows in Zigzag Canyon. **0.2**

45.0 Turn right toward Timberline Lodge. For the next 2.6 miles the road passes through small cuts of Fraser-age till and cone-building andesite. **1.0**

47.6 Andesite flow overlying andesitic breccia. **2.6**

49.1 STOP 64. Cut in interlayered fine pink ash and coarse grayish-pink diamict. A few clasts are radially jointed. Possibly the diamict was a lahar generated by collapse of a hot dome onto snow. The age is unknown but probably Polallie. **1.5**

49.6 Roadcut exposes diamict with scattered hot blocks capped by ash or wind-blown sand. Diamict overlies a cone-building andesite flow. **0.5**

50.2 Andesite flow capped by thin tephra layers. **0.6**

50.5 Turn left into **bus** parking lot just before Timberline Lodge. **STOP 65.** Restrooms in building nearest the lot. Walk back down road 100 m to overflow parking lot and then up dirt road and cross country to Timberline (Skyline) trail contouring on hill slope. Follow trail east across small creek (headwaters of Salmon River). Stop on crest of next ridge at Ghost Forest Overlook. From here one can look down White River and see a buried forest engulfed by fluvial and cold-lahar deposits of Old Maid age that form a flat-topped terrace. A ^{14}C age of 250 ± 150 years B.P. dates one of the buried stumps [*Crandell*, 1980]. The Old Maid deposits formed a fill whose surface is about 100 m above the White River. Upstream is White River Glacier, source of outburst floods that periodically destroy the bridge on Highway 35. The main thermal area of the volcano (Devil's Kitchen) can be seen between the head of White River Glacier and the summit ridge; temperatures as high as 92°C, which is above boiling at 3171 m, have been measured [*Nehring et al.*, 1981; *Cameron*, 1988]. Crater Rock, remnant of a three-lobed hornblende dacite dome [*Cameron and Pringle*, 1986], can be seen left (west) of Devil's Kitchen. Crater Rock was probably emplaced at least in part in Old Maid time [*Crandell*, 1980; *Cameron and Pringle*, 1986], although *Wise* [1968, 1969] ascribed it to what today is called the Timberline eruptive period. Trim lines from the neoglacial maximum (about 1740 A.D.) can be seen below Steel Cliff, to right of White River Glacier.

Many trees near Timberline Lodge, including those at this locality (the *ghost forest*), were killed around 1795-1810, probably near 1800 A.D. [*Lawrence*, 1948]. The trees were killed neither by beetles (no galleries exist below the remaining bark) nor by forest fire (no charring can be found on the dead trees). The ghost forest occurs only within about 500 m of the west side of White River canyon and within 2 km of the east side [*Cameron and Pringle*, 1987]. The trees were probably killed by a hot ash cloud from the dacitic lithic pyroclastic flow at Stop 62 [*Cameron and Pringle*, 1987].

Walk to south end of lower paved parking area near bus parking lot. Several layers of Timberline-age tephra in a 35-cm-thick section overlie the weathered top of a Polallie-age lahar and underlie young soil and possibly an Old Maid-age tephra. The Timberline-age tephra here comprise two or three light-colored ash layers sandwiched in dark andesitic or muddy ash. Note the surrounding ghost forest, near its western limit. **0.3**

Return to Highway 26.

55.7 Junction with Highway 26. Turn right. **5.2**

58.0 Polallie lahars in roadcuts [*Crandell*, 1980]. **2.3**

58.2 Roadcut in quartz diorite of Laurel Hill. **0.2**

58.6 Fractured margin of Laurel Hill pluton. **0.4**

58.9 At 3 o'clock across Little Zigzag Canyon is cliff of Mount Hood andesite flows. **0.3**

59.0 Roadcut in Laurel Hill pluton. Many cuts in next mile expose this pluton. **0.1**

60.2 Two dikes cut the Laurel Hill pluton. We will stop here later in the day. Roadcuts expose the Laurel Hill for next mile. **1.2**

61.4 Turn right from Highway 26 onto road toward Mount Hood Kiwanis Camp. **1.2**

63.3 STOP 66. Bus can turn around 0.2 mi farther. Small quarry in three lithic pyroclastic flows of Timberline age. Wood is entirely charred in uppermost thick, gray deposit and partly charred in the underlying two flows. Prismatically jointed blocks are scarce. Below the thickest gray flow is a poorly exposed thin unit with noncharred wood that may record high energy, surgelike emplacement. Radiocarbon ages on wood in the deposits indicate that the thick gray flow is 1440 ± 155 years B.P. [*Cameron and Pringle*, 1986] and the second unit below it is 1830 ± 50 years B.P. (K. A. Cameron and P. T. Pringle, unpub. data). Tan unit on top of gray pyroclastic flow is cold lahar that is also likely of Timberline age, for no recognizable weathering zone separates it from the pyroclastic flows. **1.9**

63.5 Bus turnaround. Return to quarry. **0.2**

65.5 Highway 26. Turn left and retrace route to Laurel Hill pluton. **2.0**

66.7 STOP 67. Stop on outside of curve. The Laurel Hill pluton is plagioclase-hornblende-phyric quartz diorite with lesser quartz monzonite and patches of granophyre [*Wise*, 1969]. Its groundmass is pervasively propylitized. Similar bodies crop out 4 km to the south along Still Creek and 2 km to the west along Zigzag River. Drilling on the south side of Zigzag Mountain and in the Old Maid Flat area, 4 km and 10 km respectively northwest of here, penetrated similar quartz diorite. These bodies are likely connected at depth [*Wise*, 1969]. The quartz diorite intrudes the Grande Ronde and Wanapum Basalts, the Rhododendron Formation (about 14-11 Ma; D. R. Sherrod, written commun., 1988), and the Last Chance andesite (K-Ar whole-rock ages of 10.7 ± 0.5 and 10.5 ± 0.5 Ma) [*Priest et al.*, 1982].

For the Laurel Hill, *Bikerman* [1970] obtained K-Ar hornblende ages of 8.4 ± 0.6 and 8.0 ± 0.6 Ma, and *Keith et al.* [1985] determined K-Ar whole-rock ages of 8.6 ± 0.14 and 8.75 ± 0.18 Ma on a single sample. *Priest et al.* [1982] reported a K-Ar hornblende age of 9.3 ± 0.9 for biotite-bearing hornblende microquartz diorite in a drill hole near Old Maid Flat. *Wise* [1969] noted that a K-Ar age of 11.6 ± 1.2 Ma for the Laurel Hill was too old, for the pluton intrudes definitely younger rocks. The Laurel Hill, just as the Tatoosh pluton at Mount Rainier, cooled during a field reversal (from reverse to normal) [*Dunn et al.*, 1971].

Walk up road 200 m to see two slightly vesicular andesite dikes cutting the quartz diorite. *Bikerman* [1970] apparently dated one of these dikes as about 5 Ma (K-Ar whole-rock). **1.2**

67.2 Tom Dick and Harry Mountain at 3 o'clock, composed of upper Miocene and Pliocene andesite. **0.5**

69.9 Turn right into parking area, turn around, and continue west. **2.7**

75.6 Road is on lahar surface of post-Fraser age. Valley walls in Rhododendron Formation. **5.7**

80.0 Forest Service Station in Zigzag. Continue west on Highway 26 toward Portland or turn right on East Lolo Pass Road for optional side trip. **4.4**

82.5 Signal light in Welches, built on a Timberline lahar surface. **2.5**

End of geologic guide. To reach I-84, continue on Highway 26 to east side of Gresham. Then,

127.0 Continue straight on Burnside rather than turning left on Highway 26 toward Portland. **44.5**

127.7 Turn right at signal on SE 242nd. Drive toward I-84 and follow signs; stay on same road all the way. **0.7**

130.5 Wood Village-Gresham interchange at I-84. 2.8

Optional Trip to Mount Tabor and Boring Lava

(0.0) At mileage 127.7, continue straight on Burnside.

(0.3) At signal light, turn left onto Division. Follow Division to mile 8.9. **0.3**

(2.4) At 9 o'clock is a cone-shaped hill of Pliocene and early Pleistocene Walters Hill Formation (conglomerate, sandstone, and mudflows) [*Trimble*, 1963]. **2.1**

(6.4) At 9 o'clock is Kelly Butte, a cinder cone of the Boring Lava [*Treasher*, 1942; *Trimble*, 1963; *Allen*, 1975; *Tolan and Beeson*, 1984], which contains Pliocene and Pleistocene olivine basalt and basaltic andesite flows with vents throughout the Portland metropolitan area. **4.0**

(7.1) Pass beneath I-205. **0.7**

(7.9) Mount Tabor, another Boring cone, to right. **0.8**

(8.9) Turn right on S.E. 60th Avenue. **1.0**

(9.4) Turn right on S.E. Salmon for Mount Tabor. **0.5**

(9.7) Sharp turn to left. Parking lot here affords view of city center and the Portland Hills, which are a faulted anticline of Grande Ronde and Wanapum (Frenchman Springs) Basalts overlain in places by Boring Lava. **0.3**

(10.0) Turn into parking lot. **STOP 68.** A partial cross section of the Mount Tabor cone of olivine basalt is exposed beside the basketball court and outdoor theater. Some of the cinder beds contain very large ejecta (more than 1 m in long dimension), whereas other beds, especially high in the section, are relatively fine grained and well sorted. The general dip is northwestward, away from the breached crater that the road traverses. Several fault surfaces suggest slumping of the flank of the cone during its formation. Gravel in some of the younger beds in the middle of the exposure were probably derived from the underlying Troutdale Formation [*Trimble*, 1963], and the younger beds at the north end of the exposure are partly palagonitized. This evidence suggests interaction with water.

Allen [1975] located more than 30 Boring vents within 21 km of here and more than 90 vents (about 50 of which he classified as "certain") within 32 km of Troutdale (5 km northeast of Gresham). The Boring in this area is between about 1.3 Ma (Rocky Butte, 3 km northeast of here) and 2.1 Ma (in bluffs near Oregon City, 18 km south of here), judging from unpublished K-Ar ages (D. R. Sherrod, oral commun., 1988) and a K-Ar age of 1.56 ± 0.2 Ma on a flow at Bear Prairie 29 km east-northeast of here [*Tolan and Beeson*, 1984].

The setting of the basalt field is puzzling and not understood. The vents lie well west of the crest of the Cascades, and those such as Kelly Butte, Mount Tabor, and the vents in the Portland Hills are in and even west of the Portland basin.

End of log.

Acknowledgments

We thank Dave Sherrod and Jim Moore for providing incisive reviews; Paul Hammond for graciously contributing unpublished information about Indian Heaven, and Jim Vallance for doing likewise for Stop 58; Gordon Keating and Lyn Topinka for preparing many of the illustrations; Craig Weaver for allowing use of Figure 3; and Barbara White for helping log several legs of the trip. Prime responsibility for preparation of the guidebook is: road log from Issaquah to Cle Elum and related text (Vance); Tertiary geology near Mount St. Helens (Evarts); road log and text for Mount Hood (Cameron and Pringle); remainder of guidebook (Swanson).

REFERENCES

Allen, J. E., Volcanoes of the Portland area, Oregon, *Ore Bin*, *37*, 145-157, 1975.

Alt, D., J. W. Sears, and D. W. Hyndman, Terrestrial maria: the origins of large basalt plateaus, hotspot tracks and spreading ridges, *J. Geol.*, *96*, 647-662, 1988.

Anderson, J. L., Pomona Member of the Columbia River Basalt Group: an intracanyon flow in the Columbia River Gorge, Oregon, *Oreg. Geol.*, *42*, 195-199, 1980.

Anderson, J. L. and B. F. Vogt, Intracanyon flows of the Columbia River Basalt Group in the southwest part of the Columbia Plateau and adjacent Cascade Range, Oregon and Washington, *Wash. Div. Geol. Earth Resources Bull.*, *77*, 249-267, 1987.

Anderson, J. L., M. H. Beeson, and T. L. Tolan, Tectonic evolution of the southwest Columbia Plateau, *Geol. Soc. Am. Abst. Prog.*, *19*, 354-355, 1987.

Armstrong, R. L., Cenozoic igneous history of the U.S. Cordillera from lat 42° to 49° N, *Geol. Soc. Am. Mem.*, *152*, 265-282, 1978.

Armstrong, R. L., W. P. Leeman, and H. E. Malde, K-Ar dating, Quaternary and Neogene volcanic rocks of the Snake River Plain, Idaho, *Am. J. Sci.*, *275*, 225-251, 1975.

Atwater, T., Implications of plate tectonics for the Cenozoic tectonic evolution of western North America, *Geol. Soc. Am. Bull.*, *81*, 3513-3536, 1970.

Bailey, M. M., The stratigraphy and petrochemistry of the Picture Gorge Basalt, north-central Oregon, M.S. thesis, 156 pp., Wash. State Univ., Pullman, 1986.

Bailey, M. M., Revisions to stratigraphic nomenclature of the Picture Gorge Basalt Subgroup, Columbia River Basalt Group, in *Volcanism and Tectonism on the Columbia Plateau*, edited by S. P. Reidel, P. R. Hooper, and J. L. Anderson, *Geol. Soc. Am. Spec. Pap.*, in press.

Baksi, A. J., Estimation of lava extrusion and magma production rates for two flood basalt provinces, *J. Geophys. Res.*, *93*, 11,809-11,815, 1988.

Barnosky, C. W., Late Quaternary vegetation near Battle Ground Lake, southern Puget Trough, Washington, *Geol. Soc. Am. Bull.*, *96*, 263-271, 1985.

Barrash, W., J. Bond, and R. Venkatakrishnan, Structural evolution of the Columbia Plateau in Washington and Oregon, *Am. J. Sci.*, *283*, 897-935, 1983.

Bates, R. G., M. E. Beck, Jr., and R. F. Burmester, Tectonic rotations in the Cascade Range of southern Washington, *Geology*, *9*, 184-189, 1981.

Beck, M. E., Jr., Paleomagnetic record of plate-margin tectonic processes along the western edge of North America, *J. Geophys. Res.*, *85*, 7115-7131, 1980.

Beck, M. E., Jr., and C. D. Burr, Paleomagnetism and tectonic significance of the Goble Volcanic Series, southwestern Washington, *Geology*, *7*, 175-179, 1979.

Becraft, G. E., Definition of the Tieton andesite on lithology and structure, M.S. thesis, 26 pp., Wash. State Coll., Pullman, 1950.

Beeson, M. H., K. R. Fecht, S. P. Reidel, and T. L. Tolan, Regional correlations within the Frenchman Springs Member of the Columbia River Basalt Group: new insights into the middle Miocene tectonics of northwestern Oregon, *Oreg. Geol.*, *47*, 87-96, 1985.

Bentley, R. D., Stratigraphy of the Yakima basalts and structural evolution of the Yakima Ridges in the western Columbia Plateau, in *Geological excursions in the Pacific Northwest*, edited by E. H. Brown and R. C. Ellis, pp. 339-389, Dept. Geology, Western Wash. Univ., Bellingham, 1977.

Bentley, R. D. and J. L. Anderson, Wrench tectonics in the Yakima ridges of the Columbia Plateau, Washington and Oregon, *Geol. Soc. Am. Abst. Prog.*, *12*, 267, 1980.

Bikerman, M., K-Ar ages of Laurel Hill pluton and dike, Oregon, *Ore Bin*, *32*, 211-215, 1970.

Birch, D. C., Big Eddy-McLoughlin Line Tower 36/2 Ladd Creek washout, *Bonneville Power Administration in-house memorandum*, 9 pp., 1961.

Blackwell, D. D. and J. L. Steele, A summary of heat flow studies in the Cascade Range, *Trans. Geotherm. Resour. Counc.*, *7*, 233-236, 1983.

Blackwell, D. D., R. G. Bowen, D. A. Hull, J. Riccio, and J. L. Steele, Heat flow, arc volcanism, and subduction in northern Oregon, *J. Geophys. Res.*, *87*, 8735-8754, 1982.

Borchardt, G. A., J. A. Norgren, and M. E. Harward, Correlation of ash layers in peat bogs of eastern Oregon, *Geol. Soc. Am. Bull.*, *84*, 3101-3108, 1973.

Brantley, S. R. and R. B. Waitt, Interrelations among pyroclastic surge, pyroclastic flow, and lahars in Smith Creek valley during first minutes of 18 May 1980 eruption of Mount St. Helens, USA, *Bull. Volcanol.*, *50*, 304-326, 1988.

Broughton, W. R., The exploration of the Columbia River, in *An extract from the journal of Captain George Vancouver*, Longview Daily News Press, Longview, Wash., 39 pp., 1929.

Brown, E. H., Geology of the Shuksan Suite, North Cascades, Washington, *Geol. Soc. Am. Mem.*, *164*, 143-154, 1986.

Brown, E. H., D. L. Wilson, R. L. Armstrong, and J. E. Harakal, Petrologic, structural, and age relations of serpentinite, amphibolite, and blueschist in the Shuksan Suite of the Iron Mountain-Gee Point area, North Cascades, Washington, *Geol. Soc. Am. Bull.*, *93*, 1087-1098, 1982.

Brown, G. C., Calc-alkaline intrusive rocks: Their diversity, evolution, and relation to volcanic arcs, in *Andesites: Orogenic Andesites and Related Rocks*, edited by R. S. Thorpe, pp. 437-461, John Wiley, New York, 1982.

Brugman, M. M., Groundwater discharge within a glaciated strato-volcano: what happens to missing surface runoff?, *Geol. Soc. Am. Abst. Prog.*, *20*, A115, 1988.

Buckovic, W. A., The Eocene deltaic system of west-central Washington, in *Cenozoic Paleogeography of the western United States*, Econ. Paleon. Mineral., Pac. Coast Paleog. Symp., Los Angeles, 147-163, 1979.

Byerly, G. R. and D. A. Swanson, Invasive Columbia River basalt flows along the northwestern margin of the Columbia Plateau, north-central Washington, *Geol. Soc. Am. Abst. Prog.*, *10*, 98, 1978.

Byerly, G. R. and D. A. Swanson, The transition from subaerial to invasive lava flows, Grande Ronde Basalt, northwestern Columbia Plateau, *Geol. Soc. Am. Abst. Prog.*, *19*, 363, 1987.

Cameron, K. A., Fumarole fields and thermal features at Mount Hood, Oregon, *Northwest Sci.*, *62*, 82, 1988.

Cameron, K. A. and P. Pringle, Post-glacial lahars of the Sandy River basin, Mount Hood, Oregon, *Northwest Sci.*, *60*, 225-237, 1986.

Cameron, K. A. and P. Pringle, A detailed chronology of the most recent major eruptive period at Mount Hood, Oregon, *Geol. Soc. Am. Bull.*, *99*, 845-851, 1987.

Camp, V. E., Geologic studies of the Columbia Plateau: Part II. Upper Miocene basalt distribution, reflecting source locations, tectonism, and drainage history in the Clearwater embayment, Idaho, *Geol. Soc. Am. Bull.*, *92*, 669-678, 1981.

Carey, S. and H. Sigurdsson, The May 18, 1980 eruption of Mount St. Helens 2. Modeling of dynamics of the Plinian phase, *J. Geophys. Res.*, *90*, 2948-2958, 1985.

Carkin, B. A., Geology and petrology of the Fifes Peak Formation in the Cliffdell area, central Cascades, Washington, *Geol. Soc. Am. Abst. Prog.*, *17*, 347, 1985.

Carlson, R. W., Isotopic constraints on Columbia River flood basalt genesis and the nature of the subcontinental mantle, *Geochim. et Cosmochim. Acta*, *48*, 2357-2372, 1984.

Carlson, R. W. and W. K. Hart, Flood basalt volcanism in the northwestern United States, in *Continental Flood Basalts*, edited by J. D. Macdougall, Kluwer Academic Pub., Dordrecht, in press.

Cashman, K. V., Crystallization of Mount St. Helens 1980-1986 dacite: a quantitative textural approach, *Bull. Volcanol.*, *50*, 194-209, 1988.

Cassidy, J. J., H. A. Coombs, and W. L. Shannon, Study of effects of potential volcanic activity on Lewis River projects, unpublished report to the Pacific Power and Light Co, Portland, Oregon, 91 pp., 1980.

Catchings, R. D. and W. D. Mooney, Crustal structure of the Columbia Plateau: evidence for continental rifting, *J. Geophys. Res.*, *93*, 459-474, 1988.

Chadwick, W. W., E. Y. Iwatsubo, D. A. Swanson, and J. W. Ewert, Measurements of slope distances and vertical angles at Mount Baker and Mount Rainier, Washington, Mount Hood and Crater Lake, Oregon, and Mount Shasta and Lassen Peak, California, 1980-1984, *U.S. Geol. Surv. Open-File Rept.*, *85-205*, 96 p., 1985.

Chadwick, W. W., R. J. Archuleta, and D. A. Swanson, The mechanics of ground deformation precursory to dome-building extrusions at Mount St. Helens 1981-1982, *J. Geophys. Res.*, *93*, 4351-4366, 1988.

Christiansen, R. L. and E. H. McKee, Late Cenozoic volcanic and tectonic evolution of the Great Basin and Columbia Intermontane regions, *Geol. Soc. Am. Mem.*, *152*, 283-311, 1978.

Clayton, G. A., Geology of the White Pass area, south-central Cascade Range, Washington, M.S. thesis, 212 pp., Univ. of Washington, Seattle, 1983.

Coleman, S. M. and K. L. Pierce, Weathering rinds on andesitic and basaltic stones as a Quaternary age indicator, western United States, *U.S. Geol. Surv. Prof. Pap.*, *1210*, 56 pp., 1981.

Condie, K. C. and D. H. Swenson, Compositional variation in three Cascade stratovolcanoes: Jefferson, Rainier, and Shasta, *Bull. Volcanol.*, *37*, 205-230, 1973.

Conway, T. R., Mount Hood in eruption, *Mazama*, *6*, 40-43, 1921.

Coombs, H. A., The geology of Mount Rainier National Park, *Univ. Wash. Pubs. Geol.*, *3*, 131-212, 1936.

Couch, R. W. and B. Baker, Geophysical investigations of the Cascade Range in central Oregon, final report, grant 14-08-001-G-393, 95 pp., U.S. Geol. Surv., Reston, VA, 1979.

Cox, K., A model for flood basalt volcanism, *J. Petrol.*, *21*, 629-650, 1980.

Crandell, D. R., Surficial geology and geomorphology of the Lake Tapps quadrangle, Washington, *U.S. Geol. Surv. Prof. Pap.*, *388-A*, 84 pp. 1963a.

Crandell, D. R., Paradise debris flow at Mount Rainier, Washington, *U.S. Geol. Surv. Prof. Pap.*, *475-B*, B135-B139, 1963b.

Crandell, D.R., Surficial geology of Mount Rainier National Park, Washington, *U.S. Geol. Surv. Bull.*, *1288*, 41 pp., 1969

Crandell, D. R., Postglacial lahars from Mount Rainier Volcano, Washington, *U.S. Geol. Surv. Prof. Pap.*, *677*, 75 pp., 1971.

Crandell, D. R., Recent eruptive history of Mount Hood, Oregon, and potential hazards from future eruptions; *U.S. Geol. Surv. Bull.*, *1492*, 81 pp., 1980.

Crandell, D. R., Deposits of pre-1980 pyroclastic flows and lahars from Mount St. Helens volcano, Washington, *U.S. Geol. Surv. Prof. Pap.*, *1444*, 95 pp. 1987.

Crandell, D. R. and R. P. Hoblitt, Lateral blasts at Mount St. Helens and hazard zonation, *Bull. Volcanol.*, *48*, 27-37, 1986.

Crandell, D. R. and R. D. Miller, Quaternary stratigraphy and extent of glaciation in the Mount Rainier region, Washington, *U.S. Geol. Surv. Prof. Pap.*, *847*, 59 pp., 1974.

Crandell, D. R. and D. R. Mullineaux, Potential hazards from future eruptions of Mount St. Helens, Washington, *U.S. Geol. Surv. Bull.*, *1383-C*, 26 pp., 1978.

Crandell, D. R. and M. Rubin, Late-glacial and postglacial eruptions at Mt. Hood, Oregon, *Geol. Soc. Am. Abst. Prog.*, *9*, 406, 1977.

Crandell, D. R. and H. H. Waldron, A recent volcanic mudflow of exceptional dimensions from Mt. Rainier, Washington, *Am. J. Sci.*, *254*, 349-362, 1956.

Crandell, D. R., D. R. Mullineaux, M. Rubin, E. Spiker, and M. L. Kelley, Radiocarbon dates from volcanic deposits at Mt. St. Helens, Washington, *U.S. Geol. Surv. Open-File Rept.*, *81-844*, 15 pp., 1981.

Criswell, C. W., Chronology and pyroclastic stratigraphy of the May 18, 1980, eruption of Mount St. Helens, Washington, *J. Geophys. Res.*, *92*, 10,237-10,266, 1987.

Davis, G. A., Late Cenozoic tectonics of the Pacific Northwest with special reference to the Columbia Plateau, Final Safety Analysis Report WNP-2, Amendment 18, appendix, 69 pp., Wash. Pub. Power Supply System, Richland, 1981.

Davis, G. D., J. W. H. Monger, and B. C. Burchfiel, Mesozoic construction of the Cordilleran "Collage", central British Columbia to central California, in *Mesozoic paleogeography of the western United States*, Soc. Econ. Paleon. Mineral., Pac. Coast Paleog. Symp. 2, Los Angeles, 33-70, 1978.

Davis, J. O., Quaternary tephrochronology of the Lake Lahontan area, Nevada and California, *Nevada Archeol. Surv. Res. Pap.*, *7*, Univ. of Nevada, Reno, 29 pp., 1978.

DeGraff, J. M. and A. Aydin, Surface morphology of columnar joints and its significance to mechanics and direction of joint growth, *Geol. Soc. Am. Bull.*, *99*, 605-617, 1987.

Doukas, M. P. and D. A. Swanson, Mount St. Helens, Washington, with emphasis on 1980-85 eruptive activity as viewed from Windy Ridge, *Geol. Soc. Am. Centennial Field Guide—Cordilleran Section*, 333-338, 1987.

Driedger, C. L. and P. M. Kennard, Ice volumes on Cascade

volcanoes: Mount Rainier, Mount Hood, Three Sisters, and Mount Shasta, *U.S. Geol. Surv. Prof. Pap.*, *1365*, 28 pp., 1986.

Duncan, R. A., A captured island chain in the Coast Range of Oregon and Washington, *J. Geophys. Res.*, *87*, 10,827-10,837, 1982.

Dungan, M. A., J. A. Vance, and D. P. Blanchard, Geochemistry of the Shuksan greenschists and blueschists, North Cascades, Washington: variably fractionated and altered meta-basalts of oceanic affinity, *Contr. Mineral. Petrol.*, *82*, 131-146, 1983.

Dunn, J. R., M. Fuller, H. Ito, and V. A. Schmidt, Paleomagnetic study of a reversal of the Earth's magnetic field, *Science*, *172*, 840-845, 1971.

Dzurisin, D., R. Y. Koyanagi, and T. T. English, Magma supply and storage at Kilauea Volcano, Hawaii, 1956-1983, *J. Volc. Geotherm. Res.*, *21*, 177-206, 1984.

Easterbrook, D. J., N. D. Briggs, J. A. Westgate, and M. P. Gordon, Age of the Salmon Springs Glaciation in Washington, *Geology*, *9*, 87-93, 1981.

Eaton, G. P., R. R. Wahl, H. J. Prostka, D. R. Mabey, and M. D. Kleinkopf, Regional gravity and tectonic patterns: their relation to late Cenozoic epierogeny and lateral spreading in the western Cordillera, *Geol. Soc. Am. Mem.*, *152*, 51-91, 1978.

Eichelberger, J. C. and D. B. Hayes, Magmatic model for the Mount St. Helens blast of May 18, 1980, *J. Geophys. Res.*, *87*, 7727-7738, 1982.

Ellingson, J. A., The rocks and structure of the White Pass area, Washington, *Northwest Sci.*, *46*, 9-24, 1972.

Endo, E. T., S. D. Malone, L. L. Noson, and C. S. Weaver, Locations, magnitudes and statistics of the March 20-May 18 earthquake sequence, *U.S. Geol. Surv. Prof. Pap.*, *1250*, 93-107, 1981.

Engels, J. C., R. W. Tabor, F. K. Miller, and J. D. Obradovich, Summary of K-Ar, Rb-Sr, U-Pb, Pb-alpha, and fission-track ages of rocks from Washington state prior to 1975 (exclusive of Columbia Plateau basalts), scale 1:1,000,000, *U. S. Geol. Surv. Misc. Field Studies Map*, *MF-710*, 1976.

Erikson, E. H., Jr., Petrology of the composite Snoqualmie batholith, central Cascade Mountains, Washington, *Geol. Soc. Am. Bull.*, *80*, 2213-2236, 1969.

Evarts, R. C. and R. P. Ashley, Preliminary geologic map of the Spirit Lake quadrangle, Washington, scale 1:48,000, *U.S. Geol. Surv. Open-File Map*, *84-480*, 1984.

Evarts, R. C., R. P. Ashley, and J. G. Smith, Geology of the Mount St. Helens area: record of discontinuous volcanic and plutonic activity in the Cascade arc of southern Washington, *J. Geophys. Res.*, *92*, 10,155-10,169, 1987.

Ewart, A., The mineralogy and petrology of Tertiary-Recent orogenic volcanic rocks: With special emphasis to the andesite-basaltic andesite range, in *Andesites: Orogenic Andesites and Related Rocks*, edited by R. S. Thorpe, pp. 25-95, John Wiley, New York, 1982.

Ewing, T. E., Paleogene tectonic evolution of the Pacific Northwest, *J. Geol.*, *88*, 619-638, 1980.

Fecht, K. R., S. P. Reidel, and A. M. Tallman, Paleodrainage of the Columbia River system on the Columbia Plateau of Washington state—a summary, *Wash. Div. Geol. Earth Res. Bull.*, *77*, 219-248, 1987.

Felts, W. M., A granodiorite stock in the Cascade Mountains of southwestern Washington, *Ohio J. Sci.*, *39*, 297-316, 1939.

Fiebelkorn, R. B., G. W. Walker, N. S. MacLeod, E. H. McKee, and J. G. Smith, Index to K-Ar determinations for the state of Oregon, *Isochron/West*, *37*, 3-60, 1983.

Finn, C. and D. L. Williams, An aeromagnetic study of Mount St. Helens, *J. Geophys. Res.*, *92*, 10,194-10,206, 1987.

Fiske, R. S., Subaqueous pyroclastic flows in the Ohanapecosh Formation, Washington, *Geol. Soc. Am. Bull.*, *74*, 391-406, 1963.

Fiske, R. S., C. A. Hopson, and A. C. Waters, Geology of Mount Rainier National Park, Washington, *U.S. Geol. Surv. Prof. Pap.*, *444*, 93 pp., 1963.

Flanagan, G. and D. L. Williams, A magnetic investigation of Mount Hood, Oregon, *J. Geophys. Res.*, *87*, 2804-2814, 1982.

Foster, R. J., Tertiary geology of a portion of the central Cascade Mountains, Washington, *Geol. Soc. Am. Bull.*, *71*, 99-125, 1960.

Frank, D. G., Hydrothermal processes at Mount Rainier, Washington, Ph.D thesis, 195 pp., Univ. Wash., Seattle, 1985.

Frizzell, V. A., Jr., R. W. Tabor, D. B. Booth, K. M. Ort, and R. B. Waitt, Jr., Preliminary geologic map of the Snoqualmie Pass 1:100,000 quadrangle, Washington, *U.S. Geol. Surv. Open-File Rept.*, *84-693*, 42 pp., 1984.

Frizzell, V. A., Jr., R. W. Tabor, R. E. Zartman, and C. D. Blome, Late Mesozoic or early Tertiary melanges in the western Cascades of Washington, *Wash. Div. Mines Geol Bull.*, *77*, 129-148, 1987.

Furlong, K. P. and D. M. Fountain, Continental crustal underplating: thermal considerations and seismic-petrologic consequences, *J. Geophys. Res.*, *91*, 8285-8294, 1986.

Gallino, G. L. and T. Pierson, The 1980 Polallie Creek debris flow and subsequent dam-break flood, East Fork Hood River basin, Oregon, *U. S. Geol. Surv. Open-File Rept.*, *84-578*, 37 pp., 1984.

Gill, J., *Orogenic andesites and plate tectonics*, 390 pp., Springer Verlag, New York, 1981.

Grant, R. Y., A John Day vertebrate fossil discovered in the Keechelus Series of Washington, *Am. J. Sci.*, *239*, 590-593, 1941.

Grant, W. C., C. S. Weaver, and J. E. Zollweg, The 14 February Elk Lake, Washington, earthquake sequence, *Bull. Soc. Am.*, *74*, 1289-1309, 1984.

Greeley, R., Basaltic "plains" volcanism, in *Volcanism of the eastern Snake River Plain, Idaho*, edited by R. Greeley and J. S. King, NASA CR-154621, 23-44, 1977.

Greeley, R. and J. H. Hyde, Lava tubes of the Cave Basalt, Mount St. Helens, Washington, *Geol. Soc. Am. Bull.*, *83*, 2397-2418, 1972.

Gresens, R. L., J. T. Whetten, R. W. Tabor, and V. A. Frizzell, Part III: Tertiary stratigraphy of the central Cascade Mountains, Washington state, in *Geological Excursions in the Pacific Northwest*, edited by E. H. Brown and R. C. Ellis, Geol. Soc. Am. Field Guide, Seattle, 84-126, 1977.

Grommé, C. S., M. E. Beck, Jr., R. E. Wells, and D. C. Engebretson, Paleomagnetism of the Tertiary Clarno Formation of central Oregon and its significance for the tectonic history of the Pacific Northwest, *J. Geophys. Res.*, *91*, 14,089-14,103, 1986.

Guffanti, M. and C. S. Weaver, Distribution of late Cenozoic volcanic vents in the Cascade Range: volcanic arc segmentation and regional tectonic considerations, *J. Geophys. Res.*,

93, 6513-6529, 1988.

Hague, A., Glaciers of Mount Hood, *Am. J. Sci.*, *3rd series*, *1*, 165-167, 1871.

Hammond, P. E., Part IV: Stratigraphy, structure, and plutonism along the upper Yakima and South Fork Snoqualmie River valleys, central Cascade Range, Washington, in *Geological Excursions in the Pacific Northwest*, edited by E. H. Brown and R. C. Ellis, Geol. Soc. Am. Field Guide, Seattle, 292-308, 1977.

Hammond, P. E., A tectonic model for evolution of the Cascade Range, in *Cenozoic paleogeography of the western United States*, Soc. Econ. Paleon. Mineral., Pac. Coast Paleog. Symp. 3, Los Angeles, 219-237, 1979.

Hammond, P. E., Reconnaissance geologic map and cross sections of southern Washington Cascade Range, *Pubs. Dept. Earth Sci.*, Portland State Univ., Portland, Oreg., 1980.

Hammond, P. E., Indian Heaven, S. Washington Cascade Range: a basaltic volcanic field supplied by a central magma system?, *Geol. Soc. Am. Abst. Prog.*, *16*, 528, 1984.

Hammond, P. E., Lone Butte and Crazy Hills: Subglacial volcanic complexes, Cascade Range, Washington, *Geol. Soc. Am. Centen. Field Guide—Cordilleran Section*, 339-344, 1987.

Hammond, P. E. and M. A. Korosec, Geochemical analyses, age dates, and flow-volume estimates for Quaternary volcanic rocks, southern Cascade Mountains, Washington, *Wash. Div. Geol. Earth Resources Open-File Rept.*, *83-13*, 36 pp., 1983.

Hammond, P. E., S. A. Petersen, K. D. Hopkins, D. Aiken, D. S. Harle, Z. F. Daneš, D. L. Konicek, and C. R. Stricklin, Geology and gravimetry of the Quaternary basaltic volcanic field, southern Cascade Range, Washington, in *Proc. Second U.N. Symp. on Development and Use of Geothermal Resources*, *1*, San Francisco, 397-405, 1976.

Harris, B. L., Genesis, mineralogy, and properties of Parkdale soils, Oregon, Ph.D. thesis, 174 pp., Oreg. State Univ., Corvallis, 1973.

Hart, W. K. and R. W. Carlson, Tectonic controls on magma genesis and evolution in the northwestern United States, *J. Volcanol. Geotherm. Res.*, *32*, 119-136, 1987.

Heliker, C. C., Inclusions in the 1980-1983 dacite of Mount St. Helens, *Eos Trans AGU*, *64*, 894, 1983.

Heliker, C. C., Inclusions in the 1980-83 dacite of Mount St. Helens, Washington, M.S. thesis, 185 pp., West. Wash. Univ., Bellingham, 1984.

Heller, P. L., R. W. Tabor, and C. A. Suczek, Paleogeographic evolution of the United States Pacific Northwest during Paleogene time, *Can. J. Earth Sci.*, *24*, 1652-1667, 1987.

Helz, R. T., The petrogenesis of the Ice Harbor Member, Columbia Plateau, Washington—a chemical and experimental study, Ph.D. thesis, 284 pp., Penn. State Univ., University Park, 1978.

Hildreth, E. W., and J. Fierstein, Mount Adams: Eruptive history of an andesite-dacite volcano at the focus of a fundamentally basaltic volcanic field, *U. S. Geol. Surv. Open-File Rept.*, *85-521*, 44-50, 1985.

Hildreth, W., J. Fierstein, and M. S. Miller, Mineral and geothermal resource potential of the Mount Adams Wilderness and contiguous roadless areas, Skamania and Yakima Counties, Washington, *U.S. Geol. Surv. Open-File Rept.*, *83-474*, 50 pp., 1983.

Hoblitt, R. P. and C. D. Miller, Comment, *Geology*, *12*, 692-693, 1984.

Hoblitt, R. P., D. R. Crandell, and D. R. Mullineaux, Mount St. Helens eruptive behavior during the past 1,500 yr, *Geology*, *8*, 555-559, 1980.

Hodge, E. T., Geology of the Lower Columbia River, *Geol. Soc. Am. Bull.*, *49*, 831-930, 1938.

Hooper, P. R., The Columbia River basalts, *Science*, *215*, 1463-1468, 1982.

Hooper, P. R., Physical and chemical constraints on the evolution of the Columbia River basalt, *Geology*, *12*, 495-499, 1984.

Hooper, P. R., W. D. Kleck, C. R. Knowles, S. P. Reidel, and R. L. Thiessen, The Imnaha Basalt, Columbia River Basalt Group, *J. Petrol.*, *25*, 473-500, 1984.

Hopson, C. A. and J. M. Mattinson, Ordovician and Late Jurassic ophiolitic assemblages in the Pacific Northwest, *Geol. Soc. Am. Abst. Prog.*, *5*, 57, 1973.

Janda, R. J., K. M. Scott, K. M. Nolan, and H. A. Martinson, Lahar movement, effects, and deposits, *U.S. Geol. Surv. Prof. Pap.*, *1250*, 461-478, 1981.

Johnson, S. Y., Eocene strike-slip faulting and nonmarine basin formation in Washington, *Soc. Econ. Paleon. Mineral. Spec. Pub.*, *37*, 283-302, 1985.

Jones, J. G., Intraglacial volcanoes of the Laugarvatn region, southwest Iceland—I, *Quar. J. Geol. Soc. Lond.*, *124*, 197-211, 1969.

Keith, T. E. C., M. H. Beeson, and K. E. Bargar, Geologic map of the Mount Hood Wilderness, Clackamas and Hood River Counties, Oregon, *U.S. Geol. Surv. Misc. Field Studies Map*, *MF-1379A*, 1982.

Keith, T. E. C., J. M. Donnelly-Nolan, J. L. Markman, and M. H. Beeson, K-Ar ages of rocks in the Mount Hood area, Oregon, *Isochron/West*, *42*, 12-16, 1985.

Kieffer, S. W., Blast dynamics at Mount St. Helens on 18 May 1980, *Nature*, *291*, 568-570, 1981a.

Kieffer, S. W., Fluid dynamics of the May 18 blast at Mount St. Helens, *U.S. Geol. Surv. Prof. Pap.*, *1250*, 379-400, 1981b.

Kjartansson, G., The Moberg Formation, in *On the geology and geomorphology of Iceland*, edited by S. Thorarinsson, T. Einarsson, and G. Kjartansson, *Geografisk Annaler*, *41*, 135-169, 1959.

Kleinspehn, K. L., Cretaceous sedimentation and tectonics, Tyaughton-Methow Basin, southwestern British Columbia, *Can. J. Earth Sci.*, *22*, 154-174, 1985.

Korosec, M. A., Geologic map of the Mount Adams quadrangle, Washington, *Wash. Div. Geol. Earth Resources Open-File Rept.*, *87-5*, 33 pp., 1987a.

Korosec, M. A., Geologic map of the Hood River quadrangle, Washington and Oregon, *Wash. Div. Geol. Earth Resources Open-File Rept.*, *87-6*, 41 pp., 1987b.

Kuntz, M. A., P. D. Rowley, N. S. MacLeod, R. L. Reynolds, L. A. McBroome, A. M. Kaplan, and D. J. Lidke, Petrography and particle-size distribution of pyroclastic-flow, ash-cloud, and surge deposits, *U.S. Geol. Surv. Prof. Pap.*, *1250*, 525-539, 1981.

Lachenbruch A. H. and J. H. Sass, Heat flow in the United States and the thermal regime of the crust, *Am. Geophys. Union Geophys. Monog.*, *20*, 626-675, 1977.

Lander, E. B., A review of the Oreodonta (Mammalia, Artiodactyla), Ph.D. thesis, 474 pp., Univ. of California, Berkeley, 1977.

Laubscher, H. P., Models of the development of Yakima deformation, *Final Safety Analysis Report WNP-2*, Amendment

18, appendix, 69 pp., Washington Public Power Supply System, Richland, 1981.

Law, L. K., D. R. Auld, and J. R. Booker, A geomagnetic variation anomaly coincident with the Cascade volcanic belt, *J. Geophys. Res.*, *85*, 5297-5302, 1980.

Lawrence, D. B., The "floating island" lava flow of Mount St. Helens, *Mazama*, *36*, 41-44, 1941.

Lawrence, D. B., Mount Hood's latest eruption and glacier advances, *Mazama*, *30*, 22-29, 1948.

Le Bas, M. J., R. W. Le Maitre, A. Streckeisen, and B. Zanettin, A chemical classification of volcanic rocks based on the total alkali-silica diagram, *J. Petrol.*, *27*, 745-750, 1986.

Lipman, P. W. and D. R. Mullineaux (Eds.), The 1980 eruptions of Mount St. Helens, Washington, *U.S. Geol. Surv. Prof. Pap.*, *1250*, 844 pp., 1981.

Lipman, P. W., J. G. Moore, and D. A. Swanson, Bulging of the north flank before the May 18 eruption—geodetic data, *U.S. Geol. Surv. Prof. Pap.*, *1250*, 143-155, 1981a.

Lipman, P. W., D. R. Norton, J. E. Taggart, E. L. Brandt, and E. E. Engleman, Compositional variations in 1980 magmatic deposits, *U.S. Geol. Surv. Prof. Pap.*, *1250*, 631-640, 1981b.

Long, P. E., and R. A. Duncan, ^{40}Ar-^{39}Ar ages of Columbia River basalt from deep boreholes in south-central Washington, *Eos*, *64*, 90, 1983.

Long, P. E. and B. J. Wood, Structures, textures and cooling histories of Columbia River basalt flows, *Geol. Soc. Am. Bull.*, *97*, 1144-1155, 1986.

Lowry, W. D. and E. M. Baldwin, Late Cenozoic geology of the lower Columbia River Valley, Oregon and Washington, *Geol. Soc. Am. Bull.*, *63*, 1-24, 1952.

Luedke, R. G. and R. L. Smith, Map showing distribution, composition, and age of late Cenozoic volcanic centers in Oregon and Washington, *U.S. Geol. Surv. Misc. Invest. Map*, *I-1091D*, 1982.

Mackin, J. H., A stratigraphic section in the Yakima Basalt and the Ellensburg Formation in south-central Washington, *Wash. Div. Mines Geol. Rept. Invest.*, *19*, 45 pp., 1961.

MacLeod, N. S., G. W. Walker, and E. H. McKee, Geothermal significance of eastward increase in age of upper Cenozoic rhyolitic domes in southeastern Oregon, in *Proc. United Nations Symposium on Development and Use of Geothermal Resources 2nd*, *1*, 465-474, 1975.

Magill, J., A. Cox, and R. Duncan, Tillamook volcanic series: further evidence for tectonic rotation of the Oregon Coast Range, *J. Geophys. Res.*, *86*, 2953-2970, 1981.

Majors, H. M., Three newly discovered accounts of activity on Mount St. Helens, 1898, 1903 and 1921, *Northwest Discovery*, *1*, 36-41, 1980.

Malone, S. D. and Swanson, D. A., Twentieth century unrest at Cascades volcanoes, *Inter. Volcanol. Cong. Abst. Vol.*, Auckland-Hamilton-Rotorua, New Zealand, 260, 1986.

Mangan, M. T., Wright, T. L., Swanson, D. A., and Byerly, G. R., Regional correlation of Grande Ronde Basalt flows, Columbia River Basalt Group, Washington, Oregon, and Idaho, *Geol. Soc. Am. Bull.*, *97*, 1300-1318, 1986.

Mankinen, E. A. and G. B. Dalrymple, Revised geomagnetic polarity time scale for the interval 0-5 m.y. B.P., *J. Geophys. Res.*, *84*, 615-626, 1979.

Mattinson, J. M., Emplacement history of the Tatoosh volcanic-plutonic complex, Washington: ages of zircons, *Geol. Soc. Am. Bull.*, *88*, 1509-1514, 1977.

McBirney, A. R., Volcanic evolution of the Cascade Range, *An. Rev. Earth Planet. Sci.*, *6*, 437-456, 1978.

McKee, E. H., D. A. Swanson, and T. L. Wright, Duration and volume of Columbia River basalt volcanism, Washington, Oregon, and Idaho, *Geol. Soc. Am. Abst. Prog.*, *9*, 463-464, 1977.

McKee, E. H., P. R. Hooper, and W. D. Kleck, Age of the Imnaha Basalt—oldest flows of the Columbia River Basalt Group, northwest United States, *Isochron/West*, *31*, 31-33, 1981.

McMillan, K., R. W. Cross, and P. E. Long, Two-stage vesiculation in the Cohassett flow of the Grande Ronde Basalt, south-central Washington, *Geology*, *15*, 809-812, 1987.

Mellors, R. A., R. B. Waitt, and D. A. Swanson, Generation of pyroclastic flows and surges by hot-rock avalanches from the dome of Mount St. Helens volcano, USA, *Bull. Volcanol.*, *50*, 14-25, 1988.

Melson, W. G., Monitoring the 1980-82 eruptions of Mount St. Helens: compositions and abundances of glass, *Science*, *221*, 1387-1391, 1983.

Michaelson, C. A. and C. S. Weaver, Upper mantle structure from teleseismic P-wave arrivals in Washington and northern Oregon, *J. Geophys. Res.*, *91*, 209-250, 1986.

Miller, R. B., The Ingalls ophiolitic complex, north-central Cascade Mountains, Washington, *Geol. Soc. Am Bull.*, *96*, 27-42, 1985a.

Miller, R. B., The pre-Tertiary Rimrock Lake inlier, southern Cascades, Washington, *Wash. Div. Geol. Earth Resources Open-File Report*, *85-2*, 16 pp., 1985b.

Minor, R., Dating the Bonneville landslide in the Columbia River Gorge, *Heritage Research Associates Report No. 31*, Eugene, Oreg., 19 pp., 1984.

Misch, P., Tectonic evolution of the northern Cascades of Washington state: a west-Cordilleran case history, *Can. Inst. Min. Metal. Spec. Vol.*, *8*, 101-148, 1966.

Misch, P., Dextral displacements at some major strike faults in the North Cascades, *Geol. Assoc. Canada Prog. Abstracts*, 37, 1977.

Moen W. S., St. Helens and Washougal mining districts of the southern Cascades of Washington, *Wash. Div. Geology Earth Resources Info. Cir.*, *60*, 71 pp., 1977.

Morgan, W. J., Hotspot tracks and the opening of the Atlantic and Indian oceans, in *The Sea*, *7*, 443-487, John Wiley, New York, 1981.

Morgan, W. J., Flood basalts and mass extinctions, in *Global catastrophes in Earth history: an interdisciplinary conference on impacts, volcanism, and mass mortality*, Lunar Planet. Inst. Contrib. 673, 126, 1988.

Moyer, T. C. and D. A. Swanson, Secondary hydroeruptions in pyroclastic-flow deposits: examples from Mount St. Helens, *J. Volc. Geotherm. Res.*, *32*, 299-319, 1987.

Mullineaux, D. R., Pumice and other pyroclastic deposits in Mount Rainier National Park, Washington, *U.S. Geol. Surv. Bull.*, *1326*, 83 pp., 1974.

Mullineaux, D. R., Summary of pre-1980 tephra-fall deposits erupted from Mount St. Helens, Washington state, USA, *Bull. Volcanol.*, *48*, 17-26, 1986.

Mullineaux, D. R. and D. R. Crandell, The eruptive history of Mount St. Helens, *U.S. Geol. Surv. Prof. Pap.*, *1250*, 3-15, 1981.

Mullineaux, D. R., R. E. Wilcox, W. F. Ebaugh, R. Fryxell, and M. Rubin, Age of the last major scabland flood of the

Columbia Plateau in eastern Washington, *Quat. Res.*, *10*, 171-180, 1978.

Mundorff, M. J., Geology and ground-water conditions of Clark County, Wash., with a description of a major alluvial aquifer along the Columbia River, *U.S. Geol. Surv. Water Supply Pap.*, *1600*, 268 pp., 1964.

Mundorff, M. J. and A. A. Eggers, Tumtum Mountain, a late Pleistocene volcanic dome in southwestern Washington, *Northwest Science*, *62*, 10-15, 1988.

Myers, C. W. and S. M. Price (Eds.), *Geologic studies of the Columbia Plateau*, Rockwell Hanford Operations, Richland, WA, RHO-BWI-ST-4, III-23-III-32, 1979.

Nehring, N. L., H. A. Wollenberg, and D. A. Johnston, Gas analyses of fumaroles from Mt. Hood, Oregon, *U.S. Geol. Surv. Open-File Rept.*, *81-236*, 9 pp., 1981.

Niem, A. R. and W. A. Niem, Oil and gas investigation of the Astoria Basin, Clatsop and northernmost Tillamook Counties, northwest Oregon, *Oreg. Dept. Geol. Min. Indust. Oil and Gas Invest. Rept.*, *OGI-14*, 1985.

Okazaki, R., H. W. Smith, R. A. Gilkeson, and J. Franklin, Correlation of West Blacktail Ash with pyroclastic layer T from the 1800 A.D. eruption of Mt. St. Helens, *Northwest Sci.*, *46*, 77-89, 1972.

Phillips, W. M., M. A. Korosec, H. W. Schasse, J. L. Anderson, and R. A. Hagen, K-Ar ages of volcanic rocks in southwest Washington, *Isochron/West*, *47*, 18-24, 1986.

Phillips, W. M., Geologic map of the Mount St. Helens quadrangle, Washington and Oregon, *Wash. Div. Geol. Earth Resources Open-File Rept.*, *87-4*, 59 pp., 1987.

Pierson, T. C., Initiation and flow behavior of the 1980 Pine Creek and Muddy River lahars, Mount St. Helens, Washington, *Geol. Soc. Am. Bull.*, *96*, 1056-1069, 1985.

Pierson, T. C. and K. M. Scott, Downstream dilution of a lahar: transition from debris flow to hyperconcentrated streamflow, *Water Resources Res.*, *21*, 1511-1524, 1985.

Porter, S. C., Pleistocene glaciation in the southern part of the North Cascade Range, Washington, *Geol. Soc. Am. Bull.*, *87*, 61-75, 1976.

Power, S. G., C. W. Field, R. L. Armstrong, and J. E. Harakal, K-Ar ages of plutonism and mineralization, western Cascades, Oregon and southern Washington, *Isochron/West*, *31*, 27-29, 1981.

Price, E. H., Structural geometry, strain distribution, and mechanical evolution of eastern Umtanum Ridge, and a comparison with other selected localities within Yakima fold structures, south-central Washington, Rockwell Hanford Oper., Richland, Wash., RHO-BWI-SA-138, 109 pp., 1982.

Priest, G. R., M. H. Beeson, M. W. Gannett, and D. A. Berri, Geology, geochemistry, and geothermal resources of the Old Maid Flat area, Oregon, *Oreg. Dept. Geol. Min. Indus. Spec. Pap.*, *14*, 16-30, 1982.

Qamar, A., W. St. Lawrence, J. N. Moore, and G. Kendrick, Seismic signals preceding the explosive eruption of Mount St. Helens, Washington on 18 May 1980, *Bull. Seis. Soc. Am.*, *73*, 1797-1813, 1983.

Rampino, M. R. and R. B. Stothers, Flood basalt volcanism during the past 250 million years, *Science*, *241*, 663-668, 1988.

Rasmussen, J. and E. Humphreys, Tomographic image of the Juan de Fuca plate beneath Washington and western Oregon using teleseismic P-wave travel times, *Geophys. Res. Lett.*, *15*, 1417-1420, 1988.

Reidel, S. P., The Saddle Mountains—the evolution of an anticline in the Yakima fold belt, *Am. J. Sci.*, *284*, 942-978, 1984.

Reidel, S. P., P. E. Long, S. W. Myers, and J. Mase, New evidence for greater than 3.2 km of Columbia River basalt beneath the central Columbia Plateau, *Eos Trans AGU*, *64*, 173, 1982.

Reidel, S. P., T. L. Tolan, P. R. Hooper, M. H. Beeson, K. R. Fecht, and J. L. Anderson, The Grande Ronde Basalt, Columbia River Basalt Group: stratigraphic descriptions and correlations in Washington, Oregon, and Idaho, in *Volcanism and Tectonism on the Columbia Plateau*, edited by S. P. Reidel, P. R. Hooper, and J. L. Anderson, *Geol. Soc. Am. Spec. Pap.*, in press.

Rite, A. and H. M. Iyer, July 1980 Mt. Hood earthquake swarm, *U.S. Geol. Surv. Open-File Rept.*, *81-48*, 21 pp., 1981.

Riddihough, R., Recent movements of the Juan de Fuca plate system, *J. Geophys. Res.*, *89*, 6980-6994, 1984.

Rosenbaum, J. G. and R. B. Waitt, Jr., Summary of eyewitness accounts of the May 18 eruption, *U.S. Geol. Surv. Prof. Pap.*, *1250*, 53-67, 1981.

Rutherford, M. J., H. Sigurdsson, S. Carey, and A. Davis, The May 18, 1980, eruption of Mount St. Helens 1. Melt composition and experimental phase equilibria, *J. Geophys. Res.*, *90*, 2929-2947, 1985.

Sager, J. W. and D. R. Chambers, Design and construction of the Spirit Lake outlet tunnel, Mount St. Helens, Washington, *Am. Soc. Civ. Eng. Geotech. Spec. Pub.*, *3*, 42-58, 1986.

Scandone, R. and S. D. Malone, Magma supply, magma discharge and readjustments of the feeding system of Mount St. Helens during 1980, *J. Volc. Geotherm. Res.*, *23*, 239-262, 1985.

Schmincke, H.-U., Petrology, paleocurrents, and stratigraphy of the Ellensburg Formation and interbedded Yakima Basalt flows, south-central Washington, Ph.D thesis, 425 pp., Johns Hopkins Univ., Baltimore, 1964.

Schmincke, H.-U., Stratigraphy and petrography of four upper Yakima basalt flows in south-central Washington, *Geol. Soc. Am. Bull.*, *78*, 1385-1422, 1967a.

Schmincke, H.-U., Graded lahars in the type sections of the Ellensburg Formation, south-central Washington, *J. Sed. Petrol.*, *37*, 438-448, 1967b.

Schmincke, H.-U., Fused tuff and pépérites in south-central Washington, *Geol. Soc. Am. Bull.*, *78*, 319-330, 1967c.

Schmincke, H.-U., Flow directions in Columbia River basalt flows and paleocurrents of interbedded sedimentary rocks, south-central Washington, *Geologische Rundschau*, *56*, 992-1020, 1967d.

Schuster, R. L., Effect of the eruptions on civil works and operations in the Pacific Northwest, *U.S. Geol. Surv. Prof. Pap. 1250*, *1250*, 701-718, 1981.

Scott, K. M., Lahars and lahar-runout flows in the Toutle-Cowlitz River system, Mount St. Helens, Washington, *U.S. Geol. Surv. Prof. Pap. 1447A*, 74 pp., 1988.

Shaw, H. R. and D. A. Swanson, Eruption and flow rates of flood basalts, in *Proc. Second Columbia River Basalt Symp.*, Eastern Wash. State Coll. Press, Cheney, 271-299, 1970.

Sherrod, D. R. and J. G. Smith, Preliminary map of upper Eocene to Holocene volcanic and related rocks of the Cascade Range, Oregon, scale 1:500,000, *U.S. Geol. Surv. Open-File Rept.*, in press.

Siebert, L., H. Glicken, and T. Ui, Volcanic hazards from

Bezymianny- and Bandai-type eruptions, *Bull. Volcanol.*, *49*, 435-459, 1987.

Simpson, R. W. and A. Cox, Paleomagnetic evidence for tectonic rotation of the Oregon Coast Range, *Geology*, *5*, 585-589, 1977.

Smith, D. R., The petrology and geochemistry of High Cascade volcanics in southern Washington: Mount St. Helens volcano and the Indian Heaven basalt field, Ph.D. thesis, 409 pp., Rice Univ., Houston, Texas, 1984.

Smith, D. R. and W. P. Leeman, Petrogenesis of Mount St. Helens dacitic magmas, *J. Geophys. Res.*, *92*, 10,313-10,334, 1987.

Smith, G. A., Coarse-grained nonmarine volcaniclastic sediment: terminology and depositional process, *Geol. Soc. Am. Bull.*, *97*, 1-10, 1986.

Smith, G. A., Neogene synvolcanic and syntectonic sedimentation in central Washington, *Geol. Soc. Am. Bull.*, *99*, 1479-1492, 1988.

Smith, G. A., Campbell, N. P., Deacon, M. W., and Shafiquallah, M., Eruptive style and location of volcanic centers in the Miocene Washington Cascade Range: reconstruction from the sedimentary record, *Geology*, *16*, 337-340, 1988.

Smith, G. O., The rocks of Mount Rainier, *U.S. Geol. Surv. Ann. Rept.*, *18*, 416-423, 1897.

Smith, G. O., Description of the Mount Stuart quadrangle, Washington, *U.S. Geol. Surv. Geol. Atlas Mount Stuart, Folio 106*, 10, 1904.

Smith, G. O. and F. C. Calkins, Description of the Snoqualmie quadrangle, Washington, *U.S. Geol. Surv. Geol. Atlas Snoqualmie, Folio 139*, 14, 1906.

Smith, H. W., R. Okazaki, and C. E. Knowles, Electron microprobe analysis of glass shards from tephra assigned to set W, Mt. St. Helens, Washington, *Quat. Res.*, *7*, 207-217, 1977.

Smith, J. G., New compilation geologic map of the Cascade Range in Washington, *Trans. Geotherm. Resour. Counc.*, *11*, 309-314, 1987.

Smith, J. G., Geologic map showing upper Eocene to Holocene volcanic and related rocks in the Cascade Range of Washington, *U.S. Geol. Surv. Misc. Invest. Map*, in press a.

Smith, J. G., Geologic map of the Yakima 2-degree sheet, *U.S. Geol. Surv. Misc. Invest. Map*, in press b.

Southwick, D. L., Geology of the alpine-type ultramafic complex near Mount Stuart, Washington, *Geol. Soc. Am. Bull.*, *85*, 391-402, 1974.

Stanley, W. D., Tectonic study of the Cascade Range and Columbia Plateau in Washington State based upon magnetotelluric soundings, *J. Geophys. Res.*, *89*, 4447-4460, 1984.

Stanley, W. D., C. Finn, and J. L. Plesha, Tectonics and conductivity structures in the southern Washington Cascades, *J. Geophys. Res.*, *92*, 10,179-10,193, 1987.

Stout, M. L., Geology of a part of the south-central Cascade Mountains, Washington, *Geol. Soc. Am. Bull.*, *75*, 317-334, 1964.

Swanson, D. A., The middle and late Cenozoic volcanic rocks of the Tieton River area, south-central Washington, Ph.D. thesis, 330 pp., Johns Hopkins Univ., Baltimore, 1964.

Swanson, D. A., Tieton volcano, a Miocene eruptive center in the southern Cascade Mountains, Washington, *Geol. Soc. Am. Bull.*, *77*, 1293-1314, 1966.

Swanson, D. A., Yakima Basalt of the Tieton River area, south-central Washington, *Geol. Soc. Am. Bull.*, *78*, 1077-1100, 1967.

Swanson, D. A., Magma supply rate at Kilauea Volcano, 1952-1971, *Science*, *175*, 169-170, 1972.

Swanson, D. A., Geologic map of the Tieton River area, Yakima County, south-central Washington, scale 1:48,000, *U.S. Geol. Surv. Misc. Field Studies Map, MF-968*, 1978.

Swanson, D. A. and G. A. Clayton, Generalized geologic map of the Goat Rocks wilderness and roadless areas (6036, Parts A, C, and D), Lewis and Yakima Counties, Washington, *U.S. Geol. Surv. Open-File Rept.*, *83-357*, 10 pp., 1983.

Swanson, D. A. and R. T. Holcomb, Regularities in growth of the Mount St. Helens dacite dome, 1980-86, in The Mechanics of Lava Flow Emplacement and Dome Growth, *IAVCEI Proceedings in Volcanology*, *2*, Springer Verlag, Heidelberg, in press.

Swanson, D. A. and T. L. Wright, Paleogeography of southeast Washington during the middle and late Miocene based on the distribution of intracanyon basalt flows, in *Cenozoic Paleogeography of the western United States*, Pac. Div. Soc. Econ. Paleon. Mineral., Pac. Coast Paleog. Symp., Los Angeles, 31, 1979.

Swanson, D. A., T. L. Wright, and R. T. Helz, Linear vent systems and estimated rates of magma production and eruption for the Yakima Basalt on the Columbia Plateau, *Am. J. Sci.*, *275*, 877-905, 1975.

Swanson, D. A., T. L. Wright, P. R. Hooper, and R. D. Bentley, Revisions in stratigraphic nomenclature of the Columbia River Basalt Group, *U.S. Geol. Surv. Bull.*, *1457-G*, 59 pp., 1979a.

Swanson, D. A., J. L. Anderson, R. D. Bentley, G. R. Byerly, V. E. Camp, J. N. Gardner, and T. L. Wright, Reconnaissance geologic map of the Columbia River Basalt Group in eastern Washington and northern Idaho, *U.S. Geol. Surv. Open-File Rept.*, *79-1363*, 26 pp., 1979b.

Swanson, D. A., J. C. Brown, J. L. Anderson, R. D. Bentley, G. R. Byerly, J. N. Gardner, and T. L. Wright, Structure contour maps on the top of the Grande Ronde and Wanapum Basalts, eastern Washington and northern Idaho, *U.S. Geol. Surv. Open-File Rept.*, *79-1364*, 1979c.

Swanson, D. A., T. L. Wright, V. E. Camp, J. N. Gardner, R. T. Helz, S. M. Price, S. P. Reidel, and M. E. Ross, Reconnaissance geologic map of the Columbia River Basalt Group, Pullman and Walla Walla quadrangles, southeast Washington and adjacent Idaho, *U.S. Geol. Surv. Misc. Invest. Map*, *I-1139*, 1980.

Swanson, D. A., J. L. Anderson, V. E. Camp, P. R. Hooper, W. H. Taubeneck, and T. L. Wright, Reconnaissance geologic map of the Columbia River Basalt Group, northern Oregon and western Idaho, *U.S. Geol. Surv. Open-File Rept.*, *81-797*, 32 pp., 1981.

Swanson, D. A., T. J. Casadevall, D. Dzurisin, S. D. Malone, C. G. Newhall, and C. S. Weaver, Predicting eruptions at Mount St. Helens, June 1980 through December 1982, *Science*, *221*, 1369-1376, 1983.

Swanson, D. A., T. J. Casadevall, D. Dzurisin, R. T. Holcomb, C. G. Newhall, S. D. Malone, and C. S. Weaver, Forecasts and predictions of eruptive activity at Mount St. Helens, 1975-1984, *J. Geodynamics*, *3*, 397-423, 1985.

Swanson, D. A., D. Dzurisin, R. T. Holcomb, E. Y. Iwatsubo, W. W. Chadwick, Jr., T. J. Casadevall, J. W. Ewert, and C. C. Heliker, Growth of the lava dome at Mount St. Helens, Washington (USA), 1981-83, *Geol. Soc. Am. Spec. Pap.*,

212, 1-16, 1987.

Tabor, R. W. and D. F. Crowder, Of batholiths and volcanoes: Intrusion and eruption of Late Cenozoic magmas in the Glacier Peak area, north Cascades, Washington, *U. S. Geol. Surv. Prof. Pap.*, *604*, 67 pp., 1969.

Tabor, R. W., V. A. Frizzell, Jr., D. B. Booth, J. T. Whetten, R. B. Waitt, Jr., and R. E. Zartman, Preliminary geologic map of the Skykomish River 1:100,000 quadrangle, Washington, *U. S. Geol. Surv. Open-File Map*, *82-747*, 1982a.

Tabor, R. W., R. B. Waitt, Jr., V. A. Frizzell, Jr., D. A. Swanson, G.R. Byerly, and R.D. Bentley, Geologic map of the Wenatchee 1:100,000 quadrangle, central Washington, *U.S. Geol. Surv. Misc. Invest. Map*, *I-1311*, 26 pp., 1982b.

Tabor, R. W., V. A. Frizzell, Jr., J. A. Vance, and C. W. Naeser, Ages and stratigraphy of lower and middle Tertiary sedimentary and volcanic rocks of the central Cascades, Washington: application to the tectonic history of the Straight Creek fault, *Geol. Soc. Am. Bull.*, *95*, 26-44, 1984.

Tabor, R. W., R. E. Zartman, and V. A. Frizzell, Jr., Possible tectonostratigraphic terranes in the North Cascades crystalline core, Washington, *Wash. Div. Geol. Earth Res. Bull.*, *77*, 107-127, 1987.

Taubeneck, W. H., Dikes of Columbia River basalt in northeastern Oregon, western Idaho, and southeastern Washington, in *Proc. Second Columbia River Basalt Symp.*, East. Wash. State Coll. Press, Cheney, 73-96, 1970.

Taylor, E. M., Field geology of the northwest quarter of the Broken Top 15' quadrangle, Deschutes County, Oregon, *Oreg. Dept. Geol. Mineral Indust. Spec. Pap.*, *21*, 20 pp., 1987.

Thiruvathukal, J. V., J. W. Berg, and D. F. Heinrichs, Regional gravity of Oregon, *Bull. Seis. Soc. Am.*, *81*, 725-738, 1970.

Tolan, T. L. and M. H. Beeson, Exploring the Neogene history of the Columbia River: Discussion and geologic field trip guide to the Columbia River Gorge. Part I. Discussion, *Oreg. Geol.*, *46*, 87-97, 1984a.

Tolan, T. L. and M. H. Beeson, Exploring the Neogene history of the Columbia River: Discussion and geologic field trip guide to the Columbia River Gorge. Part II. Road log and comments, *Oreg. Geol.*, *46*, 103-112, 1984b.

Tolan, T. L. and M. H. Beeson, Intracanyon flows of the Columbia River Basalt Group in the lower Columbia River Gorge and their relationship to the Troutdale Formation, *Geol. Soc. Am. Bull.*, *95*, 463-477, 1984c.

Tolan, T. L., S. P. Reidel, M. H. Beeson, J. L. Anderson, K. R. Fecht, and D. A. Swanson, Revisions to the areal extent and volume of the Columbia River Basalt Group (CRBG), *Geol. Soc. Am. Abst. Prog.*, *19*, 458, 1987.

Tolan, T. L., S. P. Reidel, M. H. Beeson, J. L. Anderson, K. R. Fecht, and D. A. Swanson, Revisions to the estimates of the areal extent and volume of the Columbia River Basalt Group, in *Volcanism and Tectonism on the Columbia Plateau*, edited by S. P. Reidel, P. R. Hooper, and J. L. Anderson, *Geol. Soc. Am. Spec. Pap.*, in press.

Treasher, R. C., Geologic history of the Portland area, *Oreg. Dept. Geol. Mineral Indust. Short Paper*, *7*, 17 pp., 1942.

Trimble, D. E., Geology of Portland, Oregon and adjacent areas, *U.S. Geol. Surv. Bull.*, *1119*, 119 pp., 1963.

Turner, D. L., V. A. Frizzell, Jr., D. M. Triplehorn, and C. W. Naeser, Radiometric dating of ash partings in coals of the Eocene Puget Group, Washington: implications for paleobotanical stages, *Geology*, *11*, 527-531, 1983.

Vallance, J. W., Late Quaternary volcanic stratigraphy of the southwestern flank of Mount Adams volcano, Washington, M.S. thesis, Univ. Colorado, Boulder, 1986.

Vance, J. A., Early and middle Cenozoic arc magmatism and tectonics in Washington state, *Geol. Soc. Am. Abst. Prog.*, *11*, 132, 1979.

Vance, J. A., Cenozoic stratigraphy and tectonics of the Washington Cascades, *Geol. Soc. Am. Abst. Prog.*, *14*, 241, 1982.

Vance, J. A., Early Tertiary faulting in the North Cascades, *Geol. Soc. Am. Abst. Prog.*, *17*, 415, 1985.

Vance, J. A. and R. B. Miller, The movement history of the Straight Creek fault in Washington State, *Cordilleran Sect. Geol. Assoc. Canada, Prog. Absts.*, 39-41, 1981.

Vance, J. A. and C. W. Naeser, Fission track geochronology of the Tertiary volcanic rocks of the central Cascade Mountains, Washington, *Geol. Soc. Am. Absts. Prog.*, *9*, 520, 1977.

Vance, J. A., N. W. Walker, and J. M. Mattinson, U/Pb ages of early Cascade plutons in Washington state, *Geol. Soc. Am. Abst. Prog.*, *18*, 194, 1986.

Vance, J. A., G. A. Clayton, J. M. Mattinson, and C. W. Naeser, Early and middle Cenozoic stratigraphy of the Mount Rainier-Tieton River area, southern Washington Cascades, *Wash. Div. Geol. Earth Res. Bull.*, *77*, 269-290, 1987.

Vine, J. D., Preliminary geologic map of the Hobart and Maple Valley quadrangles, King County, Washington, scale 1:24,000, *Wash. Div. Mines Geol. Map*, *GM-1*, 1962.

Vine, J. D., Geology and coal resources of the Cumberland, Hobart, and Maple Valley quadrangles, King County, Washington, *U.S. Geol. Surv. Prof. Pap.*, *624*, 67 pp., 1969.

Voight, B., H. Glicken, R. J. Janda, and P. M. Douglass, Catastrophic rockslide avalanche of May 18, *U.S. Geol. Surv. Prof. Pap.*, *1250*, 347-377, 1981.

Voight, B., R. J. Janda, H. Glicken, and P. M. Douglass, Nature and mechanics of the Mount St Helens rockslide-avalanche of 18 May 1980, *Geotechnique*, *33*, 243-273, 1983.

Waitt, R. B., Late-Cenozoic deposits, landforms, stratigraphy, and tectonism of Kittitas Valley, Washington, *U.S. Geol. Surv. Prof. Pap.*, *1127*, 18 pp., 1979.

Waitt, R. B., Deposits and effects of devastating lithic pyroclastic density current from Mount St. Helens on 18 May 1980–field guide for northeast radial, *U.S. Geol. Surv. Open-File Rept.*, *84-839*, 10 pp., 1984a.

Waitt, R. B., Comment, *Geology*, *12*, 693, 1984b.

Waitt, R. B., Case for periodic, colossal jökulhlaups from Pleistocene glacial Lake Missoula, *Geol. Soc. Am. Bull.*, *96*, 1271-1286, 1985.

Waitt, R. B. and N. S. MacLeod, Minor explosive eruptions at Mount St. Helens dramatically interacting with winter snowpack in March-April 1982, *Wash. Div. Geol. Earth Res. Bull.*, *77*, 355-379, 1987.

Waitt, R. B. and D. A. Swanson, Geomorphic evolution of the Columbia Plain and River, *Geol. Soc. Am. Centennial Spec. Vol.*, *2*, 403-423, 1987.

Waitt, R. B., T. C. Pierson, N. S. MacLeod, R. J. Janda, B. Voight, and R. T. Holcomb, Eruption-triggered avalanche, flood, and lahar at Mount St. Helens–effects of winter snowpack, *Science*, *221*, 1394-1397, 1983.

Walker, G. P. L., Lengths of lava flows, *Phil. Trans. Roy. Soc. Lond. A*, *274*, 107-118, 1973.

Walker, G. P. L. and L. A. McBroome, Mount St. Helens

1980 and Mount Pelée 1902—flow or surge?, *Geology*, *11*, 571-574, 1983.

Walsh, T. J., M. A. Korosec, W. M. Phillips, R. L. Logan, and H. W. Schasse, Geologic map of Washington—southwest quadrant, scale 1:250,000, *Wash. Div. Geol. Earth Resources Geol. Map*, *GM-34*, 28 pp., 1987.

Warren, W. C., Relation of the Yakima basalt to the Keechelus andesitic series, *J. Geol.*, *49*, 795-814, 1941.

Waters, A. C., Geomorphology of south-central Washington, illustrated by the Yakima East quadrangle, *Geol. Soc. Am. Bull.*, *66*, 663-684, 1955.

Waters, A. C., Stratigraphic and lithologic variations in the Columbia River basalt, *Am. J. Sci.*, *259*, 583-611, 1961.

Waters, A. C., The Columbia River Gorge: basalt stratigraphy, ancient lava dams, and landslide dams, *Oreg. Dept. Geol. Mineral Indust. Bull.*, *77*, 133-162, 1973.

Weaver, C. S. and G. E. Baker, Geometry of the Juan de Fuca plate beneath Washington—evidence from seismicity and the 1949 south Puget Sound earthquake, *Bull. Seis. Soc. Am.*, *78*, 264-275, 1988.

Weaver, C. S. and S. D. Malone, Overview of the tectonic setting and recent studies of eruptions of Mount St. Helens, Washington, *J. Geophys. Res.*, *92*, 10,149-10,154, 1987.

Weaver, C. S. and C. A. Michaelson, Seismicity and volcanism in the Pacific Northwest: evidence for the segmentation of the Juan de Fuca plate, *Geophys. Res. Let.*, *12*, 215-218, 1985.

Weaver, C. S. and S. W. Smith, Regional tectonic and earthquake hazard implications of a crustal fault zone in southwestern Washington, *J. Geophys. Res.*, *88*, 10,371-10,383, 1983.

Weaver, C. S., S. M. Green, and H. M. Iyer, Seismicity of Mount Hood and structure as determined from teleseismic *P* wave delay studies, *J. Geophys. Res.*, *87*, 2782-2792, 1982.

Weaver, C. S., W. C. Grant, and J. E. Shemata, Local crustal extension at Mount St. Helens, Washington, *J. Geophys. Res.*, *92*, 10,170-10,178, 1987.

Wells, R. E. and R. S. Coe, Paleomagnetism and geology of Eocene volcanic rocks of southwest Washington, implications for mechanisms of tectonic rotation, *J. Geophys. Res.*, *90*, 1925-1947, 1985.

Wells, R. E. and P. C. England, Continuum model of rotations recorded in Miocene flows of the Columbia River Basalt Group, Oregon and Washington, *Eos*, in press.

Wells, R. E. and A. R. Niem, Geology of the Columbia River Basalt Group in the Astoria Basin, Oregon and Washington: evidence for invasive flows, *Geol. Soc. Am. Abst. Prog.*, *19*, 462-463, 1987.

Wells, R. E. and R. W. Simpson, More paleomagnetic results from the Miocene Columbia River Basalt Group, Oregon and Washington: stratigraphic and tectonic implications, *Geol. Soc. Am. Abst. Prog.*, *19*, 463, 1987.

Wells, R. E., R. W. Simpson, R. D. Bentley, M. H. Beeson, M. T. Mangan, and T. L. Wright, Correlation of Miocene flows of the Columbia River Basalt Group from the central Columbia Plateau to the coast of Oregon and Washington, in *Volcanism and Tectonism on the Columbia Plateau*, edited by S. P. Reidel, P. R. Hooper, and J. L. Anderson, *Geol. Soc. Am. Spec. Pap.*, in press.

Westgate, J. A., D. G. W. Smith, and H. Nichols, Late Quaternary pyroclastic layers in the Edmonton area, Alberta, in *Pedology and Quaternary Research*, Univ. Alberta Printing, Edmonton, Alberta, pp. 179-186, 1970.

White, C., Geology and geochemistry of Mt. Hood volcano, *Oreg. Dept. Geol. Mineral Indust. Spec. Pap.*, *8*, 26 pp., 1980.

Williams, D. L. and T. E. C. Keith, Aeromagnetic and Bouguer gravity maps of Mount Hood Wilderness, Clackamas and Hood River Counties, Oregon, *U.S. Geol. Surv. Misc. Field Studies Map*, *MF-1379D*, 1982.

Williams, D. L., G. Abrams, C. Finn, D. Dzurisin, D. J. Johnson, and R. Denlinger, Evidence from gravity data for an intrusive complex beneath Mount St. Helens, *J. Geophys. Res.*, *92*, 10,207-10,222, 1987.

Williams, D. L., C. Finn, R. W. Couch, Z. F. Danes, G. S. Pitts, W. M. Phillips, and R. P. Riddihough, Residual Bouguer gravity anomaly map of the Cascade Range, California, Oregon, Washington, and British Columbia, *U.S. Geol. Surv. Geophys. Invest. Map*, *GP-973*, 1988.

Wise, W. S., The geology and mineralogy of the Wind River area, Washington, and the stability relations of celadonite, Ph.D. thesis, Johns Hopkins Univ., Baltimore, 1961.

Wise, W. S., Geology of the Mount Hood volcano, *Oreg. Dept. Geol. Mineral Indust. Bull.*, *62*, 81-98, 1968.

Wise, W. S., Geology and petrology of the Mount Hood area: a study of High Cascade volcanism, *Geol. Soc. Am. Bull.*, *80*, 969-1006, 1969.

Wise, W. S., Cenozoic volcanism in the Cascade Mountains of southern Washington, *Wash. Dept. Nat. Resources Bull.*, *60*, 45 pp., 1970.

Wright, T. L., M. Mangan, and D. A. Swanson, Chemical data for flows and feeder dikes of the Yakima Basalt Subgroup, Columbia River Basalt Group: constraints on a petrogenetic model, *U.S. Geol. Surv. Bull.*, *1821*, in press.

Yamaguchi, D. K., New tree-ring dates for recent eruptions of Mount St. Helens, *Quat. Res.*, *20*, 246-250, 1983.

Yamaguchi, D. K., Tree-ring evidence for a two-year interval between recent prehistoric explosive eruptions of Mt. St. Helens, *Geology*, *13*, 554-557, 1985.

Zoback, M. L. and G. A. Thompson, Basin and Range rifting in northern Nevada, clues from a mid-Miocene rift and its subsequent offsets, *Geology*, *6*, 111-116, 1978.

Devils Tower—Black Hills Alkalic Igneous Rocks and General Geology

Blacks Hills, South Dakota to
Bear Lodge Mountains, Wyoming
July 1–7, 1989

Field Trip Guidebook T131

Leader:
Frank R. Karner, Editor

Associate Leaders:
Don Halvorson, Gordon A. Jenner, and Stanley F. White

American Geophysical Union, Washington, D.C.

2000 Florida Ave., N.W., Washington, D.C. 20009

ISBN: 0-87590-670-2

Printed in the United States of America

COVER The Bear Lodge (mato tepee) or Devils Tower looking eastward with the low profile of the Bear Lodge Mountains in distance. Woodcut from H. A. Newton and W. P. Jenney, 1880, Report on the geology and resources of the Black Hills of Dakota: U.S. Geographical and Geological Survey of the Rocky Mountain Region.

Frank R. Karner, Editor
Department of Geology and Geological Engineering and
North Dakota Geological Survey
University of North Dakota, Grand Forks, N.D. 58202 U.S.A.

CONTENTS

IGC FIELD TRIP T131:
DEVILS TOWER — BLACK HILLS ALKALIC IGNEOUS ROCKS AND GENERAL GEOLOGY

Frank R. Karner, Editor
Department of Geology and Geological Engineering and
North Dakota Geological Survey
University of North Dakota, Grand Forks

PAHA SAPA: THE BLACK HILLS AND MATO TEPEE: THE GRIZZLY BEAR LODGE

Looming up as a dark ridgeline on the western plain, paha sapa, the Black Hills of South Dakota and Wyoming are a jewel of America, treasured by the Native American, the white explorer, the frontiersman, the cowboy, the gold prospector, the miner, the camp follower, the settler, the businessman, the tourist and also the geologist. History, folklore and legend are as conspicuous as outcrops for the geologist who travels the Black Hills. The Black Hills is a microcosm of geology. In this relatively small area is a geological section from the Archean to the Pleistocene. Precambrian igneous and metamorphic rocks, including world famous pegmatites, are overlain by an extensive Paleozoic-Mesozoic sedimentary sequence. These Precambrian-Mesozoic rocks have been uplifted and domed by Laramide deformational events and eroded to form scenic mountains, ridges and valleys, cavern systems and badlands topography. Cenozoic igneous events have formed striking volcanic-plutonic centers some of which contain significant mineral deposits.

Extending northwestward, the Black Hills uplift terminates in the Bear Lodge Mountains uplift of Wyoming with Devils Tower, named mato tepee or the grizzly bear lodge by Native Americans, and the Missouri Buttes at its western edge. The scenic Bear Lodge Mountains prominently display a carbonatite-pyroxenite alkalic rock intrusive and extrusive continental igneous association. The Bear Lodge igneous center also exhibits thorium and rare-earth mineralization, and remarkable potassium fenitization. Devils Tower is the most widely known landmark of the northern Great Plains. It has been a sacred place for Native Americans, a communal recreation site for early white settlers, and a mecca for traveling geologists since the Custer expedition of 1875. Devils Tower is recognized as the world's finest example of columnar jointing in a phonolite monolith.

Many general publications, maps and theses on the geology of the Black Hills and Bear Lodge Mountains make fascinating reading. Several are listed below and will be available for study during the trip:

Newton, H.A., and Jenney, W.P., 1880, Report on the geology and resources of the Black Hills of Dakota: U.S. Geographical and Geological Survey of the Rocky Mountain Region, p. 1-222.

O'Harra, C.C., 1902, The mineral wealth of the Black Hills: South Dakota School of Mines and Technology Bulletin 6, 88 p.

Darton, N.H., and O'Harra, C.C., 1905, Aladdin [quadrangle], Wyoming-South Dakota-Montana: U.S. Geological Survey Geologic Atlas of the United States, Folio 128, 8 p.

Darton, N.H., and O'Harra, C.C., 1907, Devils Tower [quadrangle], Wyoming: U.S.Geological Survey Geologic Atlas of the United States, Folio 150, 10 p.

Darton, N.H., 1909, Geology and water resources of the northern portion of the Black Hills and adjoining regions in South Dakota and Wyoming: U.S. Geological Survey Professional Paper 65, 105 p.

Darton, N.H., and Paige, Sidney, 1925, Central Black Hills, South Dakota: U.S. Geological Survey Geologic Atlas of United States, Folio 219, 34 p.

Page, L.R., Adams, J.W., Erickson, M.P., Hall, W.E., Hanley, J.B., Joralemon, Peter, Norton, J.J., Pray, L.C., Steven, T.A., Stoll, W.C., Stopper, R.F., and Trites, A.F., Jr., 1953, Pegmatite investigations, 1942-1945, Black Hills, South Dakota: U.S. Geological Survey Professional Paper 247, p. 293-300.

Robinson, C.S., Mapel, W.J., and Bergendahl, M.H., 1964, Stratigraphy and structure of the northern and western flanks of the Black Hills uplift, Wyoming, Montana, and South Dakota: U.S. Geological Survey Professional Paper 404, 134 p.

Harksen, J.C., and Macdonald, J.R., 1969, Guidebook to the major Cenozoic deposits of southwestern South Dakota: South Dakota Geological Survey, Guidebook 2, 103 p.

U.S. Geological Survey, 1975, U.S. Congress, Senate Committee on Interior and Insular Affairs, Mineral and water resources of South

Dakota: 94th Congress, 1st session, 313 p.
Lisenbee, A.L., 1978, Laramide structure of the Black Hills uplift, South Dakota-Wyoming-Montana in Matthews, V., III, ed., Laramide folding associated with basement block faulting in the Western United States: Geological Society of American Memoir 151, p. 165-196.
Feldmann, R.M., and Heimlich, R.A., 1980, The Black Hills, Kendall/Hunt Geology Field Guide Series, 190 p.
Halvorson, D.L., 1980, Geology and petrology of the Devils Tower, Missouri Buttes, and Barlow Canyon area, Crook County, Wyoming: Grand Forks, University of North Dakota, Ph.D. dissertation, 218 p.
Staatz, M.H., 1983, Geology and description of thorium and rare-earth deposits in the southern Bear Lodge Mountains, northeastern Wyoming: U.S. Geological Survey Professional Paper 1049-D, 52 p.
Jenner, G.A., Jr., 1984, Tertiary alkalic igneous activity, potassic fenitization, carbonatitic magmatism, and hydrothermal activity in the central and southeastern Bear Lodge Mountains, Crook County, Wyoming: Grand Forks, University of North Dakota, M.S. thesis, 232 p.
Lisenbee, A.L., 1985, Tectonic map of the Black Hills uplift, Montana, Wyoming and South Dakota: Geological Survey of Wyoming Map Series 13, scale 1:250,000.
Rich, F.J., 1985, Geology of the Black Hills, South Dakota and Wyoming, 2nd edition, 292 p.
DeWitt, E., Redden, J.A., Wilson, A.B., Buscher, D., and Dersch, J.S., 1986, Mineral resource potential and geology of the Black Hills National Forest, South Dakota and Wyoming: U.S. Geological Survey Bulletin 1580, 135 p.

This field guide contains a number of papers on general subjects along with daily field trip guides. I wish to thank the authors and recognize their special knowledge and efforts which so greatly contributed to this volume. Special thanks go to Juergen Reinhardt, Janet Evans and Penny Hanshaw of the International Geological Congress working group for support and direction, to Jack Redden of the South Dakota School of Mines and Technology for geological advice, to Ken Dorsher of the North Dakota Geological Survey for drafting, to Gloria Pederson of the University of North Dakota, Department of Geology and Geological Engineering for her willing and thorough assistance in all parts of the project, to Todd Brown of the Department for geological assistance and to Alan Cvancara and Will Gosnold of the Department for manuscript review.

IGC FIELD TRIP T131:
GEOLOGICAL FRAMEWORK OF THE BLACK HILLS -
BEAR LODGE MOUNTAINS REGION

Frank R. Karner
Department of Geology and Geological Engineering and
North Dakota Geological Survey,
University of North Dakota, Grand Forks

PRECAMBRIAN GEOLOGY

Overview

Major Precambrian orogenic belts and age provinces of North America are shown by Figure 1 (Bickford, 1988) and Figure 2 (Hoffman, 1988). A current view of the North American Precambrian record is that microcontinents of Archean age (older than 2.5 Ga or 2.5 x 10^9 years) were joined during the Early Proterozoic from about 1.81 Ga to 1.96 Ga by a network of orogenic belts to form Laurasia by an essentially plate tectonic mechanism comparable to that forming the Phanerozoic continent of Eurasia (Hoffman, 1988). The Archean provinces typically consist of belt-like granitic pluton-greenstone terranes and granitic pluton-metasedimentary gneiss terranes while the Proterozoic orogens contain highly deformed metasedimentary and metavolcanic rocks.

Black Hills-Bear Lodge Mountains Region

The Black Hills-Bear Lodge Mountains region is located at the western border of the Early Proterozoic Trans-Hudson orogen between the Archean Superior and Wyoming provinces as shown in Figure 1 (Hoffman, 1988) and Figure 3 (Reed, 1987). The Wyoming province is conventionally extended to better-exposed Archean rocks to the north which were previously placed in the problematic Churchill provinceand now subdivided into the Hearne and Rae provinces (Hoffman, 1988). Many of the Early Proterozoic orogens are transitional between collision zones containing remnants of the deformed margins of Archean microcontinents and zones of accretionary tectonics containing remnants of Early Proterozoic island arcs and associated intra-oceanic rock units. Most of the orogenic zones are asymmetric. One side is an Archean hinterland with an underlying subduction zone characterized by a magmatic arc, high grade metamorphic rock assemblages, and transcurrent shear zones. The opposite Archean foreland is characterized by an overthrust sedimentary prism that has undergone relatively low-grade metamorphism.

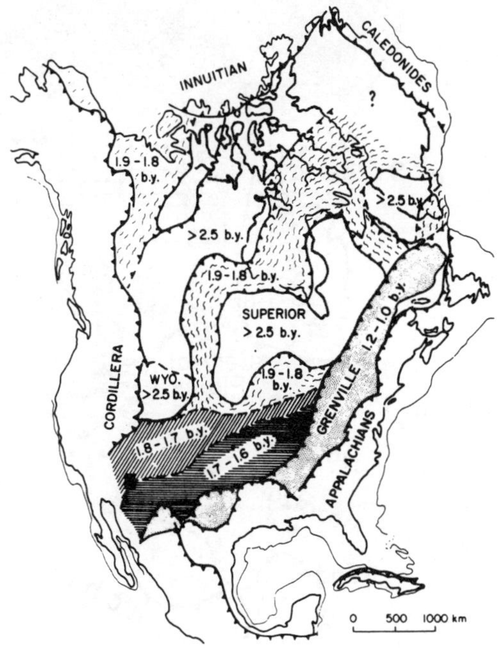

FIGURE 1 Precambrian provinces of North America (Bickford, 1988).

The Black Hills and Bear Lodge Mountains formed by the Laramide uplift of blocks of the Trans-Hudson orogen containing Early Proterozoic rocks intruded by pegmatitic granite with granitic remnants of the Archean Wyoming Province. Asymmetry in the northern part of the Trans-Hudson orogen suggests a Precambrian tectonic polarity with the Superior province foreland to the east and south and the Hearne province hinterland to the north and west. There is evidence that the Superior province has moved up relative to the Trans-Hudson orogen as is noted to be typical of the general buoyancy pattern for North American Archean provinces and Proterozoic orogens (Hoffman, 1988). The Black Hills are just

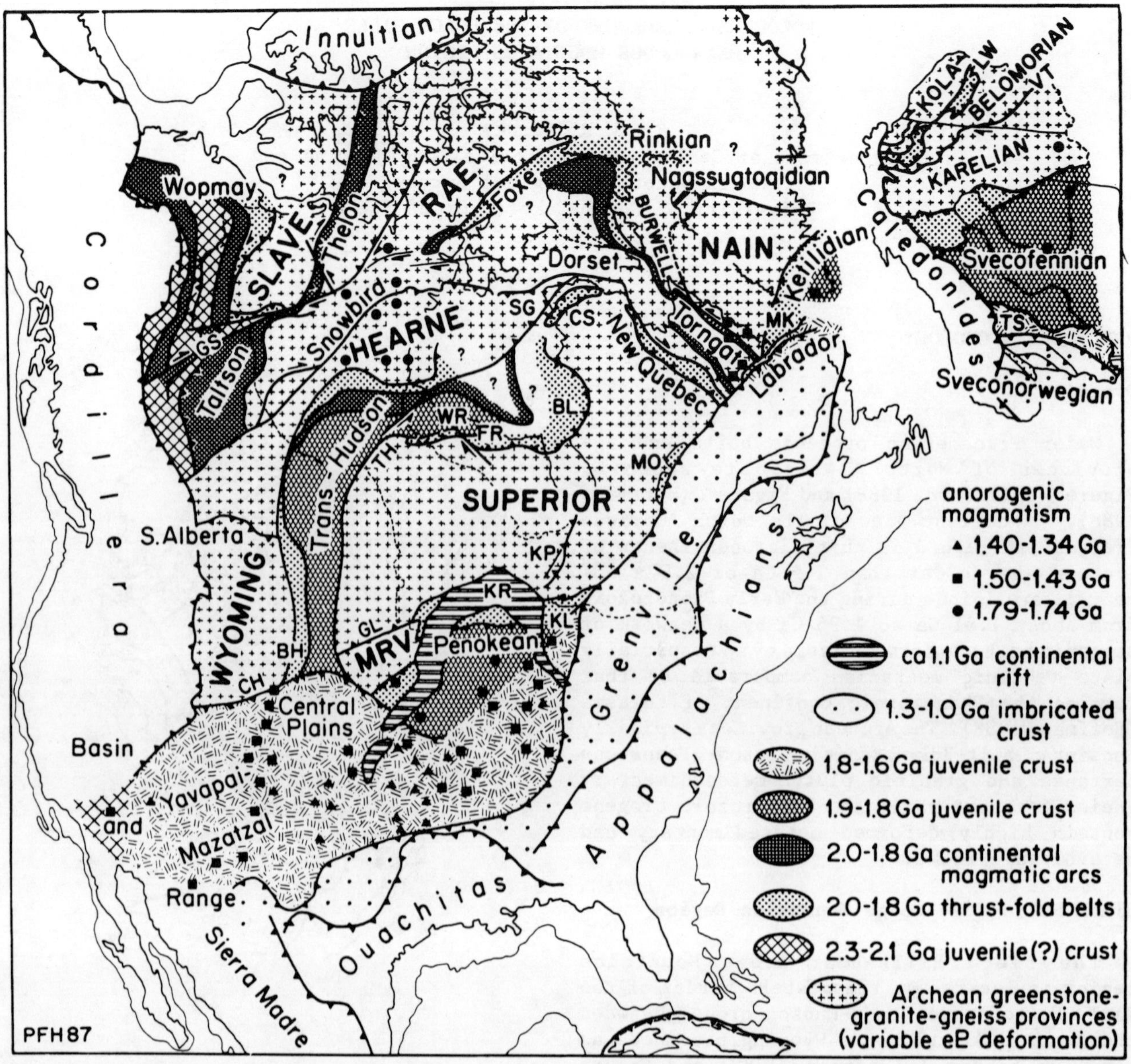

FIGURE 2 Precambrian tectonic elements of Laurentia (Hoffman, 1988). The Baltic shield is shown in a pre-Iapetus reconstruction and Greenland is restored prior to rifting from North America. Uppercase names are Archean provinces; lowercase names are Proterozoic and Phanerozoic orogens. Abbreviations: BL, Belcher belt; BH, Black Hills inlier; CH, Cheyenne belt; CS, Cape Smith belt; FR, Fox River belt; GL, Great Lakes tectonic zone; GS, Great Slave Lake shear zone; KL, Killarney magmatic zone; KP, Kapuskasing uplift; KR, Keweenawan rift zone; LW, Lapland-White Sea tectonic zone; MK, Makkovik orogen; MO, Mistassini-Otish basins; MRV, Minnesota River Valley terrane; SG Sugluk terrane; TH, Thompson belt; TS, Transscandinavian (Smaland-Varmland) magmatic zone; VT, Vetrenny tectonic zone; WR, Winisk River fault.

north of the apparent truncation of the Trans-Hudson orogen by the 1.7-1.8 Ga Central Plains orogen (Sims and Peterman, 1986) shown by Figures 1-3. The North American Central Plains electrical conductivity anomaly extends north just east of the Black Hills and may mark the subsurface eastern boundary of the Wyoming province (Hoffman, 1988).

PHANEROZOIC GEOLOGY

Overview

Western North America is dominated by the Cordilleran mountain belt along the length of the continental margin (Figures 1-2). This wide, complex belt is currently undergoing deformation. It consists of terranes added

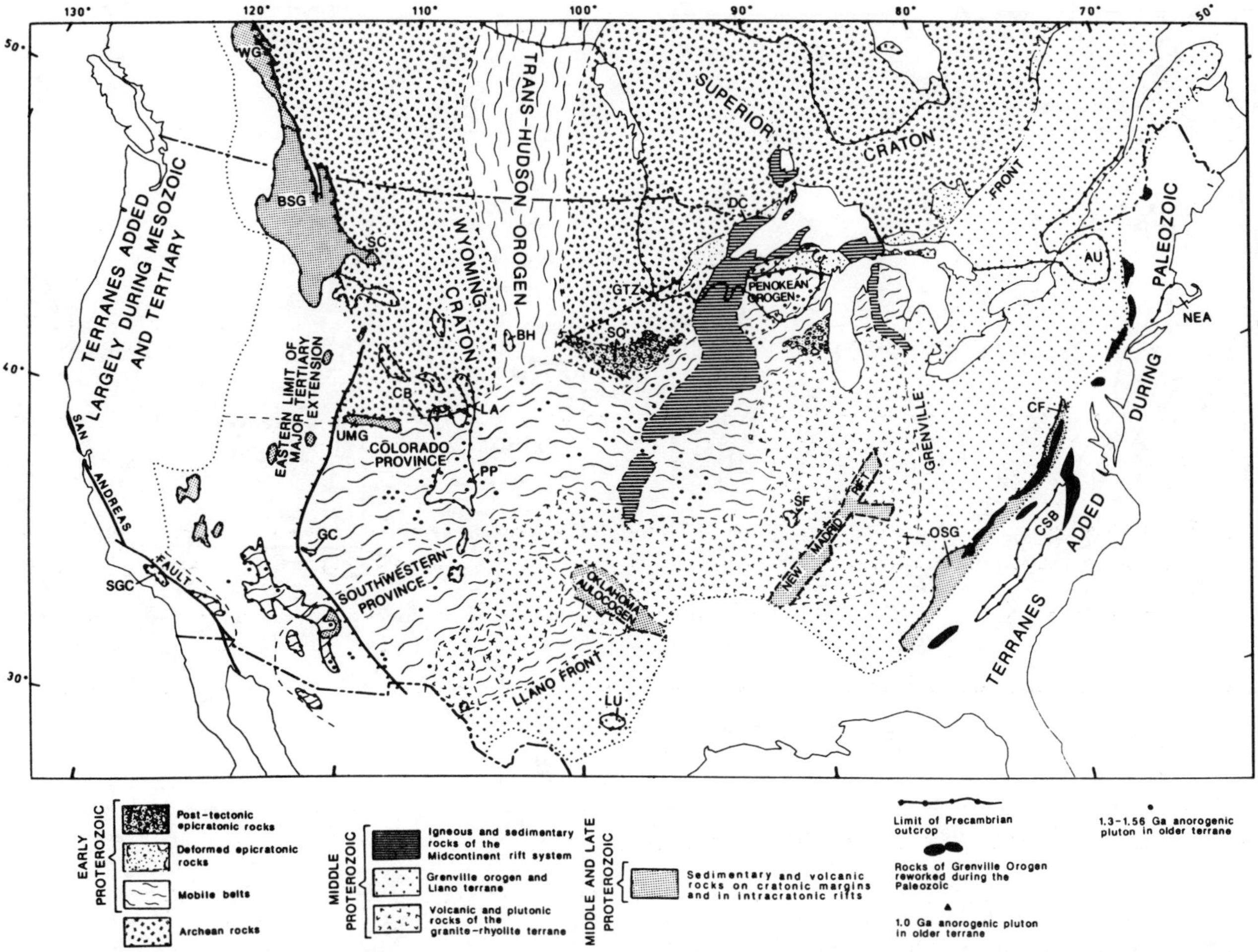

FIGURE 3 Sketch map of Precambrian geology of the United States and southern Canada (Reed, 1987). AU, Adirondack Uplift; BH, Black Hills; BSG, Belt Supergroup; CB, Cheyenne Belt; CF, Catoctin Formation; CSB, Carolina Slate Belt; DC, Duluth Complex; GC, Grand Canyon; GlZ, Great Lakes tectonic zone; LA, Laramie Anorthosite Complex; LU, Llano uplift; MEA, New England Avalon terrane; OSG, Ocoee Supergroup; PP, Pikes Peak batholith; SC, Stillwater Complex; SGC, San Gabriel Complex; SF, St. Francois Mountains; SQ, Sioux Quartzite; UMG, Unita Mountain Group; and WG, Windermere Group.

to North America (Figure 3, Reed, 1987) during the Mesozoic and Cenozoic eras as varied plate interactions modified the continent during its westward and northwestward movement. Accreted island arc and oceanic crust materials are present as fault-bounded blocks that occur with sedimentary, volcanic and plutonic rocks characteristic of extensive subduction zone-magmatic arc development. Major geological features are shown in Figure 4 (Hamilton, 1987) including the Coast Ranges dominated by accretionary tectonics, the Cascade Range volcanic arc, and the Sierra Nevada batholithic complex of granodioritic plutons.

To the east the Cordilleran mountain belt is bounded by a wide zone of Cenozoic extension (Figure 3) which is dominated by the fault block mountains and basaltic volcanism of the Basin and Range Province and the flood basalts of the Columbia Plateau (Figure 4, Hamilton, 1987).

Black Hills-Bear Lodge Mountains Region

Still further east the spectacular Rocky Mountains have formed as part of the Laramide foreland fold and thrust belt of middle Cretaceous to early Cenozoic age (Figures 4,5, Hamilton, 1987). Major structures are in a zone extending from east of the Colorado Plateau northward and northwestward into Wyoming and Montana (Figure 5). Hamilton (1987) suggests that a slowing of westward movement of North America starting in the southwest caused subsequent clockwise rotation of the Colorado Plateau. In the early stages, Proterozoic and younger sedimentary units were imbricated eastward onto the craton

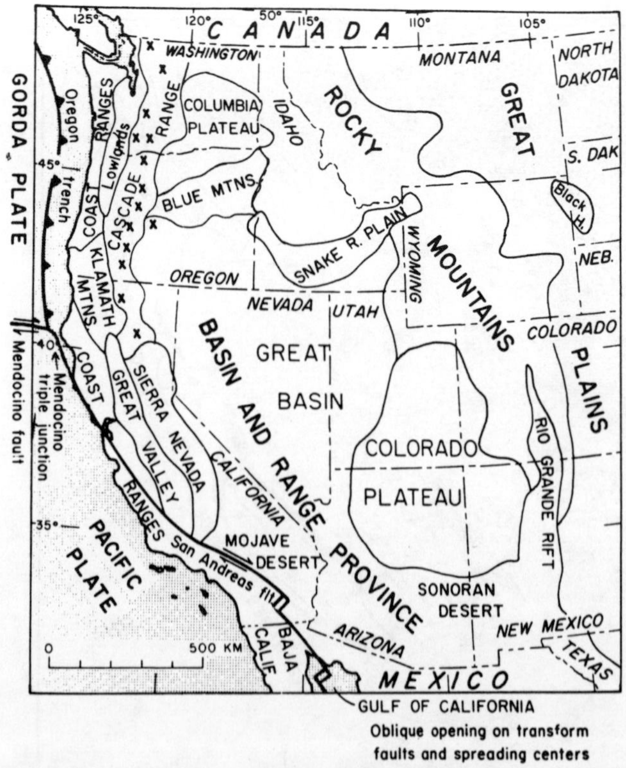

FIGURE 4 Major geological features of the western United States (Hamilton, 1987).

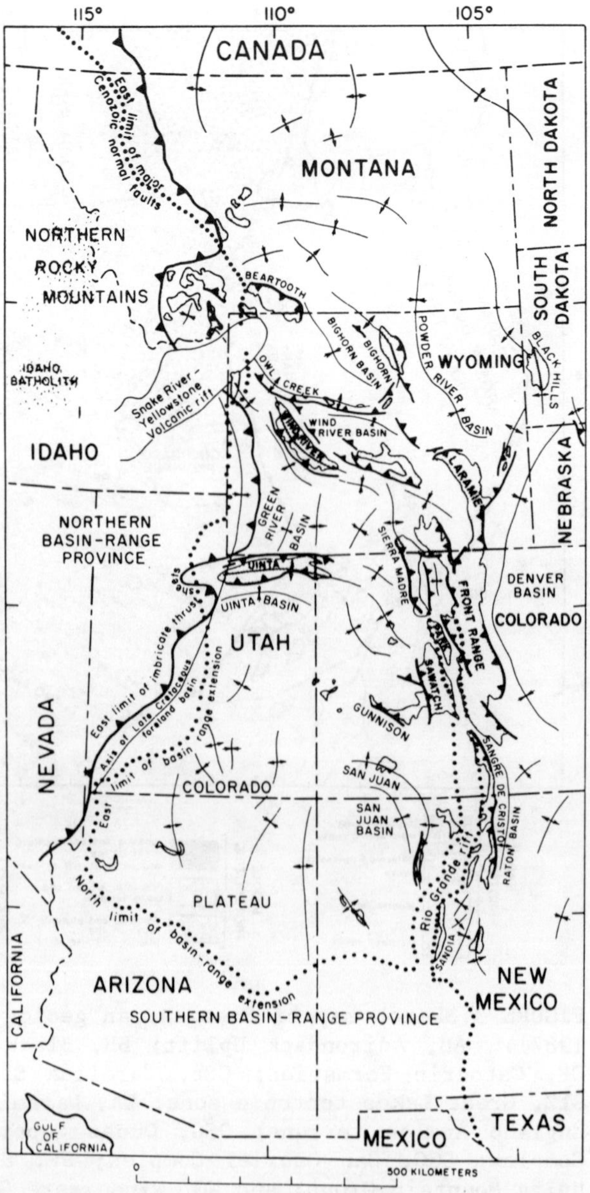

FIGURE 5 Selected tectonic elements of the western interior of the United States (Hamilton, 1987). Outcrops of Precambrian basement rocks are shaded.

along major thrusts (Figure 5) probably as a result of crustal thickening in magmatic arc terranes to the west (Hamilton, 1987). Later compression within the eastern part of the Rocky Mountains-Great Plains region (Figure 5) caused major crustal shortening with the formation of asymmetric anticlines of Precambrian basement rocks bounded by reverse faults or monoclines and adjacent intervening sedimentary basins. The Black Hills-Bear Lodge Mountains region is the easternmost of the mountainous Precambrian basement uplifts that characterize this crustal shortening event.

REFERENCES

Bickford, M.E., 1988, The formation of continental crust: Part 1. A review of some principles; Part 2. An application to the Proterozoic evolution of southern North America: Geological Society of America Bulletin, v. 100, p. 1375-1391.

Hamilton, Warren, 1987, Plate-Tectonic Evolution of the Western U.S.A.: International Union of Geological Sciences, Episodes v. 10, p. 271-6.

Hoffman, P.F., 1988, United plates of America, the birth of a cration: Early Proterozoic assembly and growth of Laurentia: Annual Reviews of Earth and Planetary Sciences, v. 16, p. 543-603.

Reed, J.C., 1987, Precambrian Geology of the U.S.A.: International Union of Geological Sciences, Episodes, v. 10, p. 271-6.

Sims, P.K., Peterman, Z.E., 1986, Early Proterozoic Central Plains orogen: A major buried structure in the north-central United States. Geology, 14:488-91.

IGC FIELD TRIP T131:
GENERAL GEOLOGY OF THE BLACK HILLS AND BEAR LODGE MOUNTAINS

Frank R. Karner
Department of Geology and Geological Engineering and
North Dakota Geological Survey
University of North Dakota, Grand Forks

Richard L. Patelke
Department of Geology and Geological Engineering
University of North Dakota, Grand Forks

REGIONAL OVERVIEW

The Black Hills and Bear Lodge Mountains are within the high plains geomorphic province. Structurally they are the easternmost of the complex assemblage of uplifts and basins of the Rocky Mountains, as described in the preceding paper (Figure 5, Karner, 1989). These uplifts are part of the Laramide foreland fold and thrust belt, and consist of Precambrian basement blocks surrounded by eroded remnants of Phanerozoic sedimentary sequences and intruded by Cenozoic igneous bodies (Figures 1-3). They, thus, illustrate many aspects of Rocky Mountain geology. The Bear Lodge Mountains are a northwest extension of the Black Hills and the center of a major Cenozoic alkalic igneous complex. This summary is a brief review of observations and ideas from selected major publications as referenced in the text.

The Black Hills-Bear Lodge Mountains uplift (Figures 1-3) consists of several distinct regions of differing physiography, structure and geological materials (Thornbury, 1965), which may be described from a central core outwards as follows:

a. the central core of Precambrian igneous and metamorphic rocks;
b. the Paleozoic dolomite and limestone plateau encircling the core;
c. the "red valley" underlain by redbeds of Permian-Triassic age;
d. the hogback ridge of Cretaceous sandstone essentially defining the Black Hills;
e. the surrounding cuestas and plains of sedimentary rocks of Mesozoic-Cenozoic age; and
f. the east-west belt of erosionally resistant Cenozoic igneous complexes extending across the northern end of the Black Hills uplift.

Precambrian Core

The rugged peaks and hills of the Precambrian core are in an elliptical area in the east-central Black Hills where a major Precambrian block has been uplifted 2-3 km relative to surrounding basement rocks. Harney Peak in the south rises to 2207 m elevation, the highest peak in the United States east of the major ranges of the Rocky Mountains. The Early Proterozoic Harney Peak Granite and related pegmatites to the south are surrounded by older deformed metasedimentary and metaigneous rocks of Early Proterozoic age and minor occurrences of Archean rocks.

Paleozoic Limestone Plateau

The plateau of Paleozoic dolomite and limestone encircles the Precambrian core. The major cliff-forming, plateau unit is the 90-185 m-thick Pahasapa Limestone (Madison-equivalent) of Early Mississippian age. Beneath the Pahasapa are 10-75 m of Ordovician-Devonian strata and the Cambrian-Ordovician, ridge-forming sandstone and dolomite of the 0-120 m-thick Deadwood Formation. The Pahasapa is overlain by Pennsylvanian-Permian strata and the 10-15 m-thick Minnekahta Formation (Permian) which lies at the surface of the plateau. Wind, Jewel and other caves in the Black Hills are formed in the massive, jointed, Pahasapa Limestone.

Red Valley

The "red valley" or "red racetrack" is a topographic depression formed on 115-245 m of easily eroded siltstone, sandstone and shale red beds of the Spearfish Formation (Permian-Triassic) just outside of the Paleozoic dolomite and limestone plateau.

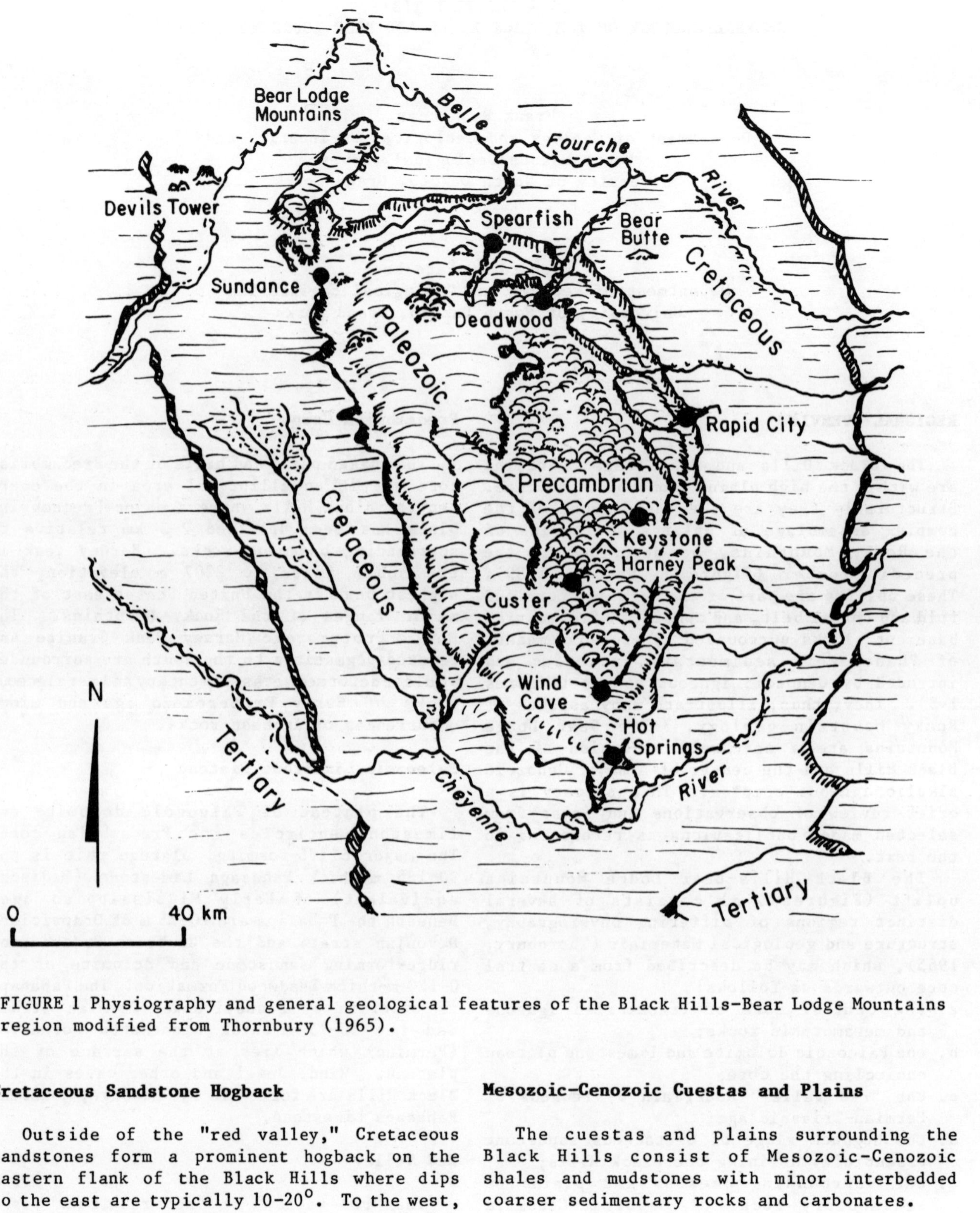

FIGURE 1 Physiography and general geological features of the Black Hills-Bear Lodge Mountains region modified from Thornbury (1965).

Cretaceous Sandstone Hogback

Outside of the "red valley," Cretaceous sandstones form a prominent hogback on the eastern flank of the Black Hills where dips to the east are typically 10-20°. To the west, dips are lower, forming a westward-dipping cuesta. Major resistant units of the Cretaceous hogback ridge are the Chilson Member (Lakota Sandstone) of the Lakota Formation, sandstone of the Fall River Formation and the Newcastle Sandstone.

Mesozoic-Cenozoic Cuestas and Plains

The cuestas and plains surrounding the Black Hills consist of Mesozoic-Cenozoic shales and mudstones with minor interbedded coarser sedimentary rocks and carbonates.

Cenozoic Igneous Centers

The east-west belt of Cenozoic igneous centers extends across the north end of the Black Hills from Bear Butte on the east to

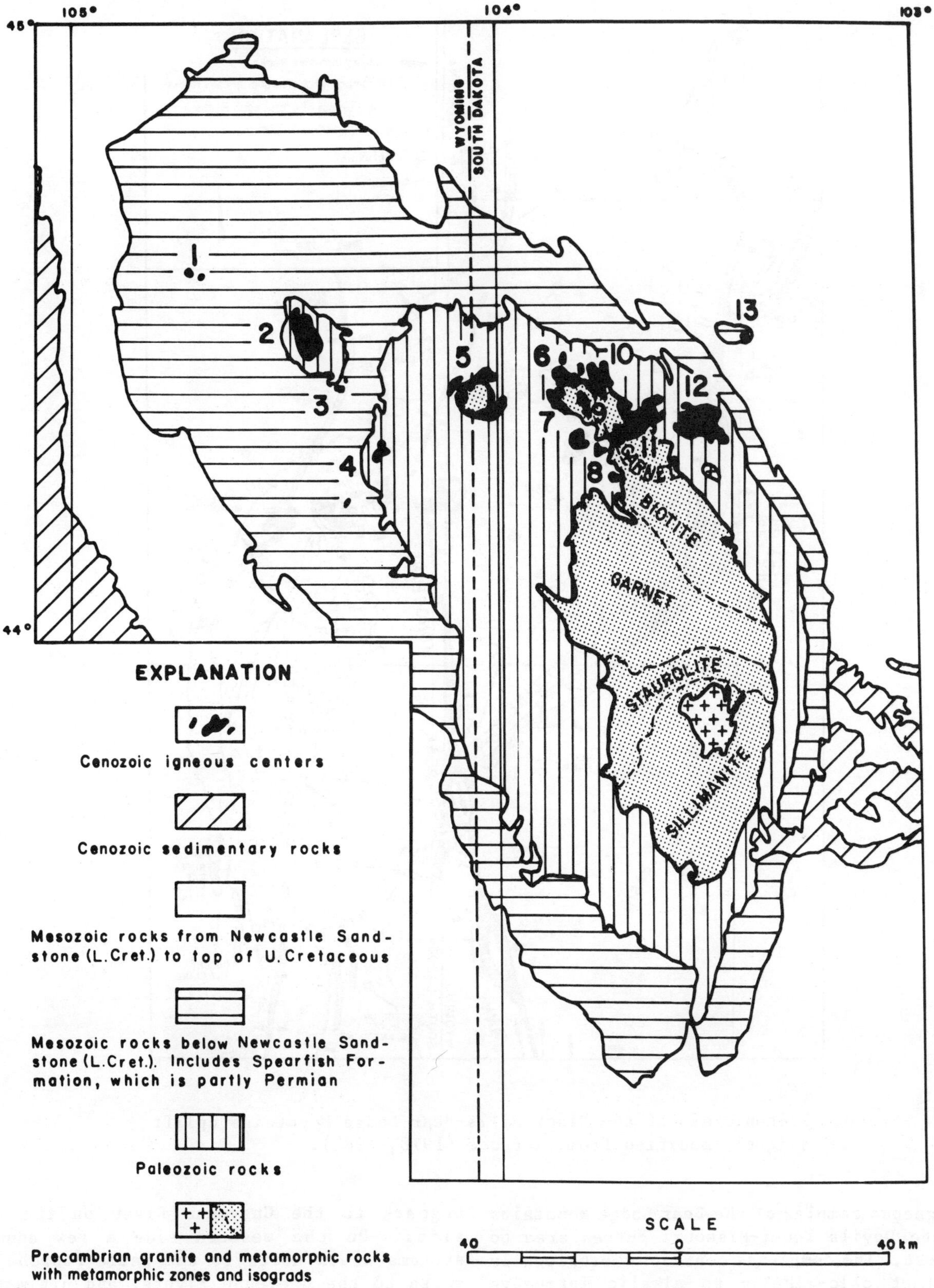

FIGURE 2 Generalized geology of the Black Hills-Bear Lodge Mountains region modified from Gries (1974), Lisenbee (1978) and Karner (1981). The 13 principal igneous centers are (1) Devils Tower - Missouri Buttes; (2) Bear Lodge Mountains; (3) Sundance Mountain-Sugarloaf; (4) Inyan Kara - Black Buttes; (5) Mineral Hill - Tinton; (6) Spearfish-Carbonate; (7) Terry Peak; (8) Strawberry Hill - Custer; (9) Cutting; (10) Mt. Theodore Roosevelt; (11) Gilt Edge - Galena; (12) Vanocker; (13) Bear Butte.

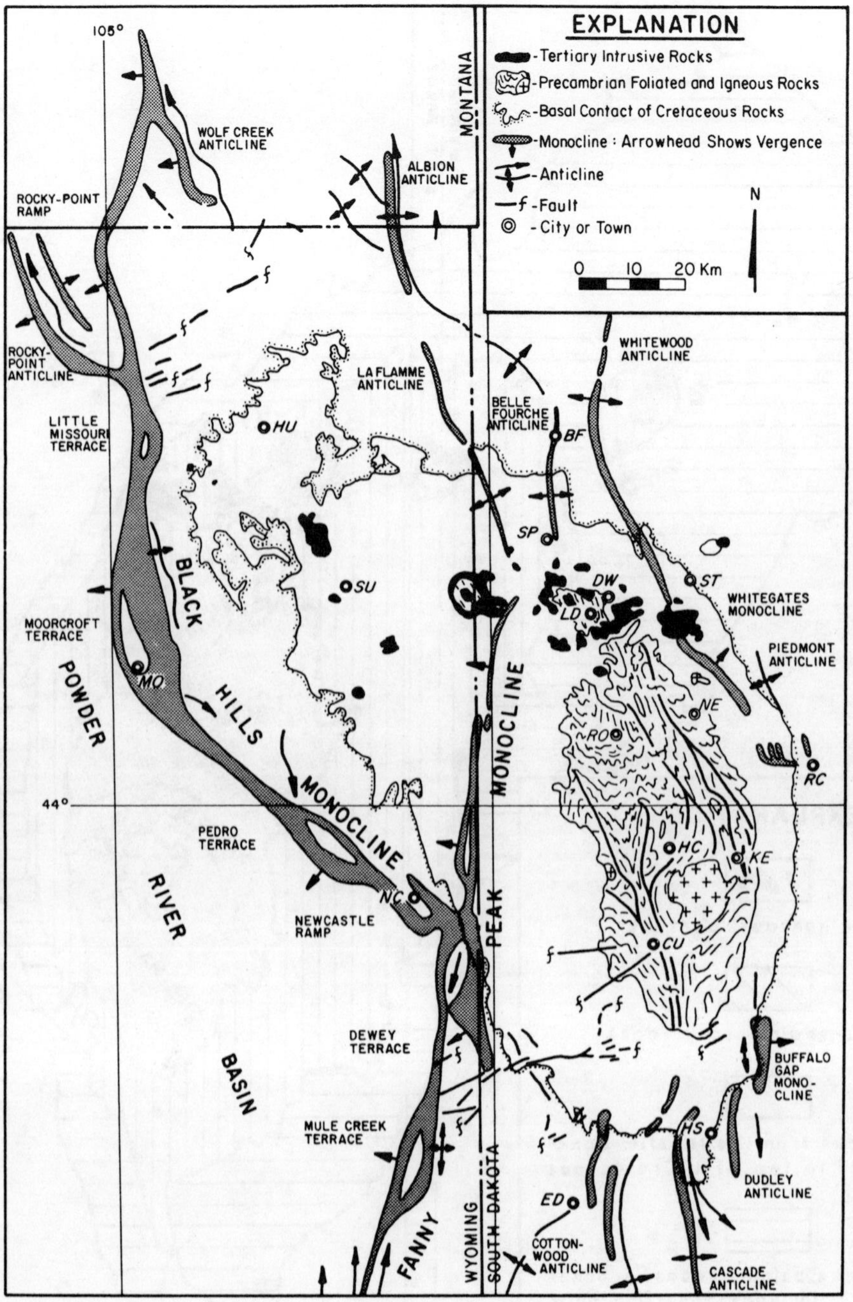

Figure 3 Tectonic map of the Black Hills-Bear Lodge Mountains uplift slightly modified from Lisenbee (1978, 1985).

the igneous complex of the Bear Lodge Mountains and the Devils Tower-Missouri Buttes area to the west. The complexes consist of erosionally resistant calc-alkalic to alkalic intrusive and extrusive rocks.

Hydrogeology

Major drainage of the Black Hills (Figure 1) is by superimposed east-west streams through the resistant ridges of the Cretaceous hogback to the Cheyenne River on the east side. On the western side a few annular streams drain north to the Belle Fourche and south to the Cheyenne rivers, and ultimately to the Missouri River 190 km to the east (Feldman and Heimlich, 1980; Thornbury, 1965). The radially draining streams of the central crystalline core become influent where they cross the Paleozoic sedimentary sequence, particularly the highly jointed, cavernous Pahasapa Limestone. This large scale diffusive infiltration makes the Black Hills a site of groundwater recharge for the surrounding plains, especially to the east (Bakalowicz and others, 1987).

Structural Geology

The structural characteristics of the Black Hills-Bear Lodge Mountains uplift are summarized below, largely following the observations and discussions by Lisenbee (1978, 1985), and DeWitt and others (1986). The uplift is the eastern-most of the major, exposed Rocky Mountain structures formed as part of the Laramide foreland fold and thrust belt described in the preceding paper (Karner, 1989). Essentially, two major blocks of Precambrian basement were uplifted in Laramide time (60-65 Ma) with Phanerozoic sedimentary strata deformed into a broad anticlinal dome of drape folding similar to the pattern of many other Rocky Mountain structures to the west.

Figure 3 (Lisenbee, 1978, 1985) illustrates many of the major structural features. The uplift consists of two major Precambrian blocks, a higher, north-south trending eastern Black Hills block and a lower, northwest-trending Bear Lodge Mountains block. Precambrian basement to the west has subsided to form the Powder River Basin (Figure 5, Karner, 1989) whereas to the east it has been somewhat elevated above its general level beneath the Great Plains. The uplifted blocks are somewhat flat-topped with monoclines of Paleozoic-Mesozoic sequences marking their fairly well-defined western boundaries (Figure 3). The westward-dipping Fanny Peak monocline separates the eastern and western blocks and is interpreted to be a drape fold over the edge of the higher eastern block. The Black Hills monocline marks the western edge of the western block. Figure 4 (Lisenbee, 1978) illustrates a possible basement configuration at the junction of the Fanny Peak and Black Hills monoclines. The structural pattern of both blocks is asymmetrical with their eastern borders marked by gentler dips of the overlying sedimentary strata. The eastern block is terminated to the north and south by north-south plunging anticlines or monoclines. These structures are often asymmetric with steeper dips on their west flanks, and paired synclines also to the west. Domal structures, many known to be cored by igneous bodies of Paleocene-Eocene age, are present in both blocks primarily in an east-west-trending zone across the northern end of the eastern block into the central portion of the western block.

The major structural features are generally interpreted to be the result of basement uplift along fault zones formed during Laramide crustal shortening and associated block faulting and drape folding of overlying sedimentary strata. Paleocene-Eocene igneous centers formed in an east-west zone during a period of 10-15 Ma following the major uplift.

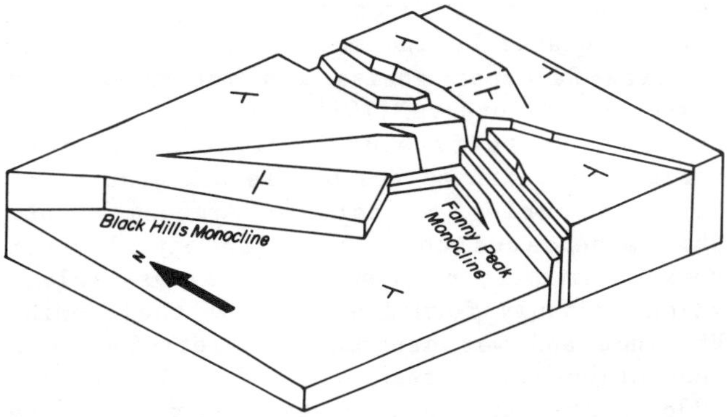

Figure 4 Block diagram illustrating possible basement configuration beneath the junction of the Fanny Peak and Black Hills monoclines illustrated in Figure 3 (slightly modified from Lisenbee, 1978).

GENERAL GEOLOGY OF THE PRECAMBRIAN ROCKS

The outcrops of Precambrian rocks in the Black Hills and Bear Lodge Mountains (Figures 2-3) expose a part of the Trans-Hudson orogen containing Early Proterozoic metasedimentary and metaigneous rocks intruded by the pegmatitic Harney Peak granite and containing granitic remnants of the Archean Wyoming Province as described in the preceding paper (Karner, 1989). This discussion of Black Hills-Bear Lodge Precambrian geology largely follows information in the summaries and maps provided by Denison and others (1984), DeWitt and others (1986), Gosselin and others (1988), and Hoffman (1988).

Archean Rocks

Archean rocks have been studied at three locations: the Little Elk Creek terrane northeast of Nemo in the eastern Black Hills; the Bear Mountain terrane northwest of Custer; and Warren Peaks in the Bear Lodge Mountains. In the Little Elk terrane (Gosselin and others, 1988), fine to medium grained, foliated, melanocratic biotite-feldspar gneiss is intruded by the medium to coarse grained, foliated, porphyroblastic, calc-alkaline, I-type, Little Elk Granite with a U-Pb intercept age of 2,549 $\pm$ 11 m.y. The biotite-feldspar gneiss has interlayers and pods of biotite-garnet schist, biotite schist, muscovite-chlorite-biotite-garnet-magnetite schist and amphibolite. Associated rock units include amphibolite interlayered with biotite gneiss and a sequence of iron formation, muscovite schist, quartzite and metaconglomerate. The biotite-feldspar gneiss was interpreted by Gosselin and others (1988) as a sequence of arkosic or tuffaceous sediments deformed during intrusion of the Little Elk Granite as well as during an earlier Archean event. This sequence and the associated and interlayered rocks form part of a supracrustal assemblage which probably formed as part of the Wyoming Province and was deformed and intruded in a subduction zone setting (Gosselin and others, 1988). At Bear Mountain, biotite-plagioclase schist (metagraywacke) is intruded by S-type granite and trondhjemite with a badly discordant U-Pb age of 2,393 ± 230 m.y. (Gosselin and others, 1988). In the Bear Lodge Mountains, granitic rock with a U-Th-Pb minimum age of 2.63 Ga (DeWitt and others, 1986) occurs as blocks associated with Cenozoic igneous rock at Warren Peaks. The Archean rocks have been further affected by Proterozoic deformational events in the Trans-Hudson orogen at about 1.9-2.0 Ga as suggested by Rb-Sr dating of the Little Elk Granite (Gosselin and others, 1988).

Early Proterozoic Rocks

The Precambrian core (Figures 2-3) of the east-central Black Hills is a north-south oriented, 40 km by 60 km area of predominantly metasedimentary rocks invaded by the Harney Peak Granite which is exposed in an irregular area 10-15 km across. The Early Proterozoic metasediments are usually strongly deformed, steeply dipping clastic to chemical rocks of intermediate to high metamorphic grade. Common rock types include schist, phyllite, metagraywacke, quartzite, metaconglomerate, iron formation and metamorphosed limestone and dolomite. Metabasalt and metagabbro are interstratified with metasedimentary strata.

Stratified Rocks. DeWitt and others (1986) divide the stratified Early Proterozoic rocks into fifteen major map units, seven of which are designated as metasedimentary units; four metaquartzite, metagraywacke and/or metaconglomerate units; and four additional units of iron formation, metachert, metabasalt, and metagabbro (Table 1).

Sequences in Major Geographic Regions. DeWitt and others (1986) summarized the Early Proterozoic geology of the Black Hills in relation to seven major geographic regions (Figure 3); the Lead-Deadwood, Nemo-Roubaix and Rochford-Pactola Reservoir areas in the north and the Southeastern, Custer, Keystone and Medicine Mountain (west of Hill City) areas in the south. The geology of the seven regions can be summarized by grouping the mapped units into the following sequences based on relative age:

(1) a sequence of older metaconglomerate-metaquartzite units;

(2) a sequence of interstratified metasedimentary, metaquartzite, and metabasalt-metachert units of intermediate age; and

(3) a sequence of younger interstratified metasedimentary and metagraywacke units.

Iron formation and metagabbro may occur within any of the three sequences.

In the Nemo-Roubaix, Keystone and Medicine Mountain areas, sequences **(1)**, **(2)**, and **(3)** are present. In the Lead-Deadwood, and Rochford-Pactola Reservoir and Custer areas sequences **(2)** and **(3)** are present and units of sequence **(1)** are not recognized. In the Southeastern area metaquartzite of sequence **(2)** predominates. In the Custer area sequence **(3)** is present with metagraywacke predominant to the southwest.

On the geologic maps of most areas (DeWitt and others, 1986) major and minor folding and faulting is obvious. Major faults commonly border the areas described above, delimiting terranes of differing structure and stratigraphy (Plate 2, DeWitt and others, 1986).

Metamorphic Grade. Metamorphic zones, marked by the occurrence in pelitic rocks of biotite, garnet, staurolite and sillimanite are shown on Figure 2. Generally, metamorphic grade increases from the greenschist facies in the north-central Black Hills to the lower amphibolite facies in the northeast and the upper amphibolite facies in the south (DeWitt and others, 1986). Metamorphic grade increases through the staurolite zone to the sillimanite zone in the pegmatite-rich aureole around the Harney Peak Granite.

Harney Peak Granite. The youngest of the Precambrian rocks in the Black Hills, the Harney Peak Granite (Figures 2-3), has a Rb-Sr whole rock age of 1.697 ± 33 Ga (DeWitt and others, 1986). It consists of gray to pink, coarse-grained to pegmatitic, muscovite granite and pegmatite. Major minerals are quartz, microcline perthite, albite and muscovite with accessory biotite, tourmaline and garnet. Texture varies with alternating, vaguely defined layers of coarse grained pegmatitic material and finer grained granitic rock (Redden and Norton, 1975). The granite is usually characterized as an igneous pluton made up of a complex of coalesced sills and dikes. As much as 20% of the area mapped as Harney Peak Granite consists of large masses of country rock (Redden and Norton, 1975). Many satellitic bodies of granite, and particularly pegmatite, surround the pluton.

TABLE 1 Major Characteristics of Early Proterozoic Stratified Rock Units of the Black Hills (DeWitt and others, 1986)

Rock Unit	Common Rock Types	Typical Source Rocks	Comments
Metasedimentary	biotite schist muscovite schist phyllite shale	shale carbonaceous shale tuffaceous shale siltstone chert	may contain coarser clastic, carbonate, and chemically precipitated sedimentary types
Metaquartzite - Metagraywacke - Metaconglomerate	siliceous schist conglomeratic schist feldspathic schist quartzite metagraywacke metaconglomerate	sandstone graywacke siltstone conglomerate shale	may contain carbonate, chemically precipitated, and volcanic rock types; may be uraniferous and/or auriferous
Iron Formation	ferruginous schist iron formation (IF) oxide IF taconite conglomeratic IF chert carbonate IF sulfide IF pyritic slate	iron formation (IF) oxide-facies IF carbonate facies IF sulfide-facies IF coarse-fine clastic rocks	often occurs as thin units within other metasedimentary units; may be auriferous
Metachert	siliceous schist cherty schist	iron-poor chert	often occurs as thin units within metabasalt; relatively minor in abundance
Metabasalt	amphibolite mafic schist tuffaceous metasediment slate chert	basalt interlayered sedimentary rock gabbro	may be pillowed; relatively minor
Metagabbro	metagabbro amphibolite serpentinite	gabbro basalt related igneous intrusive rock	relatively minor

Early Proterozoic History. As described in the previous paper (Karner, 1989) the Black Hills-Bear Lodge Mountains region is located at the western border of the Early Proterozoic Trans-Hudson orogen between the Archean Superior and Wyoming provinces. Hoffman (p. 560, 1988), using information by Redden (1987), summarizes the Early Proterozoic history of this region. "... Archean basement is overlain by syn-rift clastic sediments that were intruded by a 2.17 Ga mafic sill, then deformed and eroded, prior to deposition of younger rift- and shelf-facies clastic sediments and mafic volcanics. The shelf quartzites, which contain a 1.97 Ga tuff bed, are laterally equivalent to and overlain by thick turbidites capped by 1.88 Ga alkalic tuffs. The entire sequence was recumbently folded about easterly axes, refolded about NNW-trending axes, and intruded by the 1.72 Ga Harney Peak granite."

Pegmatites

General Occurrence, Characteristics and Use. Pegmatites in the Black Hills occur as dikes, sills, or lens-shaped bodies associated with the Harney Peak Granite. They generally have been assumed to originate from late-stage, water-rich fluids associated with magmatic emplacement. In the Black Hills about 20,000 pegmatites crop out over an area of about 870 km^2. They often stand out in relief from the enclosing, less resistant schists in an area of mature topography dominated by slopes forested with ponderosa pine (Norton, 1964). About 200 of the pegmatites are known to be suitable for mining largely because of their size and zoning. The distinction between dikes of Harney Peak Granite and unzoned pegmatites is primarily based on the size and shape of the intrusive body rather than on the texture or composition of the rock itself (Norton, 1975, Černý and others, 1982). Foliation of the enclosing schists controls the shape and attitude of most of the pegmatites with some fracture control also in evidence. Many pegmatites are approximately elipsoidal but with a rounded top, keel like bottom, and irregularities or "rolls" along contact with the schists (Norton, 1975).

The pegmatites contain such minerals as tourmaline, beryl, lepidolite, spodumene, amblygonite, tantalite-columbite, along with large crystals of feldspar, muscovite, and quartz. Zoning of the pegmatites has made selective mining and hand picking of the desired minerals both productive and practical. Pegmatite mining began in 1879 with sheet mica at the Crown mine near Custer (Norton, 1975). Major uses of the pegmatites includes the following: (1) potash feldspar for ceramics and glass; (2) sheet mica for electrical and electronic applications and scrap mica for roofing materials, paint, and rubber; (3) the lithium minerals spodumene, amblygonite and lepidolite for many uses of lithium such as lubricants, batteries, and fluxes; (4) beryl for beryllium used in copper and other alloys and; (5) columbite-tantalite for tantalum and columbium (niobium), tin-bearing minerals, white quartz, rose quartz, semi-precious stones and mineral specimens (Norton, 1975). Labor and shipping costs, along with the discontinuation of governmental programs for the purchase of strategic minerals, have so reduced the pegmatite mining industry in the Black Hills that essentially the only commercial activity in recent years has been mining of feldspar for ceramics and the collection of semi-precious stones and minerals.

Composition and Zoning. The distribution of 170 major pegmatites according to compositional features is shown in Figure 5 (Černý and others, 1982). The pegmatites are clustered around the main body of Harney Peak pegmatitic granite. The area of high concentration of pegmatites to the south and southwest coincides with areas of Early Proterozoic metagraywacke. The barren area to the southeast seems to be an unmapped area outside of the Black Hills National Forest (Plate 4, DeWitt and others, 1986). Černý and others (1982) suggested that the abundance of pegmatites to the south and southwest indicates that the Harney Peak granite may have a hidden southwest extension. According to Černý and others (1982), pegmatite distribution illustrates the normal geochemical pattern of mineralization, volatile enrichment, and Li and Cs contents all increasing with distance from the main granitic body. The

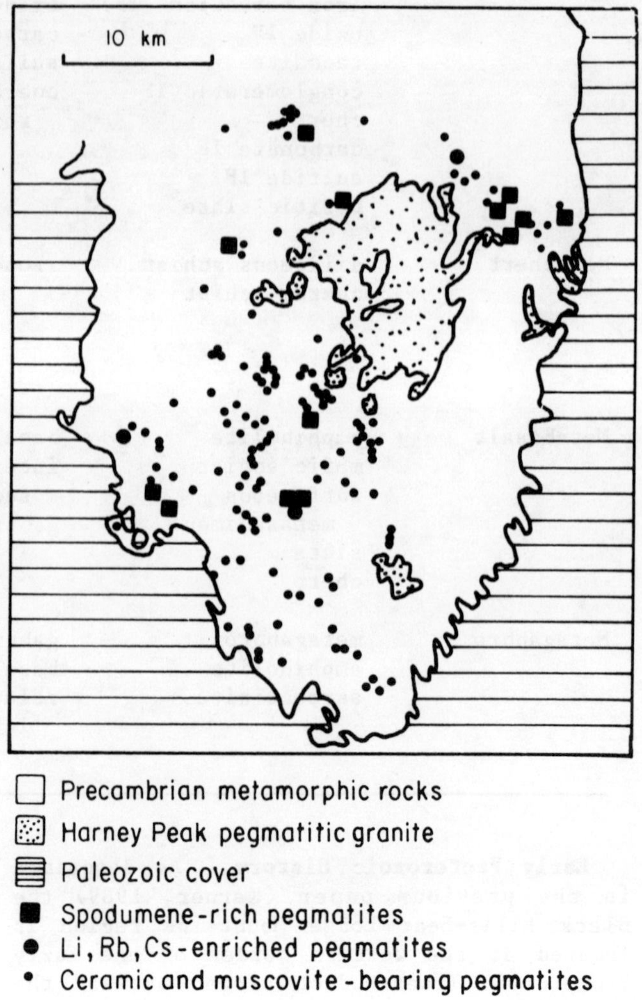

FIGURE 5 Distribution of approximately 170 major pegmatites in the southern Black Hills (modified from Černý and others, 1982).

mineralogy of Black Hills pegmatites has been summarized by Page and others (1953) as follows: "The most abundant minerals...are the plagioclase feldspars (oligoclase and albite), the potash feldspars (perthite and microcline), and quartz. In a few pegmatites spodumene, lepidolite, muscovite, and tourmaline are major constituents. The common accessory minerals are tourmaline, beryl, garnet, biotite, apatite, lithiophilite-triphylite and associated phosphates, amblygonite, spodumene, lepidolite, zinnwaldite, columbite-tantalite, cassiterite, magnetite, pyrite, uranium minerals, lollingite, and arsenopyrite. Less common are topaz, pollucite, chrysoberyl, tapiolite, microlite, siderite, galena, sphalerite, andalusite, sillimanite, graphite, and others." Table 2 summarizes the general mineral assemblages of zoned pegmatites (Norton, 1964; Cameron, 1949).

TABLE 2 Zones of Mineral Assemblages in Black Hills Pegmatites (Modified from Norton, 1964, after Cameron 1949)

Zone (from contact inward)	Assemblages
1	quartz, plagioclase, muscovite
2	quartz, plagioclase
3	quartz, plagioclase, perthite
4	quartz, perthite
5	quartz, plagioclase, with or without muscovite, perthite, amblygonite, spodumene
6	quartz, plagioclase, spodumene
7	quartz, spodumene
8	plagioclase, lepidolite, quartz
9	quartz, microcline or perthite
10	plagioclase, Li-mica, microcline, quartz
11	quartz

Geochemically, Černý and others (1982) classified the pegmatite field as a complex rare-element type with Li, Rb, Cs, Be, and Ta mineralization. Pegmatites range from barren types to types bearing Be, Nb, and Ta to types enriched in Li, Be, Ta, and Nb with abundant Li, Fe, Mn phosphates and/or spodumene (Černý and others, 1982).

Origin. The origin of the Black Hills pegmatites appears to be closely related to the emplacement of the Harney Peak Granite. Using a simple igneous model, pegmatites are formed by water-rich magma generated as a result of intrusion and crystallization of a related granitic body. However, regarding the formation of the Black Hills pegmatites, the relative importance of igneous materials and processes, metasedimentary materials and metamorphic processes, hydrological processes, and structural processes in forming the pegmatites is not clear from the general literature.

GENERAL GEOLOGY OF THE PHANEROZOIC ROCKS

Overview of Rock Sequences

The outcrops of Phanerozoic rocks in the Black Hills and Bear Lodge Mountains area (Figures 2-3) expose cratonic, sedimentary sequences of marine, continental, and transitional facies of Paleozoic, Mesozoic, and Cenozoic ages. Intrusive and extrusive igneous centers of Cenozoic age form an east-west belt across the north end of the Black Hills to the Bear Lodge Mountains. The following summary largely follows the work of Harksen and MacDonald (1969), Gries (1975), Merewether (1975), Harksen (1975), Lisenbee (1975), Lisenbee (1978) Feldman and Heimlich (1980), Gries and Martin (1985), Rich (1985), Martin (1983) and DeWitt and others (1986). Major sedimentary rock sequences and the igneous rock series are summarized in Table 3. Figure 6, based on illustrations and information in the references cited above, gives detailed information on the Precambrian to Lower Cretaceous geologic sequence which is exposed in the Black Hills and Bear Lodge Mountains. The overlying Late Mesozoic marine sequence and Cenozoic continental sequence, exposed in the surrounding Great Plains, is summarized in Table 3 and the following text. Figure 7, from Harksen and Macdonald (1969), illustrates the Oligocene, Miocene and Pliocene strata of the Badlands exposed in the region.

Paleozoic Marine Shelf Sequence

A **Cambrian-Ordovician marine shelf sandstone-conglomerate-shale-carbonate facies** of the Paleozoic North American epicontinental sea consists of massive brownish sandstone, conglomerate, shale and dolomite of the Deadwood Formation and the thinner Winnipeg Formation and Whitewood Limestone. Unconformably overlying a low-relief planar surface of Early Proterozoic rocks, these formations range in thickness up to about 175 m and are part of a transgressive sequence onto the central region of the continent. The Deadwood ranges in thickness from less than 1 m near Wind Cave to about 120 m in the vicinity of its type section at Deadwood. Deadwood sandstone and conglomerate contain paleoplacer

PRECAMBRIAN - EARLY CRETACEOUS GEOLOGICAL SECTION FOR THE BLACK HILLS REGION

ERA	PERIOD	EPOCH	AGE RANGE (Ma)	GLOBAL AGE	GROUP	FORMATION	MEMBER OR "MARKER"	THICK-NESS (m)	DESCRIPTION
MESOZOIC	CRETACEOUS	EARLY	102-104	ALBIAN	GRANEROS GROUP	NEWCASTLE SANDSTONE	"MUDDY" SS	0-30	Brown to light yellow and white sandstone.
			104-106		GRANEROS GROUP	SKULL CREEK SH		45-85	Dark gray to black shale.
			106-110		INYAN KARA GROUP	FALL RIVER FORMATION		30-60	Upper sandstone thinly bedded and reddish. Massive, tan, fine grained, calcareous ss.
			115-132	NEOCOMIAN	INYAN KARA GROUP	LAKOTA FORMATION	FUSON SHALE MINNEWASTE LS CHILSON MEM	10-220	Red to yellow silty claystone. Gray to tan limestone. Yellow to brown conglomeratic sandstone.
	JURASSIC	LATE	145-155	TITHONIAN KIMMERIDGIAN		MORRISON FORMATION		0-45	Green to maroon variegated siltstone with claystone and minor interbedded quartzose sandstone.
			156-162	OXFORDIAN		UNKPAPA SS SUNDANCE FORMATION	REDWATER SHALE	0-80	Unkpapa-white to calico, fine grained, ss. Redwater-greenish gray shale (and thin carbonates). Yellowish, glauconitic ss. near base.
		MIDDLE	163-169	CALLOVIAN BATHONIAN		SUNDANCE FORMATION	LAK MEMBER HULETT SS STOCKADE BEAVER SH CANYON SPRINGS SS	75-105	Maroon sandstone and siltstone. Tan to gray, fine grained glauconitic ss. Greenish gray shale. White, fine grained, calcareous, fossiliferous sandstone.
	TRIASSIC	EARLY	239-245	SCYTHIAN		SPEARFISH FORMATION		115-245	Red, clayey sandstone with interlayered siltstone, and shale.
PALEOZOIC	PERMIAN	LATE	245-253	TATARIAN		SPEARFISH FORMATION			Red clayey, sandy siltstone. Massive gypsum.
				KAZANIAN UFIMIAN			"GOOSE EGG"		
			255-262	KUNGURIAN		MINNEKAHTA LIMESTONE		10-15	Massive, pink to gray, clayey, finely layered limestone with interbedded shale.
		EARLY	263-267	ARTINSKIAN		OPECHE SHALE		10-40	Red shale and sandstone.
			270-284	SAKMARIAN ASSELIAN		MINNELUSA FORMATION			Yellow quartzose sandstone, beds of anhydritic dolomite. Quartzose sandstone and dolomite.
	PENNSYLVANIAN	LATE		GZELIAN			"RED" MARKER BED		
			287-300	KASIMOVIAN			"LEO" SANDSTONE	120-230	Red-black shale, local brecciated limestone. Well-sorted, quartzose sandstone.
				MOSCOVIAN					Dolomitic, anhydritic limestone.
			302-320	BASHKIRIAN			"BELL" SANDSTONE		Red, silty, sandy shale with interbedded dolomitic limestone and sandstone. Basal quartz sandstone and local red shale.
	MISSISSIPPIAN	EARLY	341-358	VISEAN TOURNAISIAN		PAHASAPA LIMESTONE (MADISON LS)		90-185	Gray, dolomite and dolomitic limestone. Gray, cherty, dolomitic limestone. Gray, cherty dolomite and dolomitic limestone.
	DEVONIAN	LATE	358-363	FAMMENIAN		ENGLEWOOD FORMATION		10-20	Pink to buff, dolomitized, clayey limestone. Shaly, sandy, dolomitic limestone.
	ORDOVICIAN	LATE	441-457	ASHGILLIAN		WHITEWOOD LIMESTONE		0-20	Buff dolomite and interlayered limestone.
				CARADOCIAN		WINNIPEG FORMATION	ROUGHLOCK MEM ICE BOX SHALE	0-15 0-20	Tan calcareous siltstone, abundant conodents. Greenish gray shale.
		EARLY	500-505	TREMADOCIAN		DEADWOOD FORMATION	"SCOLITHUS" SANDSTONE	0-120	Massive, buff sandstone with vertical borings.
	CAMBRIAN	LATE	505-518	TREMPEAL-EAUAN FRANCONIAN		DEADWOOD FORMATION			Greenish, sandy, glauconitic shales. Interlayered brown dolomite, flat pebble conglomerate, and sandstone.
		MIDDLE	521-525	DRESBACHIAN					Brown conglomeratic sandstone and dolomite.
PRECAMBRIAN	EARLY PROTEROZOIC		1700-1900?			HARNEY PEAK GRANITE		?	Younger pegmatite and Harney Peak Granite.
			1900-2500?			METASEDIMENTARY FORMATIONS AND IGNEOUS UNITS		?	Folded metasedimentary, metavolcanic and metaigneous rocks include biotite schist, muscovite schist, metagraywacke, metaquartzite and conglomeratic schist. Minor iron formation metabasalt, metachert and metagabbro.
	LATE ARCHEAN		2500-?			METASEDIMENTARY FORMATIONS AND IGNEOUS UNITS		?	Deformed, metasedimentary biotite-feldspar gneiss, biotite-plagioclase schist, mica schist, and amphibolite intruded by foliated granite. Associated interbedded schist, phyllite and iron formation in the Nemo area.

COLUMNAR SECTION

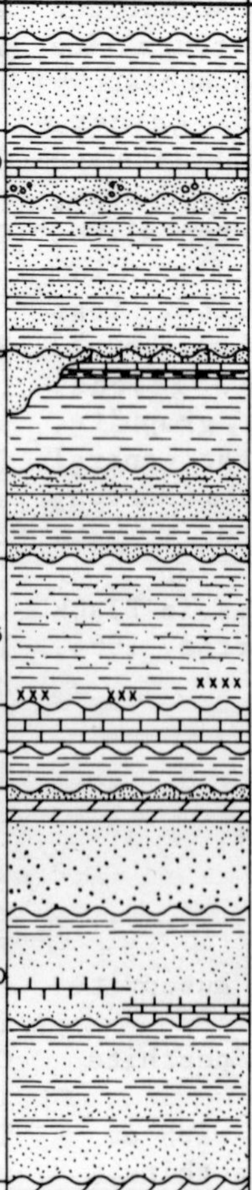

URE 6 Precambrian-Early Cretaceous Geological Section for the Black Hills region.

TABLE 3 Phanerozoic Rock Sequences of the Black Hills-Bear Lodge Region

CENOZOIC CONTINENTAL SEQUENCE
- Miocene-Pliocene-Pleistocene Great Plains continental clastic facies
- Oligocene-Miocene Badlands clastic facies
- Eocene Black Hills-Bear Lodge Mountains calc-alkalic to alkalic igneous association
- Paleocene-Eocene Great Plains transitional to fluvial and marine clastic facies

LATE MESOZOIC MARINE SEQUENCE
- Cretaceous-Paleocene Great Plains transitional marine to fluvial clastic facies
- Cretaceous Western Interior marine shale-sandstone-carbonate facies

PALEOZOIC-MESOZOIC CONTINENTAL SEQUENCE
- Jurassic-Cretaceous transitional fluvial clastic to marine sandstone facies
- Jurassic marine shelf clastic facies
- Permian-Triassic redbed facies

PALEOZOIC MARINE SHELF SEQUENCE
- Pennsylvanian-Permian marine shelf, sandstone-shale-carbonate facies
- Devonian-Mississippian marine shelf carbonate facies
- Cambrian-Ordovician marine shelf sandstone-conglomerate-shale-carbonate facies

gold deposits believed to be derived from Precambrian sources. The deposits are not expected to be a further source of gold in the foreseeable future, but could lead to the location of primary deposits in the Precambrian source areas (Norton, 1975). Tertiary replacement gold deposits are present but are of relatively small value (Gries, 1975). Deadwood fossils are mostly trilobites and ichnofossils though fish remains found near Sundance, Wyoming are among the oldest vertebrate fossils (Feldmann and Heimlich, 1980).

Silurian and Devonian rocks are apparently absent in the Black Hills, though they are widespread in the subsurface outside of the uplift (Gries, 1975). Since other formations are thinner than their counterparts in the surrounding basins, the Black Hills may have been a positive feature throughout much of the Paleozoic (Feldmann and Heimlich, 1980).

Above a major unconformity, a **Devonian-Mississippian marine shelf carbonate facies** consists of 100-205 m of the Englewood Formation and the major cliff-forming unit of the Black Hills, the light-colored limestone of the Pahasapa Formation. The Pahasapa is the equivalent of the Madison Group, the most productive of the oil and gas reservoirs in the Williston Basin just to the north of the Black Hills. The Pahasapa thins from 185 m in Spearfish Canyon to 90 m in the southern Black Hills and is entirely absent in the subsurface just to the southeast (Gries, 1975). The Pahasapa lies on Precambrian through Devonian rocks, and has an upper surface that shows the development of karst topography formed before the onset of late Mississippian or early Pennsylvanian deposition. The major caves of the Black Hills are developed in the Pahasapa, and it has economic importance as an aquifer as well as a petroleum reservoir (Gries, 1975, Feldmann and Heimlich, 1980).

A **Pennsylvanian-Permian marine shelf sandstone-shale-carbonate facies** of 140-285 m of the Minnelusa Formation, Opeche Shale, and Minnekahta Limestone overlies the erosional surface on the Pahasapa Formation. Outcrops of the Minnelusa are 120 to 230 m thick but the formation over 430 m thick in the subsurface to the southwest (Gries, 1975). The Opeche, 10 to 40 m of silty red shale and fine sand with thin, discontinuous beds of gypsum or anhydrite, is overlain by 10-15 m of thinly bedded, gray to purple Minnekahta limestone. The argillaceous and non-argillaceous high-calcium limestone units of the Minnekahta are the principal source in the Black Hills of crushed limestone, dimension stone, and cement (Gries, 1975). The Minnekahta typically is the resistant unit forming the surface of the Paleozoic plateau encircling the Black Hills.

Paleozoic-Mesozoic Continental Sequence

A **Permian-Triassic red bed facies** consists of 115-245 m of the Spearfish Formation. The non-resistant, bright red, Spearfish consists primarily of siltstone and gypsum beds at the base and argillaceous, fine grained, quartz sandstone and shale in the upper part. The "red racetrack" or "red valley" around the Black Hills just outside of the limestone

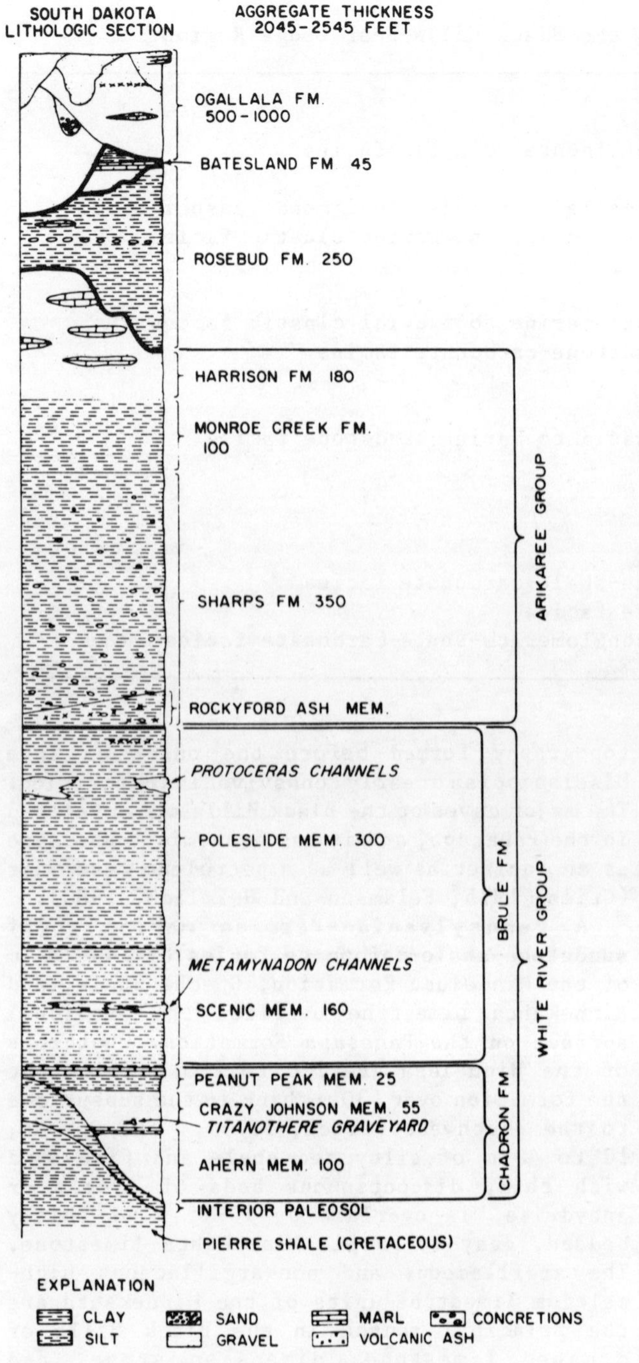

FIGURE 7 Geologic section of Tertiary strata of the Badlands and other areas of southwestern South Dakota (from Figure 3, Harkson and Macdonald, 1969). The White River Group is Oligocene in age, the Arikaree Group is Oligocene-Miocene, and the Rosebud, Batesland and Ogallala Formations are Pliocene.

plateau marks the surface exposures of the Spearfish.

A **Jurassic marine shelf clastic facies** of the Sundance Formation disconformably overlies the Spearfish Formation. The Sundance is 75-105 m thick and consists of a lower sandstone-shale transgressive sequence (Canyon Springs-Stockade Beaver Members) and a fine grained, reddish sandstone (Lak Member) unconformably overlain by a second sandstone-shale transgressive sequence of the Redwater Shale Member. Tan to reddish sandstones of the Sundance are prominent cliff-formers, particularly in the western Black Hills-Bear Lodge region.

A **Jurassic-Cretaceous transitional fluvial clastic to marine sandstone facies** is 75-520 m thick. At its base, white to reddish brown beds of the Unkpapa Sandstone form a transition from the Sundance Formation to the overlying fluviatile Morrison Formation which, in turn, is overlain by fluviatile to marine units of the Lakota and Fall River Formations of the Inyan Kara Group. The greenish claystones and sandstones of the continental Morrison Formation contain abundant dinosaur remains. Lakota and Fall River sandstones form the Cretaceous sandstone hogback or Dakota hogback of the Black Hills. The hogback is also a major recharge area for the "Dakota" sandstone and other aquifers of the Great Plains.

Late Mesozoic Marine Sequence

A **Cretaceous Western Interior marine shale-sandstone-carbonate facies** consists of up to about 1400 m of dominantly Upper Cretaceous marine shales of the Mowry, Belle Fourche, Carlile and Pierre formations. Important carbonate units within the shale sequence include the Greenhorn and Niobrara Formations. The units surround the Black Hills-Bear Lodge region and represent the partial filling of a seaway bordering the initially rising Rocky Mountain uplifts to the west.

A **Late Cretaceous Great Plains transitional marine to fluvial clastic facies** consists of the Fox Hills Sandstone and the overlying Hell Creek Formation. The marine Fox Hills is exposed north of the Black Hills where it consists of up to about 110 m of marine and brackish-water sandstone, clayey siltstone, shale and claystone documenting the regression of the Cretaceous epicontinental sea of the Western Interior seaway. Cretaceous deposits to the west are continental clastic equivalents of the Fox Hills and overlying Hell Creek. Deposited to the east of major Laramide uplifts, the Hell Creek consists of up to about 140 m of terrestrial clayey siltstone, mudstone, sandstone, lignite and related rocks. Calcareous concretions are common. The Hell Creek contains abundant fossil vertebrates including major dinosaur remains. These rocks are also believed to contain a continental record of Cretaceous-Tertiary boundary events.

Cenozoic Continental Sequence

A **Paleocene-Eocene Great Plains transitional fluvial and marine clastic facies** consists of the Fort Union Group of up to 420 m of transitional strata in the lignite-bearing Ludlow and Tongue River members. The marine sandstone and mudstone of the Cannonball Formation in central North Dakota and northwestern South Dakota include the last marine strata to be deposited in the North American interior.

An **Eocene Black Hills-Bear Lodge Mountains calc-alkalic to alkalic igneous association** consists of thirteen major igneous centers in a west-northwest-trending zone across the northern Black Hills to the Devils Tower-Missouri Buttes region at the western margin of the Bear Lodge Mountains. Sills, dikes, small stocks and related subvolcanic bodies are the most common observable evidence of major Eocene igneous activity. To the east, calc-alkaline, porphyritic, rhyolitic to trachytic bodies of various types are commonly cut by younger phonolitic dikes. Alteration and mineralization of various types are common. To the west, particularly in the Tinton (Mineral Hill) and Bear Lodge Mountain areas, alkalic rocks, including feldspathoidal syenites and pyroxenites, occur as a plutonic series accompanying the calc-alkaline, subvolcanic series. In the Bear Lodge Mountains, carbonatite cores the alkalic plutonic series. Alteration and mineralization of the subvolcanic series is, in large part, related to fenitization accompanying intrusion of the carbonatite.

An **Oligocene-Miocene Badlands clastic facies** consists primarily of up to about 300 m of the Oligocene White River Group and Oligocene-Miocene Arikaree Group spectacularly exposed in the South Dakota Badlands. The lower White River unit, the Chadron Formation, consists of about 55 m of fluvial, reddish to greenish clayey siltstone, claystone, sandstone and conglomerate deposited on the "Interior Zone," a lateritic paleosol formed on the Pierre Shale. Eroded and weathered to form rounded haystack hills, the Chadron contains a spectacular fauna including abundant titanothere bones and the remains of other mammals. The overlying Brule Formation consists of about 140 m of terrestrial sandstone, clayey siltstone, and claystone, often with nodular calcareous zones, volcanic ash zones and soil horizons. Eroded to form steep cliffs, narrow ridges and pinnacles, the Brule contains a fauna of vertebrates including abundant turtle and oreodont remains along with many other mammalian types. Overlying the White River Group in the Badlands is about 110 m of mudstone and volcanic ash of the Sharps Formation of the Arikaree Group.

A **Miocene-Pliocene-Pleistocene Great Plains continental clastic facies** consists mainly of Miocene units with only scattered evidence of Pliocene and Pleistocene erosion and deposition. The Miocene of South Dakota is a thick fossiliferous sequence of seven formations bounded by unconformities. The Arikaree Group includes the Miocene part of the Harrison Formation, about 50 m of silty sandstone and calcareous siltstone. The Rosebud Formation of uncertain age is about 90 m of pink siltstone. Both early and late Arikareean faunas have been collected from this unit and it may be equivalent to formations from the Sharps through the late Arikareean Rosebud as described near Wounded Knee, South Dakota. Channel sandstone, siltstone, claystone, and calcareous siltstones make up 14 m of the Miocene Batesland Formation exposed near the Little White River east of the Black Hills. The Fort Randall formation is about 40 m of claystone and interbedded sandstone. Uppermost of the Tertiary units are about 90 m of the discontinuous Pliocene Ogallala Group, including sandstone and siltstone and the well-cemented, calcareous sandstone and sandy limestone of the Ash Hollow Formation.

Phanerozoic History

The Phanerozoic rock record documents four major geologic events in the Black Hills-Bear Lodge Mountains:

1. A Paleozoic flooding of central North America by an epicontinental sea forming thick carbonate and clastic deposits followed by the initial development of positive features such as the Black Hills and basins such as the Williston Basin.
2. A late Paleozoic-early Mesozoic marine withdrawal with formation of redbeds, followed by a Mesozoic transitional stage with deposition of marine-fluvial facies.
3. A Cretaceous marine invasion resulting in thick shale-sandstone-carbonate deposits in the Cretaceous Western Interior seaway.
4. A late Mesozoic-Cenozoic Laramide uplift resulting in structural deformation, igneous intrusion and extrusion, erosion, and deposition of transitional marine and fluvial facies.

REFERENCES

Bakalowicz, M.J., Ford, D.C., Miller, T.E., Palmer, A.N., Palmer, M.V., 1987, Thermal genesis of dissolution caves in the Black Hills, South Dakota: Geological Society of America Bulletin, v. 99, p. 729-738.

Cameron, E.N., Johns, R.H., McNair, A.H., and Page, C.R., 1949, Internal Structure of granitic pegmatites: Economic Geology Monograph 2, p. 1-115.

Cerny, P., 1982, The pegmatite field of the southern Black Hills in Granitic Pegmatites of the Black Hills, South Dakota and Front Range Colorado: Geological Association of Canada Field Trip 12, p. 3-24.

Denison, R.E., Lidiak, E.G., Bickford, M.E., and Kisvarsanyi, E.B., 1984, Geology and geochronology of Precambrian rocks in the central interior region of the United States: U.S. Geological Survey Professional Paper 1241-6, p. C1-C20.

DeWitt, E., Redden, J.A., Wilson, A.B., Buscher, D., and Dersch, J.S., 1986, Mineral resource potential and geology of the Black Hills National Forest, South Dakota and Wyoming: U.S. Geological Survey Bulletin 1580, 135 p.

Feldman, R.M. and Heimlich, R.A., 1980, The Black Hills, Kendall/Hunt Geology Field Guide Series: Dubuque, Ia., Kendall/Hunt, 190 p.

Gosselin, D.C., Papike, J.J., Zartman, R.E., Peterman, Z.E., Laul, J.C., 1988, Archean rocks of the Black Hills, South Dakota: Reworked basement from the southern extension of the Trans-Hudson orogen: Geological Society of America Bulletin, v. 100, p. 1244-1259.

Gries, J.P., 1975, Limestone, in U.S. Congress, Senate Committee on Interior and Insular Affairs, Mineral and water resources of South Dakota: U.S. 94th Congress, 1st session, p. 115-122.

Gries, J.P. and Martin, J.E., 1985, Composite outcrop section of the Paleozoic and Mesozoic strata in the Black Hills and surrounding areas, in Rich, F.J., ed., Geology of the Black Hills, South Dakota and Wyoming: Field Trip Guidebook for G.S.A. Rocky Mountain Section Meeting, American Geological Institute, p. 261-292.

Harksen, J.C., 1975, Tertiary sedimentary rocks, in U.S. Congress, Senate Committee on Interior and Insular Affairs, Mineral and water resources of South Dakota: 94th Congress, 1st session, p. 43-45.

Harksen, J.C., and Macdonald, J.R., 1969, Guidebook to the major Cenozoic deposits of southwestern South Dakota: South Dakota Geological Survey, Guidebook 2, 103 p.

Hoffman, P.F., 1988, United plates of America, the birth of a craton: Early Proterozoic assembly and growth of Laurentia: Annual Reviews of Earth and Planetary Sciences, v. 16, p. 543-603

Karner, F.R., 1981, Geological relationships in the western centers of the northern Black Hills Cenozoic igneous province, in Rich, F.J., ed., Geology of the Black Hills, South Dakota and Wyoming: Field Trip Guidebook for G.S.A. Rocky Mountain Section Meeting, American Geological Institute, p. 126-133.

Karner, F.R., 1989, Geological framework of the Black Hills-Bear Lodge Mountains region, in Karner, F.R., ed. Devils Tower--Black Hills alkalic igneous rocks and general geology: 28th International Geological Congress Field Trip Guidebook T131, p. 3-6.

Lisenbee, A.L., 1975, Structural geology, Black Hills, in U.S. Congress, Senate Committee on Interior and Insular Affairs, Mineral and water resources of South Dakota: U.S. 94th Congress, 1st session, p. 52-56.

____, 1978, Laramide structure of the Black Hills uplift, South Dakota-Wyoming-Montana, in Matthews, V., III, ed., Laramide folding associated with basement block faulting in the western United States: Geological Society of America Memoir 151, p. 165-196.

____, 1985, Tectonic map of the Black Hills uplift, Montana, Wyoming and South Dakota: Geological Survey of Wyoming Map Series 13, scale 1:250,000.

Martin, J.E., 1983, Composite stratigraphic section of the Tertiary of western South Dakota: Museum of Geology, South Dakota School of Mines and Technology Dakoterra, v. 2, pt. 1, p. 1-8.

Merewether, E.A., 1975, Mesozoic rocks, in U.S. Congress, Senate Committee on Interior and Insular Affairs, Mineral and water resources of South Dakota: 94th Congress, 1st session, p. 38-43.

Norton, J.J., 1975, Pegmatite minerals, in U.S. Congress, Senate Committee on Interior and Insular Affairs, Mineral and water resources of South Dakota: U.S. 94th Congress, 1st session, p. 132-149.

Redden, J.A., and Norton, J.J., 1975, Precambrian geology of the Black Hills, in U.S. Congress, Senate Committee on Interior and Insular Affairs, Mineral and water resources of South Dakota: U.S.A. 94th Congress, 1st session, p 21-28.

Redden, J.A., Peterman, Z.E., Zartman, R.E., DeWitt, E., 1987, U-Th-Pb zircon and monazite ages and preliminary interpretation of Proterozoic tectonism in the Black Hills, South Dakota: Geological Association of Canada Programs Abstract 12:83.

Rich, F.J., 1985, Geology of the Black Hills, South Dakota and Wyoming, 2nd Edition: Field Trip Guidebook for G.S.A. Rocky Mountain Section Meeting, American Geological Institute, 292 p.

Thornbury, W.D., 1965, Regional geomorphology of the United States, N.Y., John Wiley and Sons, 609 p.

GLOSSARY OF PLACES AND PEOPLE OF THE BLACK HILLS

Frank R. Karner
Department of Geology and Geological Engineering and
North Dakota Geological Survey
University of North Dakota, Grand Forks

Gloria A. Pederson and Marilyn R. Shultz
Department of Geology and Geological Engineering
University of North Dakota, Grand Forks

In this glossary many of the places and people which figure prominently in a visit to the Black Hills are highlighted. The sources used are varied, but documented wherever possible. Some glossary items are quoted directly from the 1980 edition of the Encyclopedia Brittanica Micropaedia and are identified by (EBM) at the end of the quote or paraphrase. Most entries are derived from current leaflets available widely in the Black Hills area and are identified by (CL) at the end of the quote or paraphrase. Many thanks to Kris Brummer, Chloe Cermak, Jane Geiger, and Joan Karner who aided in this compilation.

Badlands National Park, fifty miles east of Rapid City in the north central Black Hills. The Badlands region (mako sica) was made a national park in 1928. Following scenic highways with numerous overlooks, early summer mornings are the best time to see the sun's rays slant into deep gorges and gullies, lighting up skies of pink, yellow, maroon, cream, gray and red on the spires and pinnacles. Buttresses and cliffs stand out in relief. After the short, infrequent rainstorms, typical of a dry climate, the colors and shapes are different.

The visitors center has video programs on the history of the Ogallala Sioux people and cultural exhibits. Self-guiding trails, guided nature walks and naturalist programs are held nightly in the campground amphitheater.

The Badlands have sparse vegetation. A gray to white terrain of ridges, pinnacles, and steep valley cut in soft sedimentary rock is beautiful and desolate. The region was fascinating yet formidable to the Dakota Indians; it was "ma-koo-si-tcha" ("evil" or "badlands"). The French trappers who crossed it to get to their hunting ground called the area "les mauvaises terres a traverser," meaning "a bad place to travel across." Many settlers avoided the Badlands as no good for farming or hunting. And yet, from this hostile terrain comes spectacular fossils of many of the mammals that lived in North America between 25 and 40 million years ago. (CL)

Bear Butte (mato paha). Bear Butte northeast of Sturgis, is not a butte, but a mountain with an elevation of 1,500 feet above the grassy plains. It is the result of molten rock that rose from great depth causing a large bulge in the earth's surface. Thousands of years of attack by wind and water have eroded the top covering exposing the underlying solidified igneous rock.

The Cheyenne religion is said to have originated at Bear Butte two or three thousand years ago. It is said to have begun when Sweet Medicine, an orphan raised by an elderly Indian couple, went on his first hunt at twelve years of age and killed a buffalo. While admiring his kill, an older Indian appeared and claimed the animal. In the scuffle that ensued, the older man was killed. Murder in self-defense against another of the tribe was not allowed and Sweet Medicine was sent away. The boy wandered to the mountain where he lived for four years. During this time, an Indian maiden, who had also been sent from her home, met Sweet Medicine and in their tradition were married. At the end of four years the gods of the mountain called the pair to a small cave on the north side and told them that the people were becoming sinful and were scattered about. A prophet was needed to bring them together for a better life. Four sacred arrows were presented to Sweet Medicine: two points were dark and were to be used for war; two were light colored and were for hunting and for peace. Four commandments were presented: Thou shalt not kill... not steal...not commit adultery...not marry your own relatives. Then Sweet Medicine was instructed to stay four more years on the mountain, after which he could go on a mission to the people. The Cheyenne religion is based on the number "4" - four sacred arrows; four commandments; four day and night fasts with four men who hope to receive visions and predictions. To the Cheyenne, "mato paha" means "The Good Mountain." Mato means "bear" and paha means "mound or hill" in the Sioux language. It is "Holy Mountain" to the Sioux and Cheyenne, who still make pilgrimages there. It has attracted such notable Indian leaders as Red Cloud, Crazy Horse, Little Wolf and Sitting Bull. (CL)

Bear Country U.S.A., south of Rapid City on Hwy. 16, a drive-through wildlife park with the largest privately owned collection of black bears. Also see buffalo, grizzly bears, wolves, cougar, rocky mountain goats, and big horn sheep roaming free in 220 acres of natural Black Hills habitat. Numerous smaller animals and park offspring can be seen in the walk-through "babyland". (CL)

Bear Lodge Mountains, in eastern Wyoming, the northwestern extension of the Black Hills uplift. The Bear Lodge Mountains are a low, north-trending range extending west and north from Sundance, Wyoming. Devils Tower lies just to the west. This scenic highland is covered with western yellow pine with aspen, white birch, hawthorn, and burr oak along valley bottoms and on northern slopes. Geologically, they are the center of an igneous complex which is a classic example of a carbonatite-pyroxenite-alkalic rock association. Deposits of rare earths, thorium gold and other valuable minerals are present (see Sundance, Devils Tower).

Black Hills (paha sapa). Rising above the Great Plains are a small range of mountains and hills surrounded by a dark, forested limestone ridge, the Pahasapa Formation. The Black Hills combines great natural beauty with Western history, exciting travel attractions and spectacular geology.

Black Hills gold and silver. The allure of gold is still very much with us, attested to by the strong demand for Black Hills gold jewelry. Traditionally patterned in delicate grape and leaf designs in colors of pale pink, green, and yellow, it is sold worldwide, but the most comprehensive selections can be found in leading South Dakota jewelry and gift shops. As a registered trademark, "Black Hills Gold Jewelry" can be manufactured only in the Black Hills, and ten local firms produce the world's inventory of distinctive jewelry. First time visitors often express surprise at the reasonable prices and often have to be assured that Black Hills Gold "really is gold." Each piece of Black Hills Gold Jewelry is subject to more than 29 separate hand-crafted operations. Rigid quality control is maintained. Black Hills gold makes a superb gift or personal piece of jewelry. It is the single most popular retail item for vacationers.

As a spinoff of Black Hills gold, Jim Langer, Spearfish jeweler, provides Black Hills silver of his own design. Like the gold, the silver is mined in the Black Hills and then brought up to sterling quality. Each piece takes about 15 hand operations to complete, and often includes semi-precious gemstones. (CL)

Black Hills Mining Museum, Lead. The Homestake Mining Company is dedicated to the preservation of the rich mining heritage of the Black Hills of South Dakota. Over 20 life-size models and many exhibits plus a wealth of historic photographs depict early days up to the high technology of modern mining. Authentic settings and displays include an assay office, blacksmith shop, cage station, windlass shaft, local history display and gambling devices. Take the simulated underground mine tours or share the thrill experienced by prospectors with the museum's hands-on, gold-panning display (see Homestake Mine). (CL)

Black Hills Passion Play at Spearfish. Now widely acclaimed as the most outstanding production of its type, the Passion Play reconstructs faithfully the dramatic event of the last seven days of the life of Christ. Endowed with all the riches, splendor, and pageantry of the Middle East, this great play speaks eloquently of the love and fury of a people 2,000 years ago: the Roman domination over Palestine, the devotion of the disciples and friends, and the supreme sacrifice of Christ. (CL)

Black Hills rock collecting. It has been said that South Dakota is a "rockhound's delight" and rightfully so. Rock shops--some in continuous business for more than 40 years--abound, and selections vary. For inexpensive beauty, South Dakota's native rose quartz (the state mineral) is a good first purchase. Fist-sized, rough pieces of rose quartz start at about a half dollar, and rise in price depending on quality and form. The much-prized Fairburn agate is also a native of South Dakota (it's the state gem stone) and is considered (not entirely out of local civic pride) to be one of the most distinctive agates. Fairburn agates are very rarely cut and shaped but instead are polished and left to display their own natural beauty. Pegmatite mines provide spectacular mineral collecting. (CL)

Boot Hill, Deadwood. Mount Moriah, a National Historic Landmark high above the city of Deadwood, reflects the hope and heritage of the frontier amidst the pine-covered beauty of the Black Hills. At Boot Hill, the cemetery of one of the wildest and most famous gold camps in the Old West, rest Wild Bill Hickok, Calamity Jane, Potato Creek Johnny, Preacher Smith and many other legendary characters of Black Hills' history. (CL)

Borglum, Gutzon (see Mount Rushmore National

Memorial).

Broken Boot Gold Mine, on Hwy. 14A, just outside of Deadwood, S.D. Gold fever brought thousands of men to the Black Hills of Dakota in 1876 to seek their fortunes. Early miners found riches in the streams, needing little more than a pick, shovel, and gold pan to separate the gold from worthless sand. Eventually, however, the "easy pickings" were gone and the miners were forced to dig in the hillsides in search of gold-bearing veins, or "lodes," from which the flakes and nuggets in the streams had come.

The Broken Boot Gold Mine opened up to the public in 1954 after lying dormant for over fifty years. It was named after an old miner's boot found in a pile of debris. (CL)

Calamity Jane. Her real name was Martha Cannary or Martha Burk, a popular heroine of the American West. In 1875 she settled in Deadwood. A former scout, bull-whacker and soldier, Calamity drank her whiskey neat and swore worse than the miners. She rode for the Pony Express between Deadwood and Custer in the days when it took a sharp trigger finger to stay alive. She had her tender side too. She is said to have solicited groceries for needy families at the point of a six-shooter. During the smallpox scourge of 1878 she nursed many residents of Deadwood back to health.

In dime novels of the 1870's and 80's, she was transferred into a beautiful, buckskin-clad crusader for justice, who, with her sweetheart "Deadwood Dick," battled against frontier badmen in the Black Hills of South Dakota. Calamity Jane appeared in the Wild West shows of the 1890's and the Pan American Exposition of 1901. (EBM & CL)

Chapel in the Hills, on Hwy. 44 west of Rapid City, a replica of the famous 830-year-old Stavkirke (Stave Church) in Borgund, Norway. Its wood carving, dragon heads and unusual construction make it a most interesting example of church architecture. (CL)

Cosmos, Nature's Mystery Area, Hwy. 16 southwest of Rapid City. Here the world is different, topsy-turvy, with nothing in its normal position, including yourself. In the mystery house, the laws of nature seem to have been changed. A physical experience which you will not forget. (CL)

Crazy Horse (ta sunko-witko), a Sioux Indian chief of the Ogallala tribe who was one of the most determined warriors in the Sioux resistance to the white man's invasion of the northern Great Plains. One of the chiefs who defeated General Custer at the Battle of Little Big Horn, Crazy Horse never surrendered, never signed a treaty and never went on a reservation. He represents the spirit of man, symbolic of the pride and dignity of the North American Indian (see Crazy Horse Memorial).

Crazy Horse Memorial, Hwy. 385 north of Custer. In 1939, Sioux Chief Henry Standing Bear invited Sculptor Korczak Ziolkowski to carve an Indian memorial in the Black Hills. The Indians picked Crazy Horse because they felt he was an Indian's Indian. Nearly eight million tons of rock have been blasted away. Work continues on the 563 foot high, gigantic statue of the Indian hero. Korczak died in 1982, but his work is being carried on by his family. His humanitarian goals for the Crazy Horse Memorial included a medical training center and university. The Indian Museum of North America at Crazy Horse houses one of the finest collections of Indian art and artifacts anywhere.

Boston-born of Polish descent, Korczak Ziolkowski was a self taught sculptor-engineer. He never had a formal lesson in sculpture or engineering, yet in 1939, his bust of Paderewski won first prize by popular vote at the New York World's Fair. Then he embarked on the biggest sculpting task in history. Before beginning Crazy Horse at the age of forty in 1949, Korczak's only mountain experience was briefly assisting Gutzon Borglum at Mount Rushmore. Mountains are not carved by weak men. He began his task with $174, living in a tent at the base of Thunderhead Mountain, site of Crazy Horse. With his wife, Ruth, and their ten children he struggled to carve the heritage of a nation in stone. With Korzcak's death in 1982, his family carries on the project using the scale models and detailed plans he left for that purpose. His story is one of determination, perseverance, and patience made more incredible when one realizes he did not begin the project until he was 40. He died at the age of 74 (see Crazy Horse). (CL)

Custer, George Armstrong, born Dec. 5, 1839. A U.S. Cavalry officer who distinguished himself in the U.S. Civil War (1861-1865) but later led his men to death at Little Big Horn, in one of the most controversial battles in U.S. history. In the summer of 1874, a 1,000 man force led by Major General George A. Custer had ridden three weeks south and west on an expedition to explore and establish the truth about the rumors of gold deposits in the Black Hills. In this expedition, as well as in a few earlier, there began the very long line of prospectors and geologists who have visited the Black Hills to see and examine its geologic features (see Sitting Bull). (EBM)

Custer, in the southern Black Hills on French Creek. The town was laid out in 1875 and named for Lieutenant Colonel George A. Armstrong Custer, whose expedition to the Black Hills (1874) included discovery of gold in French Creek. Custer has extensive mineral industries, lumbering, and livestock markets. Local museums depict frontier and gold mining memorabilia. It is the center of a major pegmatite district. (CL)

Custer State Park, in the southern Black Hills (see Needles Highway and Harney Peak). The natural beauty and unusual attractions provide an adventurous, primitive, and rewarding experience. Wildlife Loop Road is an 18 mile road designed to guide visitors through primary wildlife range, where one of the nation's largest buffalo herds is free to roam. The burros, originating from the Sylvan Lake herd that once hauled visitors to Harney Peak, are another attraction. Needles Highway Drive consists of 20 miles of hairpin curves and narrow tunnels threading through sky-piercing granite spires. The route includes the historic replica of Gordon Stockade, a fort built originally to protect whites from Sioux warriors. (CL)

Deadwood, in a canyon of the north central Black Hills. The town is built on the steep woodland inclines of Deadwood Gulch and was founded during the 1876 gold rush when about 25,000 miners swarmed the surrounding hills. Its turbulent reputation as a wide-open outpost of frontier violence was magnified by the "Deadwood Dick" dime novels. Wild Bill Hickok, soldier, scout and marshall was killed in a Deadwood saloon. Deadwood was almost completely abandoned at the news of another gold strike at Lead, a few miles southwest. (CL)

Deadwood Dick. Nat Love was a black cowboy who won the award for wild horse riding, pistol shooting and roping during the Deadwood celebration of July 4, 1876. Nat dubbed himself Deadwood Dick and a legend began. (CL)

Devils Tower National Monument, about 25 miles northwest of Sundance, Wyoming. Devils Tower (mato tepee or bear lodge) is an igneous monolith of phonolite rising 1,267 feet above the Belle Fourche River. It is one of the world's finest examples of columnar jointing. There is a legend about the tower that begins with seven Indian brothers. One day the wife of the oldest brother was carried off by a big bear. The man mourned her loss greatly and would go out and cry defiantly to the bear to release her.

The youngest brother, who had great powers, told the oldest to make four blunt arrows. Two arrows were painted red and set with eagle feathers, the other two black and set with buzzard feathers. They set out to rescue the oldest brother's wife. At the bear's cave the youngest turned himself into a gopher and dug a big hole into the bear's den, then he had the women crawl out through the hole. After they had gone, the bear woke up and he found the trail of the Indians and began to follow them.

The youngest boy, a holy man, always carried a little rock in his hand. He told his six brothers and the women to close their eyes and he sang a song and finished it. When the others opened their eyes the rock under them had grown. He sang four times and when he had finished the rock was just as high as Devils Tower is today. When the bear caught up to them he growled and kept jumping high up against the rock. His claws made the scars that are now on the rock. While doing this, the youngest brother shot the black arrows at the bear but did not hurt him. The first red arrow was shot, but did not enter the bear; the second red arrow entered the bear's head and killed him. The brothers climbed down, packed in wood, piled it on top of the body of the bear and lit a fire that burned the bear to ashes. (CL)

Dinosaur Park, on Skyline Drive west of Rapid City. This scenic drive loops from Quincy Street to Mt. Rushmore Road, and provides an excellent view. Seven life-sized concrete replica of dinosaurs stride the ridge overlooking the city. (CL)

1880 Train at Hill City. You can board the 1880 train for a 32-mile round trip in Dakota Territory vintage coaches behind a real steam locomotive for a ride through magnificent Old West mountain scenery. (CL)

Evans Plunge at the north edge of Hot Springs on Hwy. 385. Take the plunge, and enjoy the last remaining mineral-bath pool in Hot Springs. Since 1890, Evans Plunge has offered this unique mineral water pool. The 87-degree, spring-fed pool has an inflow of 5,000 gallons of water per minute. The springs rise out of the pebble bottom of the plunge, providing a complete change of water sixteen times daily. The 164-foot whale slide, Tarzan rings, and two shallow enclosures for the little ones are just a part of the fun. (CL)

Harney Peak (see Needles Highway).

Hill City, in the heart of the Black Hills, was originally settled during the gold rush of the 1880's, and was the hub of largely unsuccessful tin mining. Today it is the home of the 1880 Train, a steam engine and open cars that operate in the summer months. (CL & EBM)

Homestake Gold Mine, in Lead in the northern central Black Hills. The Black Hills was a rugged wilderness in 1874 when General George Custer's military expedition explored this part of Dakota Territory and found gold. The Homestake "lead" (pronounced leed), the gold bearing ore deposit, was discovered in 1876. Geologically, a sequence of Precambrian metasedimentary iron formation contains gold within the Homestake Formation. See the spectacular "open cut"; an open pit gold mine over 400 feet long, 1300 feet wide and 500 feet deep. Since the first lode was discovered here Homestake has processed over $1 billion in gold. Today the city of Lead is four-dimensional: over a mile high, a mile long, a mile wide, and a mile deep. Tour the surface workings and learn about a real gold mine operation. (CL)

Hot Springs, at the southeastern edge of the Black Hills. It is a town with victorian charm and grace left over from its heyday as a health spa. Railroad magnates and Wall Street tycoons came to soak their bones in the mineral springs. Hot Springs is home for both the Evans Plunge and the Hot Springs Mammouth Site. (CL)

Jewel Cave, Hwy. 16 west of Custer. The cave was discovered in 1900 when two prospectors, Albert and Frank Michaud, discovered air whistling through a small hole in the side of a canyon. They enlarged the opening and mined the calcite crystals lining the cave walls. The brothers named their mining claim the "Jewel Lode" for the sparkling of the broken crystals. In 1958, spelunking seasonal park rangers rediscovered some of the "lost" passages beyond the "known" one half mile of passages. Over forty miles of intricate passageways have been explored and mapped, making Jewel Cave one of the most extensive caves known to man. Many fascinating mineral forms are found at various locations in the cave. (CL)

Keystone, in the central eastern Black Hills. At the base of Mount Rushmore discover this captivating old west town, the gateway to Mt. Rushmore, offering train and stagecoach rides, an aerial tramway and helicopter rides with views of Mount Rushmore. (CL)

Lead, in the north central Black Hills. Settled in 1876, Lead is the only town in the Black Hills to prosper from the frontier gold rush to the present on earnings from "gold diggings." Pronounced "leed," the town's name is a mining term--when prospectors would find an outcropping of gold-bearing ore, they would follow it, hoping it would "lead" them to the mother lode. The thin little houses, victorian mansions, and old-fashioned storefronts on Lead's narrow main street exemplify 112 years of mining history. It is the site of the Homestake Mine, the largest producing gold mine in the United States. (CL)

mako sica (see Badlands).

Mammoth Site, Hot Springs. The site is of great scientific importance because of the primary accumulation (actual death area) of mammoth bones at the site. Based on tusk count, 39 mammoths have been found. Other fossils include the tibia and some ribs of the giant, short-faced bear of the Ice Age, as well as bones and teeth of rabbits, minks and prairie dogs. The Mammoth Site is the only in-place display of fossil mammoths in America. (CL)

Marine Life Aquarium, Hwy. 16 southwest of Rapid City. You'll be caught up in the excitement of dolphins, seals and sea lions in action-packed shows! View colorful coral reef fish through portholes, feed the animals, experience a kiss from a sea lion, and bring your camera. (CL)

mato paha (see Bear Butte).

mato tepee (bear lodge, see Devils Tower).

Mount Moriah (see Boot Hill).

Mount Rushmore National Memorial, in the southeastern Black Hills near Keystone. This epic sculpture portrays the faces of four presidents: George Washington, Thomas Jefferson. Abraham Lincoln, and Theodore Roosevelt. Carved in granite, with the central Black Hills as the backdrop, the four heads, each about 18 meters high and 150 meters above the cliff's base represent, respectively, the founding political philosophy, preservation, expansion, and conservation of the United States of America.

Work began in 1927 under Gutzon Borglum and was finished in 1941. Creation of the shrine to democracy took 14 years and cost a mere $1 million, financed partly by penny drives of American school children in the 1930's. Now, the monument is priceless. Gutzon Borglum was an American original. The son of Danish immigrants, he embodied the nation's pioneer spirit. He established an international reputation as a painter and sculptor by the early 1900's. The sculptor's studio, constructed in 1939, displays Borglum's original models of the four presidents. Visitors will also view tools used in the carving and a colorful history is outlined in a 20-minute movie on Borglum and his mountain. (CL)

Museum of Geology, in Rapid City. In 1885 the territorial legislature established the Dakota School of Mines in Rapid City, where it served the frontier communities of the Black Hills as a mining college and a prospector's analytical laboratory. Now the South Dakota School of Mines and Technology contains the fascinating Museum of Geology. A talking dioramas tells the story of the Big Badlands and the strange animals that once lived there. View saber-toothed cats and three-toed horses. Additional displays include: a systematic collection of minerals from all over the world; gold samples; polished agates; meteorites; fluorescent minerals; and prehistoric fossil plants and marine vertebrates. Outstanding exhibits! (CL)

Needles Highway and Harney Peak, in the

southern Black Hills (see Custer State Park). This highway curves and winds through 14 miles of solid granite tunnels and through bizarre but beautiful rock formations and granite spires called "needles." Look for the fascinating "Needles Eye," a gap in one peak, only 3 to 4 feet wide, but spanning 50 to 60 feet vertically. The road leads you past 7,242-foot Harney Peak, highest point elevation between the Rockies and the Alps. (CL)

Pactola Dam, in the central Black Hills on Hwy. 385 west of Rapid City. Built in 1956, the reservoir is a favorite recreational spot, and is designed as a flood control project and to maintain summer discharge for irrigation and municipal water supply in Rapid City.

paha sapa (see Black Hills).

Petrified Forest on Hwy. 14 northwest of Rapid City at Piedmont. This 22-acre natural forest is unique in the Black Hills. The Petrified Forest shows the rocks, fossils and overall geographic history of the Black Hills. The museum has hundreds of pieces of polished petrified wood, dozens of animals fossils and minerals galore...all from this area. A recent addition of Indian artifacts includes the largest arrowhead collection in the Black Hills. Over 3,000 pieces make this collection one of the finest ever assembled. (CL)

Poker Alice, as a young girl, married a rancher east of Sturgis. One winter her husband died. Legend has it that she didn´t get him into town for burial until spring, selling her wedding band to pay for the funeral. Destitute, she went to work and became Deadwood´s most notorious poker dealer. On the side, she became a madam, presiding over one of the bawdiest of bawdy houses in nearby Sturgis. (CL)

Potato Creek Johnny lived about 25 miles from Deadwood Gulch on Potato Creek where he discovered the largest nugget of gold ever found in the Black Hills. He used to tie it in a silk scarf, head for the saloon and offer to show it to people for drinks. He tied about a dozen knots in the kerchief, moved from table to table, and got three or four drinks before getting each knot untied. (CL)

Prairie Homestead Historical Site, at the east entrance to the Badlands National Monument on Hwy. 16A, is an original set of homestead buildings in their original location and furnished as if the homestead family was living there today. It is on the national Register of Historical Places. (CL)

Rapid City, on the eastern edge of the Black Hills. Founded in 1876, two years after gold was discovered, it is the largest city and the major gateway to the Black Hills. (CL)

Reptile Gardens, on Hwy. 16, 5 miles south of Rapid City. It is the home of hundreds of rare and unusual alligators, crocodiles, snakes and giant tortoises, as well as extremely rare snakes and lizards that are a part of the largest privately-owned collection of reptiles in the world. Make an exciting safari in the giant sky dome where colorful birds sit among many different habitats that range from deserts to gardens with great sprays of orchids and luxurious tropical foliage. Rock gardens, flower beds, tropical parrots, pygmy goats and miniature horses may be seen. (CL)

Rockerville, on Hwy. 16 southwest of Rapid City. The town gets its name from gold rockers, cradle-like boxes used to extract gold from placer deposits where water is limited. One of the earliest mining camps in the Black Hills, it is rebuilt and furnished in keeping with the period 1876. Try sarsaparilla at the old time bar. (CL)

Rushmore Aerial Tramway at Keystone. See Mount Rushmore from a different point of view. Take a relaxing ride to the top of the tramway for a view of Mount Rushmore and the Black Hills area. (CL)

Sitting Bull, (tatanka iyotake), famous Teton Dakota Indian Chief under whom the Sioux Tribes united in their struggle for survival on the Great Plains. Sitting Bull early in life became a leader of the powerful Strong Heart warrior society and later participant in the Silent Eaters, a secret group concerned with tribal welfare. Sitting Bull´s qualities of leadership, courage and wisdom were recognized when he was made head chief of the entire Sioux Nation about 1867. Sitting Bull fought many battles, but the most famous was the battle at Little Big Horn. After the Sun Dance, Sitting Bull emerged from a trance induced by self torture. He reported he had seen soldiers falling into his camp like grasshoppers from the sky. His prophecy was fulfilled on June 25, when Lt. Col. George A. Custer rode into the valley and he and all the men under his immediate command were annihilated. The Sioux emerged victors in their battle with U.S. troops; though they might win battle after battle, they could never win the war (see Custer, George Armstrong). (EBM)

South Dakota Trivia

State Gemstone - Fairburn Agate
State Mineral - Rose Quartz
State Fish - Walleye
State Tree - Black Hills Spruce
State Bird - Ringneck Pheasant
State Insect - Honey Bee
State Song - "Hail South Dakota"
State Nickname - Sunshine State
State Size - 77,515 square miles

State Motto - "Under God the People Rule"
State Flower - Pasque, also called "May Flower"
State Animal - Coyote
State Drink - Milk
State Capital - Pierre
South Dakota became a state in 1889.

Spearfish, in the northwestern Black Hills. Home of the world famous Passion Play and gateway to beautiful Spearfish Canyon, it is proud of its quality of life and called the Queen City of the Black Hills. (CL)

Stamper's Black Hills Gold Jewelry, Hwy. 16 south of Rapid City. Black Hills jewelry is a unique American art form. Hand crafted jewelry of heirloom quality and ageless beauty is produced here. Each design incorporates the traditional motif of grapes, leaves and tendrils, in combinations of yellow, pale green and rose gold. The "Jug S" or the "SB HG" stamped on every piece are registered trademarks. View the procedures used to make the jewelry and buy the pieces you choose. (CL)

Stavkirke (see Chapel in the Hills).

Stratobowl Lookout, Hwy. 16 southwest of Rapid City. In 1935, a balloon was launched from the valley below and reached a record altitude of over 21,280 meters.

Sturgis, at the northeastern edge of the Black Hills. Home of the Black Hills Motorcycle Classic, each August over 50,000 "bikers" and spectators celebrate this week long, internationally famous motorcycle racing extravaganza. Other attractions are Black Hills National Cemetery, "The Arlington of the West," Fort Meade Museum and Bear Butte State Park. (CL & EBM)

Sundance, in Crook County Wyoming, is the gateway to the Bear Lodge Mountains and Devils Tower. The surrounding country was once considered sacred by the American Indians. Here the Indians danced their famous dance to the Sun God on Sundance Mountain.

Many people have traveled this route. Some have traveled it straight into the pages of history: General Custer on his way to the Battle of the Little Big Horn in 1874; trappers and traders; stagecoaches and covered wagons filled with brave pioneers; and many of the famous and notorious characters of Old West.

Thunder Head Falls, open to the public on Hwy. 44, west of Rapid City. While hundreds of gold mines and caves penetrate the Black Hills, this underground spectacle of stream flow diverted into a tunnel for sluicing gold from sand and gravel has several interesting differences. Gushing water falls vertically at eight cubic feet of water per second for over 30 feet, and living stalactites stud the ceiling of the ancient tunnels in sparkling and colorful array. Gold-bearing quartz veins the walls in this oldest gold mine, dating back to 1880. (CL)

Trial of Jack McCall, Deadwood. In a street pageant every summer night, Wild Bill Hickok meets his maker as he did while playing poker in Deadwood's Saloon #10, August 2, 1876. Hickok liked a good card game in the saloon and for the first time took a chair with his back to the door. He never heard the boom of Jack McCall's vengeful six-gun. Hickok's cards, black aces and eights, known ever since as the "Dead Man's Hand," dropped to the floor and the famous victim slumped dead. Relive the trial of Jack McCall, presented as a vivid reenactment in the restored Old Town Hall (see Wild Bill Hickok). (CL)

Wall Drug, fifty miles east of the Black Hills in Wall, the Gateway to the Badlands. In December 1931, Ted and Dorothy Hustead bought the only drug store in a town called Wall. Only open for a few days, business was so bad that they gave themselves a limit of five years to make a go of it. They needed to come up with an idea to make the passing traffic want to stop. "What would travelers driving across the hot prairie want? They'd want water! Ice Cold Water!! They found plenty of ice and water!" Known for its free ice water, Wall Drug estimates that over 5,000 glasses are given away daily during the summer. Wall Drug is internationally famous for its incredible variety of western shops, museums, cowboy orchestra, and art gallery. Its advertising signs are placed all over the world. The town "Wall" was named because it was built on the edge of the northern extension of the "Badlands Wall". (CL)

Wild Bill Hickok, James Butler Hickok, also known as "Wild Bill" Hickok, was a famous scout and lawman who helped bring order to the frontier. His reputation as "the fastest gun in the West" made him a legend in his own time. Recognized early in life as a crack pistol shot, he left the midwest in 1855 and worked in Kansas for a time, later driving a stagecoach over the Santa Fe and Oregon Trails. For a brief period he displayed his extraordinary marksmanship for audiences of "Buffalo Bill's Wild West" exhibition. Playing poker at Deadwood in 1876, he was shot from behind and mortally wounded by Jack McCall who was convicted and hanged for the deed (see Trial of Jack McCall). (EBM)

Wild West Historical Wax Museum at Wall. Witness the wild west as it really was in the late 1800's to the early 1900's. Here are life-size wax figures of famous legends of our past, Wild Bill Hickok, Jesse James, Calamity Jane, along with the recreated historical adventure of the first trappers and miners in the west. A gun collection, historical photos, a barbed wire display and much more can be seen in the museum. (CL)

Wind Cave National Park, on Hwy. 385 in the southern Black Hills north of Hot Springs. The Sioux Indians called the cave the "sacred cave of winds." They had several legends about it, one telling that this was the cave through which their ancestors arrived at the Black Hills. Another legend told that wokan tanka (the "Great Mystery") sent the buffalo out into their hunting grounds from the cave. Other stories claimed that notorious outlaws or cowpunchers discovered the cave. Variation in rock type, jointing pattern and ground water chemistry result in a wide diversity of forms and a broad spectrum of caves and cave deposits. Wind Cave is world famous for its boxwork and ribbon deposits. (CL)

Wyoming Trivia

State Flower - Indian Paintbrush
State Bird - Meadowlark
State Tree - Plains Cottonwood
State Gemstone - Jade
State Capital - Cheyenne
State Motto - Equal Rights
State Nickname - Equality State, Cowboy State
State Size - 97,914 sq. miles, 9th largest state
State Admitted to the Union - July 10, 1890- 44th state

FIELD GUIDE DAY 1: GEOLOGY OF THE PRECAMBRIAN ROCKS OF THE KEYSTONE REGION

Frank R. Karner
Department of Geology and Geological Engineering and
North Dakota Geological Survey
University of North Dakota, Grand Forks

DAY 1 Saturday, July 1, 1989
- **Precambrian Metasedimentary and Metaigneous Rocks of the Keystone Region**
- **Harney Peak Granite (Figure 1) and Mount Rushmore National Memorial**
- **Museum of Geology at South Dakota School of Mines and Technology**

FIGURE 1 Scenery of the central Black Hills. Head of Golden Valley with a granite ridge in the distance (woodcut in Newton and Jenney, 1880).

Cumulative Miles

0.0

0.0 8.00 A.M. Depart Alex Johnson Hotel Rapid City, SD, **Turn west on Main Street.**

0.2

0.2 Join Hwy. 16. **Turn south on Hwy. 16.**

0.5

0.7 Turnoff to Dinosaur Park.

1.6

2.3 Climb onto resistant, buff-colored strata of the Cretaceous sandstone hogback which encircles the Black Hills.

4.4

6.7 Marine Life Aquarium.

5.2

11.9 Stamper Black Hills Gold Jewelry factory and show room.

6.0

17.9 Descent into the "Red Valley" or "Red Racetrack" formed in the easily eroded siltstone, sandstone, and shale red beds of the Spearfish Formation (Permian-Triassic).

6.8

24.7 Spearfish Formation (Permian-Triassic) and overlying yellow sandstone of the Sundance Formation (Jurassic) to east.

7.2

31.9 Reptile Gardens.

7.4

39.3 Spring Creek.

8.1

47.4 Gray, finely layered limestone of the Minnekahta Limestone (Permian).

8.8

56.2 Red shale and sandstone of the Opeche Shale (Permian).

9.4

65.6 Exit to Bear Country USA.

0.9

66.5 Contact of Pahasapa Limestone (Mississippian) and overlying Minnelusa Formation (Pennsylvanian-Permian).

0.1

66.6 Turnoff to Sitting Bull Crystal Cave.

1.1

67.7 Turnoff to Stratobowl Lookout.

0.3

68.0 Pahasapa Limestone (Mississippian). Large exposure of the massive, cliff-forming limestone in road cut.

0.2

68.2 Enter Black Hills National Forest. Overview.

0.5

68.7 Exit to Rockerville.

0.3

69.0 Rockerville.

0.5

69.5 Silver Mountain-Storm Mountain Center Road.

0.2

69.7 **STOP 1. SILVER MOUNTAIN: INTRODUCTION TO EARLY PROTEROZOIC PRECAMBRIAN ROCKS OF THE BLACK HILLS.**

Exposures of Early Proterozoic rocks of the central Black Hills usually contain strongly deformed, steeply dipping clastic to chemical sedimentary rocks of intermediate to high metamorphic grade. Typical rock types include schist, metagraywacke, phyllite, quartzite, metaconglomerate, and iron formation. Many exposures consist largely of metagraywacke and schist of middle to upper amphibolite facies. Generally, bedding is obvious and the rock units are steeply dipping, showing considerable evidence of tight folding. Rock units may be cut by quartzo-feldspathic and other types of veins. Shearing, alteration and mineralization may be common in some units. The rocks at this exposure on the south side of Silver Mountain are mapped as metagraywacke, metachert, metaconglomerate and metabasalt rock units of Early Proterozoic age (DeWitt and others, 1986). Road cuts continuing to the southwest expose similar rocks including iron formation northwest of Eagle Mountain (DeWitt and others, 1986). Several northwest trending faults are mapped in this area. **Important considerations at this stop are: (1) observation of typical metasedimentary Early Proterozoic rocks of the Black Hills, (2) evidences of original sedimentary characteristics; (3) evidences of metamorphism, mineralization and alteration; (4) deformational features; and (5) tectonic interpretations.**

0.4

70.1 Precambrian metasedimentary rocks. Large exposure in road cut.

0.8

70.9 Precambrian metasedimentary rocks (including iron formation?). Large exposure in road cut.

0.8

71.7 Precambrian metasedimentary rocks (including iron formation?) and amphibolite. Large exposure in road cut.

0.6

72.3 Precambrian metasedimentary rocks. Large exposure in road cut.

0.7

73.0 Precambrian metasedimentary rocks. Large exposure in road cut just northeast of 16A exit sign contains garnet-bearing, arenaceous metasedimentary rock.

1.1

74.1 Jcn. Hwy. 16A and Hwy. 16. **Turn left (south) on 16A (to Keystone)**

0.4

74.5 **STOP 2. ROCKY GULCH: ALTERED EARLY PROTEROZOIC METASEDIMENTARY ROCKS AND BASALTIC PILLOW LAVAS.**

These exposures and those to the south are mapped as schists and metagraywackes of metaconglomerate and iron formation rock units (DeWitt and others, 1986). The rocks, particularly those of the iron formation unit, contain zones of alteration and mineralization. Metabasalt with evidence of pillow structure is also present. Many of the features of Stop 1 can also be seen here.

The rocks illustrated at these first stops are believed to be products of deposition followed by metamorphism and deformation in an Early Proterozoic mobile zone between the Superior Province Archean rocks to the northeast and the Wyoming Province Archean rocks to the west. The rocks of the mobile zone are variously viewed as representing some combination of spreading/rifting, subduction, or collisional/accretional tectonic events. A current trend of thought in interpreting the North American Precambrian record postulates a tectonic regime of continental growth dominated by the joining of older Archean nucleii by Early Proterozoic mobile zones involving spreading, subduction and accretionary processes. **Important considerations at this stop are: (1) observation of typical metasedimentary Early Proterozoic rocks of the Black Hills; (2) evidences of original sedimentary characteristics; (3) evidences of metamorphism, mineralization and alteration; (4) deformational features; and (5) tectonic interpretations.**

0.2

74.7 Precambrian rock exposures in road cuts extensive as 16A descends to Keystone.

1.5

76.2 **STOP 3. KEYSTONE TUNNEL: EARLY PROTEROZOIC GARNET- AND STAUROLITE-BEARING SCHISTS.**

In the exposures of metasedimentary rock along the tunnel bypass are some fairly good examples of garnet-rich and staurolite-rich schists. These rocks show an increase of metamorphic grade from the first two outcrops. Pegmatites are also abundant in this area of higher metamorphic grade. **Important considerations at this stop are (1) the occurrence of garnet and staurolite; and (2) the metamorphic grade of the rocks in the Keystone area.**

0.6

76.8 Jcn. Hwy. 40 and Hwy. 16A at north edge of Keystone.

0.9

77.7 Jcn. Hwy. 244 and Hwy. 16A. **Follow Hwy. 244 to right (east) to Mt. Rushmore.**

1.0

78.7 **STOP 4. MOUNT RUSHMORE NATIONAL MONUMENT: EARLY PROTEROZOIC HARNEY PEAK GRANITE. LUNCH STOP.**

Note: No rock collecting at the National Monument. Harney Peak Granite and pegmatite can be collected on tomorrow's trip and to some extent at Stop 5. The Harney Peak Granite is a coarse grained muscovite granite containing muscovite, biotite, tourmaline and garnet and has a Rb-Sr emplacement age of 1,697 + 33 Ma (DeWitt and others, 1986). Pegmatitic phases and associated pegmatites are common. The variability of Harney Peak Granite; occurrence as a series of multiple coalesced dikes, sills, and other bodies; intimate association with metasedimentary rock; high $^{87}Sr/^{86}Sr$ ratios; and unusual mineralogy all suggest an origin of at least partial derivation from adjacent Early Proterozoic metasedimentary material. Surrounding satellitic bodies of granite and pegmatite may have a similar origin or may be extensions of the main body of Harney Peak Granite. **Important considerations at this stop are (1) characteristics of the Harney Peak Granite; and (2) the Mount Rushmore Memorial.**

0.2

78.9 **Leave Mt. Rushmore parking area. Turn right (north and west) on Hwy. 244.**

1.4

80.3 Leave Mt. Rushmore National Memorial.

0.1

80.4 **STOP 5. GUTZON BORGLUM MONUMENT: EARLY PROTEROZOIC PEGMATITES IN SCHIST AT CONTACT WITH HARNEY PEAK GRANITE.**

Note the monument and spend some time examining the road cut exposures in this area at the northeastern contact of the Harney Peak Granite with schistose Early Proterozoic metagraywackes. **Important considerations at this stop are (1) the evidences of igneous, metamorphic and structural effects in this contact zone; and (2) characteristics of pegmatitic bodies.**

0.0

80.4 **Return to Hwy. 244. Turn left (east) to Keystone.**

3.7

84.1 Jcn. Hwy. 16A and Hwy. 244. **Follow 16A east and north.**

2.6

86.7 Jcn. Hwy. 16 and 16A. **Turn right (northeast) on 16 to Rapid City.**

17.8

104.5 Jcn. St. Joseph St. and Hwy. 79 with Hwy. 16. **Turn right (east) on St. Joseph St. and Hwy. 795.**

0.2

104.7 Alex Johnson Hotel, Rapid City.

1.0

105.7 South Dakota School of Mines and Technology on right.

0.1

105.8 **STOP 6. SOUTH DAKOTA SCHOOL OF MINES AND TECHNOLOGY: GEOLOGY MUSEUM.**

0.1

105.9 **Return to Hwy. 79. Turn left following Main Street (west).**

1.0

106.9 Alex Johnson Hotel.

REFERENCES

DeWitt, Ed, Redden, J.A., Wilson, A.G., and Buscher, D., Dersch, J.S., 1986, Mineral resource potential and geology of the Black Hills National Forest, South Dakota and Wyoming: U.S. Geological Survey Bulletin 1580, 135 p.

Newton, H.A., and Jenney, W.P., 1880, Report on the geology and resources of the Black Hills of Dakota: U.S. Geographical and Geological Survey of the Rocky Mountain Region, p. 1-222.

FIELD GUIDE DAY 2: GEOLOGY OF THE PRECAMBRIAN ROCKS OF THE CUSTER REGION, HOT SPRINGS MAMMOTH SITE AND WIND CAVE

Frank R. Karner
Department of Geology and Geological Engineering and
North Dakota Geological Survey
University of North Dakota, Grand Forks

Richard L. Patelke
Department of Geology and Geological Engineering
University of North Dakota, Grand Forks

DAY 2: Sunday, July 2, 1989

- **Overview of Paleozoic-Mesozoic Sedimentary Geology of Black Hills along Rimrock Highway**
- **Scenic Views of Central Black Hills Including Pactola Dam and Crazy Horse Memorial**
- **Tin Mountain and Bull Moose Pegmatite Mines, Custer Region**
- **Harney Peak Granite at Sylvan Lake and the Needles Highway, Custer State Park (Figure 1).**
- **Hot Springs Mammoth Site**
- **Wind Cave**
- **State Game Lodge, Custer State Park**
- **Mount Rushmore Lighting Ceremony**

FIGURE 1 Granite needles near Harney Peak (woodcut in Newton and Jenney, 1880).

Cumulative Miles	
	0.0
0.0	8:00 A.M. Depart Alex Johnson Hotel, Rapid City, S.D. **Turn east on St. Joseph St.**
	0.8
0.8	**East Blvd., turn left (north).**
	0.2
1.0	**Omaha St., turn left (west).** Omaha St. parallels Rapid Creek which flows east from the central Black Hills. We will proceed upstream along the course of Rapid Creek on Omaha St. and then Rimrock Hwy. (Hwy. 44). A major flash flood on Rapid Creek in 1972 caused extensive damage and loss of life. Evidences of the destruction are uncommon but there has been much replanning and reconstruction including the "floodway" along which we are traveling. Some businesses remain but many buildings have been removed (Rahn and others, 1981).
	1.0
2.0	West Blvd. Continue on Omaha St. and Rimrock Hwy. (Hwy. 44).
	0.5
2.5	**STOP 1. RAPID CREEK WATER GAP: CRETACEOUS SANDSTONE HOGBACK** The stratigraphic section exposed to the north is identified in Figure 2 which is obtained from a photograph and related text in Gries and Martin (1985). The Cretaceous sandstone hogback is composed mainly of massive sandstone of the Fall River Formation and the underlying sandstone of the Lakota Formation. The sandstones dip about 20° to the east. The "Dakota Sandstone" aquifer is a major source of water in the Great Plains to the east of this recharge area. The Fall River sandstone cliff above Rapid Creek "is where, in 1876, the first white settlers camped and laid out the streets of Rapid City" (Rahn and others, 1981). **Major features at this stop are the stratigraphy, lithology, and structure of the Cretaceous sandstone hogback.**
	0.3
2.8	Unkpapa Sandstone (Jurassic) and possibly Sundance Formation (Jurassic) in hills at base of the Cretaceous sandstone hogback.
	0.1
2.9	Mountain View Road. **Turn left (south) on Mtn. View Road and continue on Rimrock Hwy. (Hwy. 44).**
	0.2
3.1	Main St. and Jcn. 79 and I90 (Business Loop).
	1.6
4.7	Rapid Creek. "The June 9, 1972 flood was particularly severe at this point, washing many houses away and killing scores of people. The area to the left, originally homes, is now the new Rapid City Municipal Golf Course" (Rahn and others, 1981).
	1.0
5.7	Spearfish Formation (Permian-Triassic) siltstone, sandstone and shale red beds above finely layered limestone of the Minnekahta Limestone. (Permian).
	0.2
5.9	Turnoff to Stavkirke (Stave Church).
	0.2
6.1	Minnelusa Formation (Pennsylvanian-Permian) variegated sandstone, limestone and shale.
	0.3
6.4	Rapid Creek
	0.6
7.0	Rapid Creek
	0.5
7.5	Dark Canyon Rd.
	0.2
7.7	Minnelusa Formation (Pennsylvanian-Permian) overlain by quartzite gravel.
	0.8
8.5	Minnekahta Limestone (Permian)?
	0.2
8.7	Opeche Formation (Permian) red shale and sandstone?
	0.1
8.8	Minnekahta Limestone (Permian)?
	0.4
9.2	**Turn left (south) on dirt road.**
	0.4
9.6	**STOP 2. DARK CANYON: OVERVIEW OF PALEOZOIC STRATIGRAPHIC SEQUENCE.** The canyon formed by entrenched meanders of Rapid Creek exposes Paleozoic strata dipping about 6° to the east. The red-brown Deadwood Formation (Cambrian-Ordovician) of sandstone, dolomite, conglomerate and shale is overlain by the massive gray Madison Formation (Mississippian) which in turn is overlain by the red and yellow Minnelusa Formation (Pennsylvanian-Permian). The overlying erosional surface veneered by Cenozoic sediments results from Laramide uplift followed by deposition of alluvium. Harney Peak (elevation 2201 m) is

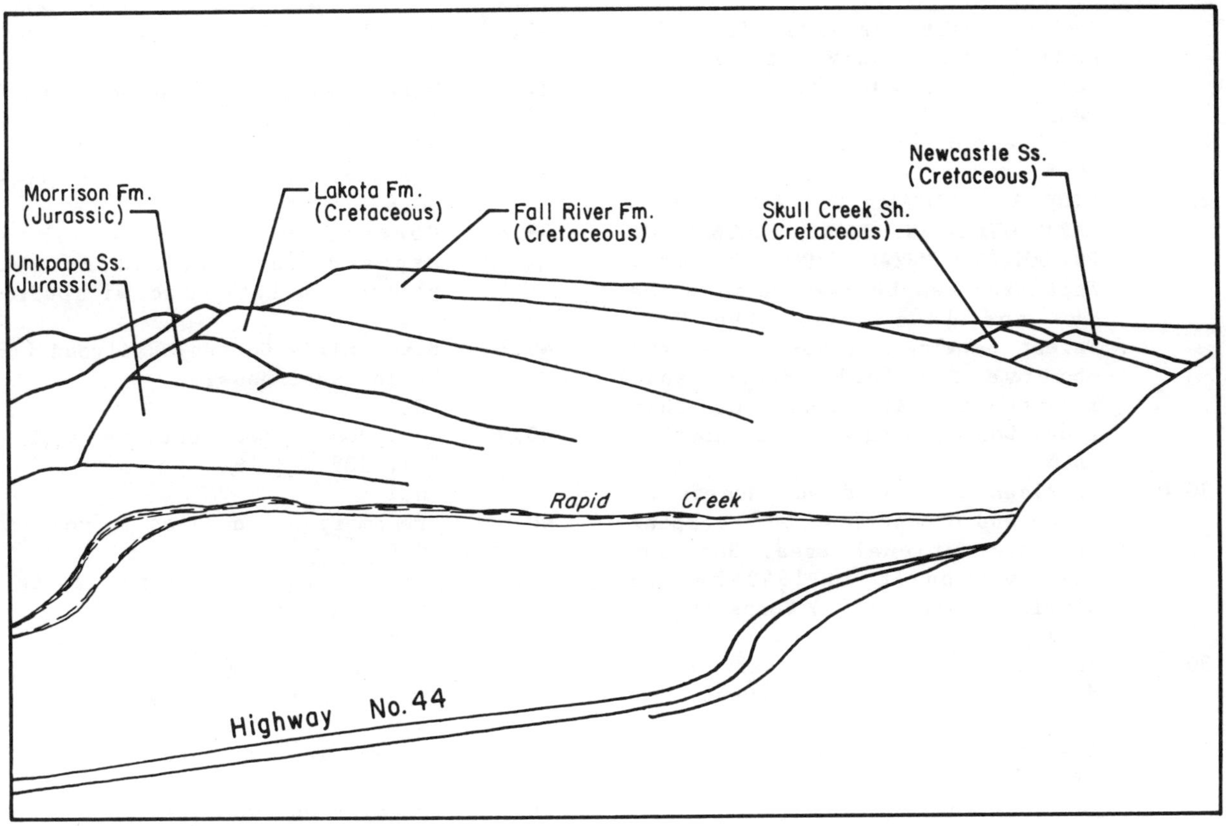

FIGURE 2 Lower Mesozoic stratigraphic section in water gap of Rapid Creek west of Rapid City. Sketch from photograph in Gries and Martin (1985).

visible to the south-southwest, Rahn and others (1980). **Major considerations at this stop are the section of Paleozoic strata and the evidence of deposition, uplift, erosion, alluvial deposition and the present-day erosional surface.**

0.4

10.0 **<u>Return to Hwy. 44 and turn left (west).</u>**

0.3

10.3 Minnelusa Formation (Pennsylvanian) overlying Pahasapa Limestone (Mississippian) with red sediment (terra rossa) at the boundary.

0.1

10.4 Pahasapa Limestone (Mississippian).

0.5

10.9 Pahasapa Limestone (Mississippian)?

0.4

11.3 Enter Black Hills National Forest.

1.0

12.3 Pahasapa Limestone (Mississippian) overlying Englewood Formation (Devonian) pink, clayey limestone.

0.2

12.5 Deadwood Formation (Cambrian-Ordovician).

0.4

12.9 Deadwood Formation (Cambrian-Ordovician).

0.2

13.1 Hisega Rd.

0.1

13.2 Precambrian metasedimentary rock.

0.5

13.7 Road to Thunderhead Falls.

2.5

16.2 Johnson Siding

4.5

20.7 Jcn. Hwy. 385 with Hwy. 44. **<u>Turn left (south) on Hwy. 385.</u>**

1.1

21.8 Exposures of Early Proterozoic metasedimentary rock. Schists contain sulfides and are gold-bearing.

0.6

22.4 North end of Pactola Dam on east side of Pactola Lake. The description of the dam follows that of Rahn and others (1981). The dam was built in 1956 for flood control, irrigation and as a water supply for Rapid City. It is 68.7 m high and 398 m long at the crest. The embankment is 3,288,200 m^3. The

dam is built across sulfide-rich schist and amphibolite in the valley of Rapid Creek.
0.5

22.9 South end of Pactola Dam.
0.5

23.4 **STOP 3. PACTOLA DAM AND PACTOLA LAKE OVERVIEW: EARLY PROTEROZOIC PRECAMBRIAN METASEDIMENTARY ROCKS.** This stop can be used to view the dam and lake, visit the U.S. Forest Service Visitor Center and observe the Early Proterozoic metasedimentary rocks of this area. Continue on Hwy. 385 south.
6.6

30.0 Sheridan Lake Road and access to north end of Sheridan Lake camping and recreational area. Sheridan Lake was built in 1940 by the Civilian Conservation Corps (CCC).
0.7

30.7 Precambrian metasedimentary rock. Major exposures in road cuts.
0.5

31.2 Sheridan Lake and major exposure of Precambrian metasedimentary rocks in road cuts.
0.3

31.5 Sheridan Lake Campground.
0.3

31.8 Precambrian metasedimentary rocks. Large exposures in road cuts.
0.4

32.2 To the left (east) about 30 m up the side of the hill across Spring Creek is the remains of an old aqueduct built in 1880 to carry water 22 km to Rockerville for placer gold mining (Rahn and others, 1981).
1.0

33.2 Jcn. Hwy. 16 with Hwy. 385. Continue on Hwy. 385 and 16 south.
0.6

33.8 Hines sawmill on left, largest in the Black Hills.
1.1

34.9 Precambrian metasedimentary rocks. Large exposures in road cuts.
0.1

35.0 Mitchell Lake built by CCC in 1937.
0.5

35.5 Precambrian metasedimentary rocks. A large recumbent fold can be observed.
0.8

36.3 Harney Ranger Station to right (northwest). Precambrian metasedimentary rocks.
0.5

36.8 Major Lake built in 1937 by CCC.
0.4

37.2 Jcn. Deerfield Rd. in Hill City.
0.4

37.6 Mateen Mine in lithium pegmatite to left (east).
0.8

38.4 Precambrian metasedimentary rocks.
0.6

39.8 Coarse, gold-bearing alluvium exposed in railroad cut with placer tunnel at contact with bedrock.
1.2

40.1 Black Hills Central Railroad (1880 Train) underpass.
0.3

40.5 Jcn. Hwy. 244. Continue south on Hwy. 385 and 16.
0.1

40.6 Precambrian exposures in roadcut.
0.1

40.7 Jcn. Hwy. 87. Continue south on Hwy. 385 and 16.
1.6

42.3 Oreville Campground. Precambrian exposures in roadcuts.
1.0

43.3 Precambrian exposures in roadcuts.
1.4

44.7 Precambrian exposures in roadcuts.
2.0

46.7 Entrance Rd. to Crazy Horse Monument.
0.5

47.2 **STOP 4. CRAZY HORSE MEMORIAL SCENIC VIEW.**
Continue south on Hwy. 385 and 16.
4.0

51.2 Custer Jcn. Hwy. 16 and Hwy. 385 and 16. **Turn right (west) on Hwy. 16.**
0.4

51.6 Jcn. Hwy. 385 (south) and Hwy. 16. Continue west on Hwy. 16.
1.9

53.5 Tourmaline-bearing pegmatite exposed in road cut on right (north).
1.7

55.2 Pleasant Valley Rd.
1.7

56.9 Harry Mills Picnic Ground.
0.3

57.2 Comanche Park Campground.
1.1

58.3 Tin Mountain Rd. (287). **Turn right (north) on Tin Mountain Rd.**
0.1

58.4 Gravel Road. Turn right (east) on gravel road.
0.5

58.9 **STOP 5. TIN MOUNTAIN PEGMATITE MINE.** The Tin Mountain pegmatite is an excellent example of a spodumene-rich, zoned pegmatite of the southern Black Hills. The country rock is garnetiferous biotite schist and amphibolite. The pegmatite has been well exposed by the mining operation. It has a wall zone of albite, quartz and muscovite enclosing a zone of perthite, quartz, albite and minor muscovite and a core of quartz, spodumene and mica with minor albite, beryl and amblygonite. Other pegmatite minerals are present in minor amounts. Large, log-like crystals of spodumene are spectacular and fine examples of other pegmatite minerals may be collected. **The major feature of this stop is the well-exposed zoned pegmatite containing large spodumene crystals and other pegmatite minerals.**

0.0

Return west on gravel road.

0.5

59.4 Jcn. Tin Mountain Rd. Turn left (south) on Tin Mtn. Rd.

0.1

59.5 Jcn. Hwy. 16. Turn left (east) on Hwy. 16 to Custer.

7.1

66.6 Jcn. Hwy. 385 (north) and 16 with Hwy. 16. **Continue east to Hwy. 16A in downtown Custer.**

0.3

66.9 Stop Light. Jcn. route to 793 south.

1.2

68.1 Jcn. Hwy. 89 with 16A. **Turn left (north) on Hwy. 89.**

5.9

74.0 Jcn. Hwy. 87 (Needles Highway) with Hwy. 89. **Turn right (east) to Sylvan Lake and Needles Highway.**

0.1

74.1 Entrance to Custer State Park. **Pay toll and continue on Hwy. 87.**

0.1

74.2 **STOP 6. SYLVAN LAKE AND THE NEEDLES: SCENIC VIEWS AND GEOLOGY OF THE HARNEY PEAK GRANITE. LUNCH STOP.**

The Harney Peak Granite, with a 1.7 Ga age, forms a scenic landscape in this area of Custer State Park. The rock consists of gray to pink, coarse-grained to pegmatitic, muscovite granite and associated pegmatite. Major minerals are quartz, microcline perthite, albite, and muscovite with accessory biotite tournamline and garnet. Textural variations and intimately associated metasedimentary rock suggest that the granite is made up of a series of smaller intrusions separated by masses of country rock or has formed by other processes such as partial melting in place or metamorphic/metasomatic processes. **The major considerations of this stop are the characteristics of the Harney Peak Granite, its unusual weathering forms, and its origin.**

3.0

77.2 Jcn. Hwy. 89 with Hwy. 87. **Turn left (south) on Hwy. 89.**

5.9

83.1 Jcn. Hwy. 16A with Hwy. 89. **Turn right (west) on Hwy. 16A.**

1.2

84.3 Stop Light. **Turn left (south) and follow truck route east and south to join 793 at Park.**

0.7

85.0 **Park Street becomes 793 and continues south.**

10.0

95.0 Jcn. 343 with 793. **Turn left (east) on 343 toward French Creek.**

2.7

97.7 Gravel Road. **Turn right (south) on gravel road (Bank 34A).**

0.4

98.1 **STOP 7. BULL MOOSE PEGMATITE MINE.**

Park at road to right (west) just north of home of James Bland. Follow gravel road west and south to the pegmatite mine. The Bull Moose pegmatite is an irregularly exposed body rich in rose quartz. The wall rock is sillimanite-grade muscovite-biotite-sillimanite schist and quartzite. Inclusions of schist are present in the outer part of the crudely zoned pegmatite which grades inward from moderately coarse grained muscovite and perthite to a coarser grained, perthite-rich zone to a core rich in very coarse grained muscovite crystals up to 30 cm across and rose quartz masses up to 2 m across (Feldmann and Heimlich, 1980). Tourmaline, columbite-tantalite and other pegmatite minerals are common. **Major considerations at this stop are the general characteristics of the Bull Moose pegmatite, its wall rock, and the occurrence of rose quartz and other pegmatite minerals.**

0.0

98.1 **Return to gravel road and turn left (west).**

3.1

101.2 Jcn. 793 with 343. **<u>Turn left (south) on 793.</u>**
3.0

104.2 Jcn. 336.
4.0

108.2 Jcn. 385 with 793. **<u>Turn left (south) on 385.</u>**
4.0

112.2 Pringle. Jcn. Hwy. 89 with Hwy. 385. Continue east on Hwy. 385.
5.8

118.0 Jcn. Hwy. 87 with Hwy. 385.
0.5

118.5 North Entrance to Wind Cave. **<u>Continue on Hwy. 385 south.</u>**
2.2

120.7 South Entrance to Wind Cave.
1.6

122.3 Roadside marker - concretions.
0.4

122.7 Roadside marker - buffalo and antelope.
1.1

123.8 South boundary of Wind Cave National Park.
0.2

124.0 Exposure of Minnekahta Limestone (Permian).
1.8

125.8 Trout Haven Rd. Exposure of Spearfish Formation (Permian-Triassic).
3.8

129.6 Hot Springs.
0.4

130.0 Exposures of Spearfish Formation (Permian-Triassic) in road cuts and along river.
0.1

130.1 Evan's Plunge, the major mineral-bath pool in Hot Springs.
0.5

130.6 Exposures of conglomeratic sandstone of Fall River Formation (Cretaceous).
0.4

131.0 Jcn. Hwy. 18 with Hwy. 385.
0.8

131.8 Jcn. Hwy. 18 Bypass (west) and Hwy. 71 with Hwy. 385. **<u>Turn right (southwest) on Hwy. 18 Bypass and Hwy. 71.</u>**
0.7

132.5 Jcn. Hwy. 18 Bypass (west) with Hwy. 71 (south). **<u>Turn right (west) on Hwy. 18 Bypass.</u>**
0.4

132.9 **STOP 8. HOT SPRINGS MAMMOTH SITE.** The Mammoth Site of Hot Springs, South Dakota is an incorporated, nonprofit organization dedicated to the preservation, through in-situ display, of this major Pleistocene paleontological find. The organization's public information is used here to summarize characteristics of the site. "More than 26,000 years ago, large Columbian mammoths were trapped and died in a spring-fed pond near what is now the southwest edge of Hot Springs, South Dakota. As many as 100 mammoths may have died here, experts say.

In 1974, during excavation for a housing project, earth-leveling equipment revealed the white bones and tusks of these huge animals. Local townspeople summoned assistance from the geology department of nearby Chadron (Nebraska) State College, and a group was formed with the initial purpose of protecting the discovery."

"The model of the geologic environment proposed for the site at the time of accumulation of the mammoth and associated fauna is a sinkhole. Walled by Spearfish shale and floored by a spring fed pond, the depression contained a standing body of water. The sediments which enclose the faunal remains are a buff to gray sand, silt, and clay....supplied to the deposit either by artesian pressure from underlying geologic units, or by over bank flow into the sink, or both."

Remains of the large Columbian mammoth are the most impressive fossils of the site. Remains of the giant short-faced bear, camel, wolf, coyote, a large carnivorous bird, rodents, fish, gastropods, and a pelecypod have also been identified. **Major considerations at this stop are the in-situ display of Pleistocene mammoths and other fossils and the geological history of their preservation.**
0.2

133.1 **<u>Return to Hwy. 18 Bypass and turn left (east).</u>**
1.1

134.2 Jcn. Hwy. 385 with Hwy. 18 Bypass and Hwy. 71. **<u>Turn left (northwest) on Hwy. 385 (north).</u>**
11.1

145.3 South Turnoff to Wind Cave. **<u>Take turnoff to visitor center.</u>**
0.7

146.0 **STOP 9. WIND CAVE.** Located in Wind Cave National Park, Wind Cave is developed in the Pahasapa Limestone (Mississippian) and is noted for its well developed calcite boxwork in 70 km of passages in a rectilinear network beneath an area of 1.8 km^2 (Bakalowicz and others, 1987). Jewel Cave, west of Custer, and Wind Cave are the world's fourth and tenth longest known caves with the orientation of their passages (Figure 3) showing clear evidence of downdip movement of groundwater during their formation. Wind Cave contains permanent lakes at its southeastern lower end. Bakalowicz and others (1987) believe that Wind Cave was formed largely in the Quaternary by regional, CO_2-rich, ascending, thermal waters envisioned as related to a basement heat source. The remarkable wall and ceiling boxwork is believed to result from solution by CO_2-rich waters combined with near-contemporaneous deposition of calcite after degassing. **Major considerations at this stop are the rectilinear passages, boxwork and other features of Wind Cave and their geological origin.**

0.0

146.0 **Continue north on park road.**

1.0

147.0 Jcn. Hwy. 385 at N entrance to Wind Cave. **Follow Hwy. 385 northwest.**

0.4

147.4 Jcn. Hwy. 87 with Hwy. 385. **Turn right (north) on Hwy. 87.** Roadside marker-prairie dogs.

0.5

147.9 Exposure of Pahasapa Limestone (Mississippian).

0.8

148.7 Precambrian pegmatite outcrop at bridge.

0.3

149.0 Roadside marker-Wind Cave.

0.8

149.8 Roadside marker-pegmatite. Exposure of Precambrian pegmatite.

0.3

150.1 Exposures of Precambrian metasedimentary rocks and pegmatite.

1.0?

151.1? Roadside marker-East meets West.

1.0?

152.1 Roadside marker-silent struggle of forest and prairie. Prairie grazing land of bison herd.

2.4

154.5 Entrance to Custer State Park.

4.5

159.0 Valley grazing land of bison herd.

0.6

159.6 Entrance to park area requiring permit.

0.9

160.5 Blue Bell Lodge.

3.1

163.6 Turnout to view large area of 1988 forest fires.

1.6

165.2 Jcn. Hwy. 87-Hwy. 16A. **Turn right (east) on Hwy. 16A and Hwy. 87.**

0.8

166.0 Legion Lake and Legion Lake Lodge.

0.6

166.6 Jcn. Hwy. 16A with Hwy. 87. **Continue east on Hwy. 16A.**

5.5

172.1 **STOP 10 STATE GAME LODGE AND VISITOR CENTER: DINNER**

0.0

172.1 **Follow western route through Custer or eastern route through Hermosa to Mount Rushmore.**

----- **STOP 11 MOUNT RUSHMORE NATIONAL MEMORIAL: LIGHTING CEREMONY**

----- **Return to Alex Johnson Hotel in Rapid City.**

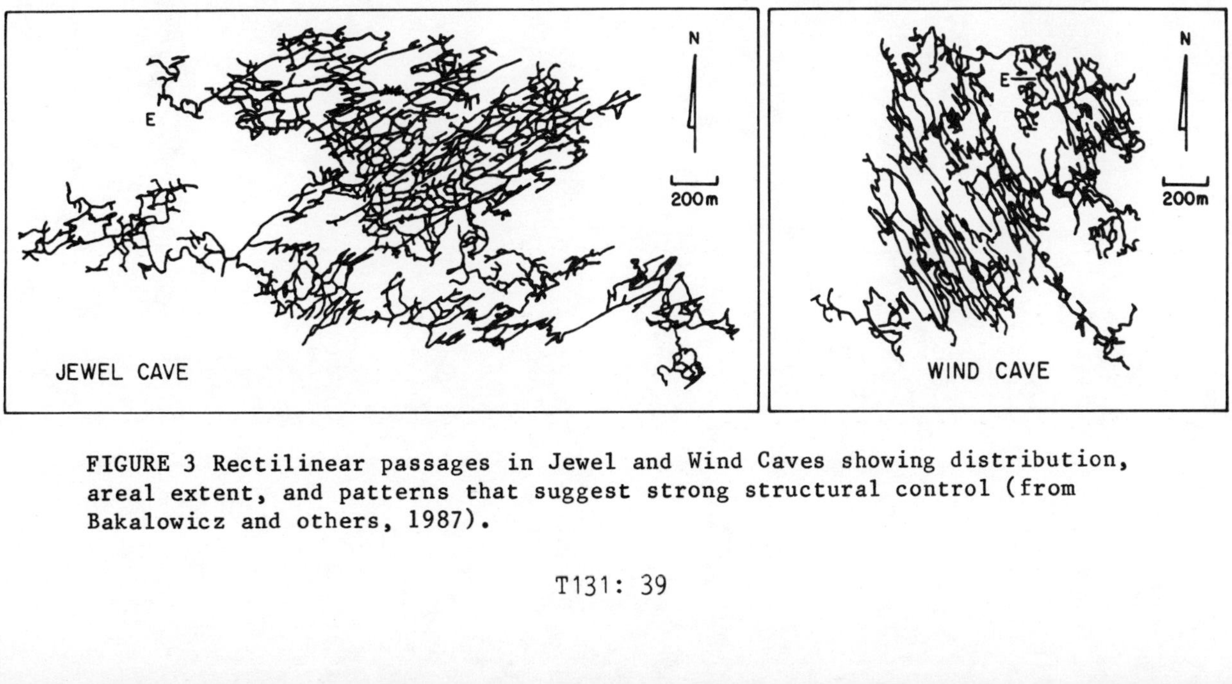

FIGURE 3 Rectilinear passages in Jewel and Wind Caves showing distribution, areal extent, and patterns that suggest strong structural control (from Bakalowicz and others, 1987).

REFERENCES

Bakalowicz, M.J., Ford, D.C., Miller, T.E., Palmer, A.N., Palmer, M.V., 1987, Thermal genesis of dissolution caves in the Black Hills, South Dakota: Geological Society of America Bulletin, v. 99, p. 729-738.

Gries, J.P. and Martin, J.E., 1985, Composite outcrop section of the Paleozoic and Mesozoic Strata in the Black Hills and surrounding areas, <u>in</u> Rich, F.J., ed., Geology of the Black Hills, South Dakota and Wyoming: Field Trip Guidebook, 2nd Ed., for G.S.A. Rocky Mountain Section Meeting, American Geological Institute, p. 261-292.

Rahn, P.H., Bump, V.L. and Steece, F.V., 1981, Engineering geology of the central Black Hills, South Dakota, Field Trip 3A: <u>in</u> Rich, F.J., ed., Geology of the Black Hills, South Dakota and Wyoming: Feild Trip Guidebook for G.S.A. Rocky Mountain Section Meeting, American Geological Institute, p. 135-153.

FIELD GUIDE DAY 3: GEOLOGY OF THE BADLANDS REGION AND THE MESOZOIC-PALEOZOIC ROCKS OF BOULDER CANYON

Frank R. Karner
Department of Geology and Geological Engineering and
North Dakota Geological Survey
University of North Dakota, Grand Forks

Stanley F. White
Maxus Energy Corporation, 717 North Harwood St.
Dallas, Texas

DAY 3: Monday, July 3, 1989

- **Badlands National Park**
- **Wall Drug**
- **Mesozoic-Paleozoic Geology from Sturgis to Deadwood on Boulder Canyon Highway**

FIGURE 1 The Cambrian-Ordovician Deadwood Formation, Deadwood (photo in O'Harra, 1902).

Cumulative Miles	
	0.0
0.0	8:00 A.M. Depart Alex Johnson Hotel, Rapid City, S.D. **Turn east on St. Joseph St.**
	0.8
0.8	**East Blvd., turn left (north) and follow north and northwest on East Blvd. and Bus. 90.**
	2.5
3.3	Jcn. Bus. 90 and Interstate 90 at Exit 60. **Turn east on I-90.**
	5.6
8.9	Exit 66. Ellsworth Air Force Base and Air and Space Museum.
	28.4
37.3	View of Badlands peaks ahead in distance.
	1.0
38.3	Exposures of gray, marine shale of Upper Cretaceous Pierre Shale.
	2.3
40.6	Exit 99 to Wasta, S.D.
	0.5
41.1	Cheyenne River
	0.3
41.4	Rest Area.
	0.3
41.7	Exposure of Pierre Shale (Cretaceous) with scattered fossiliferous concretions often containing abundant molluscan remains.
	3.4
45.1	Exposure to left (south) of Pierre Shale (Cretaceous) with prominent light layers of bentonite (?).
	7.2
52.3	Exit 109 to Wall, S.D.
	1.2
53.5	Exit 110 to Wall and Jcn. with the Badlands Loop Road, Hwy. 240. **Continue on I-90.**
	1.7
55.2	Exit 112 Jcn. Hwy. 14.
	2.5
57.7	Exposures of strata of Badlands region.
	3.1
60.8	Views of Badlands topography to right (south).
	13.0
73.8	Exit 131 to Interior, S.D. **Turn right (south) to Badlands National Monument and Cedar Pass Visitor Center.**
	2.8
76.6	Cactus Flat, Prairie Homestead, and northeast entrance to Badlands National Park.
	5.4
82.0	**STOP 1. BADLANDS NATIONAL MONUMENT: CEDAR PASS VISITOR CENTER.** Exhibits and information in the visitor center provide an excellent background on Badlands geology and natural history. The Cedar Pass Lodge is nearby. **Continue through Badlands following Badlands Loop (Hwy. 240) west.**
	0.0
82.0 TO 104.0	**STOP 2. BADLANDS LOOP: GEOLOGY OF BADLANDS AT NUMEROUS OVERLOOK SITES.** The Badlands, or les mauvaises terres a traverser of French Canadian trappers, or mako sica of Native American Dakota Indians and their neighbors, are a spectacularly eroded sequence of siltstones, claystones, sandstones, volcanic ash and associated strata. The geologic units exposed in the Badlands are described briefly in Karner and Patelke (1989) in this guidebook, in additional literature, made available on the trip, and by interpretive information at numerous overlook sites along the Badlands Loop. The fossiliferous Chadron and Brule Formations of the White River Group contain extensive mammalian and associated remains. These formations provide a detailed picture of Oligocene life in a broad depositional and erosional plain lying to the east of the Laramide uplifts of the Black Hills and Rocky Mountains. The overlying Oligocene-Miocene Arikaree Group contains similar strata only partially preserved in the Badlands.
	22.0
104.0	Jcn. Hwy. 240 with Hwy. 590. **Turn right (north) on Hwy. 240 to Wall, S.D.**
	8.0
112.0	Wall, S.D.
	0.2
112.2	**STOP 3. WALL DRUG: LUNCH STOP.** Lunch, sightseeing and shopping at Wall Drug. **Return to I-90.**
	0.2
112.4	**Turn right (west) on I-90 to Rapid City.**
	49.2
161.6	Rapid City. Exit 61.
	6.0
167.6	Exit 55. **Continue north and west on I-90.**
	4.2
171.8	Exit 51. Jcn. 79 and Bus. 90 with I-90.
	3.0
174.8	Exit 48. Stage Barn Canyon Rd. to

Stage Barn Crystal Cave.
2.3

177.1 Exit 46. Elk Creek Rd.
0.9

178.0 Piedmont, S.D.
0.7

178.7 Exit 44. Piedmont Rd. Wonderland Rd.
0.5

179.2 Little Elk Creek. Major Archean exposures to west in Little Elk Creek Canyon.
2.6

181.8 Rest Area.
1.3

183.1 Exit 40. Tilford-Bethlehem Rd.
9.7

192.8 Sturgis, S.D. Exit 30. Jcn. 14A with I-90. **Turn left (west) on 14A to Deadwood via Boulder Canyon.**
1.1

193.9 Geologic contact between gray limestone of the Permian Minnekahta Limestone and the overlying red beds of the Spearfish Formation (Permian-Triassic). As we proceed southwest in Boulder Canyon we will pass through a section of the Paleozoic sedimentary strata of the Black Hills.
0.4

194.3 Exposure of variegated sandstone, limestone and shale of the Minnelusa Formation (Pennsylvanian-Permian).
0.9

195.2 Exposure of massive limestone of Pahasapa Formation (Mississippian).
0.8

196.0 Trucker Bobs.
0.1

196.1 **STOP 4. BOULDER CANYON: RAINBOW ARCH SECTION OF MINNELUSA FORMATION.** Stop in parking area and walk upstream to examine this section of the Minnelusa. This description is from Gries and Steece (1985). About 121 m of Minnelusa is exposed in this relatively thin section which dips steeply westward into the Boulder Park syncline. A paleosol is present at the contact of the Madison Limestone with the overlying Minnelusa. The Minnelusa is dominantly sandstone at this location and is unconformably overlain by red beds of the Opeche Shale. The Opeche is overlain unconformably by the Minnekahta Limestone which is overlain unconformably by red beds of the Spearfish Formation as observed opposite the entrance to Boulder Park, 0.5 miles upstream.
0.5

196.6 Entrance to Boulder Park.
0.1

196.7 Boulder Canyon Country Club entrance. The Spearfish Formation forms the valley floor.
6.7

203.4 **Turn right just before crossing Whitewood Creek.**
0.4

203.8 **STOP 5. WHITEWOOD CANYON: GEOLOGIC SECTION OF WHITEWOOD LIMESTONE-ENGLEWOOD FORMATION-PAHASAPA LIMESTONE.** This Ordovician-Mississippian section is exposed in the canyon wall at this location.
0.3

204.1 **STOP 6. WHITEWOOD CANYON: DEADWOOD FORMATION.** We have just crossed a normal fault with approximately 100 m of throw. The Cambrian-Ordovician Deadwood Formation is exposed here.
0.2

204.3 **Turn around and return to Hwy. 14A.**
0.9

205.2 **Turn right (southwest) on Hwy. 14A.**
0.7

205.9 Jcn. Hwy. 85 with 14A. **Continue south and west on Hwy. 14A and Hwy. 85.**
0.1

206.0 **STOP 7. NORTHEAST DEADWOOD: DEADWOOD FORMATION, WINNIPEG FORMATION AND WHITEWOOD LIMESTONE.** In this highway junction roadcut described by Gries and Steece (1985) an especially thick section of Scolithus sandstone of the Upper Deadwood (Ordovician) is overlain by shale of the Winnipeg Formation (Ordovician) and the Whitewood Limestone (Ordovician).
0.5

206.5 **Turn left, cross Whitewood Creek and proceed up steep grade.**
0.2

206.7 **Turn right onto unpaved road.**
0.2

206.9 **STOP 8. WHITEWOOD CANYON: ANGULAR UNCONFORMITY-EARLY PROTEROZOIC SCHIST OVERLAIN BY DEADWOOD FORMATION. Return to Hwy. 14A.**
0.4

207.3 **Turn left (southwest) on Hwy. 14A and proceed to Franklin Hotel in downtown Deadwood.**
0.4

207.7 Franklin Hotel.

REFERENCES

Gries, J.P. and Steece, F.V., 1985, Paleozoic and Mesozoic stratigraphy of the northern Black Hills, Field Trip 4: in Rich, F.J., ed., Geology of the Black Hills, South Dakota and Wyoming: Field Trip Guidebook, 2nd Ed., for G.S.A. Rocky Mountain Section Meeting, American Geological Institute, p. 175-180.

Karner, F.R. and Patelke, R.L., 1989, General geology of the Black Hills and Bear Lodge Mountains in Karner, F.R., ed., Devils Tower--Black Hills alkalic igneous rocks and general geology: 28th International Geological Congress Field Trip Guidebook T131, p. 7-14.

O'Harra, C.C., 1902, The mineral wealth of the Black Hills: South Dakota School of Mines and Technology Bulletin 6, 88 p.

ACTIVITIES GUIDE DAY 4: UNITED STATES INDEPENDENCE DAY AND GEOLOGICAL AND CULTURAL FEATURES OF THE DEADWOOD REGION

Frank R. Karner
Department of Geology and Geological Engineering and
North Dakota Geological Survey
University of North Dakota, Grand Forks

DAY 4: Tuesday, July 4, 1989

Possible Optional Activities: Consult with field trip leader.

8:30 A.M. - 12:00 P.M.	**Bear Butte State Park hiking and sightseeing.**
9:00 A.M. - 12:00 P.M. or 2:00 P.M. - 5:00 P.M.	**Visit Black Hills Mining Museum, Lead; Broken Boot Gold Mine, Deadwood; and, Boot Hill Cemetery, Mt. Moriah.**
7:00 P.M. - 11:00 P.M.	**Attend Passion Play, Spearfish.**
9:00 P.M. - 10:00 P.M.	**Observe fireworks display, Deadwood.**

FIGURE 1 Bear Butte from the north with ridge of Pahasapa Limestone to left and slope of Sundance and Spearfish Formations to right. Photo from N. H. Darton, 1909, Geology and water resources of the northern portion of the Black Hills and adjoining regions in South Dakota and Wyoming: U.S. Geological Survey Professional Paper 65.

FIELD GUIDE DAY 5: GEOLOGY OF PRECAMBRIAN-CENOZOIC ROCKS OF THE DEADWOOD-LEAD REGION, TERRY PEAK, AND SPEARFISH CANYON

Frank R. Karner
Department of Geology and Geological Engineering and
North Dakota Geological Survey
University of North Dakota, Grand Forks

Stanley F. White
Maxus Energy Corporation, 717 North Harwood St.
Dallas, Texas

DAY 5: Wednesday, July 5, 1989

- **Precambrian-Cenozoic Geology in the Deadwood-Lead Region**
- **Open Pit at the Homestake Gold Mine, Lead**
- **Terry Peak Scenic Overview**
- **Spearfish Canyon Paleozoic Strata and Cenozoic Igneous Intrusions**
- **Introduction to the Geology of the Bear Lodge Mountains**

FIGURE 1 Cliff of the Mississippian Pahasapa Limestone, Hellgate Gulch, Spearfish Canyon (photo in Darton, 1909).

Cumulative Miles

0.0

0.0 8:00 A.M. Depart Franklin Hotel, Deadwood, S.D. **Follow Pine St. to Hwy. 14A.**

0.1

0.1 **Turn right (southwest) on Hwy. 14A to Central City and Lead.** This segment includes information from a road log by Lisenbee and others (1981).

0.5

0.6 Exposure of Early Proterozoic Grizzly Formation with Cenozoic dike cut by small fault.

0.1

0.7 Exposure of Grizzly and underlying Flagrock Formation (Early Proterozoic).

0.2

0.9 Broken Boot "gold mine" was actually mined for pyritic graphite schist used as flux in the Deadwood smelter. Folded metasedimentary rocks.

0.2

1.1 Exposure of Flagrock and underlying Northwestern Formation (Early Proterozoic).

0.4

1.5 Exposures of Northwestern and underlying Ellison Formation (Early Proterozoic).

0.2

1.7 **STOP 1. DEADWOOD GULCH HISTORIC MARKER: DEADWOOD PLACER GOLD DISCOVERY AND DEFORMED METASEDIMENTARY ROCKS.**

Two separate, initial gold discoveries were in Deadwood Gulch in November 1875. By January 1876, 50 miners were working Deadwood Gulch. By April 1876 the town of Deadwood was laid out and a city resembling "a heap of lemon boxes propped up on broomsticks" was formed (Lisenbee and others, 1981).

Lisenbee and others (1981) recognize four types of gold-bearing deposits in the Deadwood Creek basin: (1) the Precambrian Homestake Formation; (2) Cambrian placers at the base of the Deadwood Formation; Tertiary mineralization related to igneous intrusions and (4) Quaternary placers.

Note the Historic Marker and the deformed metasedimentary rocks of the Ellison, Homestake and Poorman Formations in the canyon.

0.0

1.7 **Continue on Hwy. 14A.** Exposures of Poorman Formation (Early Proterozoic) for next mile.

0.1

1.8 Bobtail Gulch on left. Exposure on right of quartz-bearing porphyry of Mt. Theodore Roosevelt stock of Cenozoic age.

0.2

2.0 Rhyolite dikes in Poorman Formation.

0.1

2.1 Central City.

0.3

2.4 Turn-off to Terraville.

0.1

2.5 Folded Poorman, Homestake and Ellison Formations.

1.4

3.9 Jcn. Hwy. 14A with Hwy. 85 in Lead, S.D. **Turn east on Hwy. 85 to Homestake open cut area and Mooney Park.**

1.0

4.9 **STOP 2. HOMESTAKE OPEN CUT: OVERVIEW OF HOMESTAKE GEOLOGY.**

The open cut illustrates many of the features of the geology of the Homestake Mine. Geologically, the gold is contained in a sequence of Early Proterozoic metasedimentary rocks. The lower unit is the barren Poorman Formation a dark gray, ankeritic phyllite. The overlying, ore-bearing, Homestake Formation is an iron formation consisting of cummingtonite-sideroplessite (iron-magnesium carbonate)-quartz schist. The upper Ellison Formation consists of phyllites and schists with abundant beds and lenses of dark, fine-grained quartzite.

Gold is disseminated in the extensively folded Homestake Formation which has also been intruded by Cenozoic rhyolite dikes. The ore occurs in chloritized bodies formed by selective mineralization and replacement in elongate zones parallel to the axes of plunging cross folds and thought to be related to zones of increased permeability. Minerals formed during the ore-forming stages include quartz, chlorite and arsenopyrite in the first stage; quartz, ankerite and pyrrhotite in the second stage; and pyrrhotite in the third stage. Gold formed in the early stages and is associated with arsenopyrite.

0.0

4.9 **Return west on Hwy. 85.**
1.0
5.9 Jcn. Hwy. 14A and Hwy. 85. **Follow Hwy. 14A and Hwy. 85 southwest.**
1.0
6.9 Jcn. Rd. in Nevada Gulch to Terry Peak chairlift. **Turn right (northwest) on Terry Peak road.**
0.2
7.1 Jcn. gravel road in Fantail Gulch. **Turn left (west) on gravel road in Fantail Gulch.**
0.8
7.9 **Turn left on to track.**
0.3
8.2 **STOP 3. FANNIE ORE BODY: MINERALIZED LOWER DEADWOOD FORMATION.**
The following description is from Lisenbee and others, (1988). "A road cut at the northern end of the Fannie ore body exposes a breccia zone consisting of angular clasts of Precambrian schist and milky quartz, and feldspar phenocrysts in a dark green to black, fine-grained, siliceous matrix.

The breccia zone is sandwiched by trachyte porphyry which forms a sill in the immediate vicinity.... The exposed upper contact of the zone is sub-horizontal. Slickensides occur near the contact, and small lenses of breccia are observed in the immediately overlying porphyry. Breccia zone development likely resulted from entrainment of Precambrian country rock along dike and sill margins during intrusion of the porphyry.

The trachyte porphyry sill rests on Precambrian schist and underlies the basal quartzite of the Deadwood Formation. A feeder dike is postulated to occur at depth based on drill intercepts. Dike emplacement presumably occurred along steep, northerly fractures which were conduits for fluids producing the mineralization in the overlying sedimentary units."
0.0
8.2 **Return to Hwy. 14A and Hwy. 85 via Fantail Gulch and Nevada Gulch.**
1.3
9.5 Jcn. Hwy. 14A and Hwy. 85. **Turn right (southwest) on Hwy. 85 and Hwy. 14A.**
0.5
10.0 Dikes of Cenozoic Cutting stock in Northwestern Formation.
0.7
10.7 **STOP 4. SUGAR LOAF LACCOLITH: PHONOLITE-TRACHYTE SILLS IN DEADWOOD FORMATION.**
Described in Lisenbee and others (1988) the exposure includes sills of phonolite-trachyte representing the base of the laccolith in the Deadwood Formation (Cambrian-Ordovician).
Continue on Hwy. 14A and 85.
2.8
13.5 **Jcn. Road to Terry Peak Summit. Turn right (north) on Terry Peak Rd.** Road crosses Deadwood Formation and the Cenozoic Terry Peak laccolith of phonolite intruded into the Deadwood.
3.4
16.9 **STOP 5. TERRY PEAK: OVERVIEW. LUNCH STOP.**
View the surrounding Black Hills and adjacent regions. **Return to Hwy. 14A and 85.**
3.3
20.2 **Turn right west on Hwy. 14A and Hwy. 85.** Exposures of Cambrian-Mississippian strata for next several miles.
2.6
22.8 Cheyenne Crossing. **Turn right (north) on Hwy. 14A to Spearfish Canyon.**
1.6
24.4 Cenozoic phonolite.
1.0
25.4 Pahasapa Limestone forms the Canyon rim.
1.7
27.1 Savoy. Latch String Inn.
1.7
28.8 Iron Creek. Narrow gauge railroad cut visible on canyon wall across Spearfish Creek.
0.4
29.2 TO 35.3 **STOP 6. SPEARFISH CANYON: PALEOZOIC STRATA, CENOZOIC PHONOLITE.**
Visit several locations to examine exposures of the lower and middle Paleozoic section of the Black Hills including the Cambrian-Ordovician Deadwood Formation, the Ordovician Winnipeg Formation and Whitewood Limestone, the Devonian Englewood Formation and the Mississippian Pahasapa Limestone. Cenozoic phonolitic intrusions are also present in this interval. Spearfish Creek has been used extensively as a source of water, diverted through a tunnel system, for the mining operations in Lead. The

stream is also used to generate hydroelectric energy.
6.1

35.3 **STOP 7. BRIDAL VEIL FALLS: CENOZOIC PHONOLITE INTRUSION.** Bridal Veil Falls has formed at the junction of Rubicon Gulch with Spearfish Canyon. Examine the canyon walls which contain exposures of a phonolite-nepheline syenite laccolithic intrusion
2.9

38.2 Leaving Black Hills National Forest.
1.0

39.2 Contact of Pahasapa Limestone and overlying Minnelusa Formation (Pennsylvanian-Permian).
1.4

40.6 Contact of Permian Minnekahta Limestone and the overlying Permian-Triassic Spearfish Formation.
0.2

40.8 Spearfish golf course.
0.4

41.2 Jcn. 14A with Bus. I-90, Main St in Spearfish, S.D. **Turn left (northwest) on Main St.**
1.1

42.3 Jcn. Main St. with Jackson Blvd. **Turn right (east) on Jackson Blvd. to I-90.**
0.8

43.1 Jcn. I-90 at Exit 12. **Turn left (west) on I-90 to Beulah and Sundance, Wyoming.** Spearfish Peak to south.
1.0

44.1 Sundance Formation (Jurassic) to right (northeast) with the underlying Gypsum Springs Formation (Jurassic) and redbeds of the Spearfish Formation (Permian-Triassic).
1.0

45.1 Exit 10. Spearfish Creek and "red valley" or "red racetrack" underlain by Spearfish Formation.
0.4

45.5 Crow Peak to left (south). To the west the red valley of the Black Hills and broader areas of exposure of the Spearfish Formation are bounded by uplifted ridges and plateaus underlain by the Minnekahta Limestone (Permian). The Bear Lodge Mountains are on the skyline to the west.
3.5

49.0 TO 54.0 **STOP 8. MINNEKAHTA LIMESTONE - OPTIONAL.** Low amplitude folds and normal faults in the Minnekahta Limestone for the next several miles.
6.0

55.0 Wyoming-South Dakota state line. Crow Peak and Cement Ridge can be seen to the south.
1.9

56.9 Sand Creek. Beulah, Wyo.
3.3

60.2 Sinkholes in Spearfish Formation.
2.1

62.3 Jcn. Hwy. 111 to Aladdin, Wyo.
6.5

67.8 A number of landmarks can be seen from this area including: Black Buttes; Green Mountain; Inyan Kara Mountain; Sundance Mountain; Warren Peaks and Sheep Mountain.
0.9

68.7 Moskee Rd.
3.0

71.7 Sundance, Wyoming. Exit 188. Sugar Loaf Mountain, a small volcaniclastic cone to the north.

73.1 Exit 187. **Follow exit to Sundance and Apache Motel.**

REFERENCES

Darton, N.H., 1909, Geology and water resources of the northern portion of the Black Hills and adjoining regions in South Dakota and Wyoming: U.S. Geological Survey Professional Paper 65, 105 p.

Lisenbee, A., Karner, F., Fashbaugh, E., Halvorson, D., O'Toole, F., White, S., Wilkinson, W., and Kirchner, J., 1981, Geology of the Tertiary intrusive province of the northern Black Hills, South Dakota and Wyoming, Field Trip 2, in Rich, F.J., ed., Geology of the Black Hills, South Dakota and Wyoming: Field Trip Guidebook for G.S.A. Rocky Mountain Section Meeting, American Geological Institute, 221 p.

Lisenbee, A.L., Paterson, C.J., Gries, J.P., Nelson, G. Emanual, K., Walsh, J., Backens, M, and Duex, T., 1988, Tertiary igneous intrusions and related gold mineralization, Northern Black Hills, South Dakota: Wyoming Geological Association Field Trip (Preliminary Manuscript).

IGC FIELD TRIP 131: EOCENE IGNEOUS ACTIVITY AND RELATED METASOMATIC AND HYDROTHERMAL EVENTS, BEAR LODGE MOUNTAINS, CROOK COUNTY, WYOMING

Gordon A. Jenner
Exxon Corporation, Midland, Texas

INTRODUCTION

In this paper the petrology and geochemistry of the alkalic-carbonatitic complex in the Bear Lodge Mountains is examined to investigate a complex sequence of early Cenozoic igneous, metasomatic, and hydrothermal events.

Fieldwork was done during the summer of 1982. Samples were analyzed at the Natural Materials Analytical Laboratory (NMAL) at the University of North Dakota and at the Isotope Laboratory (IL) at the University of Utah. Special thanks are extended to Dr. Robert Stevenson (NMAL) and Dr. John Bowman (IL) for assistance and generous provision of laboratory facilities. Jenner (1984) provides a detailed description of analytical methods.

GEOLOGIC SETTING

The Bear Lodge Mountains are at the western end of a series of early Cenozoic igneous centers that trend N 75 W across the northern Black Hills as shown in Figure 2 in the paper by Karner and Patelke (1989) in this volume. These small late Paleocene and Eocene intrusive complexes crosscut the major structural features of the Black Hills postdating the major episodes of regional uplift (Shapiro, 1971; Lisenbee, 1978). Radiometric dates from several of the centers range from 38 to 59 Ma (McDowell, 1971; Hill and others, 1975; Staatz, 1983).

The core of the Bear Lodge uplift consists of a central 9 x 4 km dome of subvolcanic igneous rocks (Figure 1). Smaller igneous plugs, dikes, and sills occur within upturned Paleozoic and Mesozoic strata around the flanks of the uplift. Although Staatz (1983) reports the occurrence of extrusive flows and pyroclastic deposits, the majority of the igneous bodies are intrusive in origin. Rock types include latite, trachyte, phonolite, syenite, carbonatite, lamprophyre, and intrusive breccia. Archean granitic rock (Karner and Patelke, 1989) occurs as large, irregularly shaped xenoliths within the central dome. The largest of the xenoliths is an elliptical body 1200 meters in length by 900 meters in width (Wilkinson, 1982).

Paleozoic and Mesozoic strata in the Bear Lodge dip radially from the central dome and

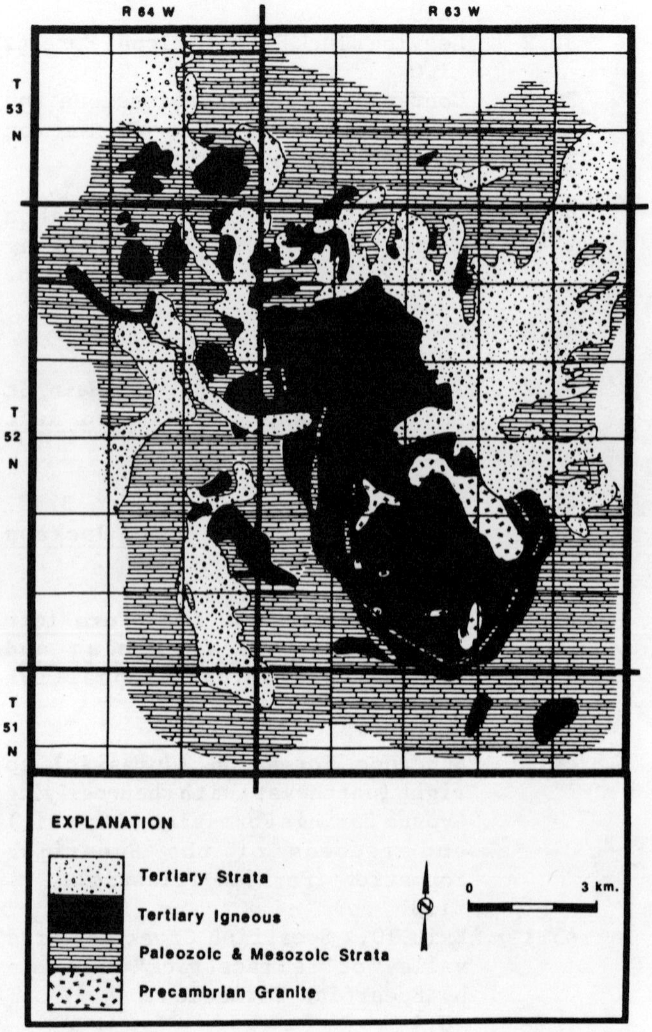

FIGURE 1 Geologic map of the Bear Lodge Mountains showing the distribution of Precambrian granitic rocks, Paleozoic and Mesozoic strata, Eocene igneous rocks, and younger Tertiary sedimentary rocks (generalized from Staatz, 1983).

have a combined thickness of approximately 1100 meters (Staatz, 1983). Cenozoic sedimentary rocks of Miocene and Oligocene age occur on the northern, eastern, and western flanks of the uplift. These conglomerates, sandstones, and siltstones postdate the structural deformation of the Bear Lodge Mountains and unconformably overlie the Paleozoic-Mesozoic strata and Eocene igneous rocks.

IGNEOUS, METASOMATIC, AND HYDROTHERMAL EVENTS

The Eocene core of the Bear Lodge records a complex sequence of igneous activity, metasomatism, and hydrothermal alteration. Five significant events from oldest to youngest are: (1) early igneous activity; (2) potassic fenitization; (3) carbonatitic magmatism; (4) hydrothermal activity; and (5) late igneous activity (Figure 2).

Early Igneous Activity

The early igneous history is revealed by a compositionally and texturally diverse suite of subvolcanic and plutonic rocks that form the bulk of the main igneous dome and many of the satellitic plugs, dikes, and sills. Potassic fenitization and hydrothermal alteration have changed the original compositions and textures especially in the central portions of the dome where the effects of these later events are most pronounced.

The rocks range from slightly oversaturated to strongly undersaturated with respect to SiO_2 (Table 1, columns A, B, C, and D). Total alkalis (Na_2O + K_2O) vary between 9 and 11 percent with all samples falling within the alkaline field proposed by Irving and Barager (1971) for volcanic rocks (Figure 3). The major rock types are latite-trachyte porphyries, phonolite-trachyte porphyries, and natrolite-melanite syenites and malignites.

Latite-trachyte Porphyries. Latite-trachyte porphyries are the most common igneous rocks exposed at the surface. Colors range from light gray to dark gray to olive-gray for fresh samples and light brown to rust-brown for hydrothermally altered samples. Major phenocryst minerals are plagioclase, sanidine, salite, and hastingsite. Accessory minerals include apatite, sphene, and titanian magnetite (Figure 4). Plagioclase phenocrysts range from oligoclase to andesine and typically show oscillatory normal zoning. Carlsbad-twinned sanidine microlites occur in an aphanitic matrix of cryptocrystalline albite, oligoclase, anorthoclase, and analcime.

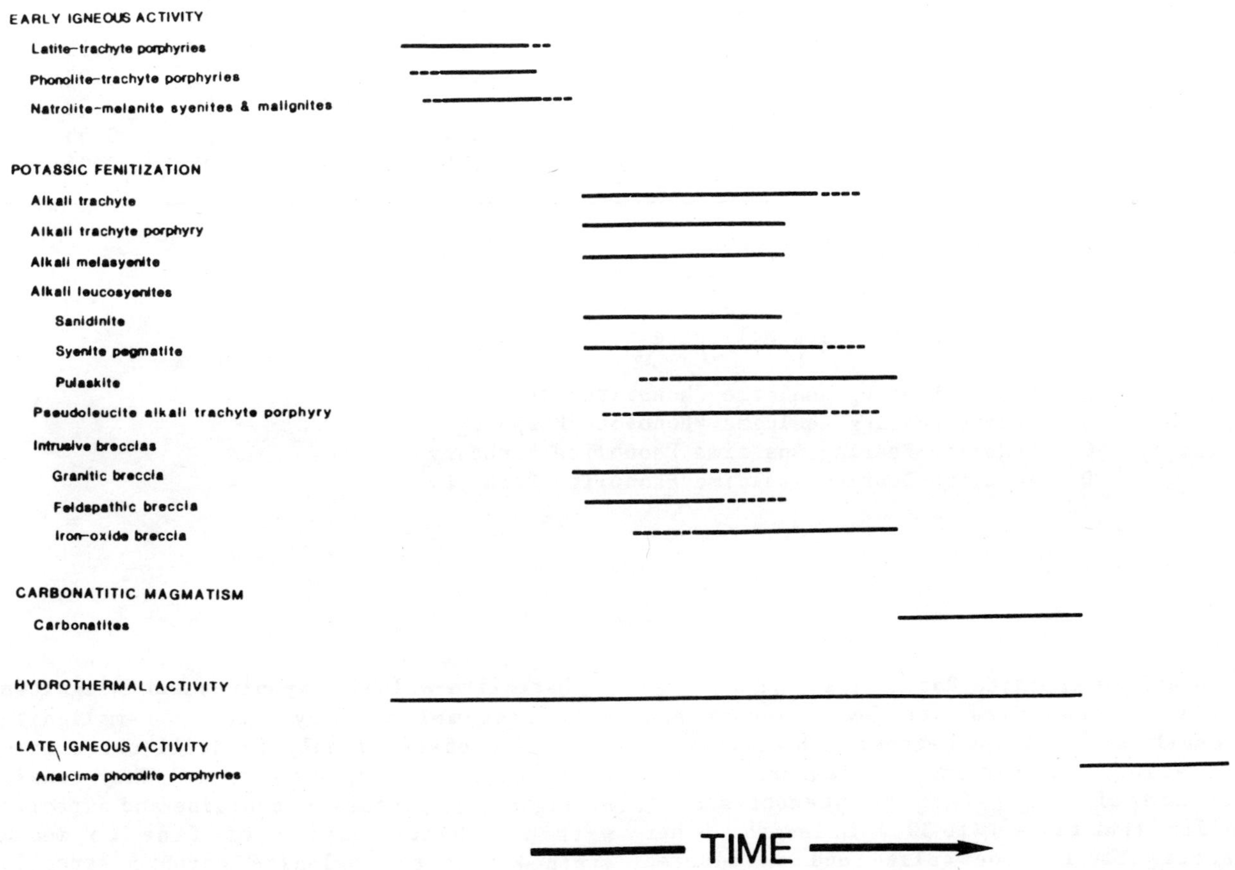

FIGURE 2 Simplified paragenesis diagram of the major petrologic events and associated rock types of the Bear Lodge Mountains. Uncertainty of relationships indicated by dashed lines.

TABLE 1 Major Element Chemical Analyses and CIPW Normative Mineralogies of Selected Samples of the Major Igneous Rock Types of the Bear Lodge Mountains

	A	B	C	D	E	F	G	H
SiO_2	56.57	51.32	46.27	43.67	58.02	55.69	50.37	44.37
Al_2O_3	17.22	18.66	17.00	15.47	18.64	19.23	20.58	19.83
Fe_2O_3*	3.42	2.63	7.93	9.02	3.28	3.40	3.16	3.24
MgO	0.79	1.62	1.53	1.97	0.54	0.44	0.19	0.17
CaO	3.62	6.03	8.90	11.51	2.67	1.78	1.64	1.70
Na_2O	5.41	1.89	2.20	0.63	5.65	6.42	8.18	4.93
K_2O	4.57	8.72	7.81	8.32	5.96	6.07	8.38	8.35
TiO_2	0.54	0.63	1.51	1.69	0.37	0.27	0.29	0.23
P_2O_5	0.18	0.12	0.27	0.32	0.10	0.03	0.03	0.02
MnO	0.14	0.16	0.27	0.30	0.12	0.16	0.14	0.17
Total	92.46	91.78	93.69	92.90	95.35	93.49	92.96	83.01
Qz	0.78	0.00	0.00	0.00	0.00	0.00	0.00	0.00
Co	0.00	0.00	0.00	0.00	0.00	0.00	0.00	0.00
Or	29.21	56.15	27.97	2.70	36.94	38.37	45.88	53.59
Ab	49.51	4.68	0.00	0.00	42.94	37.22	0.00	0.00
An	9.95	18.17	14.35	15.92	8.28	6.12	0.00	8.81
Lc	0.00	0.00	16.69	39.37	0.00	0.00	5.80	4.59
Ne	0.00	6.90	10.76	3.11	3.90	11.31	34.49	27.23
Wo	0.75	0.58	3.91	6.43	0.00	0.00	0.80	0.00
Di	5.03	9.48	17.92	23.21	3.99	2.60	5.75	1.02
Hy	0.00	0.00	0.00	0.00	0.00	0.00	0.00	0.00
Ol	0.00	0.00	0.00	0.00	0.12	1.00	0.00	1.15
Mt	3.20	0.34	4.66	4.98	2.84	2.75	0.00	3.02
Hm	0.00	2.09	0.00	0.00	0.00	0.00	0.00	0.00
Il	1.11	1.30	3.06	3.46	0.74	0.55	0.59	0.53
Ap	0.46	0.31	0.68	0.82	0.25	0.08	0.08	0.06

Early Igneous Activity
A = Latite-Trachyte Porphyry
B = Phonolite Trachyte Porphyry
C = Natrolite-Melanite Syenite
D = Natrolite-Melanite Malignite

Late Igneous Activity
A = Hauyne-Bearing Analcime Phonolite Porphyry
B = Hauyne-Bearing Analcime Phonolite Porphyry
C = Sodalite-Bearing Analcime Phonolite Porhpyry
D = Sodalite-Bearing Analcime Phonolite Porhpyry

* Total Fe as Fe_2O_3

Phonolite-trachyte Porphyries. Phonolite-trachyte porphyries are less common and generally show varying degrees of hydrothermal alteration. A distinctive feature is the presence of large tabular phenocrysts of sanidine that range up to 30 mm in length. Other phenocrysts include salite and an altered feldspathoid mineral, possibly sodalite or hauyne (Figure 5). The groundmass is similar to that of latite-trachyte porphyries.

Natrolite-melanite Syenites and Malignites. Natrolite-melanite syenites and malignites have been observed only in drill cores taken from the central igneous dome. They consist of light gray patches of sanidine and natrolite within a black matrix of fine to medium grained biotite, melanite garnet, strontian calcite, and aegirine, with accessory strontianite, pyrite, pyrrhotite, and chalcopyrite (Figure 6). Malignites are distinguished

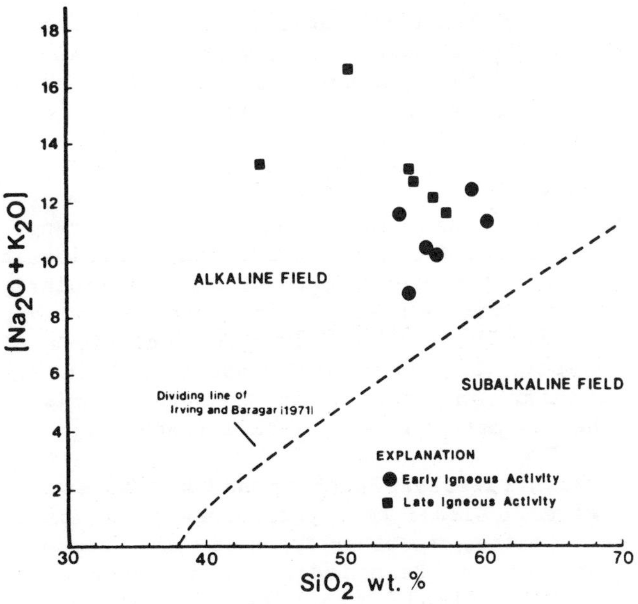

FIGURE 3 Plot of (Na_2O+K_2O) versus SiO_2 showing Irving and Baragar's (1971) dividing line for making a general distinction between alkaline and subalkaline volcanic igneous rocks. Analysis of samples associated with early and late igneous activity in the Bear Lodge Mountains plot in the alkaline field.

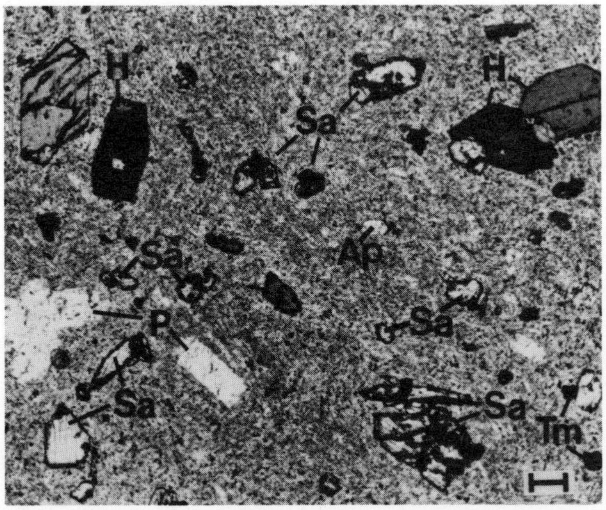

FIGURE 4 Photomicrograph of trachyte porphyry. Phenocrysts of plagioclase (P), hastingsite (H), salite (Sa), apatite (Ap), and titanian magnetite (Tm) enclosed within a microcrystalline groundmass rich in sanidine. Bar scale represents 0.25mm. Plane-polarized light.

from syenites by a higher content of mafic minerals.

Potassic Fenitization

Early igneous activity in the Bear Lodge Mountains was followed by a period of potassic fenitization, a type of metasomatic wall-rock alteration that is frequently observed in alkalic igneous complexes. Brögger (1921) first used the term to describe a process of alkali enrichment in igneous rocks of the Fen Complex in southern Norway. Inherent in Brögger's definition of fenitization is the

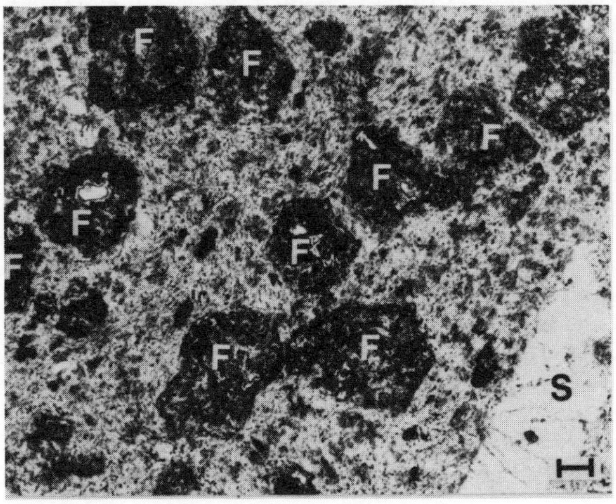

FIGURE 5 Photomicrograph of phonolite porphyry. Phenocrysts of Carlsbad-twinned sanidine (S) and an altered, rounded to hexagonal feldspathoid mineral (F) enclosed within a microcrystalline groundmass rich in sanidine. Bar scale represents 0.25mm. Crossed polars.

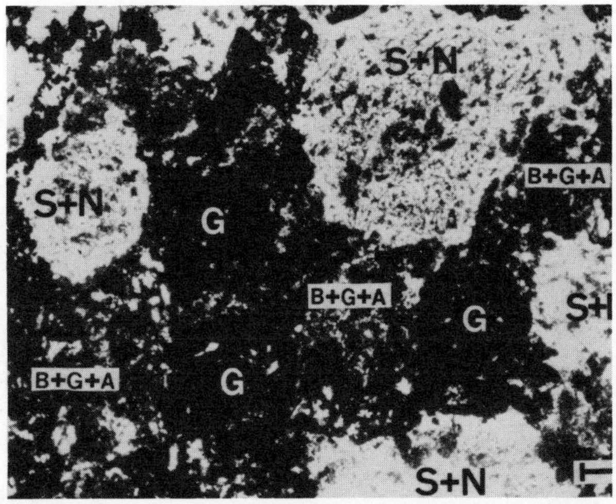

FIGURE 6 Photomicrograph of natrolite-bearing melanite syenite. Rounded to hexagonal patches of sanidine and turbid natrolite (S+N) and dark, rounded to hexagonal crystals of melanite garnet (G) enclosed within a mafic, fine-grained matrix rich in biotite, melanite garnet and aegirine (B+G+A). Bar scale represents 0.25mm. Plane-polarized light.

interpretation that it occurs without appreciable melting or mobilization of the original igneous wall rocks. Subsequent work in other complexes resulted in the recognition of igneous-looking, mobilized rocks which are commonly termed "rheomorphic fenites".

The most notable effect of the potassic fenitization is the metasomatic development of potassic alkali feldspar commonly designated as sanidine on the basis of optical and x-ray data. The feldspar occurs both in shallower subvolcanic rocks outcropping at the surface and deeper plutonic rocks observed in drill cores. Other minerals associated with the fenitization process are biotite, fluorite, strontian calcite, strontianite, and a suite of sulfide minerals that includes pyrite, pyrrhotite, chalcopyrite, sphalerite, and galena.

The fenites include both metasomatic and rheomorphic varieties. Major rock types are alkali trachyte, alkali trachyte porphyry, alkali melasyenite, alkali leucosyenite, pseudoleucite alkali trachyte porphyry, and several types of intrusive breccia. All of these rocks are characterized by high concentrations of K_2O in comparison to the surrounding alkalic igneous rocks (Figure 7).

Alkali Trachyte Fenite. Alkali trachyte is the most abundant variety of fenite exposed at the surface in the Bear Lodge Mountains. A rheomorphic type of fenite, it occurs as dikes and small plugs crosscutting older fenites and alkalic igneous rocks. Typical samples are aphanitic to very fine-grained and composed largely of sanidine (Figure 8) which ranges from Carlsbad-twinned microlites in classic trachytic textures to untwinned and irregularly shaped microcrystals in allotriomorphic granular textures. Minute inclusions of hematite are commonly disseminated in the sanidine and result in the characteristic ocherous appearance of alkali trachyte.

Alkali Trachyte Porphyry and Alkali Melasyenite Fenites. Alkali trachyte porphyry and alkali melasyenite, two other types of fenite commonly found in the Bear Lodge are metasomatic in origin. Alkali trachyte porphyry is the fenitized equivalent of latite, trachyte, and phonolite porphyries emplaced during the episode of early igneous activity while alkali melasyenite is the fenitized equivalent of natrolite-melanite syenite and malignite in the subsurface of the central dome. Both

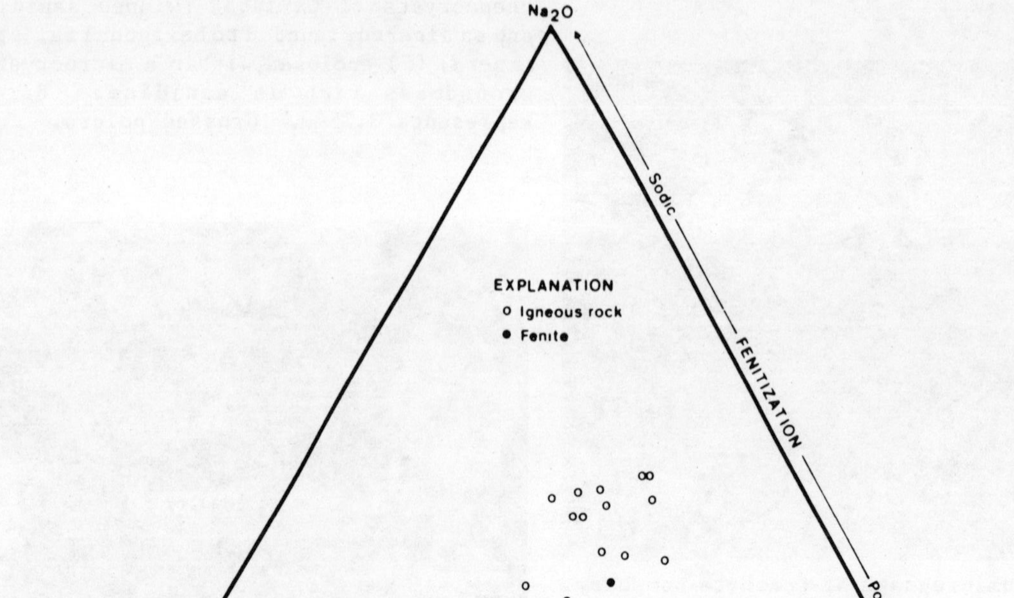

FIGURE 7 Triangular plot of ($MgO+Fe_2O_3+FeO+MnO$) versus K_2O versus Na_2O showing compositional differences between the rock types associated with the episode of early igneous activity in the Bear Lodge Mountains (open circles) and the rock types associated with potassic fenitization (closed circles).

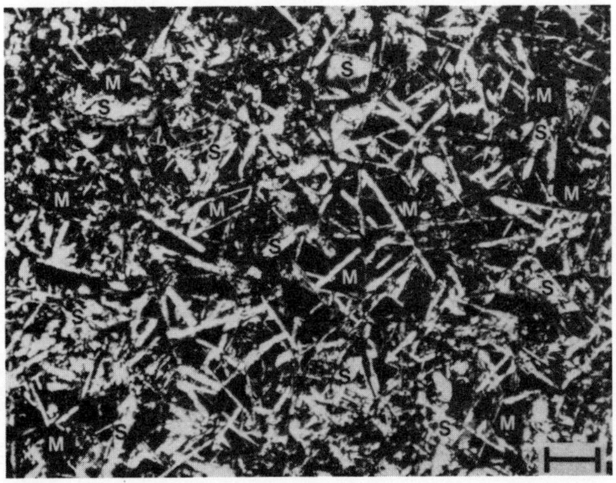

FIGURE 8 Photomicrograph of alkali trachyte. Fresh, vitreous laths of felted, Carlsbad-twinned sanidine (S) enclosing altered and oxidized mafic minerals (M). Bar scale represents 0.25mm. Crossed polars.

types of fenites contain relict igneous textures and structures (Figure 9) making megascopic distinction between fenites and igneous precursors difficult. Fenites may be recognized on the basis of their ultrapotassic chemical composition (Table 2, Columns B and C).

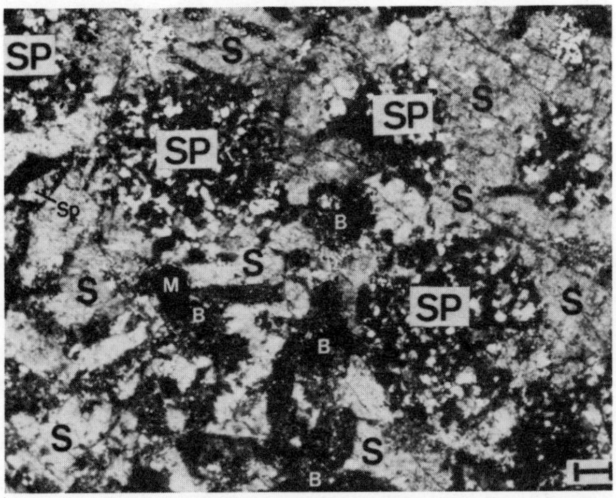

FIGURE 9 Photomicrograph of alkali melasyenite. Large poikilitic crystals of sanidine (S) of metasomatic origin enclosing patches of microcrystalline sanidine (SP), biotite (B), sphene (Sp), and magnetite (M). Bar scale represents 0.25mm. Crossed polars.

Alkali Leucosyenite Fenite. Alkali leucosyenite is another variety of rheomorphic fenite. The rock is common in the subsurface and composed largely of potassic feldspar. Three subtypes are recognized: sanidinite, syenite pegmatite, and pulaskite. Sanidinite is characterized by acicular crystals of Carlsbad-twinned sanidine, 5 to 10 mm in length, in randomly oriented to strongly trachytic fabric (Figure 10). Syenite pegmatite is very coarse-grained to pegmatitic with untwinned potassium feldspar crystals that range up to 40 mm in length. Pulaskite is characterized by blocky crystals of potassium feldspar that vary from 1 to 3 mm in size and typically occur in a fine-grained matrix of strontian calcite, aegirine, pyrite, pyrrhotite, chalcopyrite, fluorite, and strontianite (Figure 11).

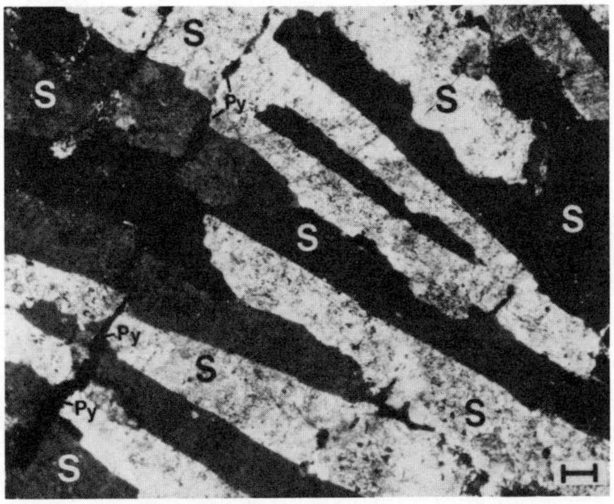

FIGURE 10 Photomicrograph of sanidinite. Acicular crystals of Carlsbad-twinned sanidine (S) exhibiting a well developed trachyte texture. Pyrite-filled microfault (Py) located in left half of photograph. Bar scale represents 0.25mm. Crossed polars.

Pseudoleucite Alkali Trachyte Porphyry Fenite. Pseudoleucite alkali trachyte porphyry is a volumetrically insignificant variety of rheomorphic fenite that occurs both at the surface and in the subsurface. The rock is distinct in appearance with light gray trapezohedral phenocrysts of pseudoleucite enclosed within a dark gray aphanitic groundmass composed largely of branching crystals of untwinned sanidine with lesser amounts of albite, analcime, biotite, and strontian calcite (Figure 12). Relict leucite with well developed lamellar twinning has been reported in several specimens of pseudoleucite alkali trachyte porphyry by White (1980).

TABLE 2 Major Element Chemical Analyses of Selected Samples of Potassic Fenites from the Bear Lodge Mountains

	A	B	C	D	E	F	G	H
SiO_2	58.64	57.16	50.60	48.39	49.96	50.53	49.01	59.26
Al_2O_3	35.53	22.79	14.64	16.21	13.22	13.55	16.78	16.67
Fe_2O_3*	6.86	2.91	4.98	4.86	3.87	5.91	4.89	2.95
MgO	0.15	0.30	1.02	0.81	1.17	1.62	1.63	0.20
CaO	0.11	0.20	6.97	8.31	5.53	5.60	7.35	0.61
Na_2O	1.59	0.23	1.76	0.37	1.01	0.41	0.83	0.97
K_2O	11.21	12.96	10.10	10.76	11.06	11.81	10.51	11.97
TiO_2	0.66	0.40	0.61	1.00	2.00	0.54	1.10	0.70
P_2O_5	0.20	0.11	0.25	0.14	0.31	0.11	0.20	0.22
MnO	0.01	0.20	0.27	0.28	0.31	0.27	0.28	0.12
Total	94.96	97.30	91.20	91.13	88.44	90.35	92.66	93.67

A = Alkali Trachyte
B = Alkali Trachyte Porphyry
C = Alkali Melasyenite
D = Sanidinite
E = Syenite Pegmatite
F = Pulaskite
G = Pseudoleucite Alkali Trachyte Porphyry
H = Feldspathic Breccia

FIGURE 11 Photomicrograph of pulaskite. Cube-shaped crystals of sanidine (S) enclosed within a finer grained matrix rich in strontian calcite (Sr-C) and pyrite (Py). Bar scale represents 0.25mm. Plane-polarized light.

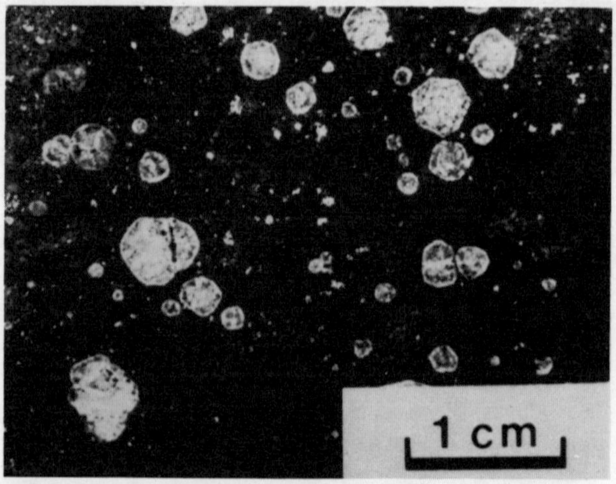

FIGURE 12 Slab photograph of pseudoleucite alkali trachyte porphyry. Trapezonedral phenocrysts of pseudoleucite enclosed within a dark, aphanitic groundmass rich in biotite, strontian calcite, and sanidine.

Intrusive Breccia Fenites. Intrusive breccias are common in the Bear Lodge and appear to have been intimately associated with the fenitization process. Like other types of fenite they are ultrapotassic and rich in potassium feldspar. Three varieties are distinguished: breccia of altered granitic rock, feldspathic breccia, and iron oxide breccia.

Breccia of altered granitic rock forms by desilicification of Precambrian granitic basement rocks during the early stages of fenitization. Desilicification of quartz-bearing wallrocks was first described by Brögger (1921) in his original monograph on the Fen complex. The granitic breccias occur as small pipes that crosscut both older Cenozoic igneous rocks and undisturbed granitic basement rocks. The breccias are composed of subangular xenoliths of Precambrian granite enclosed within a microcrystalline groundmass of granular quartz and sanidine (Figure 13). Occasional fragments of sandstone and older alkalic igneous rocks may also be present. Typically the quartz within xenoliths is granulated and optically strained grading into the matrix of the breccia at the margins of the xenoliths where it frequently contains acicular rutile inclusions.

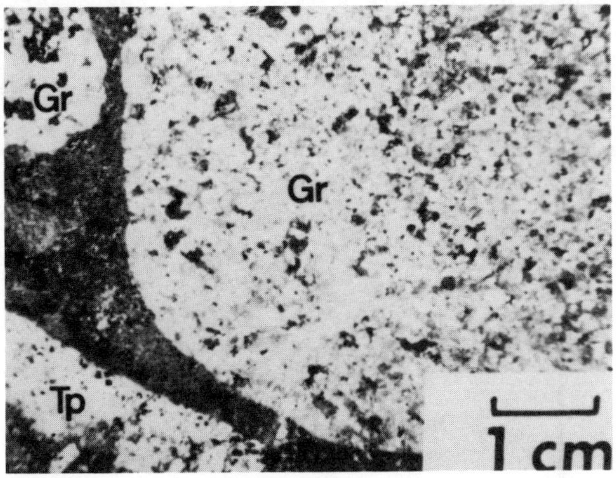

FIGURE 13 Slab photograph of granitic breccia. Subrounded to subangular xenoliths of Precambrian granite (GR) and alkali trachyte porphyry (Tp) enclosed within an aphanitic matrix rich in sanidine and quartz.

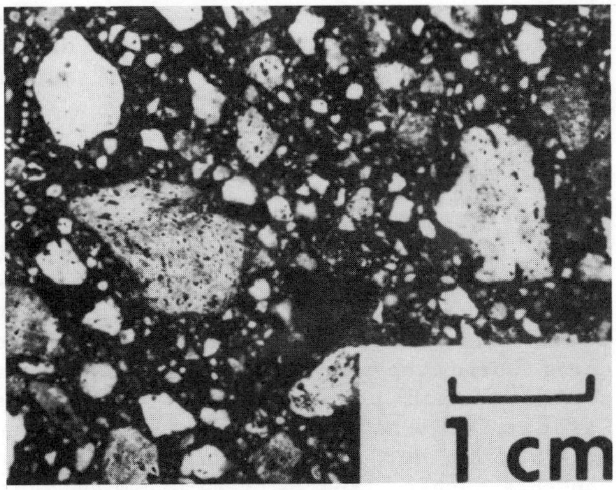

FIGURE 14 Slab photograph of iron-oxide breccia. Subrounded to subangular xenoliths of alkali trachyte enclosed within an aphanitic matrix rich in sanidine and iron oxides.

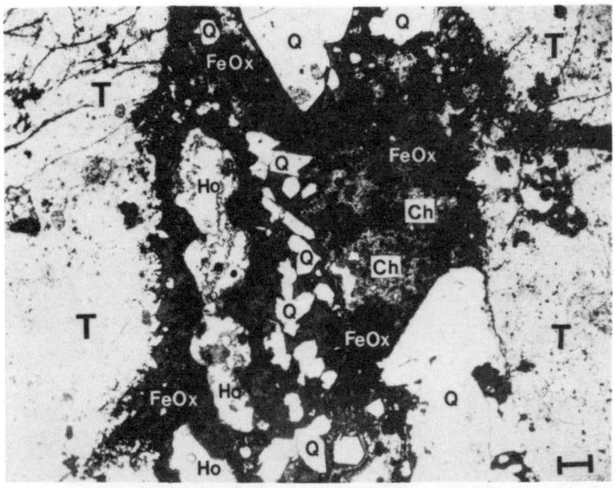

FIGURE 15 Photomicrograph of Th/REE vein crosscutting alkali trachyte. Vein composed mostly of iron oxides (FeOx) chlorite, (Ch), and quartz (Q). Holes in specimen (Ho) due to polishing. Note microfracturing with mineralization in alkali trachyte (T) adjacent to vein. Bar scale represents 0.25mm. Plane-polarized light.

Feldspathic breccia, consists of subangular clasts of older fenites and alkalic igneous rocks enclosed within a fine-grained groundmass rich in granular sanidine. The proportion of clasts to groundmass is highly variable in feldspathic breccias and the rock grades into xenolith-bearing alkali trachyte.

Iron-oxide breccia possesses a distinct rust-red to rust-orange ferruginous matrix rich in hematite and sanidine (Figure 14). The matrix encloses subangular fragments of older fenites and alkalic igneous rocks. The xenoliths constitute 30 to 50 percent of the rock and range up to 5 cm in size. The surrounding microcrystalline matrix contains abundant disseminated iron oxide.

Mineralized Veins of Fenites. The portions of the central igneous core of the Bear Lodge most affected by potassic fenitization commonly display intense fracturing and veining. The veins vary from dark brown to black in color and range from less than a centimeter to several meters in thickness (Figure 15). They contain up to 1.2 percent thorium and .8 percent total rare-earths with a complex and highly variable mineralogy that includes, in approximate order of decreasing abundance, sanidine, quartz, hematite, limonite, goethite, manganese oxides, monazite, fluorite, montmorillonite, kaolinite, cryptomelane, barite, rutile, magnetite, calcite, chlorite, brockite, bastnaesite, lepidocrocite, anatase, brookite, biotite, sphalerite, and zircon (Staatz,

1983). The veins crosscut earlier formed metasomatic and rheomorphic fenites and appear to mark the waning stages of fenitization.

Carbonatitic Magmatism

Potassic fenitization in the Bear Lodge Mountains was followed by the emplacement of dikes and veins of carbonatite near the central portions of the main igneous dome. Poorly exposed at the surface, the dikes and veins are relatively common in subsurface cores where they have steep dips and sharp contacts with the surrounding wallrocks. Carbonatite exhibits significant enrichment in thorium and rare-earths with thorium contents ranging from 54 to 1510 ppm and total rare-earth elements from 0.58 to 64 percent (Staatz, 1983). Cerium and lanthanum account for more than 95 percent of the total rare-earth content of the rocks (Staatz, 1983).

The mineralogy of the carbonatites is highly variable. The principal constituent is strontian calcite which composes from 60 to 90 percent of the rocks. Other minerals include strontianite, siderite, fluorite, biotite, aegirine, actinolite, sanidine, quartz, apatite, barite, magnetite, pyrite, pyrrhotite, chalcopyrite, galena, sphalerite, ancylite, bastnaesite, synchysite, burbankite, and carbocernaite (Figure 16). Two additional rare-earth minerals, one a carbonate and the other a phosphate, have yet to be identified and possibly represent new mineral species, (Jenner, 1984).

FIGURE 16 Photomicrograph of carbonatite. Radiating intergrowth of strontianite and ancylite (S+A) enclosed within a fine-grained matrix of strontian calcite (Sr-C). Bar scale represents 0.25mm. Plane-polarized light.

The isotopic compositions of carbon and oxygen in calcites from the carbonatites of the Bear Lodge exhibit a limited range for both carbon (-7.0 to -10.6 per mil relative to PDB) and oxygen (+7.5 to +9.9 per mil relative to SMOW). The values of δC^{13} and δO^{18} fall within or near the field proposed by Taylor and others (1967) for primary igneous carbonatites. They differ significantly from values determined for samples of sedimentary calcite from the Mississippian Pahasapa Limestone which outcrops around the perimeter of the uplift (Figure 17). Because the values of δC^{13} are similar to values observed in calcites from mantle-derived diamonds and kimberlitic calcites (Deines and Gold, 1973; Deines, 1980) and values of δO^{18} approach the values exhibited in mantle-derived ultramafic and mafic rocks (Taylor, 1968), it would seem likely that the carbonatites of the Bear Lodge were derived from mantle-derived CO_2 of relatively uniform isotopic composition.

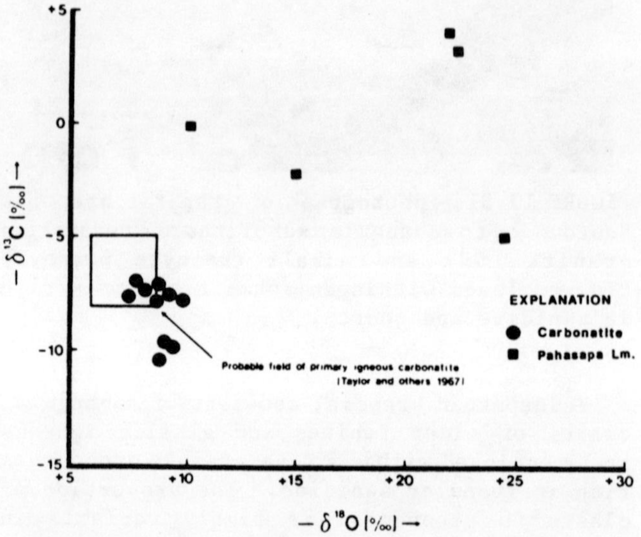

FIGURE 17 Plot of $\delta^{13}C$ versus $\delta^{18}O$ comparing compositions of calcites from carbonatites with calcites from the Early Mississippian Pahasapa Limestone. Also shown is the probable field of primary igneous carbonatite proposed by Taylor and others (1967).

Hydrothermal Activity

A hydrothermal system, active during the episodes of early igneous activity, potassic fenitization, and carbonatitic magmatism, accounts for the highly altered and oxidized rocks exposed at the surface of the Bear Lodge dome. Highly altered dikes of latite porphyry crosscut by less altered dikes of latite porphyry, and altered mafic minerals enclosed within fresh, vitreous crystals of fenitic sanidine, indicate that the hydrothermal system was active during the early igneous history of the complex (Figure 18). Alteration and oxidation of potassic fenites and carbonatites show that hydrothermal activity continued through the episodes of potassic fenitization and carbonatitic magmatism.

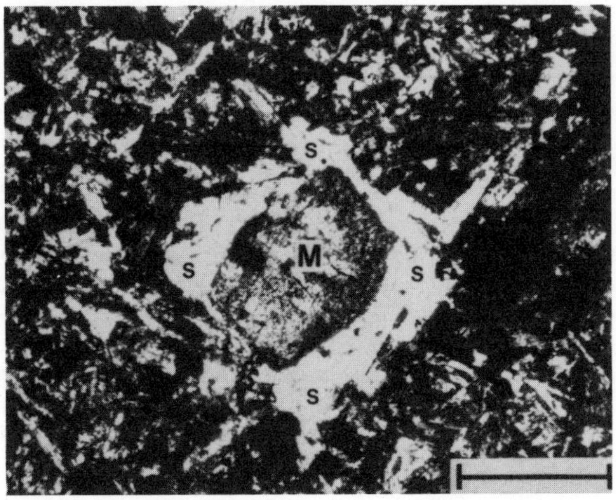

FIGURE 18 Photomicrograph of alkali trachyte. Fresh, vitreous sanidine of metasomatic origin mantling an altered mafic mineral (M). Bar scale represents 0.25mm. Crossed polars.

Mineralogical changes associated with hydrothermal activity are best observed in latite porphyries. The porphyries range from relatively unaltered to highly altered and oxidized. Unaltered porphyries contain relatively pristine crystals of zoned plagioclase, sanidine, hastingsite, salite, apatite, sphene, and titanian magnetite enclosed within a microcrystalline groundmass rich in sanidine (Figure 19a). With moderate alteration the more calcic cores of the plagioclase phenocrysts are replaced by a cryptocrystalline mixture of calcite and clays while sanidine phenocrysts and groundmass sanidine exhibit incipient sericitization. Hastingsite and salite phenocrysts are partially to completely replaced by a fine grained mixture of biotite and calcite while sphene and titanian magnetite are partially altered to iron and titanium oxides (Figure 19b). With continued exposure to hydrothermal solutions, hastingsite, salite, sphene, and titanian magnetite crystals are completely replaced by oxides of iron and titanium or, more commonly, are apparently totally dissolved with only prismatic casts as remnants. Plagioclase phenocrysts show further replacement by calcite and clays while sanidine phenocrysts and groundmass sanidine display more advanced stages of sericitization (Figure 19c).

The effects of hydrothermal activity are most pervasive in the central portions of the uplift where the zone of oxidation and alteration extends to depths of several tens of meters. The effects diminish with depth suggesting that the hydrothermal system involved the circulation of heated meteoric groundwaters, probably similar to circulation in modern geothermal areas such as the Yellowstone region of Wyoming or the Salton Sea area of California. A high geothermal gradient, related to igneous activity in the Bear Lodge during Eocene time, would account for the continuing circulation of meteoric groundwaters.

Late Igneous Activity

The final stage of igneous activity in the Bear Lodge Mountains is represented by hauyne-bearing and sodalite-bearing analcime phonolite porphyries. These younger porphyries are undersaturated with respect to SiO_2 and occur as dikes and small plugs. Their characteristically fresh appearance is in sharp contrast to the altered and oxided igneous rocks and fenites which they crosscut. Like the older latite, trachyte, and phonolite porphyries emplaced during early igneous activity, the analcime phonolite porphyries possess high percentages of total alkalis ($Na_2O + K_2O$) and fall within the alkaline field proposed by Irving and Baragar (1971) for volcanic rocks (Figure 3).

The analcime phonolite porphyries are distinctly porphyritic. The hauyne-bearing varieties are dark green with large phenocrysts of tabular sanidine and smaller phenocrysts of rounded to hexagonal hauyne. The sodalite-bearing varieties are dark green to olive green and contain small phenocrysts of hexagonal sodalite (Figure 20), aegirine, melanite garnet, and biotite. Accessory minerals include apatite, sphene, and titanian magnetite.

The groundmasses of analcime phonolite porphyries are aphanitic to very fine-grained. Sanidine, anorthoclase, and aegirine occur as microlites and exhibit trachytic to subtrachytic textures. The microlites are enclosed within an isotropic matrix rich in analcime. Natrolite occurs as an alteration product of the analcime and often forms spectacular clusters of radially arranged crystals (Figure 21).

The late analcime phonolite porphyries appears to be related to a regional pattern of phonolitic magmatism that occurred in the northern Black Hills during Eocene time. Phonolite porphyries, compositionally and texturally similar to those of the Bear Lodge, occur at other igneous centers including Devils Tower, Missouri Buttes, and Barlow Canyon in Wyoming (Halvorson, 1980) and Rubicon Gulch, Little Crow Peak, Ragged Top

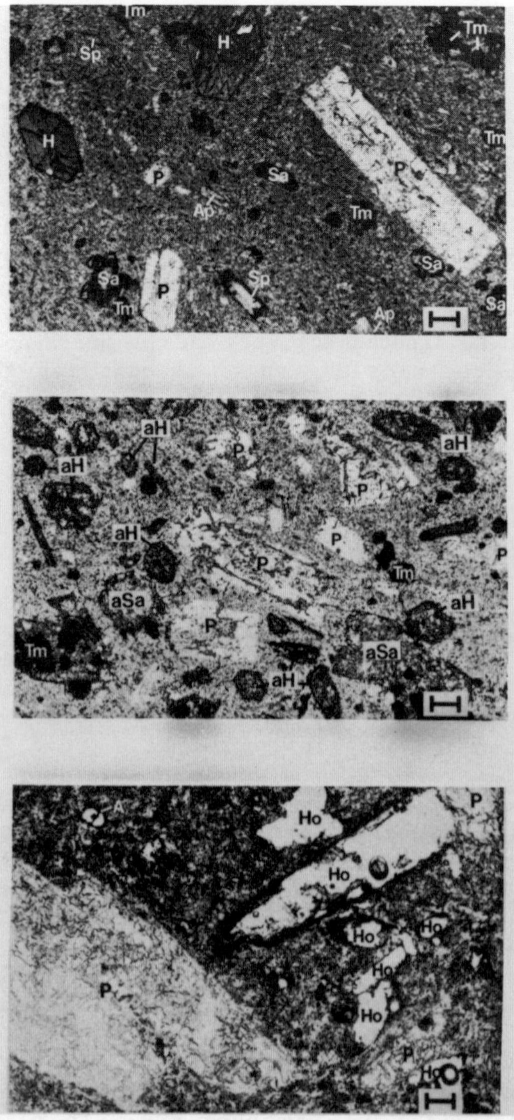

FIGURE 19 Photomicrographs showing mineralogic changes associated with hydrothermal activity in the Bear Lodge Mountains.

(a) Photomicrograph of relatively unaltered trachyte porphyry. Phenocrysts of plagioclase (P), hastingsite (H), salite (SA), apatite (Ap), sphene (Sp), and titanian magnetite (Tm) enclosed within a microcrystalline groundmass rich in sanidine. Bar scale represents 0.25mm. Plane-polarized light.

(b) Photomicrograph of moderately altered latite porphyry. Plagioclase phenocrysts (P) altered to a mixture of calcite and clays, especially in the more calcic crystal cores and along growth bands. Original hastingsite (aH) and salite (aSa) are completely replaced by a microcrystalline mixture of biotite and calcite. Titanian magnetite (Tm) exhibits incipient alteration ot iron and titanian oxides. Bar scale represents 0.25mm. Plane-polarized light.

(c) Photomicrograph of highly altered trachyte porphyry. Plagioclase phenocrysts show more advanced alteration to calcite and clays. Hastingsite, salite, and titanian magnetite are completely destroyed with only prismatic holes (Ho) as remnants. Bar scale represents 0.25mm. Plane-polarized light.

Mountain, Spearfish Canyon, Spearfish Peak and other sites in South Dakota (Kirchner, 1971). At several of the igneous centers in South Dakota the phonolite porphyries reportedly crosscut older hydrothermally altered igneous rocks that include quartz latites, latites, and trachytes (Kirchner, 1971), a situation that is remarkably similar to the relationships observed in the Bear Lodge Mountains.

PETROGENETIC MODEL

Mantle Source of Magma

The proposed petrogenetic model for the alkalic igneous rocks, potassic fenites, and carbonatites of the Bear Lodge Mountains begins with partial melting of upper mantle material metasomatically enriched in alkalis, strontium, cerium, lanthanum, fluorine, carbon dioxide, water, and other incompatible elements and volatiles (Figure 22). Such metasomatically enriched zones of the upper mantle have been proposed on the basis of petrographic and isotopic evidence to be the source region of alkali basalts, nephelinites, kimberlites, and similar types of alkali-rich ultrabasic magmas (Sun and Hanson, 1975; Cox and others, 1976; Boettcher and O'Neil, 1980; Menzies and Murthy, 1980). The metasomatic enrichment of these zones, which may have occurred billions of years ago, accounts for the characteristically high concentrations of incompatible elements in alkali-rich ultrabasic rocks and also provides the concentration of the fluxes and radiogenic heat sources necessary to initiate mantle fusion.

Although comprehensive isotopic and trace element studies are needed to show that the parent magma of the alkalic igneous rocks, potassic fenites, and carbonatites of the Bear Lodge originated from partial melting of an enriched portion of the upper mantle, the extremely high concentrations of cerium,

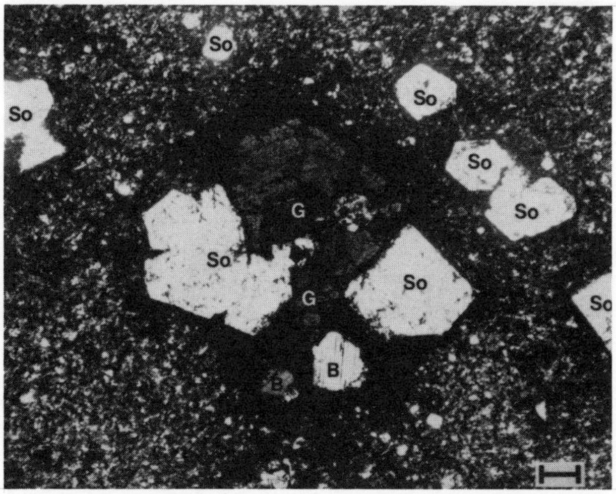

FIGURE 20 Photomicrograph of sodalite-bearing analcime phonolite porphyry hexagonal phenocrysts of sodalite (So) occurring isolated and cummulophyrically arranged wih zoned melanite garnet (G) and biotite (B). Bar scale represents 0.22mm. Plane-polarized light.

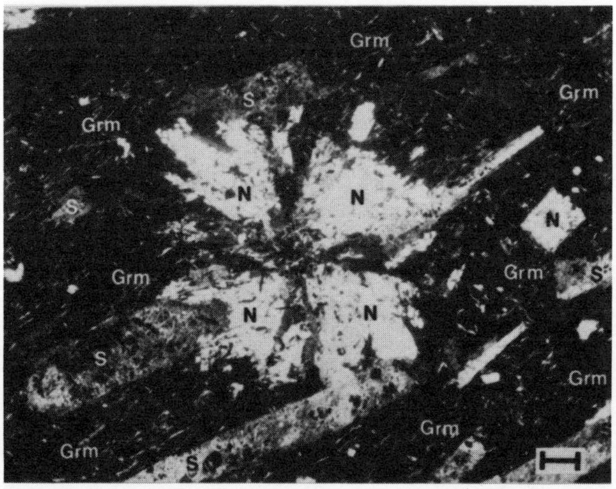

FIGURE 21 Photomicrograph of haüyne-bearing analcime phonolite porphyry. Radiating natrolite (N) replacing analcime-rich groundmass (Grm) and sanidine phenocryst (S). Bar scale represents 0.25mm. Crossed polars.

lanthanum, thorium, and other incompatible elements in the fenites and carbonatites provide preliminary evidence. The carbonatites are probably genetically related to the surrounding silicate rocks and, as discussed earlier, carbon and oxygen isotopic data indicate a mantle source for the carbonatites.

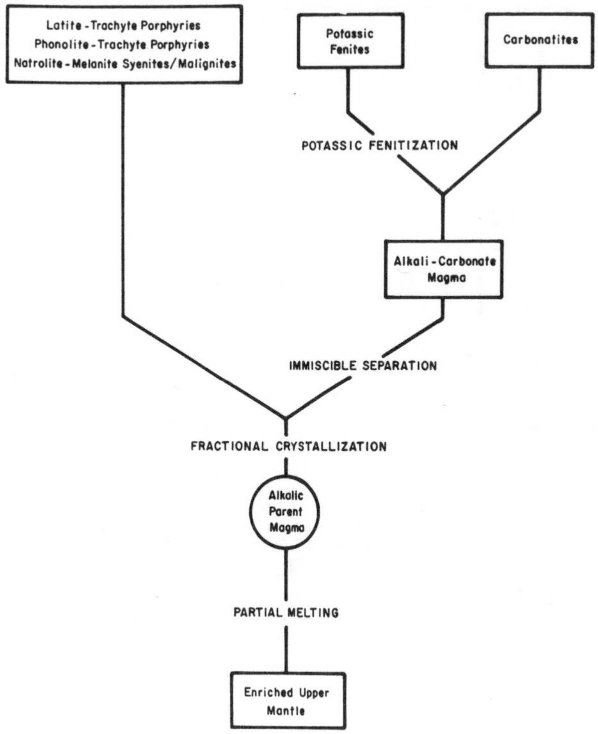

FIGURE 22 Simplified petrogenetic scheme proposed for the alkalic igneous rocks, potassic fenites, and carbonatites of the Bear Lodge Mountains.

Alkali-rich Mafic Parental Magma

An alkali-rich mafic magma is proposed as the parental magma of the alkalic igneous rocks of the Bear Lodge Mountains. A plot of SiO_2 versus differentiation index which compares analyses of unaltered igneous rocks from the Bear Lodge with generalized alkali basalt and tholeiitic basalt differentiation lineages suggests that the parental magma was compositionally similar to an alkali basalt (Figure 23). An alkali basalt composition is also suggested by the trend defined when the same analyses are plotted on a standard AFM diagram (Figure 24). Note that the projected origin of the trend delineated on the AFM diagram approaches the composition of the average alkali basalt reported by Nockolds (1954).

Early Magmatic Differentiation by Fractional Crystallization

The episode of early igneous activity in the Bear Lodge Mountains shows that the parental magma initially generated latitic, trachytic, and phonolitic differentiates. Petrographic evidence such as zoned pyroxene and plagioclase phenocrysts and the occasional occurrence of pyroxenitic xenoliths suggests

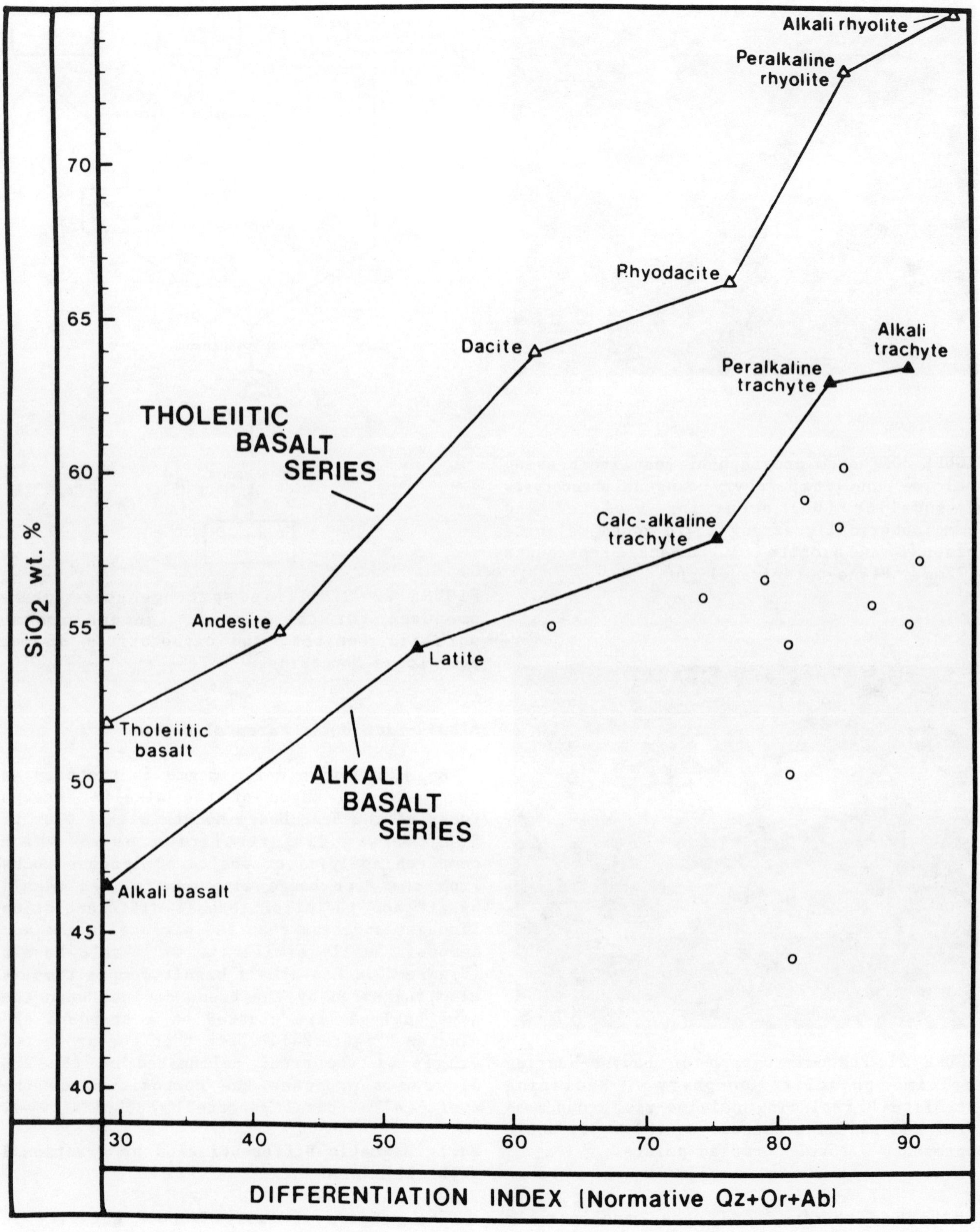

FIGURE 23 Plot of SiO_2 versus differentiation index showing trends of tholeiitic basalt series rocks (open triangles) and alkali basalt series rocks (closed triangles) as determined from analyses presented by Nockolds (1954). Samples of unaltered igneous rocks from the Bear Lodge Mountains plotted as open circles.

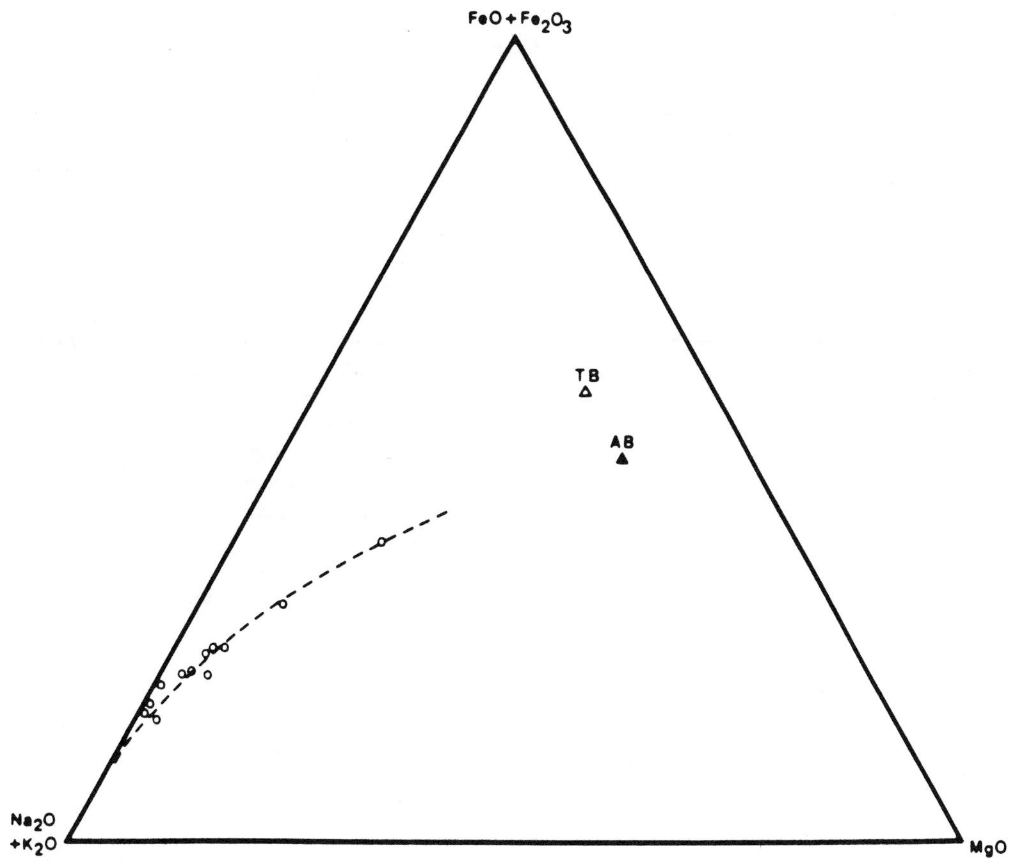

FIGURE 24 Triangular plot of (Na_2O+K_2O) versus (FeO+Fe_2O_3) versus MgO. Samples of unrelated igneous rocks from the Bear Lodge Mountains plotted as open circles. Average thoeliitic basalt (TB-open triangle) and average alkali basalt (AB-closed triangle) plotted from analyses presented by Nockolds (1954).

processes of fractional crystallization. Additional evidence for fractional crystallization is provided by a Larson (1938) variation diagram which indicates a progressive enrichment of SiO_2, Al_2O_3, Na_2O, and K_2O and a corresponding depletion of CaO, MgO, and total iron during differentiation (Figure 25). Such trends would be consistent with the early fractionation of magnesian olivene, magnesian pyroxene, and calcic plagioclase during crystallization.

Role of Liquid Immiscibility in the Origin of Alkali-rich Carbonatitic Magma

Differentiation by a process of liquid immiscibility is believed to have produced an alkali-rich carbonatitic magma in the Bear Lodge Mountains that was simliar to, although more potassic than, the natrocarbonatitic lavas erupted in Tanzania by the volcano Oldoinyo Lengai (Dawson, 1962, 1966). Immiscibility between alkali-rich carbonatitic liquids and CO_2-rich silicate magmas has been confirmed experimentally in a number of systems (Koster van Gross and Wyllie, 1966, 1968, 1973; Wendlandt and Harrison, 1979; Freestone and Hamilton, 1980). The process has also been substantiated in nature. Rankin and LeBas (1974a) report immiscible silicate and carbonate liquids coexisting in fluid inclusions in apatites from ijolite pegmatites of the Wasaki complex of western Kenya. According to the authors (p. 208) the inclusions represent the "simultaneous entrapping of a portion of a carbonate-rich melt and a silicate-rich melt".

Experimental studies by Freestone and Hamilton (1980) have shown that the miscibility gap between an alkali-rich carbonatitic liquid and coexisting silicate magma expands with increased P_{CO2} and decreased temperature and that it closes away from Na_2O-rich compositions, indicating that more felsic magmas are more likely to exsolve a carbonatitic liquid than more mafic magmas. Thus in the Bear Lodge the alkali-rich carbonatitic liquid probably separated from a phonolitic or compositionally similar differentiate of the original parent magma. The analcime phonolite porphyries emplaced during the episode of late igneous activity may have formed by

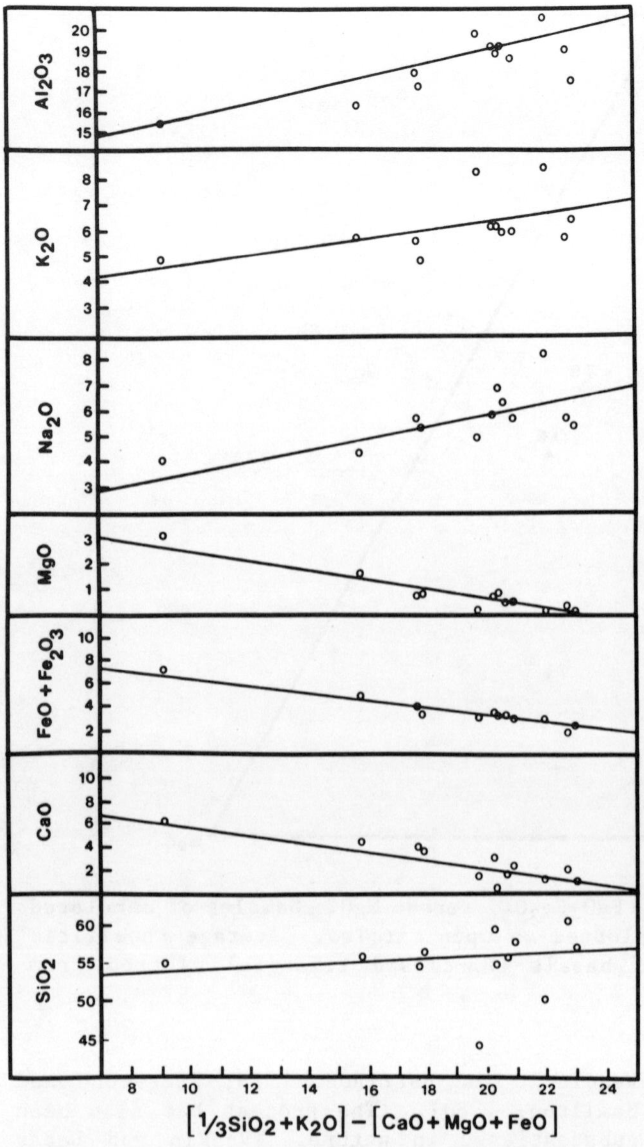

FIGURE 25 Larson (1938) variation diagram. Samples of unaltered igneous rocks from the Bear Lodge Mountains plotted as open circles. Lines best fitted by least square linear regression technique.

crystallization of the silicate magma from which the carbonatitic liquids were exsolved.

If immiscible differentiation did produce an alkali-rich carbonatitic magma in the Bear Lodge, then the extreme enrichment of cerium and lanthanum in the carbonatites suggests that light rare-earth elements were preferentially partitioned into the carbonatitic liquid during the process. Experimental studies by Wendlandt and Harrison (1979) have shown that the partition coefficients for light rare-earth elements are in favor of the carbonatitic liquid relative to the silicate liquid by a factor of 2-3 during immiscible differentiation.

Thus the rare-earth enrichment of the carbonatites can be attributed to two separate events: (1) metasomatic enrichment of light rare-earth elements in the source region of the alkali-rich mafic parent magma and (2) partitioning of rare-earth elements into the carbonatitic liquid during immiscible differentiation.

Origin of Carbonatite and Fenite

The alkali-rich carbonatitic magma is thought to have exsolved an alkali-rich aqueous saline solution, resulting in the potassium-rich fluids responsible for fenitization and also producing an alkali-depleted carbonatitic liquid that later crystallized to form carbonatite in the Bear Lodge Mountains. The same mechanism has been proposed by a number of authors to account for the intimate association between carbonatite and fenite (von Eckermann, 1948; Dawson, 1964; Wooley, 1982; Rubie and Gunter, 1983). The process has also been proposed to account for the common occurrence of alkali-rich aqueous fluid inclusions in minerals of alkali-depleted intrusive carbonatite (Rankin and LeBas, 1974b; Rankin, 1975, 1977; Nesbitt and Kelley, 1977; Metzger and others, 1977; Aspden, 1980). In the Bear Lodge the high concentrations of rare-earth elements in both potassic fenite and carbonatite suggests a genetic connection between the two types of rock and supports the hypothesis that the potassium-rich fluids responsible for fenitization were derived from a carbonatitic liquid.

SUMMARY

Although speculative, the petrogenetic model outlined above is consistent with the observed chronology of igneous and metasomatic events in the Bear Lodge Mountains. It also accounts for many of the petrographic and geochemical characteristics of the rocks such as the extreme enrichment of rare-earth elements in the carbonatites and potassic fenites. Additional isotopic, trace element, and fluid inclusion studies are needed to solve the numerous petrologic problems presented by the Eocene rocks of the Bear Lodge. The major conclusions of this study are as follows:

1. The Eocene rocks of the Bear Lodge record a complex history of multiple alkalic igneous events, potassic fenitization, carbonatitic magmatism, and hydrothermal alteration.
2. The potassic fenite and carbonatite exhibit significant enrichment in rare-earth elements and thorium.
3. The isotopic compositions of carbon and oxygen from carbonatitic calcite suggests a mantle source of CO_2.

4. A hydrothermal system involving the circulation of heated meteoric groundwater was active throughout most of the igneous history of the Bear Lodge.
5. As at other Cenozoic igneous centers in the northern Black Hills, the final stage of magmatism in the Bear Lodge was emplacement of phonolite dikes.
6. The igneous rocks of the Bear Lodge Mountains were probably derived by the differentiation of an alkali-rich parent magma generated by partial melting of upper mantle material enriched in potassium, rubidium, iron, titanium, cerium, lanthanum, fluorine, carbon dioxide, water and other incompatible elements and volatiles. Latitic, trachytic, and phonolitic differentiates were generated by fractional crystallization and an alkali-rich carbonatitic liquid was produced by later immiscible separation, probably from a phonolitic differentiate. Separation of an alkali-rich aqueous solution from the carbonatitic magma provided fenitizing potassium-rich fluids and also a residual alkali-depleted carbonatitic magma that later crystallized to form the dikes and veins of carbonatite found within the core of the igneous complex.

REFERENCES

Aspden, J.A., 1980, the mineralogy of primary inclusions in apatite crystals extracted from the Alno ijolite: Lithos, v. 13, p. 263-268.

Boettcher, A.L., and O'Neil, J.R., 1980, Stable isotope, chemical, and petrographic studies of high-pressure amphiboles and micas: evidence for metasomatism in the source regions of alkali basalts and kimberlites: American Journal of Science, v. 280-A, p. 594-621.

Brögger, W.C., 1926, Die eruptivgesteine des Kristianiagebietes. IV. Das Fengebiet in Telemarken, Norwegen: Norske Videns. Skrift. Mat. - Naturv. Kl., 1920, no.9.

Cox, K.G., Hawkesworth, C.J., and O'Nions, R. K., 1976, Isotopic evidence for the derivation of some Roman region volcanics from anamalously enriched mantle: Contributions to Mineralogy and Petrology, v. 56, p. 173-180.

Dawson, J.B., 1962, The geology of Oldoinyo Lengai: Bulletin of Volcanology, v. 24, p 349-387.

_____, 1964, Reactivity of the cations in carbonate magma: Geological Association of Canada Proceedings, v. 15, p. 103-113.

_____, 1966, The kimberlite-carbonatite relationship: International Mineralogical Association, 4th meeting, India, p. 1-4.

Deines, Peter, 1980, The carbon isotopic composition of diamond: relations to diamond shape, color, occurrence, and vapor composition: Geochimica et Cosmochimica Acta, v. 44, p. 943-961.

Deines, Peter, and Gold, D.P., 1973, The isotopic composition of carbonatite and kimberlite carbonates and their bearing on the composition of deep-seated carbon: Geochimica et Cosmochimica Acta, v. 37, p. 1709-1733.

Freestone, I.C., and Hamilton, D.L., 1980, The role of liquid immiscibility in the genesis of carbonatites - an experimental study: Contributions to Mineralogy and Petrology, v. 73, p. 105-117.

Gries, J.P., and Tullis, E.L., 1955, Geologic history of the Black Hills, in Black Hills Field Conference, North Dakota Geologic Society Guidebook, p. 32.

Halvorson, D.L., 1980, Geology and petrology of the Devils Tower, Missouri Buttes, and Barlow Canyon area, Crook County, Wyoming: unpublished Ph.D. dissertation, University of North Dakota, 163 p.

Hill, D.J., Izett, G.A., and Naeser, C.W., 1975, Early Tertiary fission trackages of sphene from Devils Tower and Missouri Buttes, Black Hills, northeastern Wyoming: Geological Society of America Abstracts with Programs, Rocky Mountain section 1975 Annual Meeting, p. 613-614.

Irving, F.N. and Barager, W.R.A., 1971, A guide to the chemical classification of common volcanic rocks: Canadian Journal of Earth Sciences, V. 8 p. 523-548.

Jenner, G.A., 1984, Tertiary alkalic igneous activity, potassic fenitization, carbonatitic magmatism, and hydrothermal activity in the central and southeastern Bear Lodge Mountains, Crook County, Wyoming: unpublished M. S. thesis, University of North Dakota, 232 p.

Kirchner, J.G., 1971, The petrography and petrology of the phonolite porphyry intrusions of the northern Black Hills, South Dakota: unpublished Ph.D. dissertation, University of Iowa, 199 p.

Koster van Groos, A.F., and Wyllie, P.J., 1966, Liquid immiscibility in the system Na_2-Al_2O_3-CO_2 at pressures to 1kb.: American Journal of Science, v. 264, p. 234-255.

_____, 1968, Liquid immiscibility in the join $NaAlSi_3O_8$-Na_2CO_3-H_2O and its bearing on the genesis of carbonatites: American Journal of Science, v. 266, p. 932-967.

_____, 1973, Liquid immiscibility in the join $NaAlSi_3O_8$-$CaAl_2Si_2O_2$-Na_2CO_3-H_2O: American Journal of Science, v. 273, p. 465-487.

Lisenbee, A.L., 1978, Laramide structure of the Black Hills uplift, South Dakota-Wyoming - Montana, in Matthews, Vincent, III, ed., Laramide folding associated with basement block faulting: Geological Society of America Memoir 151, p. 165-196.

McDowell, F.W., 1971, K-Ar ages of igneous rocks from the western United States: Isochron/West, No. 2, p. 1-20.

Menzies, Martin, and Murthy, V.R., 1980, Mantle metasomatism as a precursor to the genesis of alkaline magmas: isotopic evidence: American Journal of Science, v. 280-A, p. 622-638.

Metzger, F.W., Kelly, W.C., Nesbitt, B.E., and Essene, E.G., 1977, Scanning electron microscopy of daughter minerals in fluid inclusions: Economic Geology, v. 72, p. 141-152.

Nesbitt, B.E., and Kelly, W.C., 1977, Magmatic and hydrothermal inclusions in carbonatite of the Magnet Cove complex, Arkansas: Contributions to Mineralogy and Petrology, v. 63, p. 271-294.

Nockolds, S.R., 1954, Average chemical compositions of some igneous rocks: Geological Society of America Bulletin, v. 65, p. 1007-1032.

Rankin, A.H., 1975, Fluid inclusion studies in apatites from carbonatites of the Wasaki area, western Kenya: Lithos, v. 8, p. 123-136.

_____, 1977, Fluid inclusion evidence for the formation conditions of apatite from the Tororo carbonatite complex of eastern Uganda: Mineralogical Magazine, v. 41, p. 155-164.

Rankin, A.H., and Le Bas, M.J., 1974a, Liquid immiscibility between silicate and carbonatite melts in naturally occurring ijolite magma: Nature, v. 250, p. 206-209.

_____, 1974b, Nahcolite ($NaHCO_3$) in inclusions in apatite from some E. African ijolites and carbonatites: Mineralogical Magazine, v. 39, p. 564-570.

Rubie, D.C., and Gunter, W.D., 1983, The role of speciation in alkaline igneous fluids during fenite metasomatism: Contributions to Mineralogy and Petrology, v. 82, p. 165-175.

Shapiro, L.H., 1971, Structural geology of the Black Hills region and implications for the origins of the uplifts of the Middle Rocky Mountain Province: unpublished Ph.D. dissertation, University of Minnesota, 213 p.

Staatz, M.H., 1983, Geology and descriptions of thorium and rare-earth deposits in the southern Bear Lodge Mountains, northeastern Wyoming: U.S. Geological Survey Professional Paper # 1049-D, 52 p.

Sun, S.S., and Hanson, G.N., 1975, Evolution of the upper mantle: geochemical evidence from alkali basalt: Geology, v. 3, p. 297-302.

Taylor, H.P., 1968, The oxygen isotope geochemistry of igneous rocks: Contributions to Mineralogy and Petrology, v. 19, p. 1-71.

Taylor, H.P., Frechen, Josef, and Degens, E.F., 1967, Oxygen and carbon isotope studies of carbonatites from the Laacher See district, West Germany and the Alno district, Sweden: Geochimica et Cosmochimica Acta, v. 31, p. 407-430.

von Eckermann, H., 1948, The alkaline district of Alno Island: Sveriges Geol. Undersok. Ser. Ca., v. 36.

Wendlandt, R.F., and Harrison, W.J., 1979, Rare-earth partitioning between immiscible carbonate and silicate liquids and CO_2 vapor: results and implications for the formation of light rare-earth enriched rocks: Contributions to Mineralogy and Petrology, v. 69, p. 409-413.

White, S.F., 1980, Petrology of the Cenozoic igneous rocks of the Lytle Creek area, Bear Lodge Mountains, Wyoming; unpublished M.S. thesis, University of North Dakota, 69 p.

Wilkinson, Michael, 1982, Petrology and alteration in the core of the Bear Lodge Tertiary intrusive complex, Bear Lodge Mountains, Crook County, Wyoming: unpublished M.S. thesis, University of North Dakota, 127 p.

Wooley, A.R., 1982, A discussion of carbonatite evolution and nomenclature, and the generation of sodic and potassic fenites: Mineralogical Magazine, v. 46, p. 13-17.

IGC FIELD TRIP T131: DEVILS TOWER AND THE MISSOURI BUTTES

Frank R. Karner
Department of Geology and Geological Engineering and
North Dakota Geological Survey
University of North Dakota, Grand Forks

Don L. Halvorson
Department of Geology and Geological Engineering
University of North Dakota, Grand Forks

INTRODUCTION

This paper provides a brief introduction to the geology of the Devils Tower-Missouri Buttes area derived mainly from Halvorson (1980), Karner (1981), Lisenbee and others (1981) and Karner and Halvorson (1987). Other sources of information are cited as used. Devils Tower is the most widely known landmark of the northern Great Plains located at the northwestern edge of the Black Hills-Bear Lodge Mountains uplift as shown in Figures 1-3 of the paper by Karner and Patelke (1989) in this volume. It has been a sacred place for Native American peoples, a communal recreation site for early white settlers, and a must for traveling geologists since the Custer expedition of 1875 described in Newton and Jenney (1880). Devils Tower is recognized as the world's finest example of columnar jointing in a phonolite monolith. The Missouri Buttes are a cluster of five trachytic to phonolitic buttes west of Devils Tower.

GEOLOGICAL DESCRIPTIONS

Cenozoic Igneous Associations

Devils Tower and the Missouri Buttes are major features in the westernmost of a series of Cenozoic igneous centers that extend westward from the northern Black Hills (Figure 2 in Karner and Patelke, 1989). They are adjacent to, and may be satellitic to, the Bear Lodge Mountains center (Jenner, 1989). At Devils Tower and the Missouri Buttes, shallow intrusive and extrusive rocks form a continental phonolite-trachyte-quartz latite igneous association. This association is also present in the central Bear Lodge Mountains where carbonatite intrusions, altered pyroxenites, and igneous and metasomatic alkalic rocks form an intriguing, second igneous association. This deep-seated, second association is only partly observable at the surface but has been extensively revealed in the subsurface by explorational drilling. Potassium fenites, together with thorium and rare-earth mineralization, complete this classic suite of alkalic igneous and metasomatic rocks.

Devils Tower

Devils Tower is a steep-sided igneous body that shows spectacular columnar jointing. It consists of phonolite porphyry gradational to trachyte porphyry. The Tower is surrounded by talus and the remnants of a large volume of alloclastic breccia. It rises 382 m above the Belle Fourche River, to a maximum elevation of 1,545 m. Devils Tower is surrounded by a talus sheet, which radiates outward some 335 m. Talus is about 46 m thick near the base of the Tower, but thins rapidly. Three possible outcrops of phonolite have been noted within the talus sheet; these could be radiating dikes, or part of the original uneroded igneous body.

An elliptical knoll of alloclastic breccia is located northwest of the Tower. This breccia may at one time have been plastered against the phonolite of the Tower, somewhat resembling the volcanic necks in the Mount Taylor region New Mexico which have phonolite interiors sheathed in agglomerate.

The Devils Tower phonolite and trachyte are coarsely porphyritic with a gray to olive gray aphanitic groundmass. Phenocrysts consist of anorthoclase, aegirine-augite, and sphene in a trachytic groundmass of albite, microcline, analcime, aegirine, nepheline, and nosean. Common alteration or replacement products are calcite, zeolites, hematite, clay, and analcime. The principal mafic mineral is a green aegirine-augite, often found as phenocrysts with a dark rim of aegirine, and as groundmass needles. Analcime is one of the major constituents of the groundmass. Both nosean and nepheline occur as microphenocrysts; nosean appears as very small euhedral to subhedral grains that range in color from pale bluish gray to brown.

The Missouri Buttes

The Missouri Buttes are smaller bodies of columnar-jointed trachyte and phonolite. Five peaks rise to a maximum height of 241 m. They are arranged in a roughly rectangular pattern which tapers southward, with three buttes along the western edge and two along the eastern edge. Alloclastic breccia occurs south of the northwest butte, between the two northern buttes and on a plateau in the triangular area between the three southern buttes. The Missouri Buttes consist of trachyte porphyry and phonolite porphyry which are very similar to Devils Tower porphyries.

Sedimentary Geology

The sedimentary rocks exposed in the Devils Tower-Missouri Buttes area range in age from Triassic to Quaternary; total thickness is about 290 m. The oldest rocks are the red sandstones of the (Permian-Triassic) Spearfish Formation. The best exposures are east of Devils Tower, where 46 m of reddish brown interbedded sandstone, siltstone, and claystone are present. Gypsum occurs throughout this sequence as thin stringers along bedding planes and in fractures. The late Jurassic Sundance Formation is 60 to 130 m thick and is overlain by the Cretaceous Lakota and Fall River sandstones and siltstones 75 to 125 m thick, the Skull Creek Shale 12 m thick, the Newcastle Sandstone 6 m thick, and the Mowry Shale 9 m thick. The youngest unit is the Oligocene White River Group, consisting of a coarse feldspathic sandstone and gray calcareous claystone unit that caps large areas west and north of the Missouri Buttes. The sedimentary beds all dip westward at a shallow angle of less than 1/2 degree.

AGE, EMPLACEMENT AND FORMATION OF JOINT PATTERNS

Both Devils Tower and the Missouri Buttes are interpreted as the erosional remnants of volcanic necks by Halvorson (1980). Eocene ages of 40 Ma for Devils Tower and 50 Ma for the Missouri Buttes have been determined (Halvorson, 1980).

Emplacement

Toward the end of the Cretaceous, after the marine Fox Hills Formation was deposited, Laramide uplift occurred. Its major expression in the Black Hills was the uplift of two large blocks of Precambrian basement (Karner and Patelke, 1989). The western edge of the northern block is the Black Hills monocline just west of the Missouri Buttes. The uplift of this northern block is significant because it probably produced deep fractures in the crust which provided a route for magma to reach the surface. Before the emplacement of the Missouri Buttes and Devils Tower, probably in late Cretaceous or Paleocene time, movement of magma resulted in the formation of several domes in this area. At about the end of the Eocene, magma in the vicinity of the domes approached or perhaps reached the surface. Remnants of this volcanism are the Missouri Buttes and Devils Tower.

A maximum of approximately 2000 m of sedimentary material could have been deposited on top of the present Fall River erosional surface prior to the major Laramide uplift. This estimate is based on thickness of a vertical section from the Skull Creek Shale to the Fox Hills monocline about 10 km to the west. With the uplift, deposition virtually ceased and erosion began. By late Oligocene time, most of the present surface was exposed, as is shown by the unconformable contact of the late Oligocene White River Formation with the underlying Fall River and Skull Creek Formations. Using radiometric ages and estimated erosion rates, the range of cover of Devils Tower at the time of emplacement was 200 m to 1000 m and Missouri Buttes 800 m to 1200 m. Both of these igneous bodies lie within a depression in otherwise nearly horizontal strata. The depressions probably formed by collapse of the igneous masses back into partially evacuated magma conduits. The collapse around Devils Tower is uniform, but at the Missouri Buttes irregularities are caused by a dome to the north. The Missouri Buttes depression is deepest at the southern end where it reaches 91 m in depth, and shallowest at the northern end where it overrides the dome. The Skull Creek, Newcastle, and Mowry Formations appear to have been pulled into the depression at the southern end where dips are as high as 48°. Faults in the area of Devils Tower and the Missouri Buttes show vertical displacement and are apparently related to this collapse around the igneous bodies. The collapse would also reduce the calculated amount of sedimentary cover at the time of emplacement by as much as 100 m.

Joint Formation

The Devils Tower and Missouri Buttes volcanoes are thought to have originally vented to the surface through more than a hundred meters of sediments. Most of the pyroclastic material was blown out through these vents; some remained as a lining for the vent walls. After the more fluid phases of the eruption had ceased, cooling began from the surface downward, and from the periphery inward; this cooling history is responsible for the series of joints that parallels the circumference of the igneous mass, and for the set of vertical radial cracks which are normal to the peripheral jointing. Continued marginal cooling formed horizontal columnar joints which eventually turned upward as vertical temperature gradients became greater than horizontal temperature gradients.

REFERENCES

Halvorson, D.L., 1980, Geology and petrology of the Devils Tower, Missouri Buttes, and Barlow Canyon area, Crook County, Wyoming: Grand Forks, University of North Dakota, Ph.D. dissertation, 218 p.

Jenner, G.A., 1984, Tertiary alkalic igneous activity, potassic fenitization, carbonatic magmatism, and hydrothermal activity in the central and southeastern Bear Lodge Mountains, Crook County, Wyoming (M.S. thesis): Grand Forks, University of North Dakota, 232 p.

Jenner, G.A., 1989, Eocene igneous activity and related metasomatic and hydrothermal events, Bear Lodge Mountains, Crook County, Wyoming, in Karner, F.R., ed. Devils Tower--Black Hills alkalic igneous rocks and general geology: 28th International Geological Congress Field Trip Guidebook T131, p. 50-66.

Karner, F., 1981, Geologic relationships in the western centers of the northern Black Hills Cenozoic igneous province, in Rich, F.J., ed. Geology of the Black Hills, South Dakota and Wyoming: Field Trip Guidebook for G.S.A. Rocky Mountain Section Meeting, American Geological Institute, p. 126-133.

Karner, F.R. and Halvorson, D.L., 1987, The Devils Tower, Bear Lodge Mountains, Cenozoic igneous complex, northeastern Wyoming in Geological Society of America Centennial Field Guide-Rocky Mountain Section, p. 161-164.

Karner, F.R. and Patelke, R.L., 1989, General Geology of the Black Hills and Bear Lodge Mountains, in Karner, F.R., ed. Devils Tower--Black Hills alkalic igneous rocks and general geology: 28th International Geological Congress Field Trip Guidebook T131, p. 3-6.

Lisenbee, A., Karner, F., Fashbaugh, E., Halvorson, D., O'Toole, F., White, S., Wilkinson, W., and Kirchner, J., 1981, Geology of the Tertiary intrusive province of the northern Black Hills, South Dakota and Wyoming, Field Trip 2, in Rich F.J., ed., Geology of the Black Hills, South Dakota and Wyoming: Field Trip Guidebook for G.S.A. Rocky Mountain Section Meeting, American Geological Institute, 221 p.

Newton, H.A. and Jenney, W.P., 1880, Report on the geology and resources of the Black Hills of Dakota: U.S. Geographical and Geological Survey of the Rocky Mountain Region, p. 1-222.

FIELD GUIDE DAY 6: GEOLOGY OF THE DEVILS TOWER AND MISSOURI BUTTES REGION

Frank R. Karner
Department of Geology and Geological Engineering and
North Dakota Geological Survey
University of North Dakota, Grand Forks

Don L. Halvorson
Department of Geology and Geological Engineering
University of North Dakota, Grand Forks

DAY 6: Thursday July 6, 1989

- **The Eocene Phonolite Monolith of Devils Tower National Monument (Figure 1)**
- **Eocene Trachyte and Alloclastic Breccia of the Missouri Buttes**
- **Sedimentary Strata of the Devils Tower-Missouri Buttes**
- **Introduction to the Geology of the Bear Lodge Mountains**

Cumulative Miles

0.0

0.0 Aro Restaurant, downtown Sundance, Wyoming. **Follow Hwy. 14 west.** This field guide is based on material in Halvorson (1980), Lisenbee and others (1981), and Karner and Halvorson (1987). Additional information is contained in the introductory paper in this volume by Karner and Halvorson (1989). Sundance Mountain is the prominent hill just south of Sundance. This large exposure of quartz latite is interpreted to be a dissected volcanic cone of flow and fragmental units. It is apparently related to the igneous center of the Bear Lodge Mountains just to the northwest. Leaving Sundance, bluffs to the south expose sandstones and shales of the Jurassic Sundance Formation. These are underlain by red beds of the Permo-Triassic Spearfish Formation. Looking to the northeast, north of the town of Sundance, limestone quarries in the Permian Minnekahta Formation can be seen.

2.4

2.4 Intersection Forest Service Road 838. Rd. 838 leads north to the core of the Bear Lodge Mountains. **Continue west on Hwy. 14.** Spearfish Formation exposed along road.

1.5

3.9 Rupe Hill. Exposures in road cuts in Sundance Formation, Redwater? Member. Road continues through exposures of the upper Sundance Formation.

2.9

6.8 Southern flank of Houston Creek Dome centered about 1 km to the north. Lower Sundance Formation is exposed in the center of the dome. Road continues west for about 3 mi. in the Redwater Member of the upper Sundance Formation. As road turns northwest, beds are of successively younger age (Jurassic Morrison Formation, Cretaceous Lakota and Fall River Formations).

4.6

11.4 Terrace gravels are exposed at the top of Sunny Divide.

2.7

14.1 **STOP 1. WONDERVIEW: SCENIC VIEW OF DEVILS TOWER AND THE MISSOURI BUTTES.**

Devils Tower (Figure 1) and the Missouri Buttes are visible from this area. They form the major features in this westernmost igneous center of the northern Black Hills province. They consist of phonolite, trachyte and alloclastic breccia.

Early explorations in the Devils Tower area were made by army expeditions in the late 1850's.

FIGURE 1 Devils Tower on west bank of Belle Fourche River, south of Hulett, Wyoming. Columnar igneous rock rising above Sundance and Spearfish Formations (photo in Darton, 1909).

In 1875, Newton served as geological assistant to a United States Geological Survey party to the Black Hills, and it was he who first described the Tower and its rock. In his published report (Newton and Jenney, 1880), he gave this description of Devils Tower: "It is a great rectangular obelisk of trachyte, with a columnar structure, giving it a vertically striated appearance, and it rises 625 feet, almost perpendicular, from its base."

The Missouri Buttes are situated about 5 km northwest of Devils Tower. Five peaks rise above the Fall River Formation plateau to a maximum height of 241 m. Newton also examined the Missouri Buttes, identifying the rock there as trachyte and the agglomerate as volcanic tuff.

Theories on the mode of emplacement of Devils Tower soon began to develop. Carpenter (1888) proposed that it was the neck of an extinct volcano. Jaggar (1901) worked out a rather complete general geology of the Devils Tower area, proposing that Devils Tower and the Missouri Buttes were originally both part of a large tabular igneous body thickened in the center, with a relatively flat floor, and vented under the Missouri Buttes. Darton (1909) offered a third theory of origin. He proposed that the Missouri Buttes and Devils Tower represent two separate laccoliths with individual vents. Robinson (1956) constructed a detailed map of the Devils Tower National Monument area, and presented a fourth theory of origin:

"The evidence gathered during the present investigation...suggests that Devils Tower is a body of intrusive igneous rock, which was never much larger in diameter than the present base of the Tower, and which at depth (1000 feet or more) is connected to a sill or laccolith type body."

Halvorson (1980) concludes that Devils Tower and the Missouri Buttes are the erosional remnants of volcanic necks. **Continue west on Hwy. 14.**

1.6

15.7 Mass movement of shaly units on Odin Hill in the Inyan Kara Group causes highway instability especially on curves situated on steeper slopes. Road continues northwest down section through Morrison Formation and Redwater Member of the Sundance Formation. Road then follows plateau on Hulett Sandstone Member of the Sundance.

4.0

19.7 Carlile or Devils Tower Junction. **Turn north onto Hwy. 24 to Devils Tower.** Lak Member containing belemnites located along Hwy 14 just southeast of the junction.

0.7

20.4 Miller Creek. Road has followed valley down through the Hulett Sandstone Member into the Stockade Beaver Member, the basal unit of the Sundance Formation. Road climbs valley wall to continue on a plateau of the Redwater Member of the Sundance Formation. Along this plateau, called "Tower Divide", there are many excellent views of Devils Tower and the Missouri Buttes.

3.7

24.1 Road descends into Campstool Draw through lower Sundance Formation, Gypsum Springs Formation, and into the red beds of the Spearfish Formation.

0.7

24.8 Exposures of Hulett Sandstone of Sundance Formation.

0.3

25.1 Exposure of shales of Stockade Beaver Member of Sundance Formation.

0.3

25.4 Exposure of Spearfish Formation.

0.4

25.8 Road Junction Hwy. 24 and entrance road to Devils Tower National Monument. **Follow entrance road north and west to monument gate.**

0.0

25.8 **STOP 2. DEVILS TOWER NATIONAL MONUMENT: GEOLOGY OF DEVILS TOWER. LUNCH STOP.**

We will stop at the visitor center and follow the trail around the Tower. Other stops can be made within the monument as time allows.

President Theodore Roosevelt established Devils Tower as the first United States National Monument in 1906. No collecting is allowed within the monument.

Devils Tower is interpreted to be an erosional remnant of a volcanic neck. The landmark consists of columnar jointed phonolite porphyry surrounded by talus and the remnants of a possibly large volume of alloclastic breccia.

The phonolite porphyry of the Tower is intruded through and appears to rest upon a "platform" of the Sundance Formation, which dips gently inward toward the tower. This platform stands as a resistant hill, surrounded on all sides by streams, and capped by an armor of igneous talus, which has impeded further erosion (Figure 1). The Tower itself rises almost vertically 225 m above the Jurassic platform. The upper 180 m is characterized by distinctive columnar jointing. The lower section is up to 37 m in height and forms the massive base of the Tower. The slightly elliptical perimeter measures 225 by 300 m, the top is 55 by 90 m. From a distance, the top appears to be flat, though it is actually rounded, particularly toward the southeast. The top of the basal section (the "shoulder") is defined by outward-flaring columns, which dip toward the perimeter about 15° and merge with the blocky jointing of the base. Above the shoulder stand huge columns which taper from 2 to 5 m diameter at the base to a 1.5 m diameter at the top. The effect of the upward tapering is to yield an overall slope angle of from 75° to 85°. While most of the columns are five-sided, some have four or six sides. In many places, columns merge or split. The top 45 m of the Tower is extensively horizontally jointed. Small joint blocks that have fallen from this area show the beginnings of spheroidal weathering. The lower limit of the horizontal jointing approximates the former upper level of the Fall River Formation. These resistant beds impeded erosion so that only the top section of the Tower was exposed to weathering. When the resistant beds were finally removed, the rest of the Tower was rapidly denuded.

Devils Tower is surrounded by a talus sheet, which thins rapidly outward from the Tower base. Although most of the present talus is from the upper columns, it does not necessarily follow that the original size of the Tower is constrained by the dimensions of the present base. At Hulett, 16 km to the northeast, old stream terraces of the Belle Fourche River are littered with rock from Devils Tower; this indicates the removal of a considerable amount of talus material through the Quaternary. Evidence of four stages of talus alteration is present in the talus sheet; these, together with relict soil horizons, probably coincide with climatic changes that occurred during the Pleistocene Epoch. Such periods of talus development and removal, together with the possible outlier of phonolite and the presence of related alloclastic breccia, imply that the original igneous body may have been somewhat larger and of a different shape than the presently exposed Tower.

----- **Leave Devils Tower National Monument.**

----- **STOP 3. MISSOURI BUTTES – OPTIONAL.** The Missouri Buttes are located about five miles west of Devils Tower. They may be reached by several secondary roads depending on travel conditions and field trip plans. The geology of the Missouri Buttes is covered briefly in the preceding paper by Karner and Halvorson (1989).

----- **STOP 4. INTRODUCTION TO GEOLOGY OF THE BEAR LODGE MOUNTAINS-OPTIONAL.**

This introduction can be provided by an optional return route through the complex. The geology of the Bear Lodge Mountains is summarized in the papers in this volume by Jenner (1989a, 1989b).

----- **Return to Sundance and Apache Motel.**

REFERENCES

Carpenter, F.R., 1889, Ore deposits of the Black Hills of Dakota: American Inst. Min. Engr. Trans., v. 17, p. 587.

Darton, N.H., 1909, Geology and water resources of the northern portion of the Black Hills and adjoining regions in South Dakota and Wyoming: U.S. Geological Survey Professional Paper 65, 105 p.

Halvorson, D.L., 1980, Geology and petrology of the Devils Tower, Missouri Buttes, and Barlow Canyon area, Crook County, Wyoming: Grand Forks, University of North Dakota, Ph.D. dissertation, 218 p.

Jaggar, T. A., Jr., 1901, The laccoliths of the Black Hills, with a chapter on experiments illustrating intrusion and erosion by Ernest Howe: U.S. Geol. Survey, 21st Annual Rpt., pt. 4, p. 489-599.

Jenner, G.A., 1989a, Eocene igneous activity and related metasomatic and hydrothermal events, Bear Lodge Mountains, Crook County, Wyoming, in Karner, F.R., ed. Devils Tower--Black Hills alkalic igneous rocks and general geology: 28th International Geological Congress Field Trip Guidebook T131, p. 50-66.

Jenner, G.A., 1989b, I-Type and S-Type Carbonatites? Evidence from the Bear Lodge Mountains, Crook County, Wyoming in Karner, F.R., ed. Devils Tower--Black Hills alkalic igneous rocks and general geology: 28th International Geological Congress Field Trip Guidebook T131, p. 75-82.

Karner, F.R. and Halvorson, D.L., 1987, The Devils Tower, Bear Lodge Mountains, Cenozoic igneous complex, northeastern Wyoming in Geological Society of America Centennial Field Guide-Rocky Mountain Section, p. 161-164.

Lisenbee, A., Karner, F., Fashbaugh, E., Halvorson, D., O'Toole, F., White, S., Wilkinson, W., and Kirchner, J., 1981, Geology of the Tertiary intrusive province of the northern Black Hills, South Dakota and Wyoming, Field Trip 2, in Rich, F.J., ed., Geology of the Black Hills, South Dakota and Wyoming: Field Trip Guidebook for G.S.A. Rocky Mountain Section Meeting, American Geological Institute, 221 p.

Newton, H.A. and Jenney, W.P., 1880, Report on the geology and resources of the Black Hills of Dakota: U.S. Geographical and Geological Survey of the Rocky Mountain Region, p. 1-222.

Robinson, C.S., 1956, Geology of the Devils Tower, National Monument, Wyoming: U.S. Geological Survey Bulletin 1021-I, p. 289-302.

IGC FIELD TRIP T131: I-Type and S-Type Carbonatites? Evidence from the Bear Lodge Mountains, Crook County, Wyoming

Gordon A. Jenner
Exxon Company, Midland, Texas

INTRODUCTION

The Bear Lodge Mountains, at the western end of a series of Cenozoic igneous intrusions that extend in a N75^{o}W direction from the northern Black Hills (Figure 2, Karner and Patelke, 1989), contain a central igneous core which records a complex history of alkalic igneous activity, potassic fenitization, carbonatitic magmatism, and hydrothermal activity (Jenner, 1984 and Jenner, 1989 in this guidebook). Potassium-argon dates of sanidines indicate that the igneous activity occurred from Paleocene through early Eocene time with ages ranging from 50.5 to 38.3 Ma (McDowell, 1971, Staatz, 1983).

Two distinct types of carbonatite can be found in the Bear Lodge (Figure 1). The first, exposed in the northern portions of the main igneous dome, exhibits substantial enrichment in strontium, thorium, and rare-earths. The second lacks the enrichment of these elements and is exposed in the southeastern portions of the main igneous dome approximately 1 km from the contact between the Cenozoic igneous rocks and overlying Paleozoic strata. Field, petrographic, and geochemical evidence suggests that the enriched carbonatite is of primary igneous origin, I-type carbonatite, while the depleted type is the product of the localized fusion of the Mississippian Pahasapa Limestone, S-type carbonatite.

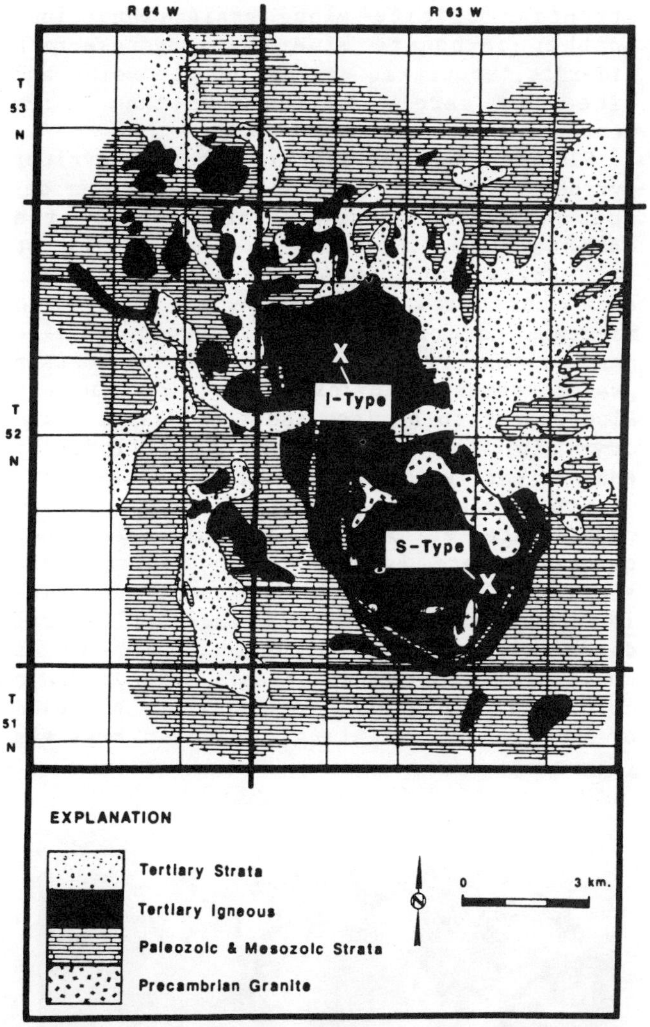

Figure 1 Geologic map of the Bear Lodge Mountains showing the distribution of Precambrian granitic rocks, Paleozoic and Mesozoic strata, Tertiary igneous rocks, and younger Tertiary sedimentary rocks (generalized from Staatz, 1983). Location of exposures of I-type and S-type carbonatites are shown by white X's.

I-TYPE CARBONATITES

Occurrence

Outcrops of I-type carbonatite are uncommon in the Bear Lodge. Two dikes are exposed in exploration trenches at the surface, one in the NW 1/4, NE 1/4, Section 20, T-52-N, R-63-W and the other in the SE 1/4, SE 1/4, Section 7, T-52-N, R-63-W (Figure 1). The dikes vary from 1 to 9 m in thickness and appear highly altered with pervasive iron-oxide staining. Surrounding alkali trachyte and intrusive breccia exhibit similar alteration and staining.

Although relatively rare at the surface, drilling in the vicinity of the exploration trenches shows that the I-type carbonatite is common in the subsurface of the Bear Lodge Mountains. Cores reveal steeply dipping veins and dikes that range from several millimeters to several tens of meters in thickness. Unlike their surface counterparts, the subsurface veins and dikes appear relatively fresh with little evidence of post-emplacement alteration.

Mineralogy

Strontian calcite is the principal mineral of I-type carbonatite, comprising from 60 to 95 percent of the rock. Strontianite occurs in variable amounts and in several specimens forms as much as 20 percent of the rock. The strontianite has a turbid appearance in plane-polarized light, commonly contains inclusions of barite, and appears to replace strontian calcite along grain boundaries. Another carbonate mineral, manganese-rich siderite, occurs in several of the veins and dikes but rarely exceeds more than a few percent.

Fluorite and the sulfide minerals pyrite, pyrrhotite, chalcopyrite, galena, and sphalerite, are common accessory minerals in I-type carbonatite. Fluorite appears to be of late-stage magmatic or hydrothermal origin and occurs with strontianite and rare-earth minerals to form irregular microcrystalline patches that replace portions of the coarser grained strontian calcite. Sulfide minerals also appear to be late and form disseminated crystals that replace and poikilitically enclose earlier formed strontian calcite.

Other minerals commonly found in I-type carbonatites are biotite, aegirine, sanidine, quartz, apatite, and magnetite. These minerals are highly variable in abundance, occurring in trace, accessory, or minor amounts. Several dikes also contain significant amounts of sodic actinolite which appears dark green in color and frequently forms clusters composed of numerous, radiating, acicular crystals (Figure 2).

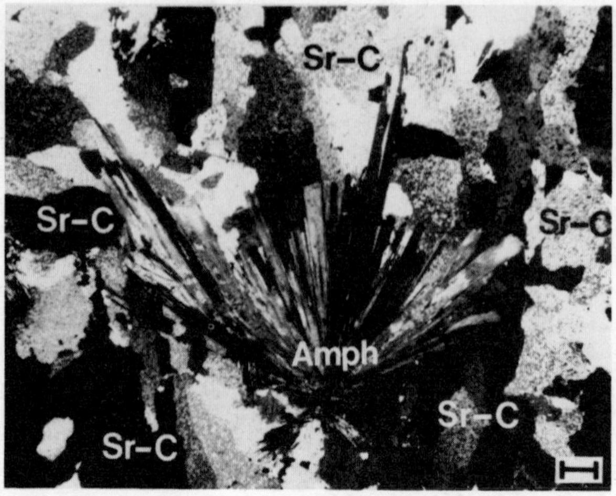

Figure 2 Photomicrograph of I-type carbonatite. Radiating crystals of sodic actinolite (Amph) enclosed within a fine to medium-grained matrix composed of strontian calcite (Sr-C). Bar scale represents 0.25 mm. Crossed polars.

The characteristically high concentration of rare-earth elements in the I-type carbonatite of the Bear Lodge is reflected by a diverse suite of rare-earth minerals. Bastnaesite [(Ce,La)FCO_3] and synchysite [(Ce,La)Ca $(CO_3)_2F$] are the principal rare-earth minerals in the altered dikes exposed at the surface (Staatz, 1983) while the fresher subsurface veins and dikes contain varying amounts of ancylite [$(Ce,La)_4(Sr,Ca)_3(CO_3)_7(OH)_4$] $3H_2O$, burbankite[$(Na,Ca,Sr,Ba,Ce)_6(CO_3)_5$], and carbocernaite [$(Na,Ca,Sr,Ce)CO_3$] (Figure 3a, b,c). Two unidentified rare-earth minerals were also observed in the core. The first of these minerals exhibits a platy to micaceous crystal habit (Figure 4) and appears to be an alteration product of anycylite. The second unidentified mineral is a rare-earth phosphate that typically occurs in clusters composed of numerous cigar-shaped crystals (Figure 5). These unidentified species probably represent new rare-earth minerals although further work is necessary to confirm this conclusion.

S-TYPE CARBONATITES

Occurrence and Evidence for Intrusive Origin

S-type carbonatites in the Bear Lodge are exposed approximately 5 km southeast of I-type carbonatite. They are best observed above the southern wall of Ogden Canyon in the SW 1/4, SW 1/4, Section 27, T-52-N, R-63-W and the SE 1/4, SE 1/4, Section 28, T-52-N, R-63-W (Figure 1). Carbonate bodies have been described by previous workers as xenoliths of Precambrian marble (Chenoweth, 1955) or as xenoliths of the Mississippian Pahasapa Limestone (Staatz, 1983). However, the exposures exhibit the following features which indicate an intrusive origin:

1) The dike-like form of the carbonate bodies and their sharp contacts with the surrounding alkalic igneous rocks (Figure 6). They are tabular in shape and trend in a NNW direction that is roughly perpendicular to the strike of the nearest Paleozoic sedimentary rocks;
2) The lack of bedding, fossils, or other features indicative of a sedimentary origin;
3) The presence in the carbonate bodies of numerous xenoliths of latite and trachyte porphyry (Figure 7). These xenoliths are frequently fractured and veined, indicating that the carbonate was in a volatile-rich, liquid state when injected into the surrounding older and obviously solidified alkalic igneous rocks;
4) The presence of sandstone xenoliths, presumably derived from the adjacent Deadwood Formation of Late Cambrian and

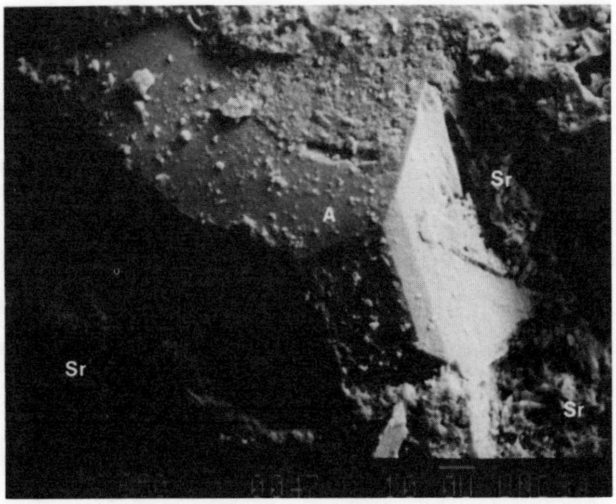

Figure 3 Rare-earth minerals from I-type carbonatites of the Bear Lodge Mountains.

(a) Scanning electron photomicrograph of euhedral ancylite crystal (A) enclosed within a matrix of microcrystalline strontianite (Sr). Bar scale represents 10 microns.

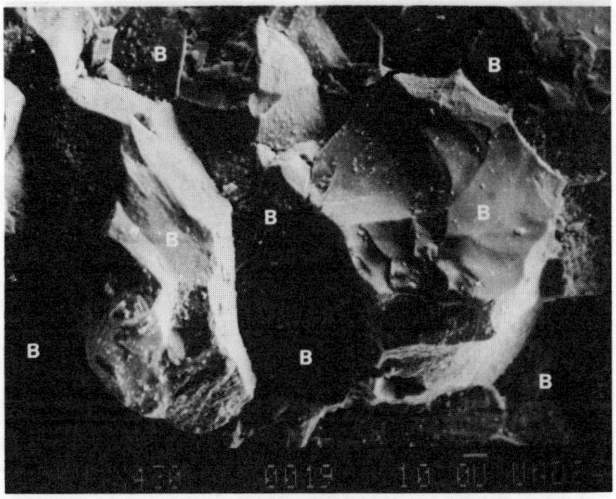

(b) Scanning electron photomicrograph of anhedral burbankite crystals (B) exhibiting pronounced conchoidal fractures. Bar scale represents 10 microns.

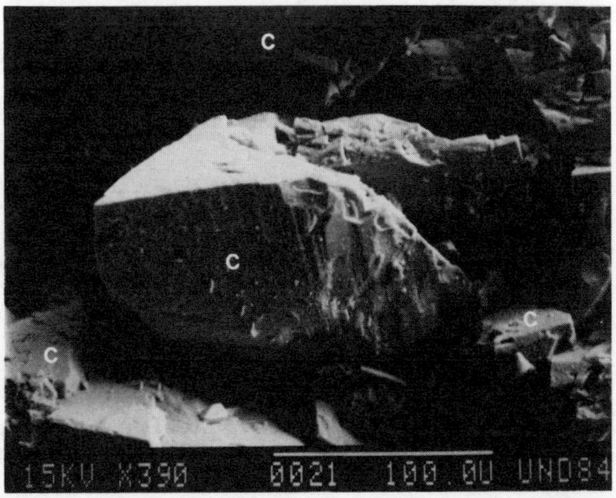

(c) Scanning electron photomicrograph of euhedral carbocernaite crystals (C). Bar scale represents 10 microns.

Early Ordovician age. Diagenetic quartz overgrowths are preserved in several of these sandstone xenoliths (Figure 8); and

5) The occurrence of small veins of calcite in the wall rocks adjacent to the outcrops. The veins "dehydrate" the sericite of the wall rocks, suggesting that the calcite crystallized from a liquid injected at relatively high temperature (Figure 9). The dikes of S-type carbonatite range up to 20 m in thickness with the longest exposed over a distance of approximately 100 m. They are less altered than the surface exposures of I-type carbonatites to the northeast and appear light gray to medium gray in color with only minor iron-oxide staining.

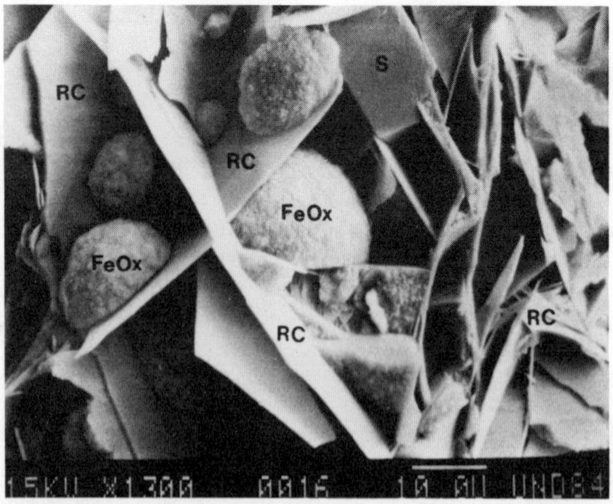

Figure 4 Scanning electron photomicrograph of unidentified rare-earth mineral (RC) exhibiting pronounced micaceous habit. The mineral is intergrown with strontianite (S) and spheres of iron oxide (FeOx). Bar scale represents 10 microns.

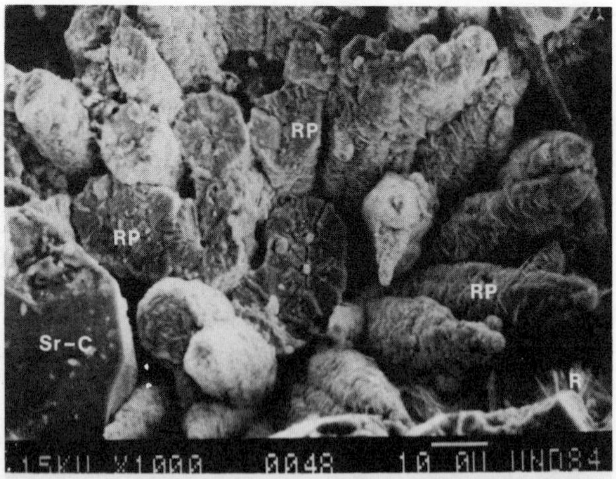

Figure 5 Scanning electron photomicrograph of unidentified rare-earth phosphate mineral (RP) intergrown with strontian calcite (Sr-C). Bar scale represents 10 microns.

Mineralogy

Calcite is the principal carbonate mineral in S-type carbonatite, composing from 90 to 95 percent of the rock. The calcite lacks the strontium enrichment characteristic of I-type carbonatite. In several of the veins and dikes 1-2 percent siderite is an accessory constituent.

Other minerals found in S-type carbonatite include sanidine, quartz, chalcedony, barite, and oxides of iron, titanium, and manganese. Most of the sanidine and quartz occurs in association with xenoliths. Chalcedony forms irregularly shaped patches, often with inclusions of barite, that replace the surrounding calcite. The oxides of iron, titanium, and manganese occur as alteration products of original pyrite and manganese-rich ilmenite. Strontianite and rare-earth minerals, characteristic components of I-type carbonatite of the Bear Lodge, are absent in S-type carbonatite.

Figure 6 Dike of S-type carbonatite outcropping above Ogden Canyon in the southeastern Bear Lodge Mountains. Note hammer for scale.

STABLE ISOTOPE GEOCHEMISTRY AND THE ORIGIN OF THE BEAR LODGE CARBONATITES

Introduction

A major school of thought on the origin of carbonatite is that it is derived through the assimilation of either sedimentary limestone or metamorphic marble by an ascending magma, a notion first proposed by Daly (1910, 1933). A second, more widely accepted hypothesis, is that carbonatite is magmatic in origin and forms from juvenile CO_2 (Heinrich, 1966, p. 272-329).

Recent studies of the isotopic compositions of carbon and oxygen in calcites from carbonatites provide a strong argument for the magmatic model. Results have shown that values of $\delta^{13}C$ and $\delta^{18}O$ are relatively uniform and significantly lower than values in calcites

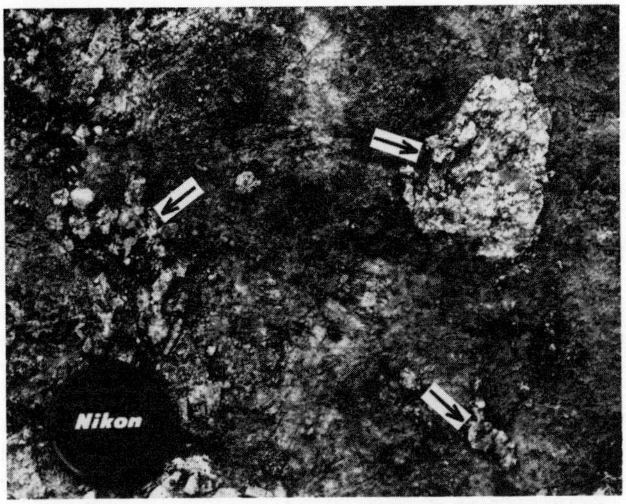

Figure 7 Brecciated xenoliths of latite and trachyte porphyry (arrows) in outcrop of dike of S-type carbonatite.

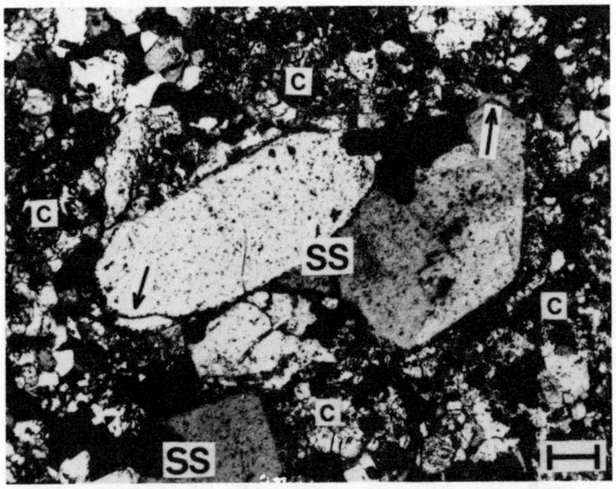

Figure 8 Photomicrograph of sandstone xenoliths (SS) in S-type carbonatite. The xenoliths are surrounded by a fine-grained matrix of calcite (C). Note preservation of diagenetic quartz overgrowths in the xenoliths (arrows). Bar scale represents 0.25 mm. Crossed polars.

from marine limestone (Baertschi, 1957; Gonfiantini and Tongorgi, 1964; Kukarenko and Dontosova, 1964; Taylor and others, 1967; Conway and Taylor, 1969; Deines, 1970; Deines and Gold, 1973). Conway and Taylor (1969, p. 618) conclude "it is improbable that limestone could be the source of carbonatitic material except through complete isotopic reconstitution at great depth".

Oxygen isotope geothermometry based upon coexisting calcite-magnetite pairs from the Oka, Quebec and Magnet Cove, Arkansas carbonatites provide additional evidence in support of the magmatic model. Equilibration temperatures range

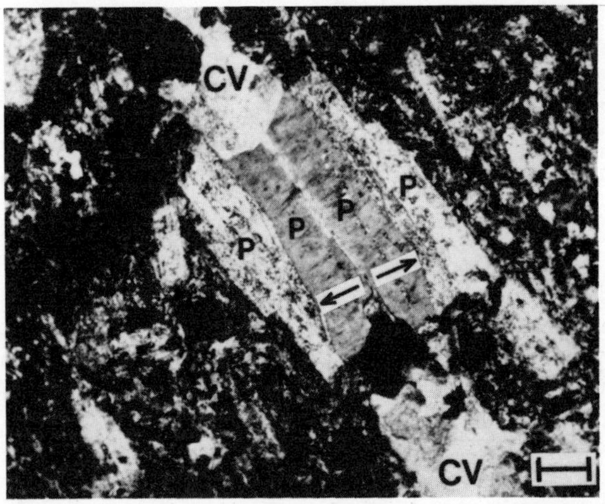

Figure 9 Photomicrograph of vein of S-Type carbonatite (Cv) crosscutting a plagioclase Phenocryst (P) in sample of latite porphyry. The plagioclase phenocryst is altered to sericite except in the area adjacent to the vein where it appears clear and devoid of sericite. Bar scale represents 0.25 mm. Crossed polars.

between 595° and 855°C, values that are comparable to estimates for the crystallization temperatures of other plutonic igneous rocks such as gabbros and granites (Conway and Taylor, 1969).

Based upon their study of the carbonatites of the Laacher See district of West Germany and the Alno Island district of Sweden, Taylor and others (1967) have proposed, as a "working hypothesis", that the isotopic range for calcites from primary igneous carbonatites is between -8.0 and -5.0 per mil relative to PDB for carbon and between +6.0 and +8.5 per mil relative to SMOW for oxygen. Because the values of $\delta^{13}C$ for carbonatitic calcites are nearly identical to values determined for mantle-derived diamonds and kimberlitic calcites (Deines and Gold, 1973; Deines, 1980) and the values of $\delta^{18}O$ are virtually the same as values determined for mantle-derived ultramafic and mafic rocks (Taylor, 1968), it has been suggested that carbonatites form from mantle-derived CO_2 of relatively uniform composition.

Stable Isotope Geochemistry and Origin of Bear Lodge Carbonatites

Isotopic compositions of carbon and oxygen have been determined for calcites from the I-type and S-type carbonatites and compared with compositions determined for sedimentary calcites from the Mississippian Pahasapa Limestone, the major carbonate unit that outcrops along the flanks of the uplift. Isotopic compositions were determined at the Stable Isotope Laboratory of the Department of Geology and Geophysics at the University of Utah under the supervision of Dr. John Bowman using procedures outlined in Jenner (1984).

I-type Carbonatite. The isotopic compositions of carbon and oxygen in calcites from the unaltered drill cores of I-type carbonatite exhibit a rather limited range for both carbon (-7.0 to -10.6 per mil) and oxygen (+7.5 to +9.9 per mil). Three of the ten analyses fall within the field proposed by Taylor and others (1967) for primary igneous carbonatites while the remaining seven fall relatively close to the proposed field (Figure 10). The results are consistent with a primary igneous origin for I-type carbonatite with the CO_2 derived from a mantle source region.

In contrast to fresh, drill-core samples, calcites from the altered and oxidized surface exposures of I-type carbonatite exhibit a significant enrichment in $\delta^{18}O$. Differences in the isotopic composition of carbon between the two types of carbonatite are negligible. The enrichment of $\delta^{18}O$ is probably related to isotopic exchange between the calcites of I-type carbonatite and the hydrothermal solutions responsible for the alteration and oxidation of the samples. One of the notable characteristics of the hydrothermal alteration in the Bear Lodge is that it appears most severe at or near the surface and gradually diminishes with depth, a feature that strongly suggests a hydrothermal system dominated by heated, meteoric groundwaters. However, meteoric groundwaters are typically +5 to +10 per mil lighter in $\delta^{18}O$ relative to SMOW (Taylor, 1971; Hoefs, 1980) and therefore, much lighter in $\delta^{18}O$ than the calcites of altered I-type carbonatite. If heated meteoric groundwaters are responsible for the $\delta^{18}O$ enrichment in the altered surface exposures, then the isotopic exchange must have occurred at relatively lower, nonmagmatic temperatures where fractionation of $\delta^{18}O$ into calcite would be greater than the fractionation of $\delta^{16}O$.

S-type Carbonatite. Calcites from altered and oxidized surface exposures of S-type carbonatite also exhibit a significant enrichment in $\delta^{18}O$ relative to the fresh drill core samples of I-type carbonatite as well as a less pronounced enrichment in $\delta^{13}C$. The enrichment in $\delta^{18}O$ is probably, in part, attributable to exchange with hydrothermal solutions in a manner similar to that proposed for the $\delta^{18}O$ enrichment in calcites from the altered and oxidized surface exposures of I-type carbonatite. However, the enrichment in $\delta^{13}C$ is anomalous and, together with the lack of strontianite and rare-earth minerals in S-type carbonatite, suggests a possible genetic difference between S-type and I-type carbonatites.

It is hypothesized that S-type carbonatite resulted from the assimilation of sedimentary limestone by an alkalic, silicate magma. The isotopic compositions of carbon and oxygen in the calcites from S-type carbonatite approach the compositions determined for sedimentary calcites from the Pahasapa Limestone with one of the analyses of S-type carbonatite virtually indistinguishable from Pahasapa Limestone. The depletion of $\delta^{13}C$ in S-type calcites relative to Pahasapa Limestone calcites is possibly attributable to loss of CO_2 by degassing during assimilation or by limited isotopic exchange with the intruding magma.

Liquid immiscibility between coexisting silicate and carbonate melts has been demonstrated experimentally in several systems (Koster van Groos and Wyllie, 1966, 1968, 1973; Wendlandt and Harrison, 1979; Freestone and Hamilton, 1980). Thus if sedimentary limestone was assimilated by a silicate magma in the Bear Lodge Mountains it is probable that the resulting carbonate melt existed immiscibly within the silicate magma. A similar interpretation has been proposed for carbonate ocelli found within diabase dikes of the Tarr albitite complex of the southeastern Sinai (Bogoch and Magaritz, 1983). The carbonate ocelli, which lack the light rare-element signature of igneous carbonatites (Gold, 1963), are believed to represent immiscible segregations formed by stoping and melting of marble wallrocks by the intruding diabase magma. It is interesting to note that the calcites of the ocelli are depleted in ^{13}C relative to the calcites of the surrounding marble wallrocks by approximately 1 to 8 per mil, a depletion that is similar in magnitude to that observed between calcites of S-type carbonatite and Pahasapa Limestone in the Bear Lodge Mountains.

CONCLUSIONS

The isotopic compositions of carbon and oxygen in calcites from Bear Lodge I-type carbonatite, S-type carbonatite, and the Pahasapa Limestone suggest: (1) I-type carbonatite is of primary igneous origin and formed from mantle-derived CO_2; (2) hydrothermal alteration, if it involved heated meteoric groundwaters, probably occurred at lower, nonmagmatic temperatures; and (3) S-type carbonatite formed through the stoping and assimilation of sedimentary limestone by an alkalic silicate magma.

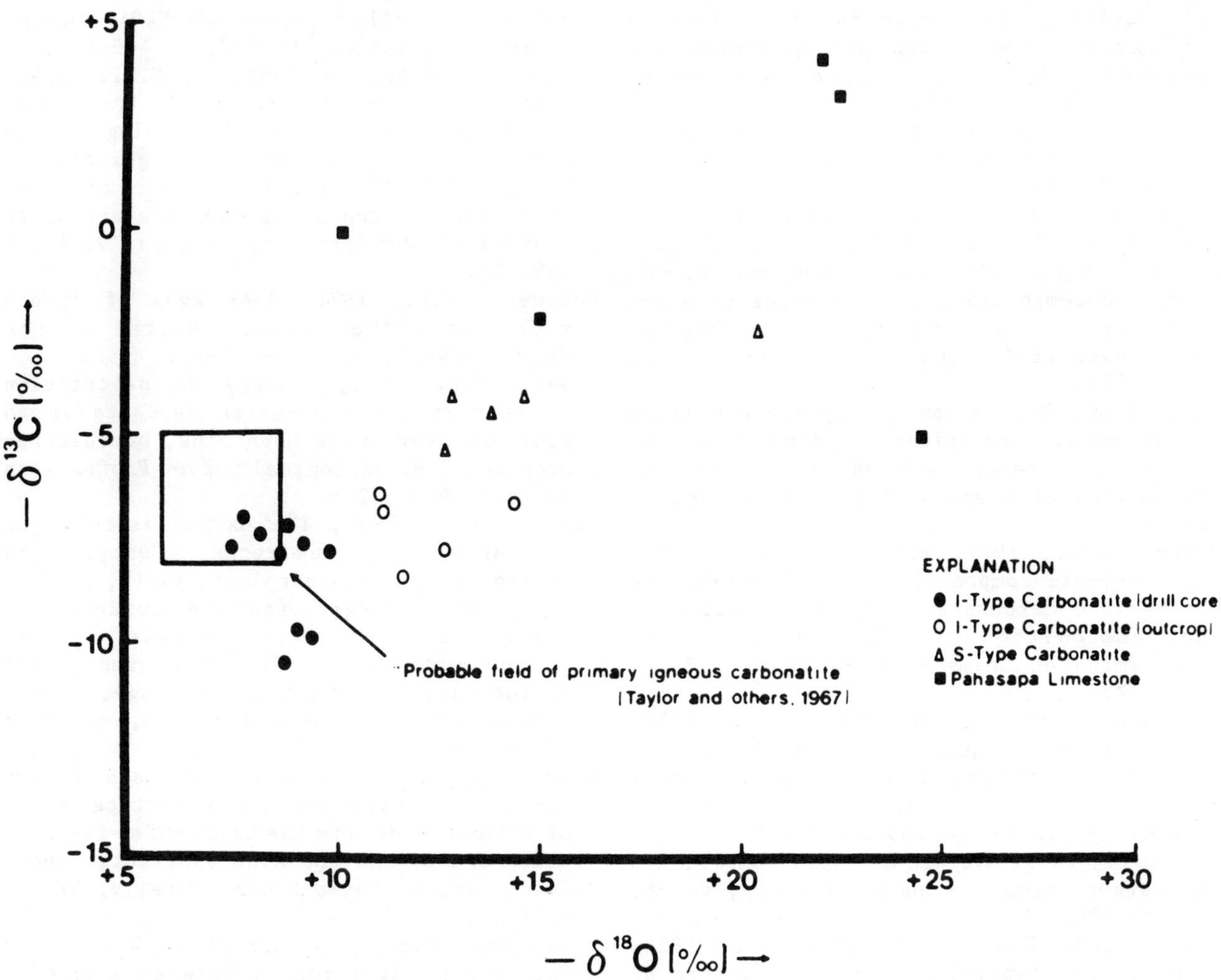

Figure 10 Plot of $\delta^{13}C$ versus $\delta^{18}O$ comparing isotopic compositions of Bear Lodge calcites from subsurface I-type carbonatites (closed circles), surface exposures of I-type carbonatites (open circles), surface exposures of S-type carbonatites (open triangles), and surface exposures of the Mississippian-age Pahasapa Limestone (closed squares). Also shown is the probable field of primary igneous carbonatite proposed by Taylor and others (1967).

ACKNOWLEDGEMENTS

I thank Dr. John Bowman and Ray Lambert at the Isotope Laboratory at the University of Utah and Dr. Robert Stevenson at the Natural Materials Analytical Laboratory (NMAL) at the University of North Dakota for their valuable assistance and generous use of laboratory facilities. I also appreciate the critical review of various aspects of this project by Dr. Frank Karner, Dr. Dexter Perkins, and Dr. Don Halvorson at the University of North Dakota. Finally, I am indebted to Exxon Company, U.S.A. for granting me permission to publish this study and to JaJuan DeGraffenreid for her valuable assistance in the preparation of this manuscript.

REFERENCES

Baertschi, P., 1957, Messung and Deutung relative Haufigkeitsvariatonen von ^{18}O and ^{13}C in Karbonatgesteinen and Mineralien: Schweizer Mineralogische and Petrographische Mitt., v. 37, p. 73-152.

Bogoch, R., and Magaritz, M., 1983, Immiscible silicate-carbonate liquids as evidenced from ocellar diabase dykes, southeast Sinai: Contributions to Mineralogy and Petrology, v. 83, p. 227-230.

Chenoweth, W.C., 1955, The sedimentary and igneous rocks, structure and mineral deposits of the southeastern Bear Lodge Mountains, Crook County, Wyoming: unpublished M.S. thesis, 220 p.

Clayton, R.N., and Mayeda, T., 1963, The use of bromine pentafluoride in the extraction of oxygen from oxides and silicates for isotopic analysis: Geochimica et Cosmochimica Acta, v. 27, p. 43-52.

Daly, R.A., 1910, Origin of alkaline rocks: Geological Society of America Bulletin, v. 21, p. 87-118.

_____, 1933, Igneous rocks and the depths of the earth: New York, McGraw Hill, 598 p.

Deines, Peter, 1970, The carbon and oxygen isotopic composition of carbonates from the Oka carbonatite complex, Quebec, Canada: Geochimica et Cosmochimica Acta, v. 34, p. 1199-1225.

_____, 1980, The carbon isotopic composition of diamond: relations to diamond shape, color, occurrence, and vapor composition: Geochimica et Cosmochimica Acta, v. 44, p. 943-961.

Deines, Peter, 1973, and Gold, D.P., 1973, The isotopic composition of carbonatite and kimberlite carbonates and their bearing on the composition of deep-seated carbon: Geochimica et Cosmochimica Acta, v. 37, p. 1709-1733.

Freestone, I.C., and Hamilton, D.L., 1980, The role of liquid immiscibility in the genesis of carbonatites - an experimental study: Contributions to Mineralogy and Petrology, v. 73, p. 105-117.

Gold, D.P., 1963, Average chemical composition of carbonatites: Economic Geology, v. 58, p. 988-996.

Gonfiantini, R., and Tongiorgi, E., 1964, composition isotipique des carbonatites du Kaiserstuhl: EURATOM pub. EUR, 1827, d, e, f, p; p. 193-199.

Gries, J.P., and Tullis, E.L., 1955, Geologic history of the Black Hills, in Black Hills Field Conference: North Dakota Geologic Society Guidebook, p.32.

Heinrich, W.E., 1966, The geology of carbonatites: New York, Rand McNally, 555 p.

Hoefs, Jochen, 1980, Stable isotope geochemistry: New York, Springer Verlag, 208 p.

Jenner, G.A., 1984, Tertiary alkalic igneous activity, potassic fenitization, carbonatitic magmatism, and hydrothermal activity in the central and southeastern Bear Lodge Mountains, Crook County, Wyoming: unpublished M.S. thesis, University of North Dakota, 232 p.

Koster van Groos, A.F., and Wyllie, P.J., 1966, Liquid immiscibility in the system Na_2O-Al_2O_3-SiO_2-CO_2 at pressures to 1kb.: American Journal of Science, v. 264, p. 234-255.

_____, 1968, Liquid immiscibility in the join $NaAlSi_3O_8$-Na_2CO_3-H_2O and its bearing on the genesis of carbonatites: American Journal of Science, v. 266, p. 932-967.

_____, 1973, Liquid immiscibility in the join $NaAlSi_3O_8$-$CaAl_2Si_2O_8$-Na_2CO_3-H_2O: American Journal of Science, v. 273, p. 465-487.

Kukharenko, A.A., and Dontsova, E.I., 1964, A contribution to the problem of the genesis of carbonatites: Economic Geology of the U.S.S.R., v. 1, p. 47-68 (in English).

McCrea, J.M., 1950, On the isotopic chemistry of carbonates and a paleotemperature scale: Journal of Chemistry and Physics, v. 18, p. 849-857.

McDowell, F.W., 1971, K-Ar ages of igneous rocks from the western United States: Isochron/West, no. 2, p. 1-20.

Staatz, M.H., 1983, Geology and descriptions of thorium and rare-earth deposits in the southern Bear Lodge Mountains, northeastern Wyoming: U.S. Geological Survey Professional Paper 1049-D, 52 p.

Taylor, H.P., 1968, The oxygen isotope geochemistry of igneous rocks: Contributions to Mineralogy and Petrology, v. 19, p. 1-71.

_____, 1971, Oxygen isotope evidence for large-scale interaction between meteoric ground waters and Tertiary granodiorite intrusions, western Cascade Range, Oregon: Journal of Geophysical Research, v. 76, p. 7855-7874.

Taylor, H.P., Frechen, Josef, and Degens, E.F., 1967, Oxygen and carbon isotope studies of carbonatites form the Laacher See district, West Germany and the Alno district, Sweden: Geochimica et Cosmochimica Acta, v. 31, p. 407-430.

Wendlandt, R.F., and Harrison, W.J., 1979, Rare-earth partitioning between immiscible carbonate and silicate liquids and CO_2 vapor: results and implications for the formation of light rare earth-enriched rocks: Contributions to Mineralogy and Petrology, v. 69, p. 409-413.

FIELD GUIDE DAY 7: GEOLOGY OF THE BEAR LODGE MOUNTAINS

Frank R. Karner[1], Gordon A. Jenner[2], Stanley F. White[3], and Don L. Halvorson[4]

DAY 7: Friday, July, 2, 1989

- **Eocene Phonolite-Trachyte-Quartz Latite Subvolcanic Igneous Association**
- **Carbonatite-Pyroxenite-Alkalic Rock Intrusive Association**
- **Fenitization, Hydrothermal Alteration, and Thorium and Rare-Earth Element and Associated Mineralization**
- **Sedimentary Strata of the Bear Lodge Region**

Cumulative Miles

0.0

0.0 Aro Restaurant, downtown Sundance, Wyoming. **Follow Hwy. 14 west**. This field guide is based on material in Lisenbee and others (1981), Staatz (1983), and Karner and Halvorson (1987). Sundance Mountain to south.

2.4

2.4 **Turn north (left) on Rd. 838 toward Warren Peaks and core of Bear Lodge Mountains.**

1.4

3.8 Exposures of redbeds of Spearfish Formation (Permian-Triassic). Dip is south away from intrusive core of Bear Lodge.

0.4

4.2 Vista West entrance.

0.5

4.7 Exposure of Permian Minnekahta Limestone above Spearfish. Dip is south.

0.2

4.9 Reuter Campground on the west side of the road is located on a plateau of the Minnekahta.

0.6

5.5 Exposure of red, sandy shale of Opeche Formation of Permian age above Minnekahta. Road continues north through exposures of the Pennsylvanian-Permian Minnelusa Formation consisting of buff-colored, limy sandstones.

0.3

5.8 Upper Minnelusa eolian sandstone with collapse breccia at top. This interval produces oil in the Powder River Basin.

0.8

6.6 Exposure of fossiliferous Mississippian Pahasapa Limestone. Dips are steep to the south.

0.2

6.8 Exposure of the Ordovician Whitewood? Limestone.

0.1

6.9 Exposure of sandstone of the Cambrian-Ordovician Deadwood Formation dipping about 65° south. A trachyte porphyry sill cuts the Deadwood here.

0.2

7.1 **STOP 1. ALTERED LATITE PORPHYRY AND TRACHYTE PORPHYRY WITH QUARTZITE AND ULTRAMAFIC XENOLITHS.**
Most of the rocks in the intrusive core of the Bear Lodge are extensively altered. Typical altered rocks are found in this road cut where latite porphyry intrudes sedimentary rocks (Deadwood?) near the southern contact of the Bear Lodge intrusion. The altered latite porphyry is a medium gray aphanitic rock with phenocrysts of altered plagioclase up to 2 cm long in an altered trachytic groundmass. Probable unaltered equivalents in other regions of the Bear Lodge are quartz-bearing and quartz free. Quartzite and pyroxenite inclusions occur within the latite and are interpreted as xenoliths.

Toward the north end of the outcrop, trachyte porphyry occurs

---

[1]Department of Geology and Geological Engineering and North Dakota Geological Survey, University of North Dakota, Grand Forks.
[2]Exxon Company, P.O. Box 1600, Midland, Texas.
[3]Maxus Energy Corporation, 717 North Harwood Street, Dallas, Texas.
[4]Department of Geology and Geological Engineering, University of North Dakota, Grand Forks.

in contact with the latite porphyry. Typical trachyte porphyry contains conspicuous alkali feldspar phenocrysts up to 4 cm long in an aphanitic to trachytic groundmass. Altered mafic clusters are probably the remains of pyroxene and amphibole phenocrysts. The groundmass consists of microscopic feldspar laths with a distinct trachytoid texture. Contacts between the trachyte and latite porphyry appear to be gradational over the distance of a few centimeters.

0.0

7.1 **Follow Rd. 838 north**. Outcrops of sills and dikes of altered felsite along road for next 0.7 miles.

0.6

7.7 Jcn. Rd. 899 to east.

0.1

7.8 **Follow logging road northwest**.

0.1

7.9 **STOP 2. HOUSTON CREEK SODALITE-BEARING PHONOLITE.** The rock exposed at this outcrop is dark green phonolite containing isolated to glomerophyric phenocrysts of sodalite. This and other phonolites in the Black Hills-Bear Lodge region rocks are believed to result from a final phonolitic phase of Cenozoic igneous activity. The exposure appears to be a dike cross-cutting trachyte porphyry rich in Precambrian granitic xenoliths. Darton (1905) first mapped the body as a tinguaite dike, one of a series of dikes cropping out in the Bear Lodge Mountains. Phenocrysts of sodalite stand out as violet to pale blue aggregates on fresh surfaces and weather to a pale brown color. A microcrystalline, trachytic groundmass is composed of alkali feldspar laths and acicular aegirine crystals. Patches of secondary analcime and calcite are abundant. Euhedral to subhedral, dodecahedroal phenocrysts 1-4 mm in size are zoned and extensively replaced by natrolite, calcite and analcime. Embayments and post-embayment exsolution rims are common. Microphenocrysts of altered nepheline and sanidine, aegirine-augite, and melanite garnet, are minor in abundance.

0.0

7.9 **Return to Rd. 838.**

0.1

8.0 **Turn left (north) on Rd. 838.** Numerous exposures of altered felsite for the next several miles.

0.8

8.8 Jcn. Rd. 839.

0.1

8.9 Cattle guard.

0.3

9.2 The Warren Peaks area is visible ahead. Hydrothermally altered felsite underlies most of this area.

0.3

9.5 Jcn. Rd. 830.

0.1

9.6 Jcn. Rd. 708. **Turn right toward the Warren Peaks Fire Tower and park.**

0.0

9.6 **STOP 3. WARREN PEAKS: ALTERED FELSITE, PHONOLITE DIKE, AND BRECCIATED PRECAMBRIAN GRANITIC ROCK.** Rock exposures at the fire tower, in road cuts and on the crests of the peaks include trachyte porphyry of the central portion of the Bear Lodge intrusion, xenoliths of granitic rock of the Archean Precambrian basement, and younger dikes of phonolite porphyry.

The flattened peak to the northwest was used as the site of a U.S. Air Force radar base. The nuclear reactor that was used for power has been dismantled but the site is still monitored for radiation leakage.

The most abundant rock type in the central core of the Bear Lodge Mountains is altered trachyte porphyry which appears to underlie much of the Warren Peaks area. The rock may be white to pinkish but most often is iron-stained and consists of sanidine phenocrysts in a trachytic, aphanitic groundmass. The trachyte porphyry also contains small, mafic inclusions which are interpreted as altered amphiboles and pyroxenes. Pyrite cubes or pseudomorphs after pyrite are also often present in accessory amounts. Excellent examples of euhedral sanidine crystals can be found in areas of weathered trachyte porphyry on top of Warren Peaks.

Brecciated granite occurs as xenolithic inclusions in many areas throughout the intrusion. One large block appears to be about 1200 x 240 m in outcrop area and forms four of the Warren Peaks. Much granitic rock occurs as clasts in breccia with other rock types and a fine grained igneous or fragmental matrix. The granites are grey to pink and weather to a yellowish color. They are medium to coarse grained with an allotriomorphic granular texture. Some clasts are gneissic. The granitic clasts consist of quartz, alkali feldspar (typically microcline perthite), oligoclase and minor biotite altered to chlorite, magnetite and accessory minerals. Feldspars may be extensively altered to sericite and epidote. Some of the granites appear to have been affected by alkali metasomatism (fenitization) which has selectively removed quartz leaving a porous, alkali feldspar-rich, syenitic rock. Chert is disseminated in the altered granites.

A dike of phonolite porphyry is exposed in the road cut at the turn-off to the fire tower. It crosscuts both trachyte porphyry and granite. The phonolite consists of phenocrysts of sanidine, aegirine-augite and altered nepheline in a trachytic groundmass. Sanidine phenocrysts are aligned subparallel to the orientation of the dike. Pyroxenitic inclusions are also present.

0.0

9.6 **Turn right (northwest) on Rd. 838.**

0.3

9.9 Jcn. with road to abandoned radar installation. **Bear left and follow road to radar base.**

0.2

10.1 Parking area near gate to radar installation. Around the perimeter of the fence many interesting examples of core rock of the Bear Lodge Mountains can be found. The exposures on the peak area just to the south show many of the same rock types. **Turn around and return to Rd. 838.**

0.2

10.3 **Turn left (northwest) on Rd. 838.**

0.1

10.4 Note many trenches and prospect pits dating from the 1950's when there was active exploration for Th and rare-earth elements (Staatz, 1983).

0.2

10.6 Junction with road to west.

0.5

11.1 Mineralized and fenitized felsitic rock.

0.3

11.4 View of Devils Tower and Missouri Buttes.

0.9

12.3 "Four Corners" intersection of Rd. 838, Rd. 851 and Rd. 847. The central Bear Lodge is traversed by roads which may not be passable at all times. The following road log to Stops 4-9 is suggested but other routes may be followed. **Turn left (southwest) on Rd. 847 toward Lytle Creek.**

1.2

13.5 Jcn. Rd. 838 and Rd. 860 to north and access south to Rd. 854. **Continue west-northwest, northwest and north along Lytle Creek on Rd. 847.**

4.0

17.5 Cross Lytle Creek. **Stop on north side of Lytle Creek and follow trail about 0.7 miles southeast.**

0.0

17.5 **STOP 4. LYTLE CREEK: CALC-ALKALI PHONOLITE-TRACHYTE AND TRACHYTE.**

Calc-alkali phonolite, calc-alkali trachyte, and trachyte crop out as small laccoliths of less than two square kilometers and as sills two to six meters thick. At this location in Section 2 only trachyte is present. At a laccolith about 2 km north in section 35, all three rock types are present. Stratigraphically, the calc-alkali phonolite-trachyte and trachyte laccoliths and sills were emplaced at or slightly above the Spearfish-Sundance contact. Silicification and/or color alteration of a zone 20-30 cm from the igneous contact are the major types of country rock alteration.

The calc-alkali phonolite-trachyte and trachyte are aphanitic gray to gray-green porphyritic rocks. Common fabrics are pilotaxitic and trachytic. Sanidine, sodian ferroaugite, and sphene occur as phenocrysts. Natrolite pseudomorphs of nepheline and melanite occur only in the calc-alkali phonolite-trachyte. The groundmass of all three rock varieties is composed of microcrystalline sanidine, acicular sodian ferroaugite, dolomite, magnetite, and apatite. Quartz is present in the groundmass of the trachyte and in some samples of calc-alkali trachyte.

Two distinct types of inclusions which are interpreted as xenoliths are common in the calc-alkali phonolite-trachyte and trachyte: Mafic inclusions are present in all three rock varieties and syenitic inclusions are present in the saturated calc-alkali trachyte and trachyte. The mafic inclusions range from approximately 1/4 mm to 4 cm in maximum dimension and are rounded to angular. Most are composed of sodian ferroaugite and alkali feldspar; but inclusions almost entirely composed of sodian ferroaugite are present. The syenitic inclusions are 1-2 mm in maximum dimension. Syenitic inclusions are irregularly shaped and composed of a hypidimorphic granular arrangement of alkali feldspar. A few alkali feldspar megacrysts contain cores of plagioclase. Mafic minerals are not present in the syenitic inclusions.

Chemically, these rocks have intermediate SiO_2 contents which range from 55 to 62 percent, are slightly peraluminous, and have K_2O/Na_2O ratios ranging from 1:1 to 2:7. The three rock varieties range from undersaturated to oversaturated in SiO_2 content. Normative nepheline abundances range from 13-18 percent and quartz values range up to 6 percent.

Syenitic inclusions in the saturated to oversaturated trachyte varieties may represent xenoliths of granitic material which are quartz-depleted as a result of metasomatic reaction with undersaturated, calc-alkaline phonolitic magma.

0.0

17.5 **Return on Rd. 847 to "Four Corners" intersection.**

5.2

22.7 "Four Corners" intersection. **Turn left (north) on Rd. 838.**

0.1

22.8 **STOP 5: BEAR LODGE TRACHYTE PORPHYRY, CARBONATITE, PYROXENITE AND MINERALIZATION.** Trachyte porphyry, fenitized trachyte porphyry, pyroxenitic xenoliths in syenite, carbonatite and copper mineralization are present in this area. The dominant rock type is trachyte porphyry similar to that described for Stop 1. Altered trachyte porphyry, Cu mineralization including turquoise?, and fenitized trachyte? occur in the prospect pit (mine?) alongside the road at this stop. The fenitized rock contains thin, monomineralic veinlets of coarse grained, bluish alkali feldspar. An exposure of fenitized trachyte occurs on Rd. 851 about 0.08 miles from the "Four Corners". I-type carbonatite (Jenner, 1989b) occurs as a dike at the east end of a ridge just northwest of the prospect pit. The carbonatite is a fine to medium grained, brown equigranular rock containing abundant calcite, minor fluorite, alkali feldspar, opaques and altered rare earth element-bearing mineral grains. Breccias, altered rocks and other calcite-bearing rocks are associated with the carbonatite.

Syenite in the vicinity of this stop commonly contains pyroxenitic xenoliths which consist primarily of green pyroxene, alkali feldspars and opaques. A good exposure for observing the pyroxenitic xenoliths is located about half way between Stop 5 and 6.

0.0

22.8 **Continue north on Rd. 838.**

0.7

23.5 Syenite with xenoliths of pyroxenite in trench at location mentioned in description of Stop 5.

0.1

23.6 Jcn. Rd. 860? to west.

0.7

24.3 Intersection of Rd. 838 and Rd. 897 to east. **Turn east on Rd. 897.**

0.1

24.4 **STOP 6: PSEUDOLEUCITE TRACHYTE PORPHYRY DIKES.** Pseudoleucite trachyte porphyry occurs as dikes 1-3 meters thick cutting trachyte. Contacts with the trachyte are well defined and marked by Fe-oxide staining. The pseudoleucite trachyte porphyry is light gray to gray. Phenocrysts of pseudoleucite range in size from approximately 0.5 mm to 12 mm in diameter and are glomerophyrically arranged. The groundmass is composed of cryptocrystalline to microcrystalline orthoclase, secondary natrolite, hematite, muscovite and dolomite. Trapezohedral, pseudoleucite phenocrysts comprise approximately 35 percent of these rocks. The pseudoleucite crystals consist primarily of orthoclase and are rimmed by radially arranged natrolite. The rock is extremely potassic and its low content of calcium and sodium is striking.

0.0

24.4 **Return west on Rd. 897.**

0.1

24.5 Junction Rd. 897 and Rd. 838. **Turn north (right) on Rd. 838.**

1.9

26.4 Junction Rd. 838, Rd. 849 and Rd. 843. **Turn east (right) on Rd. 843.**

1.6

28.0 Junction Rd. 843 and Rd. 841. **Turn south (right) on Rd. 841.**

3.0

31.0 **STOP 7: WHITE RIVER FORMATION (OLIGOCENE).**

Pebble-to cobble-size clasts of volcanic and granitic rocks are common in the White River Group (Oligocene) sandstones and gravels. The matrix is rich in calcite and quartz. The White River clast assemblages may be a major source of information on volcanic and plutonic rock types present in the Bear Lodge Mountains igneous center.

0.0

31.0 **Return to Junction Rd. 838 and Rd. 839 just south of Warren Peaks.** Best route may be Rd. 841 south to Rd. 851 (follows Whitelaw Creek) southwest and Rd. 838 south.

6.6 (approx.)

37.6 (approx.) Turn left (east) on Rd. 839.

1.4 (approx.)

39.0 (approx.) **STOP 8. OGDEN CANYON: S-TYPE CARBONATITE.**

This area is described by Jenner (1984, 1989b) and a part of his summary (1989b) is repeated here. The carbonatites are best observed above the southern wall of Ogden Canyon. An alternative explanation is that the carbonate bodies are xenoliths of sedimentary carbonate from the Precambrian basement, the Deadwood Formation or the Pahasapa Limestone. However, many features of the carbonate bodies suggest an intrusive origin:

1) The dike-like form and sharp contacts with the surrounding alkalic igneous rocks. They are tabular in shape and trend in a NNW direction that is roughly perpendicular to the strike of the nearest Paleozoic sedimentary rocks;

2) The lack of bedding, fossils, or other features indicative of a sedimentary origin;

3) The presence of numerous xenoliths of latite and trachyte porphyry. These xenoliths are frequently fractured and veined, indicating that the carbonate was in a volatile-rich, liquid state when injected into the surrounding older and obviously solidified alkalic igneous rocks;

4) The presence of sandstone xenoliths, presumably derived from the adjacent Deadwood Formation of Late Cambrian and Early Ordovician age. Diagenetic quartz overgrowths are preserved in several of these sandstone xenoliths and

5) The occurrence of small veins of calcite in the wall rocks adjacent to the outcrops. The veins "dehydrate" the sericite of the wall rocks, suggesting that the calcite crystallized from a liquid injected at relatively high temperature.

The dikes of S-type carbonatite range up to 20 m in thickness with the longest exposed over a distance of approximately 100 m. They are less altered than the surface exposures of I-type carbonatites to the northeast and appear light gray to medium gray in color with only minor iron-oxide staining.

Calcite is the principal carbonate mineral in S-type carbonatite and in several of the veins and dikes siderite is an accessory. Other minerals include sanidine, quartz, chalcedony, barite, and oxides of iron, titanium, and manganese. Most of the sanidine and quartz occurs in association with xenoliths. Chalcedony forms irregularly shaped patches, often with inclusions of barite, that replace the surrounding calcite. Strontianite and rare-earth minerals, characteristic components of I-type carbonatite of the Bear Lodge, are absent.

Jenner (1989b) suggests that S-type carbonatite formed through the stoping and assimilation of sedimentary limestone by an alkalic silicate magma.

0.0

39.0 (approx.) **Return to Jcn. Rd. 838 and Hwy. 14A via Rd. 839 and Rd. 838.**

7.6 (approx.)

46.6 (approx.) Intersection of Rd. 14 and Rd. 838. **Reset mileage and return east on Rd. 14 toward Sundance.**

1.1

47.7 (approx.) Junctions Rd. 14 and Rd. 116. **Turn south (right) on Rd. 116.**

0.1

47.8 (approx.) Junction Hwy. 116 and farm driveway. **Enter east on ranch driveway, park and walk to Stop 9.**

0.0

47.8 **STOP 9. SUNDANCE MOUNTAIN: DISSECTED VOLCANIC CONE OF QUARTZ LATITE FLOW AND FRAGMENTAL ROCKS - OPTIONAL STOP.**

Sundance Mountain is a dissected volcanic cone of quartz latite. It occupies an area of approximately 2 km^2, and shows local relief of

170 m. The most notable exposures are on the north and northwest sides of the mountain, where sheer cliffs rise above a prominent talus slope. Numerous grass- and pine-covered lobes of colluvium extend out from the base. Prominent joint sets can be identified on the northwest cliff face. Columnar jointing is most strongly developed in the lower two-thirds of the exposure. Subhorizontal joint sets, dominant near the top, are subparallel to a faint layering within the rock. The rock is a light to dark gray, slightly porphyritic quartz latite, with a predominantly cryptocrystalline groundmass consisting of alkali feldspar and quartz. Oligoclase occurs as zoned phenocrysts and microlites. The phenocrysts have distinct oscillatory zones, but have a narrow compositional range.

The quartz latite occurs in several different varieties; layered sequences, monolithic breccias, and massive types. Brecciated units are most common near the base of Sundance Mountain. Individual breccia units tend to be thin (1.5 to 5 m in thickness). The breccias consist of more than 60 percent angular to subangular clasts of quartz latite, set in a weakly isotropic groundmass. Breccias may be interbedded with massive and layered quartz latites; the contact between a given breccia unit and overlying layered rock is always a sharp disconformity. The layered sequences consist of alternating light and dark gray laminae approximately a millimetre thick. The layers tend to dip outward from the interior of Sundance Mountain, and also tend to fill in spaces between the angular breccia clasts. Structures in the layering often resemble soft-sediment deformation structures; small-scale folding may also be present. Folds are asymmetrical, and are overturned in the direction of dip. Layered sequences often grade upward into massive units. Massive quartz latite is most common in the middle part of the exposed section of Sundance Mountain. Such massively bedded units are generally 10 to 36 m in thickness.

0.0

47.8 **Travel to "The Sluice" restaurant in Spearfish, South Dakota for dinner and our final meeting. Return to the Alex Johnson Hotel in Rapid City. END OF TRIP.**

REFERENCES

Darton, N.H. and O'Harra, C.C., 1985, Aladdin [quadrangle], Wyoming-South Dakota-Montana: U.S. Geological Survey Geologic Atlas of the United States, Folio 128, 8 p.

Jenner, G.A., 1984, Tertiary alkalic igneous activity, potassic fenitization, carbonatitic magmatism, and hydrothermal activity in the central and southeastern Bear Lodge Mountains, Crook County, Wyoming: Grand Forks, University of North Dakota, M.S. thesis, 232 p.

Jenner, G.A., 1989b, I-Type and S-Type Carbonatites? Evidence from the Bear Lodge Mountains, Crook County, Wyoming in Karner, F.R., ed. Devils Tower--Black Hills alkalic igneous rocks and general geology: 28th International Geological Congress Field Trip Guidebook T131, p. 75-82.

Karner, F.R. and Halvorson, D.L., 1987, The Devils Tower, Bear Lodge Mountains, Cenozoic igneous complex, northeastern Wyoming in Geological Society of America Centennial Field Guide-Rocky Mountain Section, p. 161-164.

Lisenbee, A., Karner, F., Fashbaugh, E., Halvorson, D., O'Toole, F., White, S., Wilkinson, W., and Kirchner, J., 1981, Geology of the Tertiary intrusive province of the northern Black Hills, South Dakota and Wyoming, Field Trip 2, in Rich, F.J., ed., Geology of the Black Hills, South Dakota and Wyoming: Field Trip Guidebook for G.S.A. Rocky Mountain Section Meeting, American Goeological Institute, 221 p.

Staatz, M.H., 1983, Geology and description of thorium and rare-earth deposits in the southern Bear Lodge Mountains, northeastern Wyoming: U.S. Geological Survey Professional Paper 1049-D, 52 p.

Quaternary Volcanism of Long Valley Caldera and Mono-Inyo Craters, Eastern California

Long Valley Caldera, California
July 20–27, 1989

Field Trip Guidebook T313

Leader: *Roy A. Bailey*

Associate Leaders:
C. Dan Miller   Kerry Sieh

American Geophysical Union, Washington, D.C.

Published 1989 by American Geophysical Union

2000 Florida Ave., N.W., Washington, D.C. 20009

ISBN: 0-87590-620-6

Printed in the United States of America

COVER Long Valley caldera and Mono-Inyo Craters volcanic chain viewed from the southeast. Perspective drawing by Tau Rho Alpha, U.S. Geological Survey (from Alpha, Bailey, Lajoie, and Clark, 1983). Thin solid line from southeast to northwest is U. S. Route 395; line around north end of Mono Craters chain is California Route 120.

Leader:

Roy A. Bailey
U.S. Geological Survey
345 Middlefield Road, MS-910
Menlo Park, CA 94025

Associate Leaders:

C. Dan Miller
U.S. Geological Survey
Cascades Volcano Observatory
5400 MacArthur Blvd.
Vancouver, WA 98661

Kerry Sieh
Division of Geology and Planetary Sciences
California Institute of Technology
Pasadena, CA 91125

IGC FIELD TRIP T313
FIELD GUIDE TO LONG VALLEY CALDERA AND MONO-INYO CRATERS VOLCANIC CHAIN, EASTERN CALIFORNIA

Roy A. Bailey[1], C. Dan Miller[2], and Kerry Sieh[3]

GEOLOGIC SUMMARY

Setting

Long Valley caldera (Fig. 1) is located at the western edge of the Basin and Range Province straddling the eastern frontal fault escarpment of the Sierra Nevada, in which it forms a reentrant or offset commonly referred to as the "Mammoth embayment." The floor of the caldera ranges in elevation from 2000 m in its eastern half, where it is dominated by Lake Crowley and sage- and grass-covered Long Valley, to 2600 m in its western half, which is hillier and heavily forested. The caldera walls rise steeply to elevations of 3000-3500 m on all sides except the east and southeast, where the floor rises only 150 m before merging with the Volcanic Tableland at 2300 m elevation. The Mono-Inyo Craters volcanic chain extends from the western part of Long Valley caldera northward from Mammoth Mtn. to Mono Lake, a distance of 50 km. Although commonly described as subparallel to the Sierran front, the chain trends nearly due north at a noticeable angle to the northwest-trending Sierran faults (Figs. 1, 2). The prevolcanic basement in the area is mainly Mesozoic granitic rocks of the Sierra Nevada batholith and Paleozoic metasedimentary and Mesozoic metavolcanic rocks of the Mt. Morrison and Ritter Range roof pendants. The late Tertiary terrain upon which Long Valley volcanism was initiated was a maturely eroded upland drained by westward-flowing streams.

Volcanism

Volcanism associated with Long Valley caldera (Bailey and others, 1976; Bailey, 1989) began with widespread eruption of trachybasaltic-trachyandesitic lavas between 3.6 and 3.0 Ma. Erosional remnants of these precaldera lavas are scattered discontinuously over a 4000-km^2 area around the caldera (Fig. 2)--a distribution that suggests an extensive mantle source region for these initial mafic eruptions. Slightly younger rhyodacite domes and flows associated with these mafic lavas erupted near the north and northwest rims of the present caldera between 3.2 and 2.6 Ma. They probably represent the onset of deep-crustal magmatic accumulation and differentiation that eventually culminated in formation of the large shallow Long Valley magma chamber, from which subsequent more silicic eruptions originated. The first eruptions from this silicic chamber were on the northeast rim of the present caldera at Glass Mountain, where 1000 m of high-silica rhyolite domes, flows, and tuffs accumulated between 2.1 and 0.8 Ma (Metz and Mahood, 1985).

Catastrophic rupturing of the roof of the silicic magma chamber 0.73 Ma expelled at least 600 km^3 of rhyolite magma as plinian ash falls and incandescent ash flows. This partial emptying of the chamber caused collapse of its roof to form the 2-to-3-km-deep oval depression of Long Valley caldera. The resulting ash-flow deposits, the Bishop Tuff (Gilbert, 1938; Hildreth, 1979), inundated 1500 km^2 around the caldera and accumulated locally to thicknesses approaching 200 m on the Volcanic Tableland and to lesser thicknesses in upper Owens Valley, Adobe Valley, and Mono Basin. A large volume of the Bishop Tuff also ponded within the caldera as it collapsed; although the tuff is not exposed at the surface within the caldera, drill holes have confirmed that as much as 1500 m of the Bishop Tuff is buried beneath younger volcanic and sedimentary caldera fill. During this climactic eruption, associated plinian ash clouds drifted thousands of kilometers downwind and deposited an ash layer recognized as far east as Kansas and Nebraska (Izett, 1982). Distal ash-fall deposits equivalent to the Bishop Tuff (informally termed the Bishop ash) also are recognized in southern California (Merriam and Bischoff, 1975), as well as in deep-sea cores from the East Pacific Ocean (Sarna-Wojcicki and others, 1987).

[1] U.S. Geological Survey, Menlo Park, CA
[2] U.S. Geological Survey, Cascades Volcano Observatory, Vancouver, WA
[3] Division of Geological and Planetary Science, California Institute of Technology, Pasadena, CA

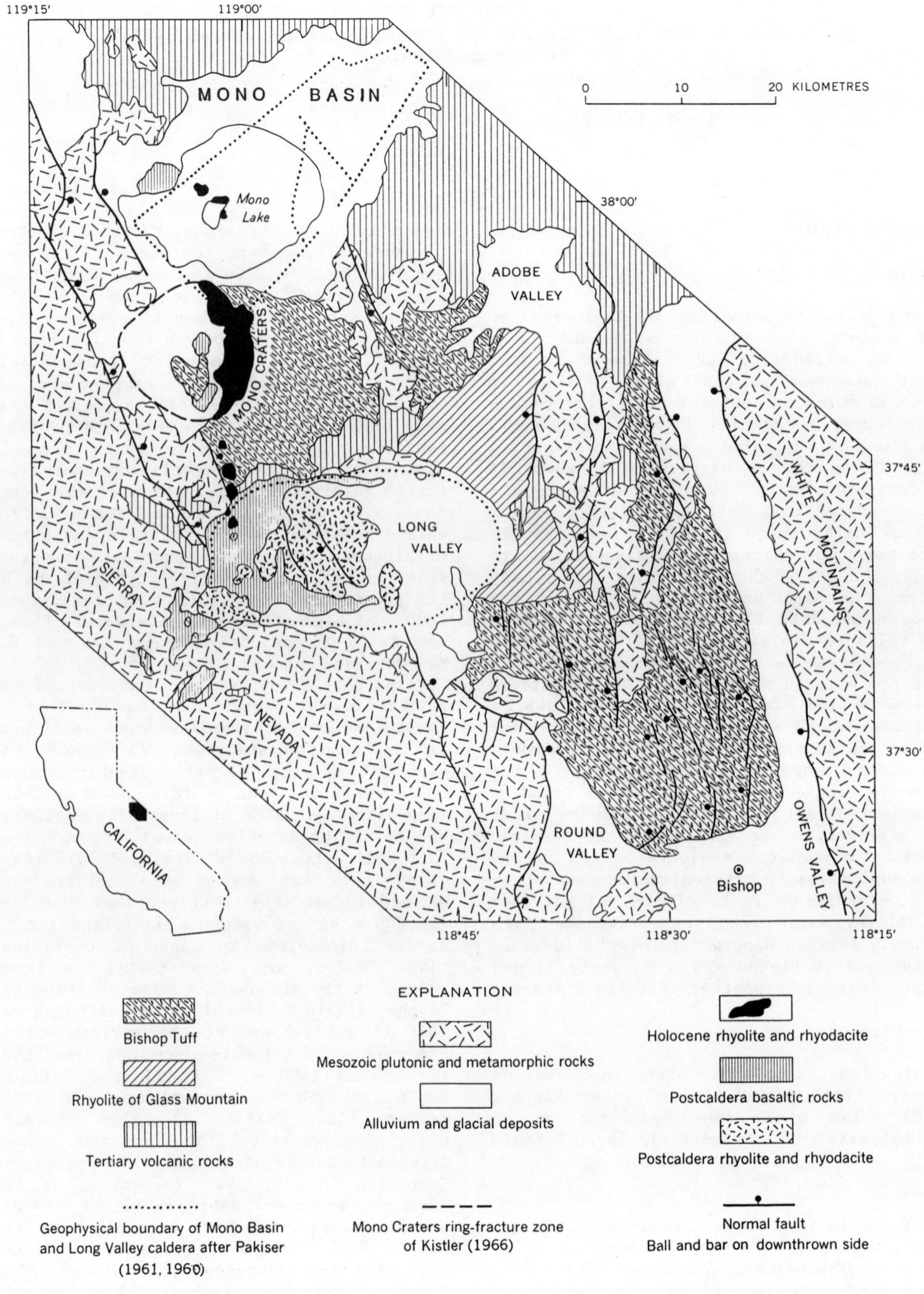

FIGURE 1 Generalized geologic map of the Long Valley–Mono Basin area, eastern California (from Bailey and others, 1976).

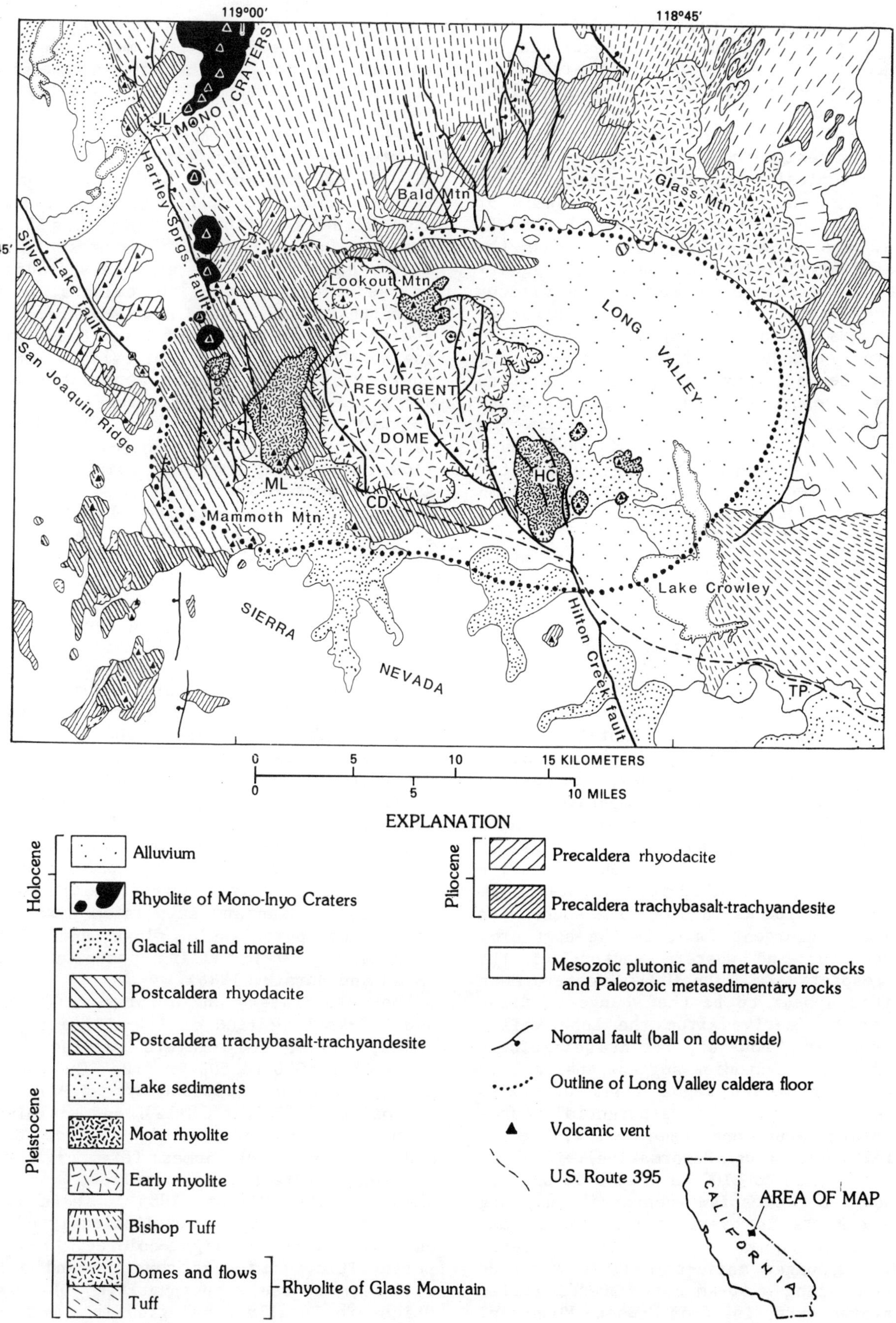

FIGURE 2 Geologic map of Long Valley caldera. JL: June Lake; ML: Mammoth Lakes; CD: Casa Diablo; HC: Hot Creek; TP: Toms Place (modified from Bailey, 1987).

After collapse of the roof of the magma chamber, rhyolitic volcanism continued on the caldera floor. Pyroclastic eruptions, followed by extrusion of thin, hot, fluid obsidian flows, produced a 100-500-m-thick sequence of postcaldera tephra and lavas, informally designated as the early rhyolite. They are typically aphyric to sparsely porphyritic, contain less than 5% crystals of plagioclase, hypersthene, biotite, and Fe-Ti oxides, and contrast strikingly with the preceding crystal-rich Bishop Tuff, suggesting a marked change in magmatic conditions after caldera collapse. Simultaneous renewal of magma pressure accompanying this early rhyolite episode uplifted, arched, and faulted the early rhyolite flows and tephra, forming a resurgent dome with a northwest-trending medial graben (Figs. 2, 5). The resurgent dome rose within 100,000 years after caldera collapse, between 730 and 650 ka as indicated by K-Ar ages of the contemporaneous obsidian flows (Bailey and others, 1976; Mankinen and others, 1986).

After a quiescent interlude of about 100,000 years, crystal-rich rhyolite again began erupting within the caldera, mainly in the moat, probably from ring fractures peripheral to the resurgent dome. This coarsely porphyritic rhyolite, informally designated as the moat rhyolite, typically contains up to 20% phenocrysts of plagioclase, quartz, sanidine, biotite, hornblende, and Fe-Ti oxides. It forms thick, steep-sided domes and flows, suggesting higher viscosity and lower temperature than the early rhyolite, and it probably signaled the onset of cooling and crystallization of the magma chamber. The moat rhyolite erupted at about 200,000-year intervals at 500, 300, and 100 ka in clockwise succession around the resurgent dome, in the northern, southeastern, and western sectors of the moat, respectively. The 100-ka western moat rhyolites appear to be the youngest extrusions so far derived from the Long Valley magma chamber. However, recent seismological and geodetic studies suggest that a body of partially molten magma still underlies the resurgent dome and is a potential source for future eruptions (see below, Recent Seismicity and Ground Deformation).

This 3.6-Ma to 100-ka mafic-to-silicic sequence of volcanism centered on Long Valley caldera is overlapped spatially and temporally by the Mono-Inyo Craters volcanic chain, a younger mafic-to-silicic sequence localized along a 50-km north-trending fissure system extending from Mammoth Mountain, on the southwestern caldera rim, through the western caldera moat to Mono Lake. This younger sequence began 300-200 ka with the eruption of trachybasaltic-trachyandesitic lavas in and near the west moat, where they accumulated to at least 250 m and poured around the resurgent dome sending lava tongues eastward into both the north and south moat (Fig. 2). Younger mafic vents erupted successively farther north near June Lake (20-40 ka) and at Black Point on Mono Lake (13,300 yr B.P.), suggesting localization along a northward-propagating fissure system. During these mafic eruptions, rhyodacite domes and flows erupted sporadically in the western part of the caldera, as well as farther north in and near Mono Lake. The greatest accumulations of these domes are on the northwest and southwest caldera margins, where the fissure system apparently intersected caldera ring fractures. On the southwest caldera rim between 200 and 50 ka, repeated extrusion of rhyodacite domes and flows built the imposing cumulovolcano of Mammoth Mountain.

Rhyolites began erupting along the Mono-Inyo fissure system about 35 ka--first at the Mono Craters chain, northwest of the caldera, and more recently at the Inyo Craters chain, which spans the northwest caldera rim and extends into the west moat. The Mono Craters form a 17-km-long, arcuate chain of 30 or more coalesced rhyolite domes, flows, and craters ranging in age from about 35 ka to 600 yr B.P. (Wood, 1977b; Sieh and Bursik, 1986). They are composed of high-silica rhyolite and consist predominantly of aphyric to sparsely porphyritic obsidian, pumiceous glass, and tephra (Kelleher and Cameron, 1988). The most recent of these eruptions occurred at the northern end of the chain from a 6-km-long line of vents that ejected about 1 km^3 of magma as plinian and subplinian pyroclastic falls and pyroclastic flows and surges, followed by vent-filling domes and flows (Sieh and Bursik, 1986).

The Inyo Craters chain forms a discontinuous 12-km-long line of low-silica rhyolite domes, flows, and craters ranging in age from about 6000 to 500 yr B.P. The youngest Inyo eruptions (600-500 yr B.P.) began explosively (Figs. 10-12) and culminated with extrusion of Obsidian, Glass Creek, and Deadman Creek domes (Fig. 9), which apparently were fed by a shallow, 8-km-long, rhyolite dike (Miller, 1985). These youngest domes consist of two distinctly different rhyolites--a light-colored, coarsely porphyritic, pumiceous rhyolite and a more silicic, sparsely porphyritic obsidian--which are locally commingled in spectacular "marbled-cake" fashion. Petrologic and geochemical studies indicate that the two types probably came from separate magma chambers

(Sampson and Cameron, 1987). The youngest Inyo eruptions succeeded the youngest Mono eruption by only a few years (Sieh and Bursick, 1986), so that the northern and southern ends of the Mono-Inyo chain were active almost simultaneously about 600 yr B.P.

Several phreatic eruptions occurred almost contemporaneously with the youngest Inyo magmatic eruptions. Shortly before the Inyo tephra eruptions, phreatic explosions broke out on the north face of Mammoth Mountain, forming six small craters subparallel to the caldera wall. Shortly after the tephra eruptions, several phreatic eruptions occurred in the west moat near Deer Mountain, a 115-ka dome of moat rhyolite, forming several large craters, two of which contain the Inyo Craters Lakes. The latter eruptions, according to radiocarbon dating and dendrochronological evidence (Wood, 1977a), occurred between 1340 and 1460 A.D. The most recent eruptive activity in the region, however, occurred in Mono Lake between 1720 and 1850 A.D. with the emergence of Paoha Island during intrusion of a rhyolite cryptodome (Stine, 1984). There is also a report of an apparent sulfurous steam eruption in Mono Lake during an earthquake in 1890 (Holden, 1892), but the original source of the report (The Homer Mining Index, 1890) was of notoriously questionable reliability and the event has not been corroborated.

Structure

The dominant structural trend in the Long Valley region is the northwest-trending Sierran frontal fault zone, which parallels the northwest elongation of the Sierran plutons and the associated roof pendants. Within the caldera this northwest trend is reflected by the medial graben on the resurgent dome and by the alignment of the vents of the early and moat rhyolites. The caldera ring fracture is not exposed, but its general location, coinciding with the annular caldera moat, is suggested by the distribution of the moat rhyolite vents peripheral to the resurgent dome.

Two major Sierran frontal faults transect the caldera -- the Hilton Creek fault on the southeast and the Hartley Springs fault on the northwest (Fig. 2). These faults, which show major pre- and postcaldera displacement outside the caldera, terminate as major escarpments at the caldera margin and continue within only as minor discontinuous splinters. This marked difference in intracaldera and extracaldera displacement suggests that the subcauldron magma chamber has tended to hydraulically dampen or absorb intracaldera fault movements and that only recently has the chamber roof thickened sufficiently, as a result of cooling and crystallization, to behave rigidly and transmit tectonic stresses across the cauldron block (Bailey and others, 1976). An alternative or possibly additional explanation for the lack of significant structural relief on intracaldera tectonic faults is that regional extension within the caldera has been compensated volumetrically by repeated injection of magma into the chamber or into fissure and faults (Bursik and Sieh, 1989).

Recent Seismicity and Ground Deformation

Based on the 200,000-year eruption periodicity of the moat rhyolite and its most recent 100-ka eruption, future eruptions from the residual Long Valley magma chamber would seem only a remote possibility, not to be expected for another 100,000 years. However, an unusual succession of earthquake swarms, including a sequence of magnitude-6 earthquakes in May 1980 and another of magnitude-5 earthquakes in January 1983, accompanied by 50 cm uplift of the resurgent dome, suggests that new magma has been injected into the chamber and possibly into the south moat ring-fracture zone (Savage and Clark, 1982; Savage and Cockerham, 1984). Although the intensity of seismicity and the rate of uplift and horizontal distention across the resurgent dome have decreased since 1984, evidence for the presence of magma beneath the caldera persists, and the potential for future eruptions remains (Hill and others, 1985). Statistically, however, the more likely site for future eruptions is along the Mono-Inyo Craters volcanic chain, where eruptions have occurred most recently about 450, 700, 1200, and 1345 A.D. (Miller, 1985; Sieh and Bursik, 1986, Sieh, unpublished data). The potential hazards associated with possible future eruptions in the area have been outlined by Miller and others (1982).

DAY 1: RENO, NEVADA, TO MAMMOTH LAKES, CALIFORNIA, VIA U.S. ROUTE 395

About 13 mi south of Bridgeport, CA (0.9 mi past Conway Summit sign), turn RIGHT off Route 395 into Scenic Viewpoint.

STOP 1. Conway Summit Overlook: Mono Basin. From this vantage point (Fig. 3; see also Fig. 19), the entire Mono Basin and most of its important geologic features can be seen. Mono Lake, in the center of the basin, is the saline remnant of a much larger Pleistocene fresh-water lake known as Lake Russell, named for Israel C. Russell, who first mapped the area in 1881 (Russell, 1889). The

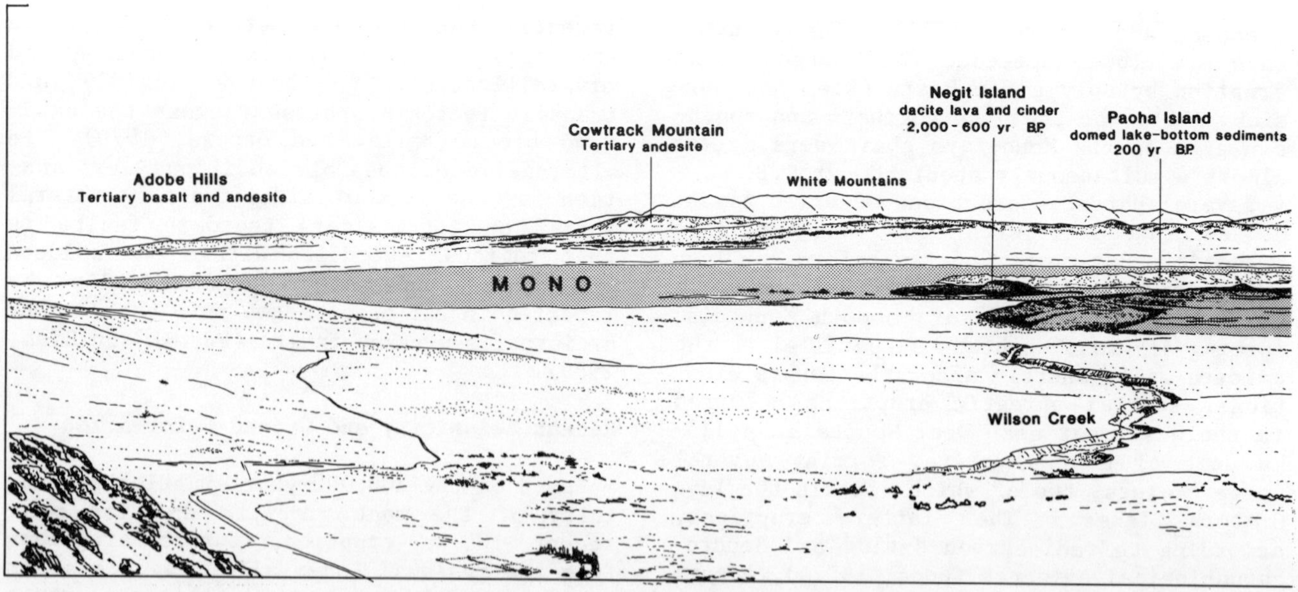

FIGURE 3 (this and facing page). Panorama of

surface of Mono Lake is presently at an elevation of about 1941 m and has been declining at a rate of about 1 m per year as a result of evaporation and diversion of most of its fresh-water inflow to the city of Los Angeles. The shore of Lake Russell once lapped against the foot of the granite slope immediately below the overlook, and its former highest strandline at 2188 m (Putnam, 1950) can be seen along the steep slope west of Mono Lake. Lake Russell overflowed a low pass to the southeast and by a circuitous route through Adobe Valley drained into Owens Valley, 80 km to the south and just east of the White Mountains on the distant southeast skyline.

The buff-colored flat-topped mound on the near shore of Mono Lake is Black Point, a terraced, sublacustrine, basaltic cinder and palagonite cone that formed 13,300 yr B.P. (Christensen and Gilbert, 1964; Lajoie, 1968). The dark hat-shaped island, Negit, and the several small islets to the east are dacitic and rhyolitic lava domes and flows that range in age from about 2000 to about 220 yr B.P. (Stine, 1984). The largest, light-colored island, Paoha, is principally a domed mass of lake-bottom sediments, apparently uplifted by intrusion of a rhyolite cryptodome (Lajoie, 1968). Lake-terrace chronology indicates that Paoha probably emerged between 1720 and 1850 A.D. (Stine, 1984). Beyond the south shore of the lake lie the Mono Craters, a 17-km-long arcuate chain of rhyolite domes and flows ranging in age from 35 ka to 600 yr B.P. (Wood, 1977b; Sieh and Bursik, 1986). The peaks on the distant skyline beyond the Mono Craters (Mt. Morgan, Mt. Morrison, and Bloody Mountain) mark the south wall of Long Valley caldera, which extends from Glass Mtn. west to Mammoth Mtn., barely visible on the skyline beyond June Mtn. The steep escarpment immediately west of Mono Lake is the Lee Vining fault, one of the main faults bounding the east front of the Sierra Nevada. It has formed over the past 3 m.y. and has at least 2000 m of structural relief (Gilbert and others, 1968). It has been active in the Holocene, displacing glacial moraines of the Tioga glaciation at the mouth of Lundy Canyon about 20 m. Displacement on the range-front faults in the vicinity of the Mono Craters appears to have diminished in the past few ten-thousand years as the Mono Craters have become active (Bursik and Sieh, 1989). This relation suggests that locally, dike intrusion beneath the Craters has replaced normal faulting along the Sierran front as the principal mechanism of accommodating extention.

Continue south on Route 395 to Mammoth Lakes via June Lake Loop.

DAY 2: LONG VALLEY CALDERA CIRCUIT (FIG. 4)

[Begins and ends in Mammoth Lakes]

From Mammoth Lakes go west on Route 203 to Mammoth Mountain Ski Lodge. Take gondola to summit of Mammoth Mtn. (3372 m); walk about 200 m SE along the crest.

STOP 2. Summit of Mammoth Mountain: Long Valley caldera overview.

The view northeast from the summit of Mammoth Mountain (Fig. 5) encompasses the entire expanse of Long Valley caldera, from the west rim on the left to the east rim

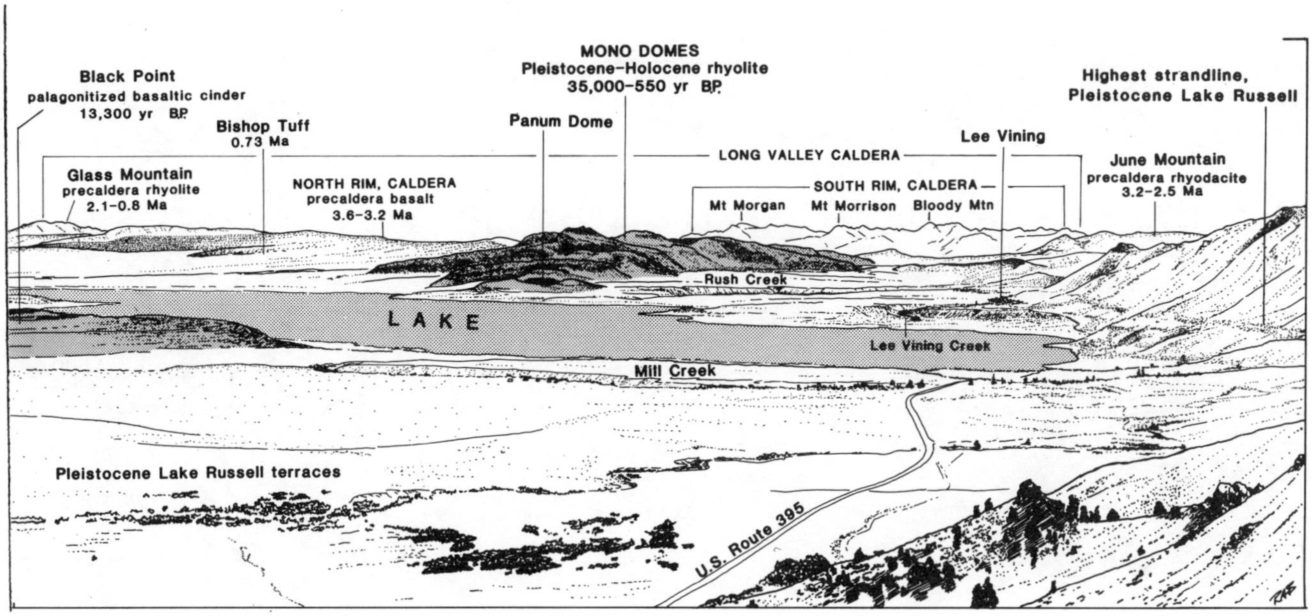

Mono Basin from Conway Summit Overlook (Stop 1).

beyond Lake Crowley, and from the south rim at Laurel Mountain on the right to the north rim at Bald Mountain. The steep humocky terrain in the immediate foreground below is composed of rhyodacite lavas of Mammoth Mountain, which was built on the southwest rim of the caldera about 200 to 50 ka. The summit dome, underfoot, is the youngest extrusive on the mountain (52 ka) and consists of coarsely porphyritic, hornblende-biotite rhyodacite, typical of other postcaldera rhyodacites in the caldera. Extending beyond the foot of Mammoth Mtn. to the left and right are the west and south caldera moat, respectively, underlain mainly by 200- to 60-ka postcaldera trachybasalt and trachyandesite flows that partially encircle the resurgent dome (Fig. 5, shaded), which is underlain by 730- to 650-ka aphyric to sparsely porphyritic flows and tuffs of the early rhyolite. At three sites around the resurgent dome are clusters of porphyritic hornblende-biotite rhyolite domes and flows of the moat rhyolite sequence; they include North dome on the north (520 ka), Hot Creek flow on the southeast (288 ka), and Mammoth Knolls on the west (106-97 ka). In the distant center is Glass Mountain, a 2.1- to 0.8-Ma precaldera high-silica rhyolite complex, which has been truncated on the southwest by caldera collapse. The White Mountains, on the distant skyline, bound the eastern side of the Owens Valley Rift and rise to 4373 m (nearly 300 m higher than the local Sierran peaks); they are underlain by Mesozoic plutons and Paleozoic roof-pendant rocks. On the left, extending from the west moat northward to Mono Lake are the rhyolite domes and flows of the Inyo and Mono Craters, the culminating effusions from the Mono-Inyo Craters volcanic chain.

Return to gondola and Ski Lodge at base of Mountain. Retrace Route 203 east toward Mammoth Lakes going 9.4 mi to Earthquake Fault; turn LEFT into visitors parking area.

STOP 3. Earthquake Fault.

Earthquake Fault is a vertical cleft 2-3 m wide, as much as 20 m deep, and about 0.7 km long crossing the surface of an eroded and pumice-mantled glassy rhyodacite flow. Sunlight does not reach the bottom of the cleft most of the year, and snow and ice usually persist at its bottom through summer. The cleft merges to the north with a fault that can be traced for an additional 2.5 km across the crest of Southwest dome, where it shows as much as 10 m west-downward displacement. The fault is one of several that define an en echelon graben subparallel to the Inyo Craters volcanic chain in the west moat. At this locality the fault shows little or no vertical displacement; irregularities in the joint faces on either side of the cleft can be perfectly matched and show mainly horizontal pull-apart movement with slight right-lateral offset. Earthquake Fault and the related en echelon system were first mapped by seimologists Benioff and Gutenberg (1939), and for several years they and their Caltech colleagues periodically measured the distance between stainless steel pins set in the walls of the cleft in an effort to measure strain in the region. Measurements made through the mid-1940s showed widening of a

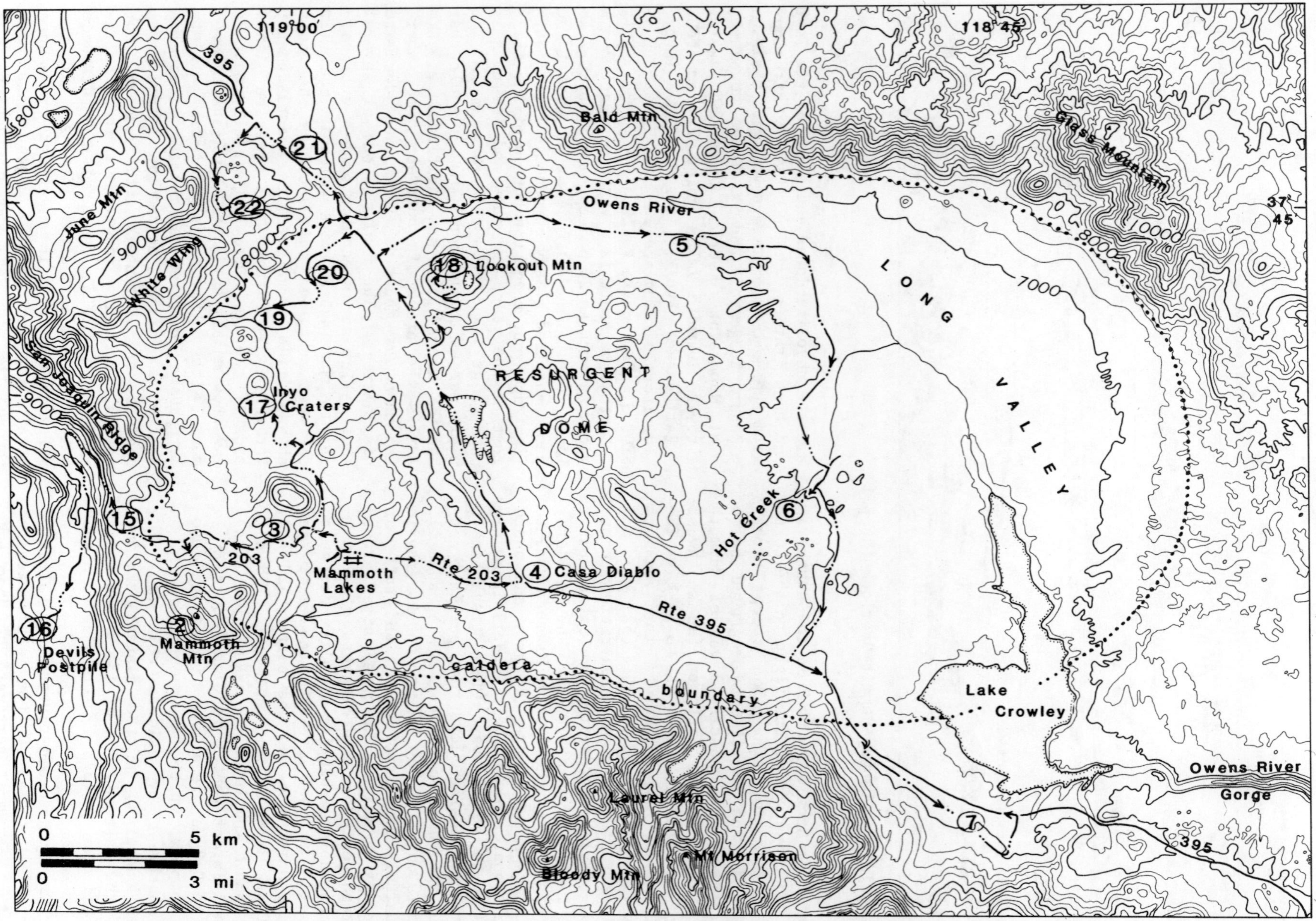

FIGURE 4 Map of Long Valley caldera showing routes and stops of Days 2, 4, and 5. (Number of dots along route indicates day: two dots is Day 2, etc.).

millimeter or so; however, after a 20-year hiatus, a measurement in 1967 revealed that the cleft had closed suffiently to prevent insertion of the steel caliper-bar between the pins, and the measurements were terminated (Francis E. Lerner, written commun., 1982). During the 1980-1983 Mammoth Lakes earthquake swarms, the ladders to the bottom of the cleft were damaged and later removed as a safety precaution. During the same swarms, a number of collapse pits as much as 5 m in diameter and equally as deep formed along the northern extension of the fault. Similar but smaller pits also formed along other faults in the Inyo en echelon system further north.

Continue east on Route 203 through Mammoth Lakes to the T beyond US Route 395; turn LEFT at the T and LEFT again just beyond the Mammoth-Pacific Geothermal Power Plant; go to the sharp bend at the top of the hill and park on the right.

STOP 4. Casa Diablo: Geothermal activity; 1980-1983 seismic epicenter.

Casa Diablo Hot Springs is located on the southern edge of the resurgent dome, on a major northwest-trending fault within the medial graben. The rock exposed in the hillsides to the north is sparsely porphyritic, biotite-bearing early rhyolite that issued from a vent on the crest of the resurgent dome, about 6 km to the north. Postcaldera trachybasalts ranging in age from 129 to 62 ka flood the south moat and lap against the rhyolite of the resurgent dome. Across the south moat rises the steep south wall of the caldera, dominated by Mount Morrison, Laurel Mountain, and Bloody Mountain, which are underlain mainly by Paleozoic metasedimentary rocks locally intruded by Cretaceous granite. At Casa Diablo, a deep 1605-m exploratory geothermal drill hole passed through 460 m of the early rhyolite, 914 m of the Bishop Tuff, and penetrated Paleozoic metasedimentary basement rock at 1377 m. The depth to basement at Casa Diablo indicates a total subsidence of about 2100 m in the south moat. In the upper part of the Bishop Tuff, a 50-m-thick granite breccia was encountered; the similarity of the granite to that exposed on the south wall suggests that the breccia is a slide block that avalanched off the south wall during the latter stages of caldera collapse.

During the seismic swarms of 1980-83, in which the resurgent dome rose 0.5 m, the south moat from Casa Diablo to Convict Creek, a distance of 8 km, was the principal locus of seismicity. Modelling of the seismicity and associated ground deformation (Savage and Clark, 1982; Savage and Cockerham, 1984) led to the conclusion that new magma had not only welled into the chamber beneath the resurgent dome but also possibly had intruded the south moat ring-fracture zone to depths as shallow as 3 km, thereby raising serious concern about the possibilty of an eruption. During that episode a previously dormant fumarole at Casa Diablo showed vigorous increase in flow, and several new hot springs broke out along the southern edge of the resurgent dome. Currently a scientific drill hole targeted to a depth of 5-7 km is being drilled on the resurgent dome by Sandia National Laboratory (Rundle and others, 1986) to explore the cause of this recent activity.

Hot Springs, fumaroles, and areas of active hydrothermal alteration are particularly prevalent in the south and southeast moat and on adjacent flanks of the resurgent dome. Most of them are localized along faults. The heat source for these features is presumably the residual magma chamber beneath the resurgent dome, but drill holes show that hot waters are confined to relatively shallow aquifers (<700 m) and that except in the west moat temperatures decline or do not increase substantially with depth. Possible explanations for this puzzling thermal distribution are that cold groundwater flow through the deeper caldera fill masks the deep thermal source or that the principal heat source is not beneath the resurgent dome but is a younger magma body related to the Inyo Craters volcanic chain in the west moat (Blackwell, 1985; Sorey, 1985). Maximum temperatures measured in drill holes in the caldera are about 220°C, but chemical geothermometry indicates deep reservoir temperatures near 240°C. In recent years substantial efforts have been made to develop and utilize this geothermal resource. At Casa Diablo Hot Springs several shallow drill holes supply hot water for two 7-megawatt binary electrical power plants; nearby, a California State fish hatchery utilizes local warm-spring waters to breed trout; and at Whitmore Hot Springs, thermal waters are used in a public swimming pool.

Return to US Route 395; turn RIGHT and go north about 13 mi to Owens River Rd. Turn RIGHT (east) and follow Owens River Rd. 6.7 mi to large gravel pit on right; park in pit and walk to its south rim.

STOP 5. Owens River Road gravel pit: North moat viewpoint.

This borrow pit, in Pleistocene periglacial gravels, is in the middle of the north moat of the caldera. The steep escarpment to the north is the northern

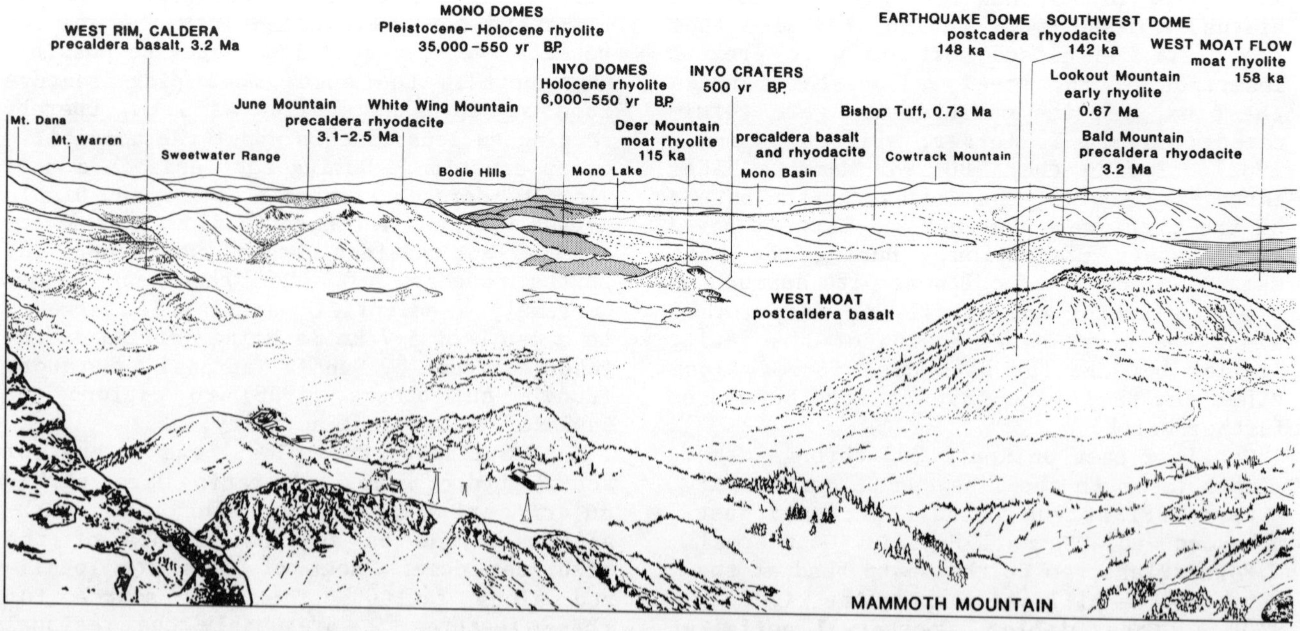

FIGURE 5 (this and facing page). Panorama of

caldera wall, underlain by Mesozoic plutonic and Paleozoic metasedimentary rocks and capped with 3.2- to 2.6-Ma precaldera trachybasalts and trachyandesites. On the rim to the northwest is Bald Mountain, composed of 3.2-Ma precaldera rhyodacite domes and flows. On the far western skyline is the crest of the Sierra Nevada and, somewhat closer and lower, the western caldera rim. On the northeast rim is Glass Mountain, a 2.1- to 0.8-Ma precaldera rhyolite complex, the southwestern half of which is downfaulted and buried within the caldera. The low flat surface south of Glass Mountain is the southeast caldera rim, beyond which is the Volcanic Tableland, underlain by the Bishop Tuff. The high Sierran peaks on the southern skyline mark the south rim of the caldera. The low forested hills in the south foreground are the resurgent dome, underlain by 730- to 650-ka tuffs and obsidian flows of the early rhyolite. In the near southwest, the high knob rising above the surface of the resurgent dome is a 520-ka lava dome of the moat rhyolite, composed of coarsely porphyritic hornblende-biotite rhyolite. On the sage-covered lower slopes of the resurgent dome, nondeformed lake terraces and strandlines of Pleistocene Long Valley Lake may be seen. On the north side of Owens River Road, the moat is floored by a 15-km-long, 108-ka, postcaldera trachybasalt flow that erupted from a vent in the west moat. The dark-brownish crags forming the southern edge of this flow can be traced from the base of Lookout Mountain on the west to near the base of Glass Mountain on the east. The flow, which shows no evidence of having flowed into water, indicates that the caldera lake was confined to the eastern moat and may have been completely drained by this time. Just beyond the toe of the trachybasalt flow, at the very base of Glass Mountain, is a small conical hill--a small postcaldera rhyodacite dome that has not been dated but appears to have erupted subaqueously in Pleistocene Long Valley Lake in early postcaldera time.

Continue on Owens River Road east and south to the T junction just beyond the small bridge across Hot Creek (7.4 mi); turn RIGHT at T and bear RIGHT at next junction (0.5 mi); continue 1 mi to Hot Creek Recreational Area parking lot on the right.

STOP 6. Hot Creek Gorge: Southeast moat rhyolite; hot-spring activity.

Hot Creek is the lower reach of Mammoth Creek where several vigorous hot springs issue from its bottom and form this popular bathing area. Hot Creek Gorge has been incised into a 288-ka rhyolite lava flow that erupted from a vent about 4 km to the south on the south shore of Pleistocene Long Valley Lake, which at that time was at an elevation of 2195 m. The flow crept northward into the lake, becoming entirely subaqueous in the vicinity of the present gorge. The initially hot glassy rhyolite was pervasively hydrated by interaction with the lake water and later was partly hydrothermally altered by local hot-spring activity. Outcrops of partly altered, perlitized flow breccia are exposed along the paved trail leading to the bottom of the gorge.

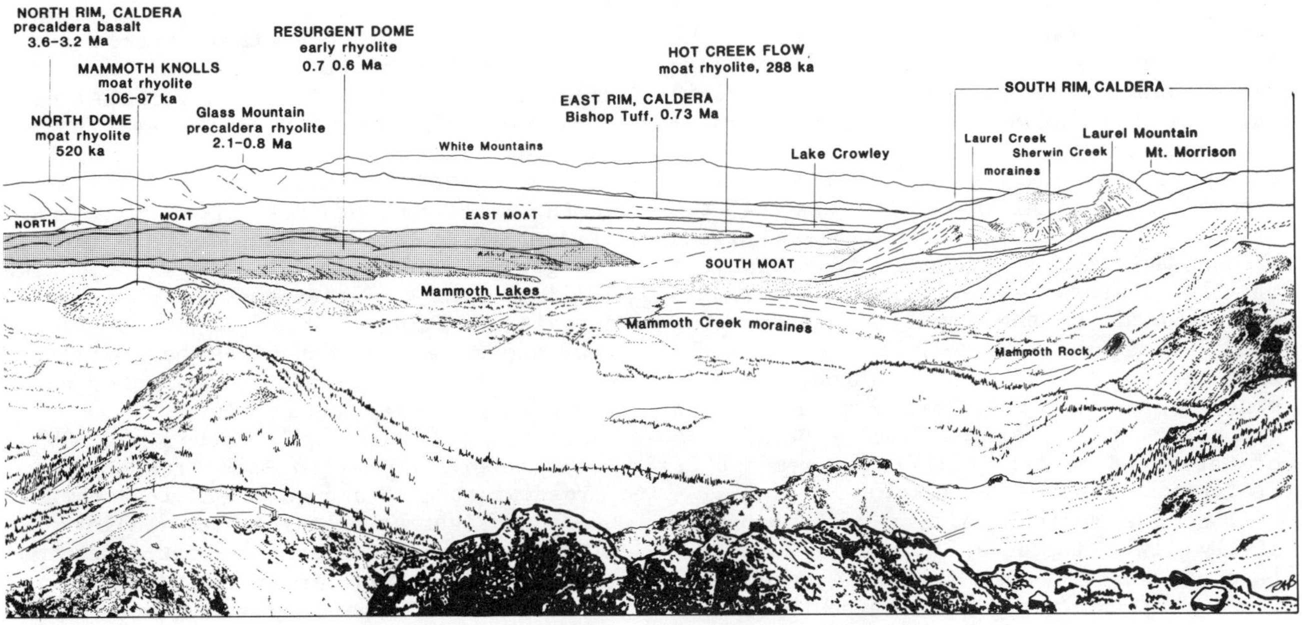

Long Valley caldera from the summit of Mammoth Mountain (Stop 2).

The Hot Creek flow is a sparsely porphyritic, sanidine-augite-bearing rhyolite of the moat sequence. Its chemistry and mineralogy differ from that of other, more coarsely porphyritic hornblende-biotite rhyolites of the moat-rhyolite sequence, and it shows a chemical fractionation pattern (relative to the older flows) similar to that of the Bishop Tuff. This chemical change about 300 ka possibly reflects thermal rejuvenation of the Long Valley magma chamber brought about by a new influx of mafic magma into the roots of the magmatic system, as shortly thereafter trachybasalts began erupting in the west moat (Bailey, 1984a). Additional uplift of the resurgent dome probably accompanied this magmatic rejuvenation, as at this time a sudden influx of coarse detritus derived from the resurgent dome built extensive flanking deltas into the caldera lake. These deltaic sediments are intensely silicified and permeated with fossil hot springs and fumaroles, suggesting that increased widespread hydrothermal activity also accompanied this magmatic and structural activity.

The largest and hottest springs in the bottom of the gorge are located on two north-trending faults that bound a shallow, 1-km-wide graben transecting the rhyolite flow (Fig. 2). The numerous smaller, cooler springs along the creek between the two faults are produced by lateral flow of hot water from the faults through the basal breccia of the rhyolite flow where mixing with cold meteoric water occurs. The several boiling pools in the gorge (93°C at this elevation) commonly change in vigor and configuration in response to local earthquakes and are also significantly affected by seasonal rise and fall of the water level in the creek. The white deposits lining the pools are calcium carbonate (travertine). Numerous older travertine deposits surround extinct hot springs at higher levels along the gorge.

Return east along Hot Creek Road and turn RIGHT (south) at first junction (1 mi); continue 2.5 mi to US Route 395; turn LEFT (east) onto Route 395 and continue 1.7 mi to Crowley Lake exit; turn RIGHT onto Crowley Lake Drive and continue 3.5 mi to Crowley Lake Community Center on left; park well clear of the adjacent fire station and walk between and about 50 m behind buildings.

STOP 7. Crowley Lake Overlook: Pleistocene Long Valley Lake; Hilton Creek fault.

During early postsubsidence time, Long Valley caldera was filled by a large lake, terraces and strandlines of which are well preserved along the eastern caldera wall and locally on the flanks of the resurgent dome. Rise of the resurgent dome gradually raised the lake level above the low southeastern rim (east of Lake Crowley), where it overflowed and cut the Owens River Gorge. Gradual downcutting, accompanied by intermittent tectonic lowering of the southeast rim along the Sierran front, eventually drained the lake. From this viewpoint in early morning or late afternoon light, the terraces of Pleistocene Long Valley Lake (Mayo, 1934) are readily visible along the eastern caldera wall beyond Lake Crowley. Although these terraces have not been dated directly, geologic relations and dated intracaldera volcanic units suggest that the higher terraces are the oldest and the lower ones the youngest, and that the lake was extant between about 600 ka and 100-50 ka. Scattered on many of the lake terraces are glacial erratics of granitic rocks that can have been derived only from plutons exposed in the High Sierra west of the caldera and which must have been carried across the lake by icebergs from glaciers on the western shore. Similar erratics were lodged on the upper slopes of the resurgent dome, which was an island in the lake at that time. (Lake Crowley is a modern, man-made reservoir impounded by a dam built in 1941 by Los Angeles Department of Water and Power; it is not a remnant of Pleistocene Long Valley Lake.)

Along the eastern caldera wall, progressively lower terraces, traceable from the base of Glass Mountain to Lake Crowley, are increasingly downwarped to the south (Christensen, 1966)--a reflection of repeated faulting along the eastern Sierran front producing reverse drag on the downthrown side, a relation that has been argued as evidence for listric faulting along the front. The imposing escarpment west of the viewpoint is the Hilton Creek fault. At the base of the escarpament south (left) of McGee Creek Canyon can be seen the light-colored jagged trace of the slickensided fault surface cutting Paleozoic Hilton Creek Marble. Structural relief on the fault exceeds 1200 m, probably more than half of which postdates formation of Long Valley caldera. Late Pleistocene (Tioga) moraines at the canyon mouth are displaced about 15 m down to the east.

Continue east on Crowley Lake Drive 0.9 mi; turn LEFT and go 0.9 mi to Route 395; turn LEFT (west) and return to Mammoth Lakes.

DAY 3. BISHOP TUFF CIRCUIT (FIG. 6)
[Begins and ends in Mammoth Lakes]

Take Route 395 south to Toms Place and continue 1 mi to the wide turnout on the right, opposite a 20-m-high roadcut on the left.

STOP 8. Big Pumice Cut: Bishop Tuff - Sherwin Till contact.

In this roadcut on the north side of U.S. Route 395, the basal, bedded pumice fall and overlying nonwelded ash flows of the 0.73-Ma Bishop Tuff rest on weathered Sherwin Till (Sharp, 1968). The contrasting modes of emplacement of the pumice fall and ash-flow members of the Bishop Tuff are well displayed here--the eastward-dipping airborne pumice fall mantling the initial sloping surface of the deeply weathered till, and the overlying fluidized ash flows banking against the slope, with their nearly horizontal pumice swarms converging with the dipping pumice fall beds. Vertically cutting the tuff and till are several north-trending clastic dikes that are subparallel to major Sierran frontal faults in the area. In the till, some of these dikes cut cleanly through large boulders weathered to incoherent gruss. The source for the clastic dike fillings is a Pleistocene glacio-fluvial deposit overlying the tuff at the top of the roadcut; note how the dikes become finer-grained downward, with only the upper parts containing pebbles and cobbles.

Carefully U-turn back onto Route 395 and return 0.1 mi to Rock Creek Canyon road; turn LEFT and go 5.5 mi to a turnout on right overlooking Round Valley.

STOP 9A. Round Valley Viewpoint: Deformation of Bishop Tuff; fumarolic mounds.

From this viewpoint the surface of the Bishop Tuff slopes about 5 degrees to the south into Round Valley, a prominent depression bounded on the west by the Wheeler Crest fault, which forms one of the more imposing scarps along the eastern Sierran front. While 5 degrees does not seem a particularly steep slope, when considered within the context that initial dips on the Bishop Tuff elsewhere (and on ash-flow tuffs in general) seldom exceed 1 or 2 degrees, it is clear that the Bishop Tuff in Round Valley has been structurally downwarped at least 600 m against the Sierran front. As a consequence of this structural oversteepening of the tuff surface, 20 m or more of the uppermost unconsolidated part of the tuff has been removed by erosion, and many fumarolic mounds (Sheridan, 1970) rise 10-20 m above the stripped surface. They form prominent hillocks on the tuff surface around Round Valley, as well as elsewhere on the Volcanic Tableland. Locally they are arranged in crude dendritic patterns that mimic the approximate courses of pre-eruption stream valleys, suggesting that the main source of volatiles for the fumaroles was streams overrun by Bishop ash flows.

Continue 0.3 mi. and park at junction with Swall Meadow Road. Walk 100 m east to rim of Rock Creek Gorge.

STOP 9B. Lower Rock Creek Gorge Overlook: Zonation in the Bishop Tuff.

Down-canyon on the east wall of the gorge, the internal zonation of the Bishop Tuff is well displayed. Here, the tuff laps around a granitic high on the Sierran frontal fault scarp and thickens southward toward Round Valley. Exposed from the base upward are (1) the lower nonwelded zone (in contact with the granite), (2) lower vitrophyre

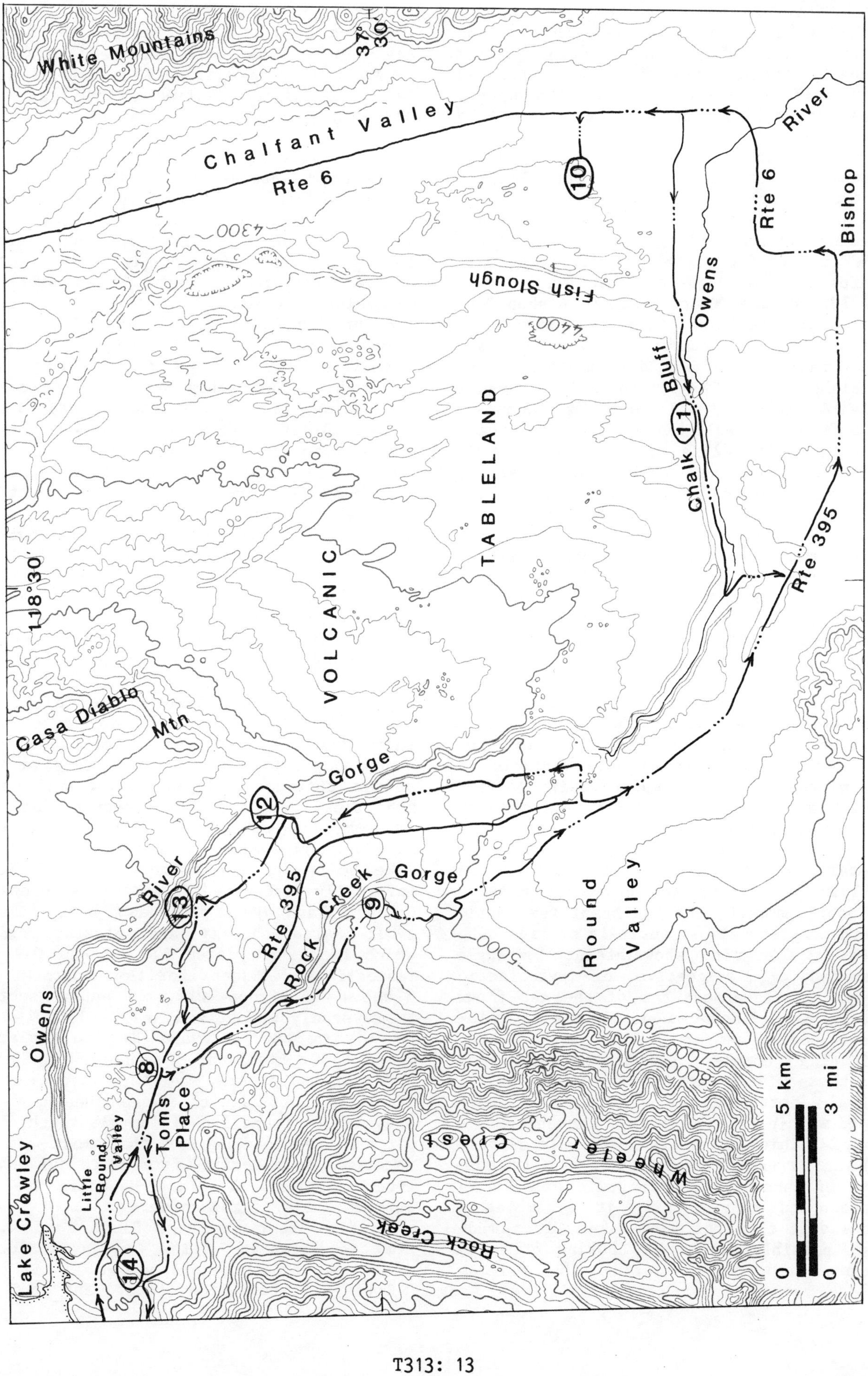

FIGURE 6 Map of Volcanic Tableland showing route and stops of Day 3.

(dark-gray layer dipping down-canyon), (3) central devitrified zone (brown zone gradually thickening down-canyon), and (4) upper buff vapor-phase zone. The upper, vitric, nonwelded zone has been removed by erosion.

Continue south 7.7 mi to junction with Route 395; turn RIGHT (SE) onto Route 395 and continue 9.5 mi to junction with US Route 6; turn LEFT (north) on Route 6 and continue 8.5 mi to Rudolf Road; turn LEFT and go 0.8 mi to Insulating Aggregates Co. pumice quarry (inactive).

STOP 10. Bishop Pumice Quarry: Bishop Tuff basal pumice fall and distal ash flows.

Exposed in this quarry are 4 m of basal pumice fall and 4-6 m of distal ash flows of the Bishop Tuff. The pumice fall is roughly divisible into (1) a lower third, composed of moderately well-bedded ash and fine pumice lapilli, (2) a middle third that is lighter-colored, less well-bedded, coarser, and fines depleted, with pumice lapilli up to 2-cm diameter, and (3) an upper third that is darker-colored, more distinctly bedded, with numerous fine-to-coarse alternations containing pumice lapilli up to 4-cm diameter. Discontinuous dark manganiferous(?) streaks occur in the upper third of the deposit. Lithic clasts, like the pumice lapilli, tend to increase in size upward through the deposit. The lack of distinct discontinuities suggests continuous deposition over a relatively short time, with eruption intensity increasing toward the middle third and becoming more variable or intermittent in the upper third. Very subtle low-angle cross-bedding in the lower part of the deposit suggests that the tephra rained to the surface from very dense, laterally drifting, eruption clouds. The overlying ash flows are generally massive and poorly sorted, but the basal few centimeters consists almost entirely of fine ash, which locally also shows faint low-angle cross-bedding, suggestive of pyroclastic surge. The lowermost 2 m of the ash flows contain many discontinuous swarms of relatively coarse pumice lapilli and blocks, indicating complex multiflow deposition near these distal parts of the formation.

At the north end of the quarry, interbedded within the pumice fall sequence, is a 1- to 2-m thick, tongue-shaped, buff deposit with an irregular upper surface. It consists of unusually coarse pumice blocks in a nonsorted ashy matrix; it resembles a pyroclastic flow but more likely is a mudflow, possibly formed by choking of local streams with tephra during the plinian episode.

At one or two localities in the quarry, east-dipping low-angle reverse faults with as much as 1 m displacement can be seen. The fault planes decline in dip eastward, becoming bedding-plane faults that are difficult to trace. In this structural setting, near the western boundary fault of the White Mountains and on the eastern edge of the Volcanic Tableland, which is broken by numerous normal faults, reverse faults seem anomalous. Possibly they are related to local compressional wedging or bending within downfaulted blocks in an otherwise extensional region.

Return to Route 6, turn RIGHT (south) and go 2.5 mi to Jean Blanc Road; turn RIGHT (west) and go 3.3 mi to Fish Slough; just beyond Fish Slough crossing, bear LEFT (southwest) onto Chalk Bluff Road; go 1 mi to where the road rises over prominent white outcrops along the base of Chalk Bluff; park beyond the rise in the meadow on the left, and walk back to the roadcut.

STOP 11. Chalk Bluff: Glass Mountain fluvio-lacustrine sediments and ash.

Chalk Bluff is the southern edge of the Volcanic Tableland, which is underlain principally by the Bishop Tuff. The cliffs at the top of the bluff are partly welded Bishop ash flows. The basal pumice fall is obscured here by talus and slope wash but is exposed in nearby steep gullies along the bluff. The tuffaceous sediments exposed in this roadcut and adjacent outcrops underlie the Bishop Tuff. They include massive nonsorted tuffaceous deposits, possibly thin ash flows or mudflows; even-bedded and cross-bedded ashy silts and sands, some showing soft-sediment deformation, probably lake beds; and two primary ash-fall beds, informally designated as the Glass Mountain-G and Glass Mountain-D ash beds (Izett, 1981; Sarna-Wojcicki and others, 1984), which have been dated 1.0 Ma and 0.9 Ma, respectively. Except for the uppermost 1-2 m below the Bishop Tuff, the entire section is reversely magnetized and is believed to be within the Matuyama Reversed-Polarity Chron (J.C. Liddicoat, as quoted in Sarna-Wojcicki and others, 1984). The chemistry, mineralogy, and age of the ash beds and the prevalence of small obsidian clasts throughout the section suggest that most of the materials were derived from precaldera rhyolite eruptions at Glass Mountain on the northeast rim of the caldera.

Continue west on Chalk Bluff Road 4.5 mi to Pleasant Valley Dam Road; turn sharp LEFT (southeast) and go 1.7 mi to Route 395; turn RIGHT (west) and go 5.8

mi to Gorge Road; turn RIGHT (east) and go 0.7 mi to T; turn LEFT (north) and go 6.0 mi; at Y bear RIGHT and go 0.2 mi to locked gate. Park on left and walk down powerhouse road to bottom of canyon and return (0.8 mi roundtrip).

STOP 12. Owens River Gorge: Section of the Bishop Tuff; rosette jointing.

The Los Angeles Department of Water and Power (LADWP) powerhouse road descends the west wall of Owens River Gorge and affords spectacular closeup views of a 150-meter-thick section of the Bishop Tuff on the east wall. At the same time, the road permits closehand examination of roadcuts and outcrops that show the vertical changes in density and texture of the entire tuff section.

On the east wall of the gorge, the Bishop Tuff displays a remarkable variety of columnar jointing. In the upper half of the cliff the joint columns are remarkably well-formed five- to six-sided columns 1 to 3 m in diameter. Locally these columns curve and converge toward common foci, forming joint rosettes. These rosettes are the locus of large fossil fumaroles. At the top of the gorge, directly above many of these joint rosettes, are indurated fumarolic mounds, like those seen near Round Valley, exhumed by erosional stripping of the upper nonwelded surface of the Bishop Tuff. The distribution of these mounds close to the present gorge suggests that the volatiles responsible for their formation were derived in part from the ancestral Owens River, which was overrun and vaporized by Bishop ash flows.

In the lower half of the gorge, in dark-gray, densely welded tuff, the joint columns are gigantic, 10-20 m in diameter, and are relatively crudely formed, probably a consequence of slower cooling about widely spaced cooling centers. Locally some of these large joint columns have secondary horizontal joint columns developed perpendicular to their surfaces. Between the upper and lower tiers of joint columns is a broadly undulating, ill-defined parting that marks the contact between two sub-cooling units of the Bishop Tuff. This parting and another like it near the top of the gorge road represent brief hiatuses in the deposition of the tuff and show that the tuff consists of multiple cooling units (Sheridan, 1967). However, the partings probably do not represent intervals of more than a few days or weeks (Hildreth, 1979) or at most 2-3 years (Snow and Yund, 1988). The prominent crystal-rich parting in the upper part of the gorge, near the gate, is the contact between the Tableland and Gorges cooling units (Hildreth, 1979); the Tableland (upper) unit contains two pyroxenes and has Fe-Ti oxides temperatures of 737-763°C, whereas the Gorges unit lacks pyroxenes and has oxide temperatures of 725-736°C.

Exposures along the left (west) side of the powerhouse road display almost the entire range of lithologic variation in the Bishop Tuff. Near the upper gate the rock is light-gray to pinkish or purplish, porous, poorly to moderately welded, vapor-phase tuff. Downward the tuff grades through brown to dark gray, becomes progressively less porous and more densely welded, and displays conspicuous eutaxitic texture. In the most densely welded, darker gray facies, eutaxitic texture is almost entirely obscured by devitrification. At the first major left bend in the road, however, eutaxitic texture becomes more prominent over a short 5-meter interval where scattered obsidian fiamme occur, suggesting a minor eruptive/cooling interval. The tuff below this level is monotonously dark gray, densely welded, and devitrified to the bottom of the gorge. The basal contact is not exposed in the gorge, but the foundation of the powerhouse was excavated in dense eutaxitic vitrophyre, indicating that the base is not far below.

From upper gate return to Gorge Road (0.2 mi); turn sharp RIGHT (north) and go 1.8 mi; bear RIGHT and go 0.3 mi; turn LEFT onto narrow dirt road and go 0.1 mi; park near rock-ring fireplace and walk to edge of gorge.

STOP 13. Upper Owens River Gorge Overlook: Gorge stratigraphy and structure.

Owens River Gorge is incised into the surface of the Volcanic Tableland and here crosses a north-trending fault zone that has been intermittently active for the past 3-4 m.y. The bottom of the gorge is cut in the Triassic Wheeler Crest Quartz Monzonite, which is overlain by a 20-m-thick sequence of 3.2-Ma precaldera trachybasalt flows that erupted from a vent about 1 km north of the gorge. Locally overlying the trachybasalt are a few meters of outwash gravel of the Sherwin glaciation. The upper 60 m of the gorge exposes partly welded ash flows of the Bishop Tuff. Down-gorge, the tuff thickens to a maximum of about 200 m, gradually becoming more densely welded and intensely jointed. Up-gorge, the tuff thins over a quartz monzonite high formed by the north-trending fault zone, multiple strands of which can be seen on the opposite canyon wall. Increasingly greater offset down-to-the-east in successively older units--the Bishop Tuff, trachybasalt, and quartz monzonite--demonstrates the growth nature of the fault zone.

Continue 0.2 mi along dirt road to aqueduct surge tank and junction with unpaved U.S. Forest Service road; continue staight ahead following unpaved road 2.5 mi to Route 395; turn RIGHT and go 2.1 mi to Toms Place exit; turn LEFT then immediately RIGHT on Crowley Lake Drive toward Toms Place and continue 3.0 mi to curve in saddle on the ridge beyond Little Round Valley; park on right shoulder beyond saddle.

STOP 14. Little Round Valley: Bishop Tuff-glacier contact; fumarolic pipes.

Exposures in and around Little Round Valley suggest that active glaciers existed in the Sierra Nevada when the Bishop Tuff erupted about 0.73 Ma. The north side of Little Round Valley is a steep semi-circular scarp that exposes three cooling units of the Bishop Tuff. On the basis of mineralogy and Fe-Ti oxide temperatures, the lower and middle units belong to the Gorges unit and the upper to the Tableland unit (Hildreth, 1979, p. 47). The lower and middle units are unusual in that they contain numerous indurated, pipe-like, fossil fumaroles, which locally stand out in relief around the northern edge of the valley and are particularly evident in outcrops and roadcuts along Route 395. Within Little Round Valley are but a few low outcrops and roadcuts of Bishop Tuff, which are poorly to nonwelded, strongly oxidized, and unusually rich in lithic clasts, commonly concentrated in irregular clots or entrained in steep pipes or chimneys. The tentative interpretation of the morphology of the valley and the distribution of the lithic-rich chimneys and fumarolic pipes is that at the time of eruption of the Bishop Tuff, Little Round Valley was occupied by a small, steep-fronted, lobate glacier that flowed from the south out of Rock Creek Valley. The northern terminus of the glacier probably coincided with the present steep, arcuate, northern margin of Little Round Valley. During eruption of the Bishop Tuff, ash flows banked against the toe of the glacier and probably also mantled the surface of the terminal lobe. Interaction between the hot tuff and the glacial ice along the steep lobate contact generated abundant steam, which streamed through the adjacent poorly consolidated tuff forming myriads of fumaroles now preserved as indurated pipe- and rib-like forms. Gradual melting of the glacial ice lobe caused settling of the thin nonwelded tuff that partially mantled the toe of the glacier, and the tuff became oxidized and intermixed with glacial debris melting out of the ice; the resulting heterogeneous deposit now thinly mantles the valley floor. Glacial till occurs locally along the steep slopes above and south of Little Round Valley, but no large terminal moraine is evident, though it could be buried beneath the Bishop Tuff ringing the valley. Complete melting of the ice lobe formed a crescent-shaped lake in Little Round Valley, which eventually overflowed northward across the surface of the Bishop Tuff and cut a small steep-sided gorge leading to the head of the Owens River Gorge, which is now drowned by Lake Crowley. If separate from the Sherwin glaciation, how significant or widespread the glaciation represented by this glacier is not known.

Continue north on Crowley Lake Drive to Route 395; turn LEFT and return to Mammoth Lakes.

DAY 4. DEVILS POSTPILE-INYO CRATERS (FIG. 4)

[Begins and ends in Mammoth Lakes]

Follow Route 203 west from Mammoth Lakes to Minaret Summit (6.4 mi). Turn RIGHT into Minaret Vista parking area.

STOP 15. Minaret Vista: Sierran Cretaceous basement and Tertiary volcanic rocks.

The view west from Minaret Vista overlooks the deep glaciated canyon of the Middle Fork of the San Joaquin River and the Ritter Range beyond, the most prominent peaks of which, viewed from south to north, are the jagged Minarets, massive Mt. Ritter, and Banner Peak. The rocks of the Ritter Range include coarse metavolcanic breccias, welded tuffs, and associated granitic plutons that are part of a deeply dissected Cretaceous caldera, probably comparable in size to Long Valley caldera but formed in a Cretaceous island arc rather than a Plio-Pleistocene continental setting (Fiske & Tobisch, 1978).

San Joaquin canyon is an ancient drainage that predates uplift of the Sierra Nevada. The river was initially established on the gradually rising western slope of the Sierra Nevada in Tertiary time, possibly as early as 25 Ma (Huber, 1981). It formerly drained a large area east of the present Sierran crest, but about 3 Ma, during initiation of Sierran frontal faulting, the upper reaches were beheaded by faulting and dammed by eruption of basaltic lavas. Blockage of this drainage formed ephemeral lakes, as many of the canyon-filling lavas are thick pillow complexes. These Tertiary lavas and associated overlying rhyodacite domes and breccias are part of the 3.6- to 2.2-Ma precaldera sequence that heralded the more silicic eruptions from the Long Valley magma chamber. They are exposed along San Joaquin Ridge, which extends north from Minaret Vista and forms the present drainage divide. Two Teats and San Joaquin Mountain, the two most prominent peaks along the divide, are dissected remnants of large rhyodacite domes and their surrounding pyroclastic cones.

During early Pleistocene time the Middle Fork was reexcavated by repeated glaciations and acquired its present broad U-shaped form. Intracanyon remnants of the Bishop Tuff show that the valley was temporarily filled by the caldera-forming eruptions but has been reexcavated by more recent glaciations.

Continue west on Route 203 to Devils Postpile National Monument (about 6 mi). Park at the Visitors Center and follow the footpath 0.3 mi to the Postpile.

STOP 16. Devils Postpile National Monument: Basalt joint columns.

Devils Postpile National Monument is renowned for its spectacular display of columnar-jointed basalt, which, although more limited in extent, rivals the displays at Fingal's Cave in Scotland and Giant's Causeway in Ireland.

The Middle Fork of the San Joaquin River in the vicinity of Devils Postpile is excavated mainly in Cretaceous porphyritic granite. In late postcaldera time the deeply glaciated valley was partially filled by the trachybasaltic lavas that form the Devils Postpile. Although not precisely dated, the (informal) Devils Postpile flow and several other related flows probably erupted 200-100 ka, about the same time similar flows were erupting in the west moat of Long Valley caldera. Local geologic relations suggest that formation of the Postpile columns was the consequence of ponding of the lava behind a recessional moraine. The Devils Postpile flow has since been eroded by subsequent glaciations, as is shown by the finely polished and striated outcrops along the trail at the top of the Postpile.

Leave Devils Postpile National Monument, returning on Route 203 about 10 mi toward Mammoth Lakes; 0.7 mi past Earthquake Fault turn LEFT (north) on Mammoth Scenic Loop and go 2.5 mi to signed Inyo Craters turnoff; turn LEFT (west) and follow unpaved road 1.1 mi to Inyo Craters parking area; park and follow trail 0.3 mi west to Craters.

STOP 17. Inyo Craters: Phreatic explosion craters and deposits.

The Inyo Craters (Fig. 7) are three, northerly aligned, phreatic explosion craters on the south flank and summit of Deer Mountain (Rinehart and Huber, 1965), a 115-ka porphyritic rhyolite dome in the west moat of the caldera. The two southernmost craters (informally designated as north crater and south crater) are about 200 m in diameter, about 60 m deep, and contain small lakes; the crater on the summit of Deer Mountain (informally referred to as summit crater) is smaller, irregular in outline, breached on its south side, and dry. The lake in south crater is yellowish green, suggesting the presence of suspended sulfur, but the water is cold (11°C) and other evidence of thermal influx is lacking. The lake in north crater is colored brown with organic matter. Despite present differences in morphology and vegetation, all three craters formed at nearly the same time, probably within hours or days. They erupted in succession from north to south, as the light-colored deposits from summit crater underlie the darker ones of north crater, which in turn overlie the darkest ones of south crater--relations that are well exposed in the northeast wall of north crater and on the east flank of Deer Mountain (Mastin and Pollard, 1988). The light-colored debris around summit crater is composed primarily of pulverized hornblende-biotite rhyolite of Deer Mountain dome, whereas the darker debris around north and south craters consists largely of fragmented trachyandesite derived from flows like those exposed in the walls of south crater. In the north wall of south crater, exposed in succession above the trachyandesite flows, are: (1) trachyandesitic cinder (10 m), (2) pumiceous rhyolite tephra from the Inyo domes magmatic eruptions, centered a few kilometers to the north (1 m), and (3) coarse, crudely bedded phreatic explosion deposits, mainly from south crater (13 m). The latter, which extend as far as 1 km from the crater (Fig. 7), consist mainly of trachybasalt-trachyandesite blocks in a grayish-brown, compact to semi-indurated, fine-grained matrix; blocks and boulders of granite and metamorphic rocks, probably derived from subsurface glacial till or Sierran basement, also are included in the debris. A battered log incorporated in the deposit of south crater yielded a radiocarbon age of 710 ± 60 yr B.P., which together with dendrochronological data gives a calendar age of between 1340 and 1460 A.D. for the phreatic eruptions (Wood, 1977a).

The terrain around the craters is broken by many north-trending faults and fissures (Fig. 7), which show up to 20 m displacement individually and define a graben 0.6 km wide and 2.5 km long. The graben probably formed as a consequence of uplift and distension above a rising dike (Mastin and Pollard, 1988), which probably generated the phreatic explosion craters. To test this hypothesis, Sandia National Laboratories (Eichelberger and others, 1988) slant drilled an 865-m

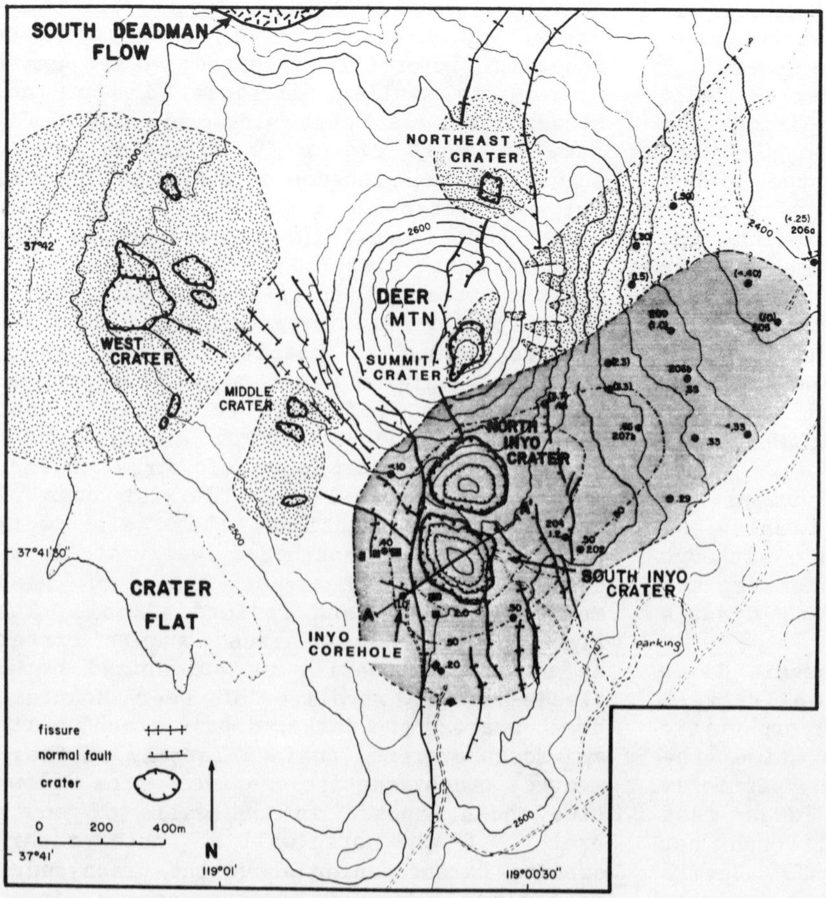

FIGURE 7 Geologic map of Inyo Craters (Stop 17), showing faults, fissures, and distribution of phreatic deposits from north and south craters (dark-shaded) and summit, middle, and west craters (light-shaded) (from Mastin, 1988). Line A-A' is section-line of Figure 8.

scientific core hole (Fig. 8) northeastward at an angle of 68 degrees from a site on the top of the fault scarp just west of south crater. It passed through a 320-m sequence of postcaldera trachyandesite and trachybasalt flows, about 50 m of gravel, 360 m of tuffs and flows of the early rhyolite, 63 m of glacial till, and terminated in 30 m of Paleozoic quartzite, presumed to be Sierran basement rock. At a depth of about 600-650 m, within the early rhyolite section and directly beneath the center of south crater, the hole penetrated an apparent vent breccia consisting primarily of pulverized early rhyolite and postcaldera trachybasalt. Included within the breccia were small pumiceous fragments of a distinctive high-silica rhyolite, apparently juvenile and considered to be the magma that generated the phreatic explosions (Eichelberger and others, 1988). Presumably a dike or conduit of this magma rising from depth encountered water-saturated tuffs of the early rhyolite, causing explosive flashing of the water to steam, which in turn reamed a vent through the overlying postcaldera trachybasalt sequence, producing the Inyo explosion craters and surrounding phreatic deposits. Surprisingly, little or no juvenile rhyolite magma reached the surface during the eruptions.

Return to parking area and retrace route to Mammoth Lakes via Mammoth Scenic Loop and Route 203.

DAY 5. INYO DOMES CIRCUIT (FIG. 4)

[Begins and ends in Mammoth Lakes]

Follow Route 395 north from Route 203 junction 5.2 mi to Mammoth Scenic Loop (on left); turn RIGHT (east) and follow gravel road 3.1 mi to summit of Lookout Mountain.

STOP 18. Lookout Mountain: Overview of Inyo domes and western caldera moat.

Lookout Mountain is a small rhyolite stratovolcano with a 2-km-diameter summit crater, which forms the tree-covered depression immediately west of the viewpoint. The volcano is situated in the northwest moat on the northern edge of the resurgent dome, and although not part of the resurgent dome, it is contructed of interstratified 0.69-Ma flows and tuffs of the early rhyolite. The aphyric obsidian underfoot is typical of the early rhyolite.

In the view to the west (Fig. 9) can be seen the west moat of the caldera, underlain mainly by 150 to 60-ka postcaldera trachybasalts and trachyandesites, and the west

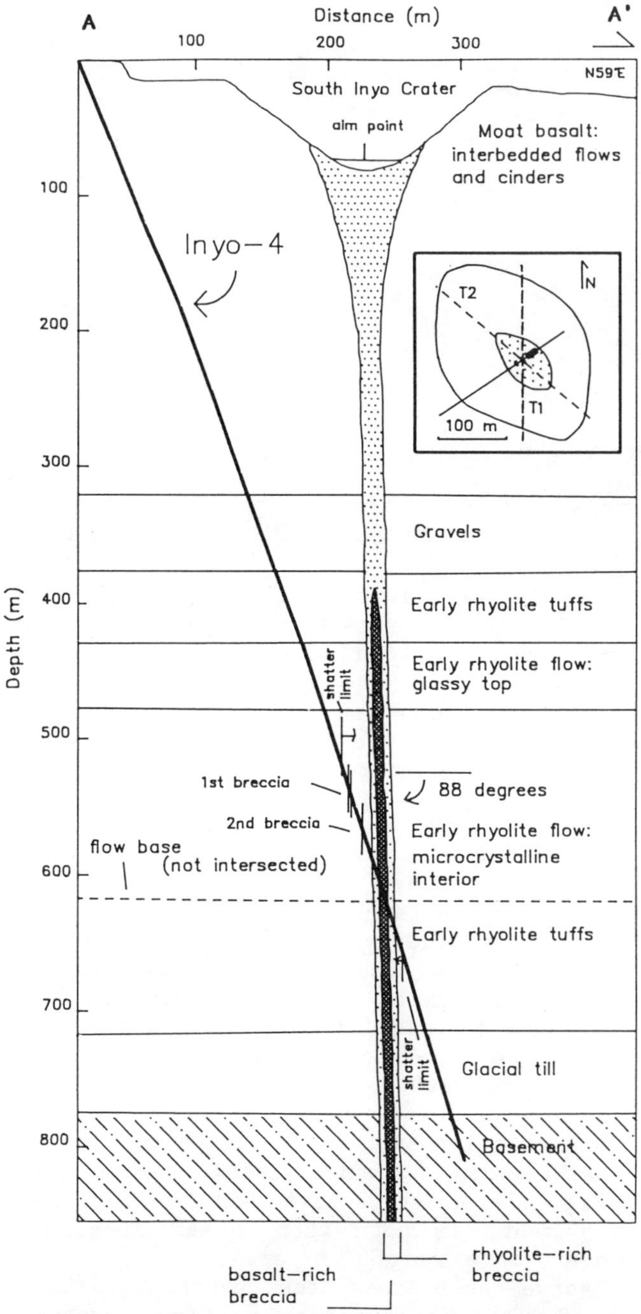

FIGURE 8 Path of Inyo Craters core hole with generalized stratigraphy and interpreted breccia structure of phreatic vent. Inset shows path of core hole (solid line) in plan view with outline of crater rim and lake; heavy bars on core-hole path in inset show location of breccia zones at depth. T1 and T2 are main N-S and subsidiary NW-SE trends of faults and fissures in area (from Eichelberger and others, 1988).

and northwest wall of the caldera, underlain by Sierran granitic and metamorphic rocks and capped by Pliocene precaldera volcanic rocks. In the middle distance are the barren, craggy, Holocene rhyolite dome flows of the Inyo Craters chain, including from north to south: Wilson Butte, Obsidian Dome, Glass Creek Dome, and Deadman Creek Dome. Wilson Butte, the older of the four (1350 yr B.P.), is chemically and petrographically similar to the Mono Craters, visible on the skyline to the north. Obsidian, Glass Creek, and Deadman Creek domes are younger (650-550 yr B.P.) and are composed of commingled sparsely and coarsely porphyritic rhyolites. Activity at these three younger vents occurred within a very short time interval and progressed from initial phreatic explosions through pyroclastic eruptions and concluded with passive lava extrusion, the latter occurring only after pyroclastic activity had ceased at all three vents. Pyroclastic eruptions began at the Deadman Creek vent, shifted to the Obsidian Dome vent and then to the Glass Creek vent. The minimum calendar age of the tephras from these three vents, based on cross dating of tree rings of Jeffrey pines growing on the deposits, is 1369, 1433, and 1469 A.D., respectively (David Yamaguchi, in Miller, 1985). The distribution of the tephras was mainly to the northeast and southwest (Figs. 10-12). North-south alignment of the Inyo phreatic and magmatic vents and their association with subparallel grabens and fissures suggest that the eruptions were fed by an 8-km-long dike (Miller, 1985) that tapped and intermixed rhyolite magmas from at least two sources (Bailey and others, 1976; Sampson and Cameron, 1987). This feeder dike transected the caldera margin, as Obsidian Dome erupted outside the caldera, Glass Creek Dome erupted on the caldera rim and flowed down the northwest wall, and Deadman Creek Dome erupted well within the caldera.

Return to Route 395; turn RIGHT (north) and go 2.8 mi to Deadman Creek Road; turn LEFT (west) and go 2.5 mi to South Deadman Creek Campground; park off road to the left of the campground entrance and walk to the bluff east of and overlooking the campground.

STOP 19. Deadman Creek Campground: Deadman Creek pyroclastic flow sequence.

From the eastern bluff of Deadman Creek, the craggy edge of Deadman Creek Dome can be seen through the trees just to the west. It is the southernmost of the Inyo domes and the vent for most of the deposits at this and the next stop. The lava dome represents the final, relatively passive, vent-filling stage of the Deadman Creek eruptions.

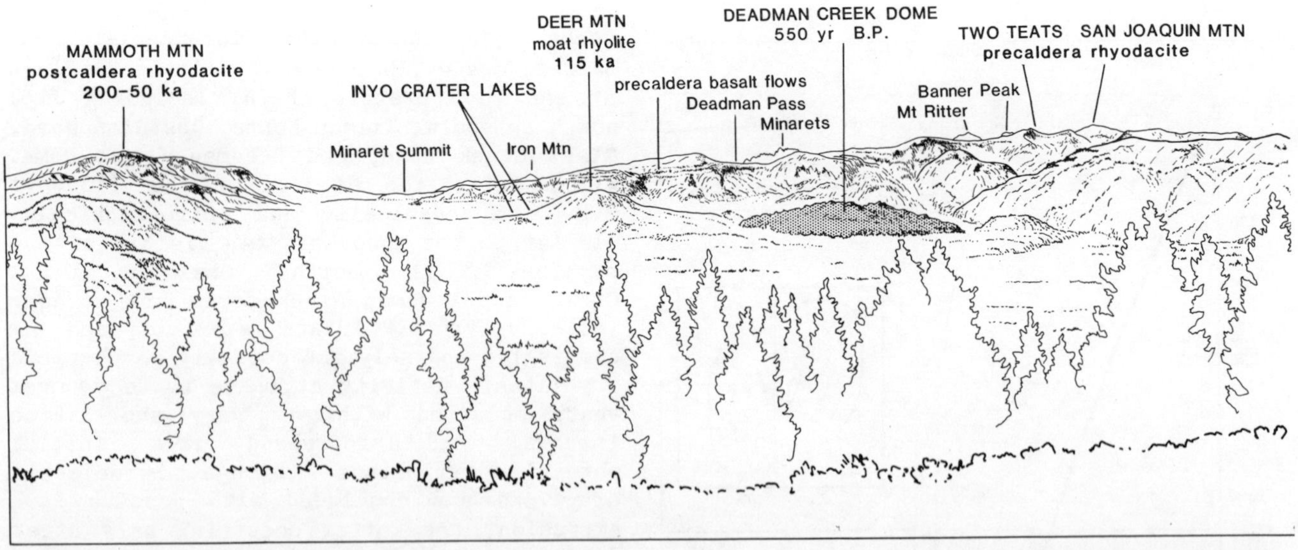

FIGURE 9 (this and facing page). Panorama from west summit of Lookout Mountain (Stop 18),

In the embankment of Deadman Creek is exposed a pyroclastic flow sequence more than 10 m thick (Fig. 13). It consists of multiple flow units interbedded with surge deposits. The flow units extend several kilometers to the west of the dome and 6 km to the northeast, beyond Route 395 (Fig. 10), covering a 15-km^2 area (Miller, 1985). The main flow unit has a pink oxidized center, which contains charcoal branches and twigs and shows reverse grading of pumice lapilli and blocks and normal grading of lithic clasts. Radiocarbon ages of the flow units range from 510 ± 100 to 770 ± 120 yr B.P, and Jeffrey pine stumps on the deposit have as many as 615 annual rings.

At this near-vent location, the pyroclastic flow is overlain by a series of thin pyroclastic flow and surge deposits that vary in thickness from 1 to 2.5 m and form dunes, hillocks, and levees on the modern ground surface. This upper sequence of deposits contains occasional bits of charcoal and shows considerable variation in bedding, grading, and sorting that records near-vent variations in eruptive mode and intensity. Such variations are not seen in pyroclastic flow deposits more distant from the vent (e.g., at Stop 20, about 2 km to the east).

Return east on Deadman Creek Road and go 1.0 mi; park on right shoulder at the top of hill.

STOP 20. Deadman Creek Road: Deadman Creek pyroclastic sequence.

The roadcut on the north side of the road exposes about 3 m of pyroclastic deposits that overlie a 100-ka postcaldera dark rhyodacite flow exposed downhill along the road. The pyroclastic sequence includes tephra deposits erupted from at least three separate vents over the last 1700 years. At the bottom of the section (Fig. 14) is 25 cm of a pyroclastic flow, with its base not exposed. The deposit contains charcoal twigs and limbs and shows no evidence of soil oxidation or erosion of its flat upper surface. Its source vent is unknown, but its radiocarbon age is 1660 ± 140 yr B.P (Miller, 1985).

Overlying the pyroclastic flow is a 1- to 1.5-cm-thick fine gray-white rhyolite ash that probably is correlative with a 600-yr-B.P. gray glassy ash sequence in the northern Mono Craters (Sieh and Bursik, 1986). Overlying this thin gray-white ash is 55 cm of bedded lapilli fall composed of varying mixtures of juvenile pumice, obsidian, and lithic clasts that erupted from the Deadman Creek vent. At this locality, this unit is east of the axis of the northeast-trending Deadman Creek tephra lobe (Fig. 10), which is more than 2 m thick near its vent, is more than 20 km long, and covers more than 80 km^2 area.

Above the Deadman Creek tephra deposit is a 2-m-thick pyroclastic flow probably correlative with the main pyroclastic flow exposed in the embankment at Deadman Creek Campground (Stop 19). Although much thinner, it retains its pink oxidized interior and contains charcoal fragments, but it shows none of the vertical or lateral variability evident near the campground.

Return to Route 395 (1.6 mi); Turn LEFT (north) and go 0.9 mi to Deadman Summit; park on right shoulder.

STOP 21. Deadman Summit: Inyo tephra deposits.

This 2.2-m-high roadcut is about 1.5 km northeast of Obsidian Dome, visible through the trees to the southwest. Like Stop 20,

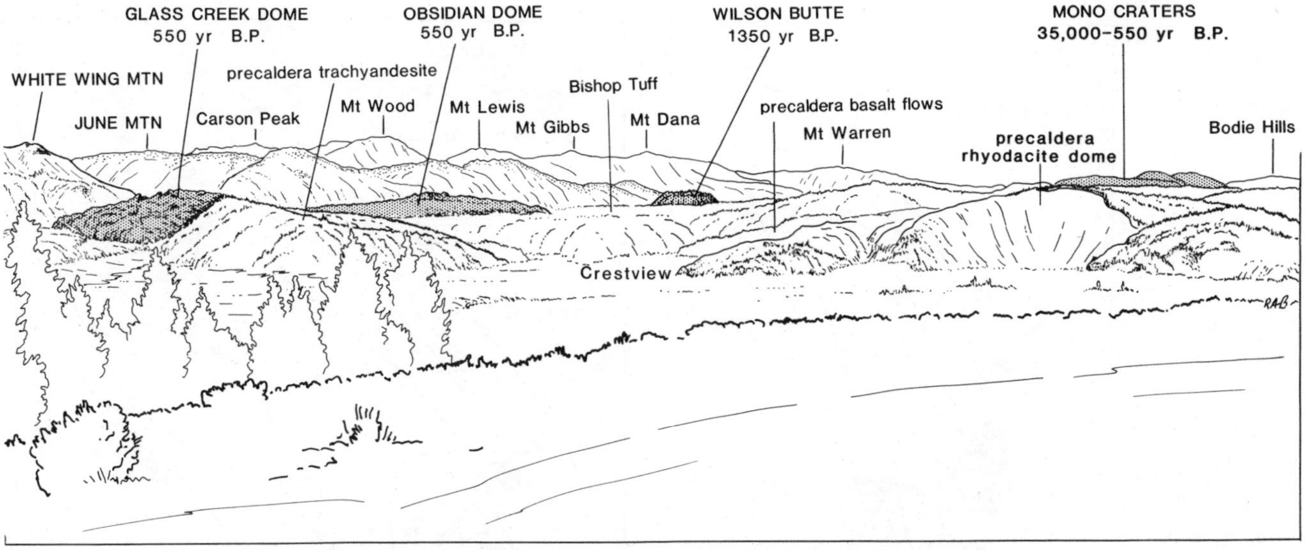

showing north-south alignment of Inyo domes (shaded).

this roadcut exposes pyroclastic units that erupted from at least three separate vents during the last 1200 yr. At the base of the section (Fig. 15) is a pyroclastic flow more than 25 cm thick, with its base not exposed. The unit contains small charcoal fragments and shows no evidence of soil oxidation or erosion of its flat upper surface. The source vent for this unit is uncertain, but a likely candidate is Wilson Butte, about 1.5 km to the northwest. The age of this unit is unknown, but it underlies the 600-yr-B.P. gray-white ash from the northern Mono Craters, which here is 3 cm thick, slightly thicker than at the previous stop because of closer proximity to its source. Overlying the gray-white ash is a 1.7-m sequence of lapilli-fall beds that erupted from the Obsidian Dome vent.

This sequence lies on the axis of a tephra lobe that can be traced 25 km northeast and covers an area of more than 140 km^2 (Fig. 11). Individual tephra layers vary in thickness from a few to 25 cm and consist of varying mixtures of juvenile pumice, obsidian, accessory lithic, and occasional accidental fragments. No radiocarbon age has been obtained from this Obsidian Dome sequence, but it overlies the Deadman Creek tephra and pyroclastic flow sequences exposed at Stops 19 and 20, and Jeffery pines growing on the Obsidian Dome sequence have as many as 525 annual rings; thus the sequence appears to be between 615 and 525 years old (Miller, 1985).

Continue 0.4 mi north on Route 395; turn LEFT onto Glass Creek Road and go 2.4 mi to Glass Creek. Park amongst the large Jeffrey pines at the south edge of Obsidian Dome and walk up the bulldozer road to the surface of the dome.

STOP 22A. Glass Creek: Obsidian Dome margin and squeeze-up.

The steep south-facing talus escarpment composed of obsidian blocks is the southern edge of Obsidian Dome (Fig. 16), which vented 1 km to the north. The morphology of the dome edge is virtually unmodified since it formed about 600 years ago. The road up the escarpment leads to a commercial pumice prospect on the dome surface and provided access to a scientific core hole drilled through the distal end of the dome flow. The core hole revealed details of the internal lithology and structure not evident from surface exposures (Eichelberger & others, 1985; Fink & Manley, 1987; Manley & Fink, 1987) and, together with data from another core hole through the conduit of the dome, has provided a better understanding of the role of volatiles during lava extrusion (Eichelberger & others, 1986).

Near the drill site on the flow surface is an unusual feature, commonly referred to as a squeeze-out or crease structure, composed of very coarsely vesicular obsidian. The feature is roughly circular in plan, about 100 m in diameter, slightly depressed below the general level of the flow surface, and bisected by a central "crease," about which flow foliation and lineation symetrically fan outward. The crease in such structures is usually oriented perpendicular to the flow margin, subparallel to the direction of flow, and appears to form as a result of lateral or tangential spreading of the flow carapace allowing more fluid gas-rich lava from the flow interior to well up to the surface (Anderson and Fink, 1987). Such features most commonly form where flows move onto gentler slopes and where forward movement slows and lateral spreading increases.

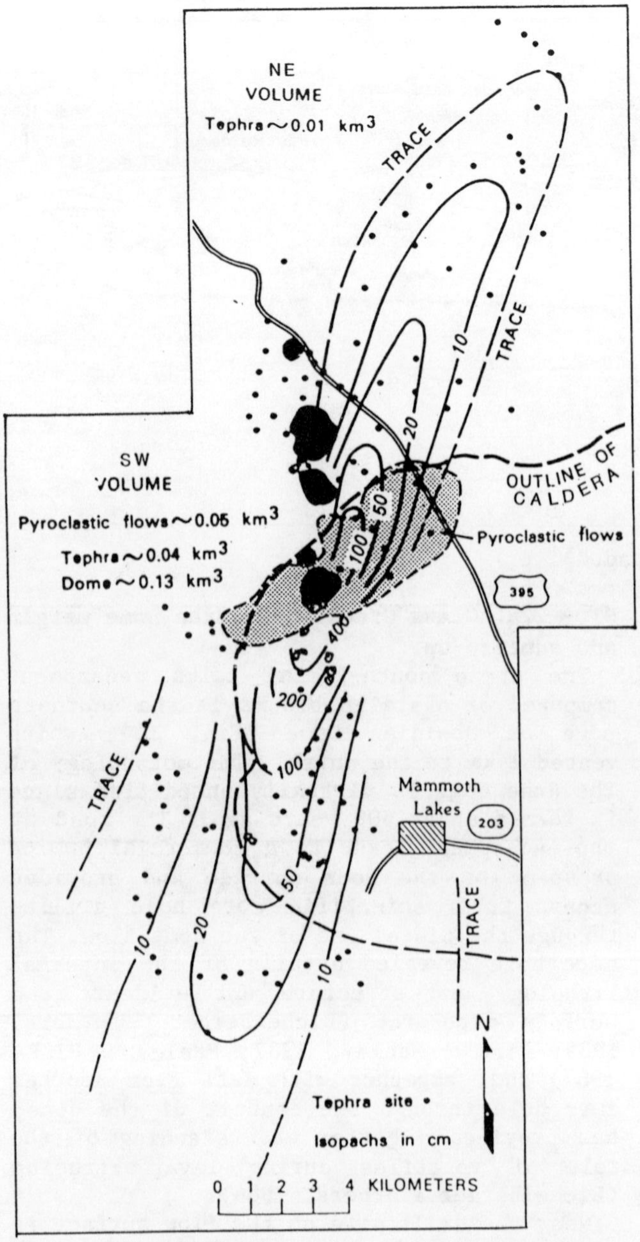

FIGURE 10 Isopach map of tephra deposits from Deadman Creek vent, Inyo volcanic chain. Northeast and southwest tephra lobes correspond to first and second eruptive pulses, respectively (from Miller, 1985).

Return to base of dome and walk to north brink of gulch cut by Glass Creek; descend to creek and walk east (left) 50 m to steep undercut south bank that exposes pyroclastic flow deposits.

STOP 22B. Glass Creek pyroclastic sequence.

Exposed along Glass Creek are poorly sorted pumiceous flowage deposits, at least one of which is incipiently welded. These deposits have an aggregate thickness of 25 m

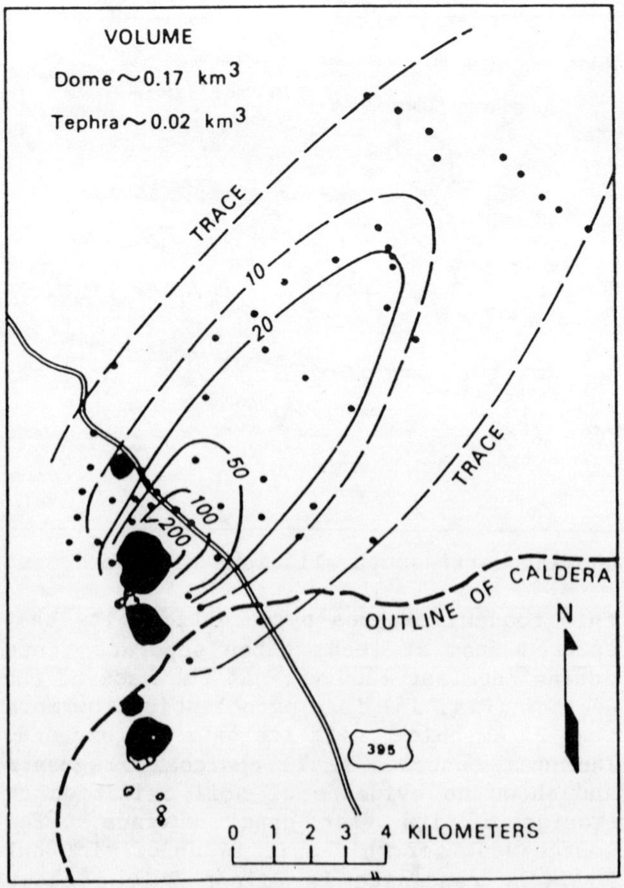

FIGURE 11 Isopach map of tephra deposits from Obsidian Dome vent, Inyo volcanic chain (from Miller, 1985).

near the source vent and can be traced for several kilometers down valley. Although they resemble deposits of pumiceous pyroclastic flows, these strata probably were generated from thick, near-vent accumulations of Glass Creek tephra deposits that avalanched down the steep slopes west of the vent (Miller, 1985). Along the creek bottom are numerous pumiceous bread-crust bombs that have been eroded out of the deposits.

Walk upstream (west) about 500 m along the fishermen's trail on north bank to road crossing the creek and follow road south (left) 200 m to base of Glass Creek Dome.

STOP 22C. Glass Creek Dome: Commingled rhyolite lava.

Directly south of Glass Creek is the steep, talus- and block-mantled north margin of Glass Creek Dome The dome is composed of two distinctive rhyolites (Bailey, 1984b; Sampson, 1987): (1) sparsely porphyritic rhyolite, typically consisting of black obsidian or vitrophyre and occurring mainly

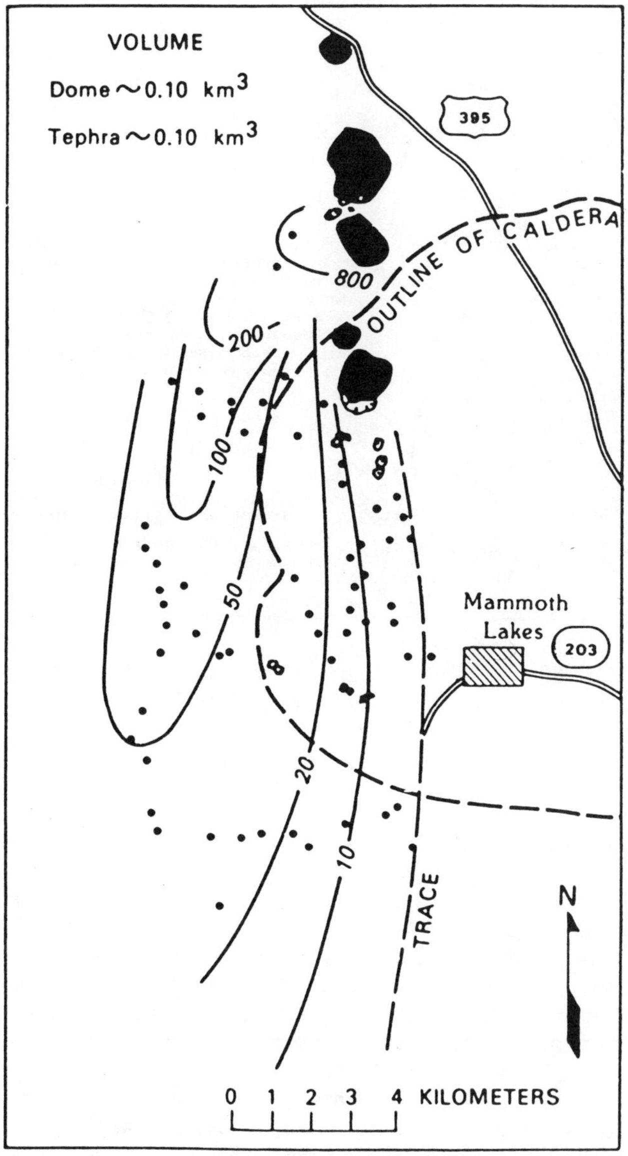

FIGURE 12 Isopach map of tephra deposits from Glass Creek vent, Inyo volcanic chain (from Miller, 1985).

on the dome margins, and (2) light-gray, pumiceous, coarsely porphyritic, hornblende-biotite rhyolite, occurring mainly in the center. Between the margin and center is an intervening zone where the two lava-types have commingled, producing fascinating "marbled-cake" structures (Fig. 17) that illustrate the contrast in viscosity between the relatively fluid obsidian and the more viscous porphyritic rhyolite. Petrologic and chemical studies (Sampson and Cameron, 1987) indicate that the two contrasting rhyolites originated and evolved in separate chambers, possibly confirming an early suggestion (Bailey and others, 1976) that the sparsely and coarsely porphyritic types had come from the Mono Craters and Long Valley magma chambers, respectively, and that they became intermixed along an interconnecting fissure just prior to eruption.

Return across Glass Creek to road on north side; walk 200 m west (left) to crest of knoll on south side of road.

STOP 22D. Glass Creek: cratered dome.

Against the nearby Sierran frontal fault scarp (Hartley Springs fault) is a small rhyolite dome about 300 m in diameter with a crater in its summit. It is informally referred to as "cratered dome." This dome is chemically and petrographically similar to Obsidian Dome and probably was extruded at about the same time, as it appears to intrude and bake (oxidize) the phreatic and pyroclastic deposits on its flanks. These flanking deposits underlie Obsidian Dome, hence both the smaller dome and Obsidian Dome postdate them. The summit crater (Fig. 18) is about 100 m in diameter, 30 to 35 m deep, and has a discontinuous inner bench about 10-15 m below the rim. This bench suggests that the crater formed by collapse rather than explosion, because crags on the bench match those on the rim, suggesting that they are subsided remnants of radial ribs that initially occupied the summit of the dome. Such small collapse craters also occur on several domes in the Mono Craters chain to the north where relationships are more obvious. Typically, no explosion debris mantles the benches and rims of such craters, affirming their non-explosive origin. Possibly they form by slow degassing and thermal contraction of the conduit or by actual magma withdrawal.

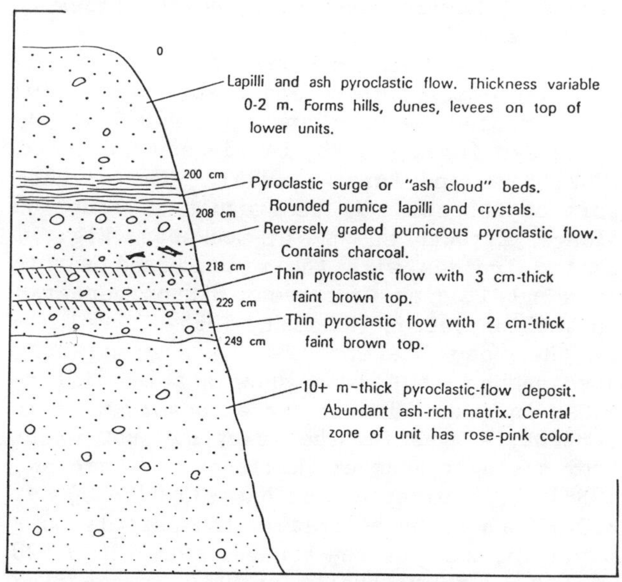

FIGURE 13 Sketch of stratigraphic section of Inyo pyroclastic flow and surge deposits exposed along entrance road to south Deadman Creek Campground (Stop 19) (from Miller, 1984).

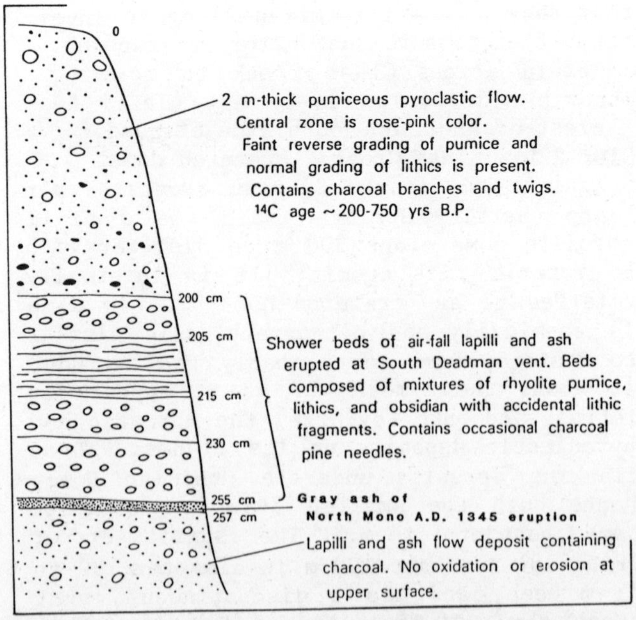

FIGURE 14 Sketch of stratigraphic section in excavation along Deadman Creek Road (Stop 20) (from Miller, 1984).

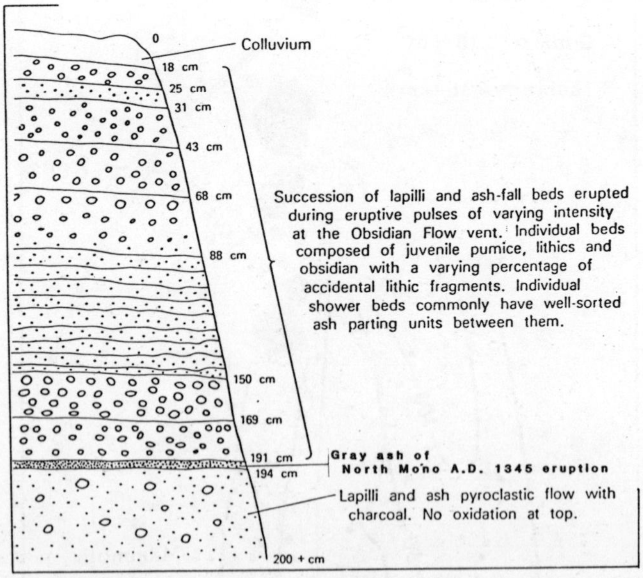

FIGURE 15 Sketch of stratigraphic section in excavation at Deadman Summit (Stop 21) (from Miller, 1984).

Return to Mammoth Lakes via Glass Creek Road and Route 395 south.

DAY 6. MONO CRATERS CIRCUIT (FIG. 19)

[Begins and ends in Mammoth Lakes]

Go north on Route 395 about 14 mi from Route 203 junction to Pumice Mine Road; turn RIGHT and go 0.4 mi to gravel road: turn LEFT and go 1.7 mi; park next to road and walk NW to crater rim.

STOP 23. Devils Punch Bowl, north crater.

The dome NW of the road is part of the Mono Craters chain and one of more than 30 rhyolite domes, flows, and craters (Fig. 20) that erupted between about 35,000 and 600 years ago (Wood, 1977b; Lajoie and Robinson, 1982; Sieh and Bursik, 1986). The arcuate form of the chain is the result of eruption along a structural zone of weakness (Fig. 1) that coincides with the mylonitized margin of a subcircular Cretaceous pluton centered on Aeolian Buttes (Kistler, 1966). Chemically the Mono Craters lavas are remarkably homogeneous, but they show subtle differences in trace element and accessory mineral content, suggesting that they did not erupt from a single chamber (Kelleher and Cameron, 1988). The older extrusions (35,000-3000 yr B.P.) tend to be moderately to sparsely porphyritic, whereas the younger ones (2000-620 yr B.P.) are virtually aphyric, suggesting that the latter erupted at higher temperature or from shallower depth, or both (Bailey, 1982). The appearance of aphyric lavas also coincides with an apparent increase in volumetric eruption rate (Wood, 1977b).

The dome northwest of the road is a body of porphyritic rhyolite that erupted more than 1200 yr ago; obsidian hydration-rind dating of a similar nearby dome suggests that the dome is late to middle Holocene in age (Wood, 1977b; Sieh and Bursik, unpub. data). The dome sits between the ridges of tephra that probably were created during the same eruption that created the dome.

In the cleft between the dome and the tephra ridge, we have open for inspection an excavation that reveals pyroclastic flow and fall deposits associated with two subsequent eruptive episodes. Yellowish-brown ash and lapilli from the 600-yr-B.P. eruption of the Inyo chain are exposed in the upper few tens of centimeters of the excavation. Underlying these Inyo pyroclastic fall beds is a thin gray ash derived from eruptions at the north end of the Mono chain. Discussion of this ash, which erupted about 1345 A.D., just prior to the Inyo eruption, will be deferred until later in the day when the near-source equivalent and associated dome are seen. Beneath the gray North Mono ash deposits are pyroclastic flow and fall deposits associated with the South Mono eruption of 700 A.D. The fall deposits near the base of the pit were erupted from sources now buried beneath and just north of South Coulee (Fig. 20). The blocky bed above the fall beds consists of juvenile and accessory debris produced during the creation of the large crater a few hundred meters northwest of the excavation.

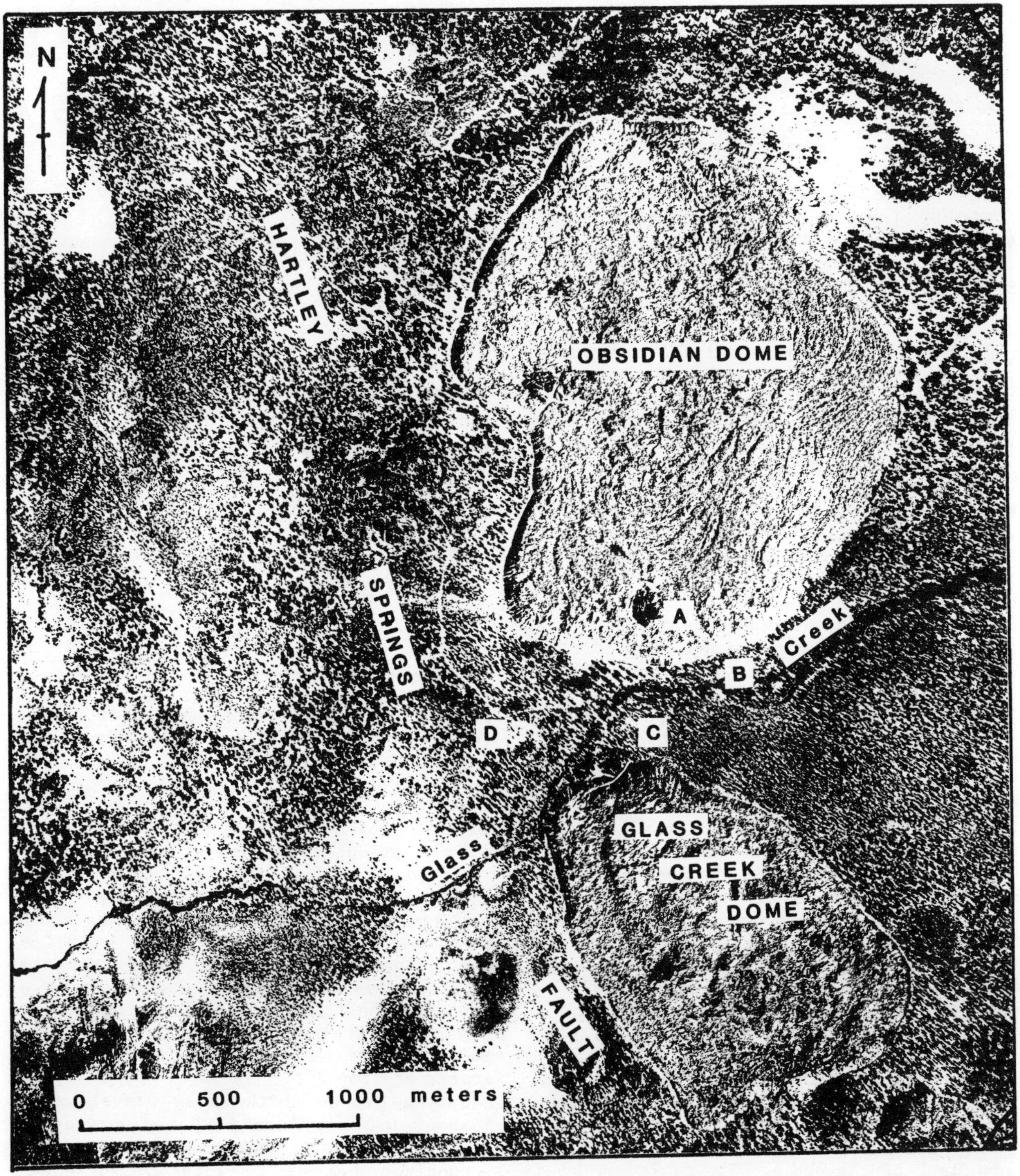

FIGURE 16 Vertical airphoto of Obsidian Dome and Glass Creek Dome, Inyo volcanic chain. A, B, C, D correspond to Stops 22A, B, C, and D.

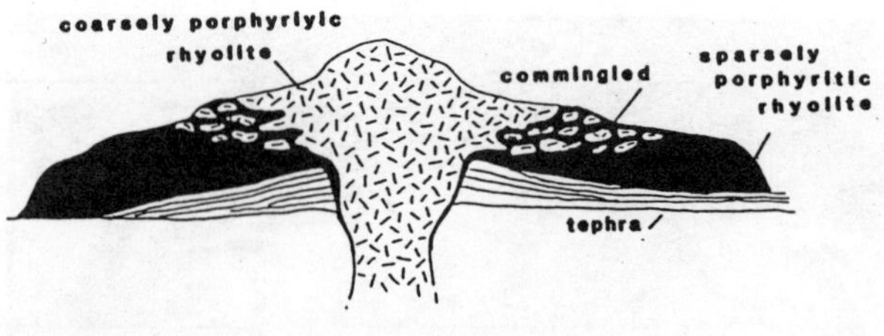

FIGURE 17 Schematic cross section of Glass Creek Dome showing distribution of marginal sparsely porphyritic rhyolite, central coarsely porphyritic rhyolite, and commingled zone, Inyo volcanic chain (Stop 22C).

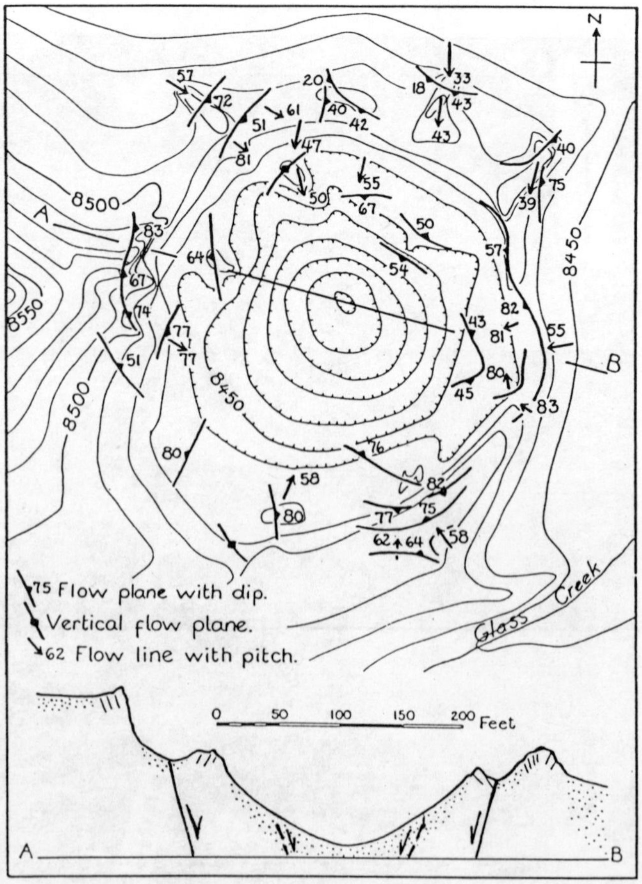

FIGURE 18 Map and cross section of Cratered dome, Inyo volcanic chain (Stop 22D) (modified from Mayo and others, 1936).

Return to road and continue northeast 0.7 mi; turn RIGHT onto another gravel road and go east about 1.5 mi; turn LEFT (north) onto a wider gravel road and go 0.3 mi to a gravel road that forms a Y with the main gravel road.

In this region, about 4 km east of the Mono Craters, an aphyric pyroclastic fall and surge series blankets the ground surface. These beds are stratigraphically beneath the 700 A.D. beds of the South Mono eruption and are constrained by radiocarbon dating to have been erupted between 345 and 535 A.D. Their source has not been identified yet but it must lie within the southern half of the Mono Craters.

Bear LEFT at the Y and continue N about 1.8 mi along the edge of the forest.

Off to the left, across the sandy plain, the prominent, barren flow is South Coulee. The volume of this aphyric rhyolite flow is about 0.5 km^3, and it flowed from the crest of the chain down both the eastern and western flanks. Radiocarbon dating constrains its emplacement to about 700 A.D. The pumiceous surface of the flow has been mined for several decades by the U.S. Pumice Company of Lee Vining and is used to face buildings and as ornamental stone in landscaping.

At the next Y bear LEFT and go about 0.5 mi to another Y; bear LEFT again and go west and then south for about 0.5 mi.

The small swale crossed by the road a few tens of meters east of the end of South Coulee is a channel that has been beheaded by a young, north-trending fault along the eastern flank of the chain. The fault may be a west-dipping normal fault that slipped during the South Mono eruption in response to the evacuation of magma through vents now beneath and immediately N of South Coulee.

Continue south on the pumice mine road 0.15 mi to the borrow pit and park.

STOP 24. South Coulee borrow pit: South Mono pyroclastic sequence.

This old, man-made cut exposes three pyroclastic flow deposits of the 700 A.D. South Mono eruption. The cut is not deep enough to expose most of the fall beds of this eruptive episode; but near the top of the cut are gray glassy fall beds whose source is beneath and/or just north of South Coulee. Some observers have suggested that these represent a phreatoplinian phase of the eruption. The dark, thin, fall bed rich in accessory clasts between the lower flow beds is from the eruption that formed the crater at Stop 23.

Retrace route north along gravel road about 0.6 mi; turn LEFT and go north about 1.8 mi; turn LEFT again and continue 1.5 mi to intersection with Route 120 (paved).

To the left (west) is the east flank of the northern half of the Mono Craters chain. The most recent eruption in this vicinity took place about 1345 A.D. and is called the North Mono eruption (Sieh and Bursik, 1986). The volume of magma erupted during this episode was about 1 km^3. The barren, rocky flow to the left and through the trees is

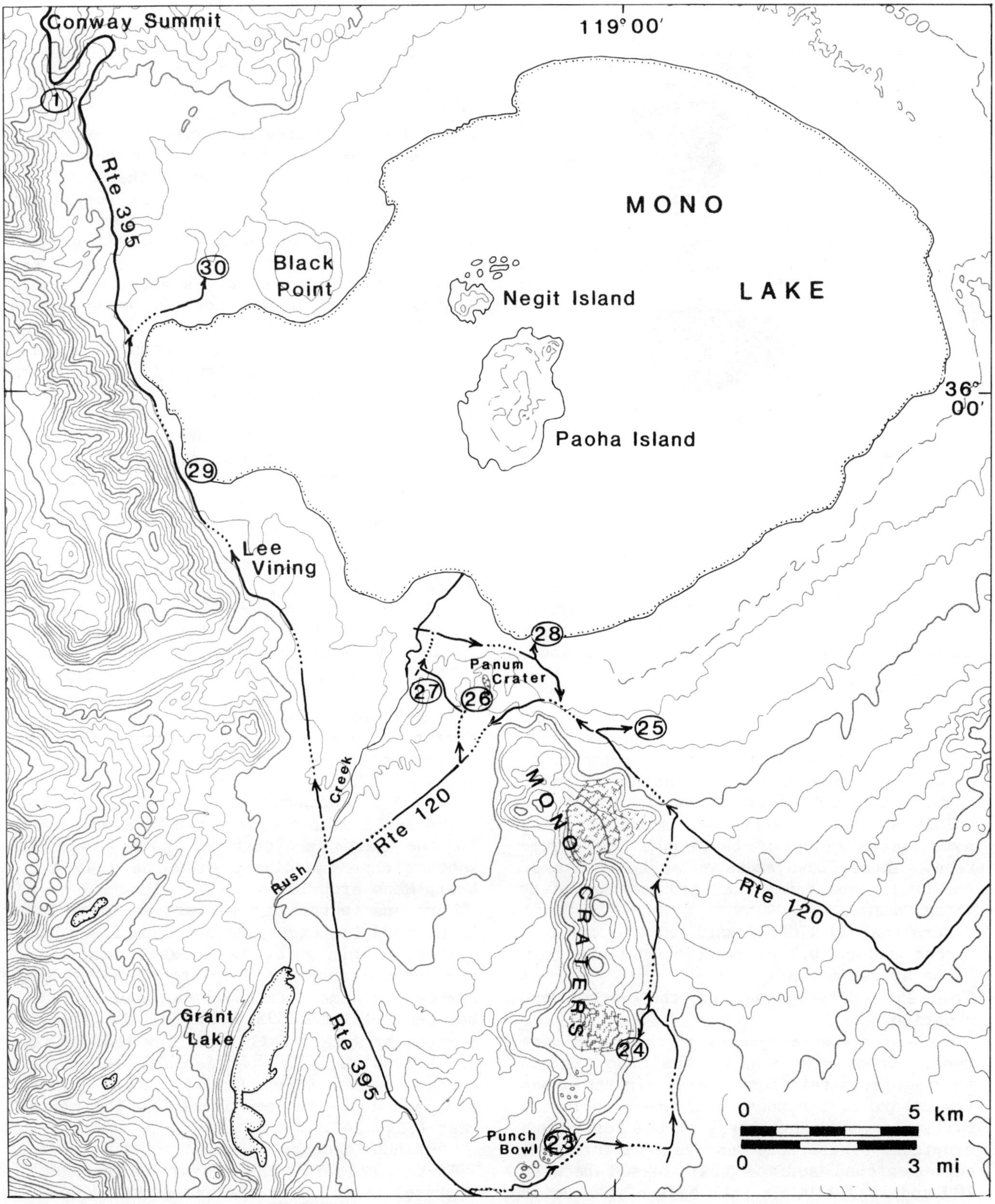

FIGURE 19 Map of Mono Craters and Mono Lake showing routes and stops of Days 1 and 6.

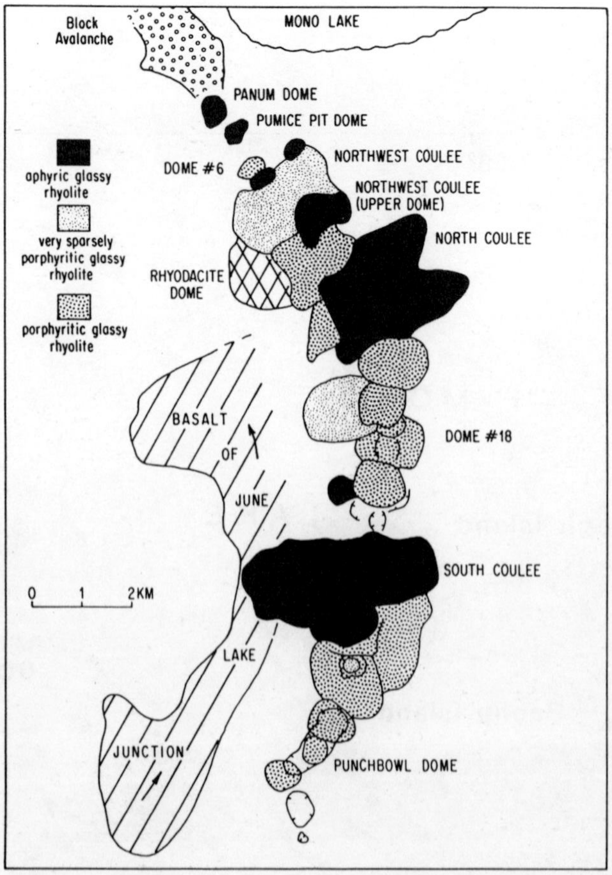

FIGURE 20 Geologic map of Mono Craters chain showing distribution of aphyric, sparsely porphyritic, and porphyritic rhyolites (from Sieh, 1984, modified from Wood, 1977b).

North Coulee; its volume represents about half that of the entire eruptive episode.

Turn LEFT onto Route 120 and go 2.2 mi until just out of the sparse forest.

This stretch of Route 120 rests on thick pyroclastic flows of the North Mono eruption. These flows emanated at the crest of the chain from vents that are now buried by North Coulee.

Turn sharply RIGHT (east) onto a gravel road and go 0.7 mi to a borrow pit in the sparse forest.

STOP 25. Stratigraphy of the North Mono eruption.

Much of the stratigraphy of the three most recent major eruptions of the Mono Craters is visible in this borrow pit. In the upper few meters of the south wall are pyroclastic fall, flow, and surge beds of the North Mono eruption. These beds rest on a bioturbated, buried ground surface that formed between 700 and 1345 A.D. on fall beds of the South Mono sequence about 1 m thick. The volumes of the two eruptions are similar, but the South Mono sequence at this site, being more distant from its source, is thinner.

Beneath the floor of the pit is a series of fall and surge beds that were deposited during an eruptive episode between 385 and 595 A.D. The sources of these beds are uncertain, but they probably underlie Pumice Pit dome and Northwest Coulee, a few kilometers to the west (Fig. 20). Recall that deposits east of the southern part of the chain also were erupted during this period.

Isopachs of the first bed of the North Mono sequence (bed 1) show that it was erupted from a linear vent now buried by North Coulee (Fig. 21). This locality is marked by the dot next to "222" on the figure. The isopach map of the second bed (bed 2) reveals that it was erupted from a source further north, now buried by Upper Dome (Fig. 22). This locality is marked by the dot next to "96" on the figure. The next several fall beds emanated from sources now beneath North Coulee and Upper Dome. Bed 7, a pyroclastic fall bed noted for its high percentage of accidental and accessory clasts, erupted from a source very near, if not under Upper Dome (Fig. 23). This locality is marked by the dot next to "130" on the figure. Above Bed 7 is a distinctive white pyroclastic surge bed. Above the surge bed is a thick sequence of planar-bedded, gray, glassy fall beds, very similar to those that cap the South Mono eruption sequence at Stop 24, near South Coulee.

Return to Route 120 and turn RIGHT (west).

The barren flow about 1 km to the left is Upper Dome, which was extruded during the 1345 A.D. North Mono eruption. The sparsely forested flow northwest of Upper Dome is Northwest Coulee, probably erupted between 385 and 595 A.D.

Go 0.9 mi northwest on Route 120.

The small barren dome a few hundred meters to the left is "cratered dome." It is another dome associated with the 1345 A.D. North Mono eruption.

Continue another 0.9 mi west on Route 120.

The road traverses the southern flank of a crater, within which is a small, circular, pumiceous, aphyric rhyolite dome called Pumice Pit dome. It may have been extruded between 385 and 595 A.D. It is thickly mantled by surge, fall, and flow beds of the North Mono sequence, particularly those that emanated from Panum Crater, whose tephra ring and dome are immediately to the northwest (Fig. 24).

Continue another 0.9 mi southwest on Route 120 to the first northbound gravel road beyond the one marked "Panum Crater" and turn RIGHT (north) toward Panum Crater.

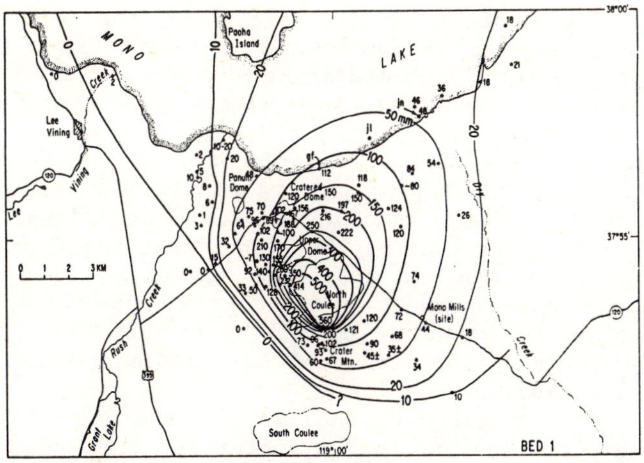

FIGURE 21 Proximal isopach map of bed 1, North Mono eruption, shows that low-level winds blew NE during eruption. Shape of 500-mm isopach shows that bed 1 tephra issued from a NW-trending linear vent or line of vents now buried beneath North Coulee (from Sieh & Bursik, 1986).

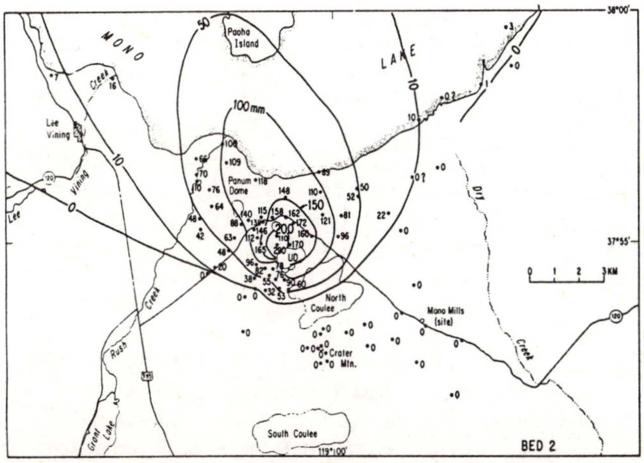

FIGURE 22 Near-vent isopach map of bed 2, North Mono eruption, shows that tephra was blown NNW from vent now buried beneath Upper Dome (UD) (from Sieh & Bursik, 1986).

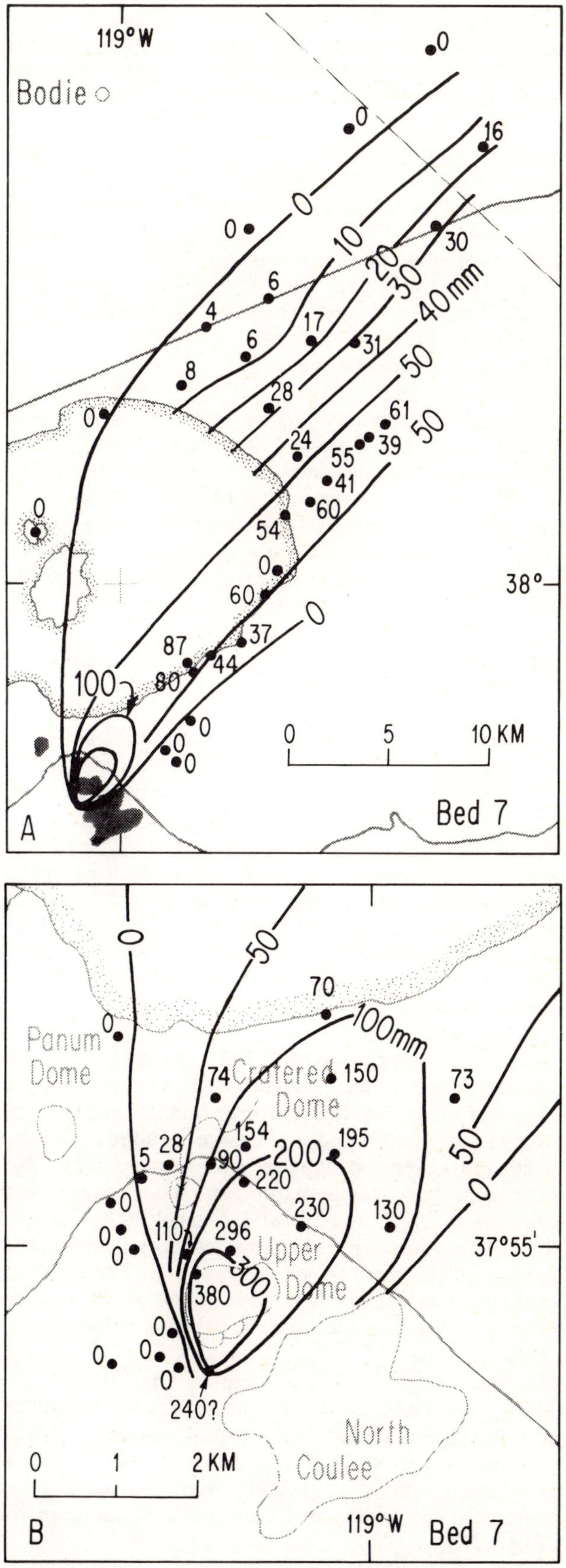

FIGURE 23 Isopach maps of bed 7, North Mono eruption: (A) Distal data show remarkable asymmetry and elongation of isopachs; (B) Near-vent data show vent was near or under Upper Dome (from Sieh & Bursik, 1986).

This road traverses a field of volcanic dunes (Fig. 24), the crests of which have amplitudes of about 1 m and wave lengths of about 5-10 m. The dunes are roughly concentric about a point just S of Panum Crater and are composed principally of subrounded lapilli and blocks of white, silvery, attenuated pumice. The deposit represents the bedload of a highly inflated eruption cloud that emanated from Panum Crater before deposition of the tephra ring and dome.

Continue 0.8 mi to the Panum Crater parking area and follow the trail to the summit of the dome.

STOP 26. Panum Crater.

Panum Crater (Figs. 20 and 24) is the northernmost and youngest vent of the North Mono eruption (Sieh and Bursick, 1986). The stratigraphic sequence of deposits emplaced during this youngest activity (Fig. 25) includes (in ascending order): (1) a vent-clearing breccia, (2) four pyroclastic flow and surge deposits varying considerably in character, (3) a tephra-ring deposit, and (4) a composite, crater-filling dome. The sequence suggests two episodes of explosive activity from the same vent, each followed by dome extrusion. The initial vent-clearing eruption produced an early tephra ring and crater, from which early pyroclastic flow and surge deposits were erupted. An early dome rising into this crater apparently collapsed or exploded laterally, breaching the northwest side of the tephra ring and sending a block avalanche or coarse pyroclastic flow northwestward into Mono Lake (Wood, 1977b; Sieh and Bursik, 1986). This outburst possibly caused momentary decompression in the conduit, and more pyroclastic flows followed. Subsequent, less energetic, pyroclastic fall eruptions partially healed the breach in the northwest side of the tephra ring, and finally the current composite dome was extruded. Panum Dome consists of three subunits (Fig. 26) extruded in succession in the east and southwest, the south, and finally the north. Both the eastern and southwestern subunits have a thin mantle of tephra on them, suggesting that they are the shouldered-aside remnants of the early dome that exploded and produced the north-west breach in the crater rim and the block avalanche to the northwest. Extrusion of the southern and northern parts of Panum Dome concluded the North Mono eruption episode, which appears to have developed above a rising rhyolite dike that broke to the surface in northward succession (Fig. 27).

From the Panum parking area follow 4-wheel-drive road north 1.0 mi; park on the left overlooking gully on the left.

STOP 27. Panum block avalanche deposit.

The surface to the right of the road northwest of Panum Crater is underlain by the Panum block-avalanche deposit of Wood (1977b). Many car-sized blocks may be seen littering its surface, and subdued longitudinal flow ridges are evident from the air (Fig. 24). The deposit is confined between opposing west- and northeast-facing linear declivities that may be either sector-graben fault scarps or erosional scarps produced during emplacement of the coarse high-energy flow by scouring of the soft substrate, which consists mainly of lake sediments and pyroclastic deposits.

In the gully to the left of the road is exposed the internal structure of the Panum block-avalanche deposit. It is about 7 m thick here but is probably several tens of meters thick near its source and thins to 1 m at its distal end near the shore of Mono Lake. The deposit is a fines-depleted, clast-supported breccia consisting of partially agglutinated lapilli and blocks (Sieh and Bursik, 1986). About 90% of the clasts are angular fragments of dense to weakly pumiceous gray glass, on the surface of which are numerous white, powdery impact marks, indicating that they were rigid during emplacement. About 10% of the clasts are oblate blocks of black to light-gray pumiceous glass with pronounced lumpy, bread-crusted surfaces, into which are impressed and agglutinated many of the smaller angular clasts. Paleomagnetic studies (Salyards, 1986) indicate that the deposit was emplaced at a temperature above 600°C. The association of contrasting rigid and viscid clasts in the deposit, together with their high-temperature emplacement, suggest an origin by collapse of a partially molten dome, most likely the early dome in Panum Crater.

Continue north and west along the 4-wheel-drive road 0.9 mi to junction with gravel road; turn RIGHT and go 0.6 mi to Rush Creek crossing; turn RIGHT (east) and go 1.9 mi; Turn LEFT to Mono Lake State Nature Preserve parking area. Follow the self-guided trail to the shore and back to the parking area.

STOP 28. Mono Lake State Nature Preserve: Tufa towers.

The delicate, castellated, white towers along the shores of Mono Lake are calcareous sublacustrine spring deposits formed where freshwater springs percolate through lake-bottom sediments and bubble up through the saline lakewater. Calcium in the freshwater combines with carbonate of the lakewater, possibly catalyzed by algae, resulting in precipitation of calcite or aragonite. The towers formed entirely

FIGURE 24 Vertical airphoto of Panum Crater and associated pyroclastic deposits: A, composite dome; B, tephra ring; C, block avalanche deposit; D, dune field of pyroclastic flow and surge deposits; E, Pumice Pit dome.

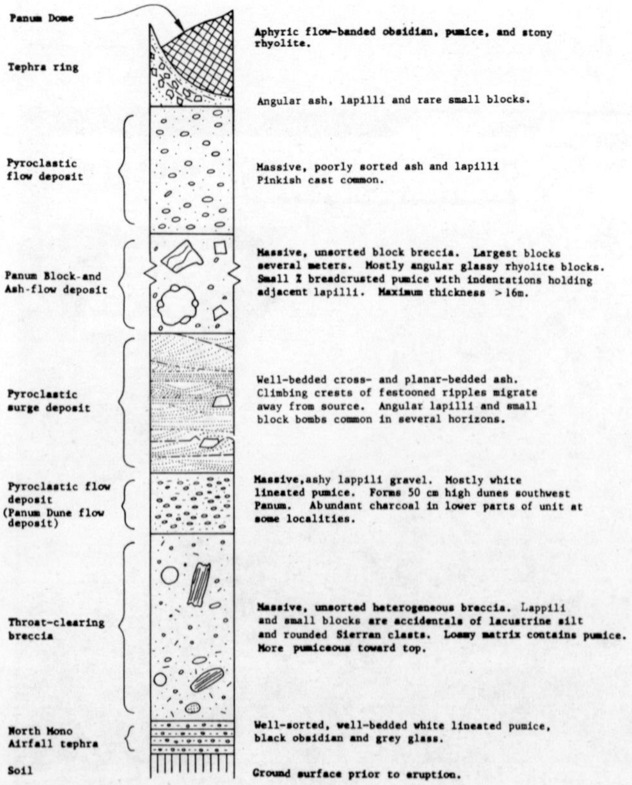

FIGURE 25 Generalized columnar section of North Mono eruption deposits in vicinity of Panum Crater (from Sieh and Bursik, 1986).

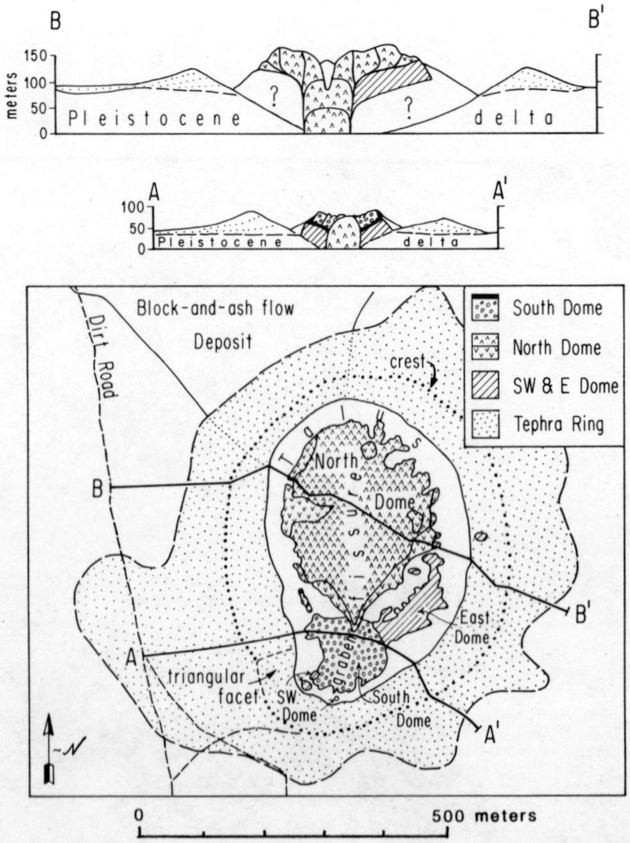

FIGURE 26 Generalized map and cross sections of Panum Dome (Stop 26) (from Sieh and Bursik, 1986).

underwater and are being exposed as the lake level declines. Numerous signs along the well-marked trail through the tufa towers and along the lakeshore describe the many fascinating and unique features of the Mono Lake habitat, including its brine shrimp, brine flies, bird life, and Indian and early settlement history.

From tufa towers parking area go south (uphill) 0.9 mi to Route 120; turn RIGHT (west) and go 4.8 mi to Route 395; turn RIGHT (north) and go 7.0 mi (through Lee Vining) to Mono Lake Marina; turn RIGHT into parking area and walk to lakeshore.

STOP 29. Mono Lake Marina: Rafted pumice blocks.

The western and northern shores of Mono Lake are strewn with thousands of large (≤2 m), subangular to subround blocks of biotite-bearing rhyolite pumice, which apparently formed during a sublacustrine eruption. The bouyant blocks probably floated to the surface of the dense saline lakewater and drifted ashore, where they lodged and later became coated with calcareous tufa. The blocks are concentrated at and below the 1948-m strandline and absent above, indicating that the lake was at that level at the time of eruption (Stine, 1984). In lower Lee Vining Creek, pine cones sandwiched between one of these blocks and overlying river gravel gave a radiocarbon age of 1495 ± 45 yr B.P., providing a minimum age for the eruption. Excavations here at the marina show the blocks to lie well above a 1990-yr-B.P. Mono Craters ash bed; thus the eruption probably occurred between about 250 and 450 A.D. The site of the eruption is not known with certainty but extrusions of similar pumiceous rhyolite occur on two of the islets northeast of Negit Island and also on Paoha Island. These or possibly other vents still submerged are likely sources, as the greatest concentration of blocks occurs along the nearby northern lakeshore.

Continue north on Route 395 2.5 mi to Danberg Park Road (just beyond Mono Inn); turn RIGHT (east) and go 2.0 mi to prominent roadcut in lake sediments and black ash on right, beyond Mill Creek crossing.

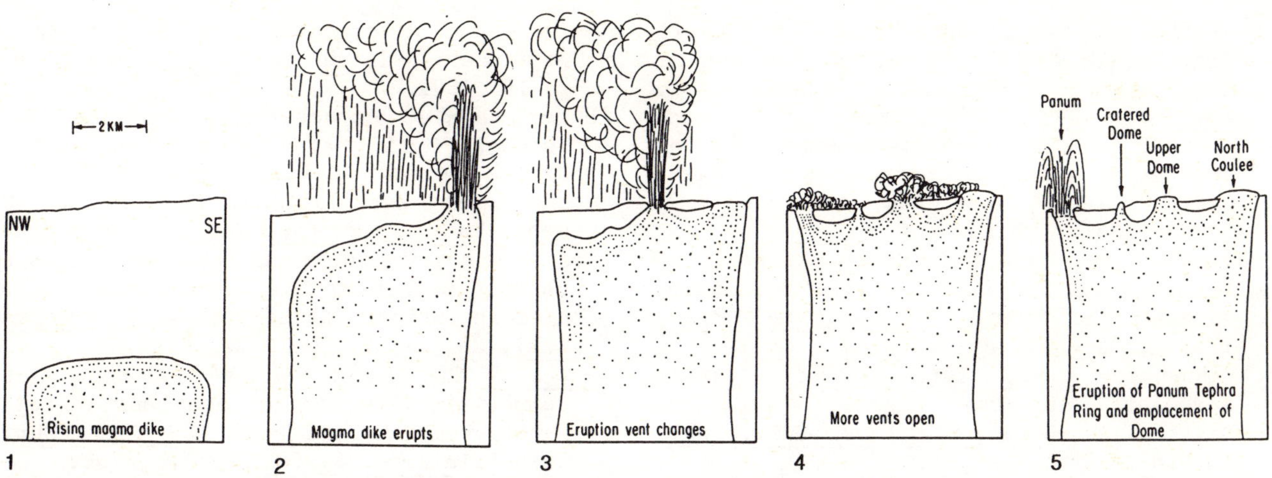

FIGURE 27 Hypothetical development of North Mono eruption episode above a rising rhyolite dike (from Sieh, 1984).

STOP 30. Mill Creek: Basaltic ash layer of Black Point.

The 2-m-thick basaltic ash layer exposed in the roadcut on the right, on the upper east side of Mill Creek, is in the uppermost part of the Wilson Creek Formation of Lajoie (1968, 1969), which consists predominantly of fine clayey silts and sands and intercalated rhyolite ash beds and deltaic gravels. The formation constitutes most of the exposed sediments around Mono Lake. On the basis of radiocarbon dating of ostracodes and calcareous tufa and estimated sedimentation rates (Lajoie and Robinson, 1982), it ranges in age from 36,000 to 11,000 yr B.P. and was approximately coeval with the latest Pleistocene glaciations and with the early eruptions of the Mono Craters.

The basaltic ash layer exposed here was derived from eruptions at Black Point about 13,300 ± 500 yr B.P. (Lajoie, 1968). The ash was deposited in at least 110 m of water and the abundant climbing ripple marks indicate emplacement by south- to southwest-directed turbid bottom currents, probably flowing from glaciated Lundy Canyon to the west. The penecontemporaneous deformation probably formed as the water-saturated sediments slumped downslope to the south, triggered either by slope instability or possibly by earthquakes associated with the Black Point eruptions (Lajoie, 1968).

Return to Route 395; turn RIGHT (south) and return to Mammoth Lakes.

END OF EXCURSION

REFERENCES

Alpha, T.R., R.A. Bailey, K.R. Lajoie, and M.M. Clarke, Physiographic diagrams of Long Valley, California, US Geol. Surv. Misc. Field Studies Map 1659, 1983.

Anderson, S.W., and J.H. Fink, Modelling crease structures on silicic lava flows (abs.), Eos Trans. AGU, 68:44, 1545, 1987.

Bailey, R.A., Other potential eruption centers in California: Long Valley-Mono Lake, Coso, and Clear Lake volcanic fields, Calif. Div. Mines Geol. Spec. Publ. 63, 17-28, 1982.

______, Chemical evolution and current state of the Long Valley magma chamber, in Hill, D.P., R.A. Bailey, and A.S. Ryall (eds.), Active tectonic and magmatic processes beneath Long Valley caldera, US Geol. Surv. Open-File Rept. 84-939, 24-40, 1984a.

______, Introduction to the Late Cenozoic volcanism and tectonism of the Long Valley-Mono Basin area, eastern California, in Lentz, J., Jr. (ed.), Guidebook for Western Geological Excursions, 2, 56-67, Dept. Geol. Sci., MacKay School of Mines, University of Nevada, Reno, 1984b.

______, Long Valley caldera, eastern California, Geol. Soc. Am. Centennial Field Guide - Cordilleran Section, 163-168, 1987.

______, Geologic map of Long Valley caldera, Mono-Inyo Craters volcanic chain, and vicinity, US Geol. Surv. Misc. Inv. Series Map 1939, scale 1:62,500, 1989.

Bailey, R.A., G.B. Dalrymple, and M.A. Lanphere, Volcanism, structure, and geochronology of Long Valley caldera, J. Geophys. Res., 81:5, 725-744, 1976.

Benioff, H., and B. Gutenberg, The Mammoth "Earthquake Fault" and related features in Mono County, California, Seismol. Soc. Am. Bull., 29, 333-340, 1939.

Blackwell, D.D., A transient model for the geothermal system of Long Valley caldera, California, J. Geophys. Res., 90:B13, 11229-11241, 1985.

Bursik, M.I., and K.E. Sieh, Range-front faulting and volcanism in the Mono Basin, eastern California, J. Geophys. Res., 1989 (in press).

Christensen, M.N., Late Cenozoic crustal movements in the Sierra Nevada of California, Geol. Soc. Am. Bull., 77, 163-182, 1966.

Christensen, M.N., and C.M. Gilbert, Basaltic cone suggests origin of some guyots, Science, 143:3603, 240-242, 1964.

Eichelberger, J.C., P.C. Lysne, C.D. Miller, and L.W. Younker, Research drilling at Inyo domes, 1984 results, Eos Trans. AGU, 66:17, 186-187, 1985.

Eichelberger, J.C., C.R. Carrigan, H.R. Westrich, and R.H. Price, Nonexplosive silicic volcanism, Nature, 323, 598-602, 1986.

Eichelberger, J.C., T.A. Vogel, L.W. Younker, C.D. Miller, G.H. Heiken, and K.H. Wohletz, Structure and stratigraphy beneath South Inyo Crater, Long Valley caldera, California, J. Geophys. Res., 93:B11, 13208-13220, 1988.

Fink, J.H., and C.R. Manley, Origin of pumiceous and glassy textures in rhyolite domes and flows, Geol. Soc. Am. Spec. Paper 212, 77-88, 1987.

Fiske, R.S., and O.T. Tobische, Paleogeographic significance of volcanic rocks of the Ritter Range pendant, central Sierra Nevada, in Howell, D.G., and K.A. McDougall (eds.), Mesozoic Paleogeography of the Western United States, Soc. Econ. Paleotol. Mineral., Pacific Sec., Los Angeles, Calif., 209-222, 1978.

Gilbert, C.M., Welded tuff in eastern California, Geol. Soc. Am. Bull., 49, 1829-1862, 1938.

Gilbert, C.M., M.N. Christensen, Y. Al Rawi, and K.R. Lajoie, Structural and volcanic history of Mono Basin, California-Nevada: Geol. Soc. Am. Memoir 116, 275-329, 1968.

Hildreth, Wes, The Bishop Tuff: Evidence for the origin of compositional zoning in silicic magma chambers, Geol. Soc. Am. Spec. Paper 180, 43-75, 1979.

Hill, D.P., R.A. Bailey, and A.S. Ryall, Active tectonic and magmatic processes beneath Long Valley caldera, eastern California: an overview, J. Geophys. Res., 90:B13, 11111-11120, 1985.

Holden, E.S., Earthquakes in California in 1890 and 1891, US Geol. Surv. Bull. 95, p. 18, 1892.

Huber, N.K., Amount and timing of Late Cenozoic uplift and tilt of the Central Sierra Nevada, California--evidence from the upper San Joaquin River Basin, US Geol. Surv. Prof. Pap. 1197, 28 pp., 1981.

Izett, G.A., Volcanic ash beds: recorders of upper Cenozoic silicic pyroclastic volcanism in western United States, J. Geophys. Res., 86:B11, 10200-10222, 1981.

______, The Bishop ash bed and some older compositionally similar ash beds in California, Nevada and Utah, US Geol. Surv. Open-File Rept. 82-582, 47 pp., 1982.

Kelleher, P.C., and K.L. Cameron, The geochemistry of the Mono Craters-Mono Lake islands complex, eastern California, J. Geophys. Res., 1988 (in press).

Kistler, R.W., Structure and metamorphism in the Mono Craters quadrangle, Sierra Nevada, California, US Geol. Surv. Bull. 1221-E, E1-53, 1966.

Lajoie, K.R., Late Quaternary stratigraphy and geologic history of Mono Basin, eastern California, Ph.D. dissertation, 271 p., Univ. California, Berkeley, 1968.

______, Late Quaternary stratigraphy and geologic history of Mono Basin, eastern California, Dissertation Abstracts, B:30:3, 1202, 1969.

Lajoie, K.R., and S.W. Robinson, Late Quaternary glacio-fluvial chronology, Mono Basin, California (abs.), Geol. Soc. Am. Abstr. Programs, 14:4, 179, 1982.

Mankinen, E.A., C.S. Gromme, G.B. Dalrymple, M.A. Lanphere, and R.A. Bailey, Paleomagnetism and K-Ar ages of volcanic rocks from Long Valley caldera, California, J. Geophys. Res., 91:B1, 633-652, 1986.

Manley, C.R., and J.H. Fink, Internal structure of rhyolite flows as revealed by research drilling, Geology, 15, 549-552, 1987.

Mastin, L.G., Phreatic eruption processes at Inyo Craters, Long Valley caldera, California, Ph.D. dissertation, 50 p., Stanford Univ., Stanford, Calif., 1988.

Mastin, L.G., and D.D. Pollard, Surface deformation and shallow dike-intrusion processes at Inyo Craters, Long Valley caldera, California, J. Geophys. Res., 93:B11, 13221-13235, 1988.

Mayo, E.B., The Pleistocene Long Valley Lake in eastern California, Science, 80, 95-96, 1934.

Mayo, E.B., L.C. Conant, and J.R. Chelikowski, Southern extention of the Mono Craters, California, Am. J. Sci., 32, 81-97, 1936.

Merriam, Richard, and J.L. Bischoff, Bishop Ash: A widespread volcanic ash extended to southern California, J. Sediment. Petrol., 45, 207-211, 1975.

Metz, J.M., and G.A. Mahood, Precursors to the Bishop Tuff eruption: Glass Mountain, Long Valley, California, J. Geophys. Res., 90:B13, 11121-11126, 1985.

Miller, C.D., D.R. Mullineaux, D.R. Crandell, and R.A. Bailey, Potential hazards from future volcanic eruptions in the Long Valley-Mono Lake area, east-central Calif. and Nevada--A preliminary assessment, US Geol. Surv. Circ. 877, 10 pp., 1982.

Miller, C.D., Chronology and stratigraphy of recent eruptions at the Inyo volcanic chain, in Stine, S., S. Wood, K. Sieh, and C.D. Miller, Holocene Paleoclimatology and Tephrachronology East and West of the Central Sierra Crest, Guidebook for the Friends of the Pleistocene, Pacific Cell, Field Trip, October 12-14, 1984, p. 88-96, 1984.

Miller, C.D., Holocene eruptions at the Inyo volcanic chain, California: Implications for possible eruptions in Long Valley, Geology, 13, 14-17, 1985.

Pakiser, L.C., Gravity, volcanism, and crustal deformation in Long Valley, California, US Geol. Surv. Prof. Pap. 424-B, B250-253, 1961.

Pakiser, L.C., F.C. Press, and M.F. Kane, Geophysical study of Mono Basin, California, Geol. Soc. Am. Bull., 71, 415-448, 1960.

Putnam, W.C., Moraine and shoreline relationships at Mono Lake, California, Geol. Soc. Am. Bull., 61, 115-122, 1950.

Rinehart, C.D., and N.C. Huber, The Inyo Crater Lakes--A blast in the past, Calif. Div. Mines Geol., 18:9, 169-172, 1965.

Rundle, J.B., C.R. Carrigan, H.C. Hardee, and W.C. Luth, Deep drilling in the magmatic environment, Long Valley caldera, Eos Trans. AGU, 67:21, 790-791, 1986.

Russell, I.C., Quaternary History of the Mono Valley, California, US Geol. Surv. Eighth Ann. Rept., 263-394, 1889.

Salyards, S.L., Thermal and depositional constraints on a block-and-ash-flow deposit from Panum Crater, Mono County, California, from paleomagnetic analysis (abs.), Geol. Soc. Am. Abstr. Progams, 18:2, 180, 1986.

Sampson, D.E., Structures and textures of the lavas of the Inyo chain, eastern California: Implications for feeder-dike and conduit geometries, Geol. Soc. Am. Spec. Pap. 212, 89-101, 1987.

Sampson, D.E., and K.L. Cameron, The geochemistry of the Inyo volcanic chain: Multiple magma systems in the Long Valley region, eastern California, J. Geophys. Res., 92:B10, 10403-10421, 1987.

Sarna-Wojcicki, A.M., H.R. Bowman, C.E. Meyer, C.E., P.C. Russell, M.J. Woodward, Gail McCoy, J.J. Rowe, Jr., P.A. Baedecker, Frank Asaro, and Helen Michael, Chemical analyses, correlations, and ages of Upper Pliocene and Pleistocene ash layers of east-central and southern California, US Geol. Surv. Prof. Pap. 1293, 40 pp., 1984.

Sarna-Wojcicki, A.M., S.D. Morrison, C.E. Meyer, and J.W. Hillhouse, Correlation of upper Cenozoic tephra layers between sediments of western United States and the East Pacific Ocean, and comparison with biostratigraphic and magnetostratigraphic age data, Geol. Soc. Am. Bull., 98, 207-233, 1987.

Savage, J.C., M.M. and Clark, Magmatic resurgence in Long Valley caldera, California: Possible cause of the 1980 Mammoth Lakes earthquakes, Science, 217, 531-533, 1982.

Savage, J.C., and R.S. Cockerham, Earthquake swarm in Long Valley caldera, California, January 1983: Evidence for dike injection, J. Geophys. Res., 89:B10, 8315-8324, 1984.

Sharp, R.P., Sherwin Till-Bishop Tuff geological relationships, Sierra Nevada, California, Geol. Soc. Am. Bull., 79, 351-364, 1968.

Sheridan, M.F., Double cooling-unit nature of the Bishop Tuff in Owens River Gorge, California (abs.), Geol. Soc. Am. Spec. Pap., 115, 351, 1967.

_____, Fumarolic mounds and ridges of the Bishop Tuff, California, Geol. Soc. Am. Bull., 81, 851-868, 1970.

Sieh, Kerry, Most recent eruptions of the Mono Craters, eastern central California, in Stine, S., S. Wood, K. Sieh, and C.D. Miller, Holocene Paleoclimatology and Tephrachronology East and West of the Central Sierra Crest, Guidebook Friends of the Pleistocene, Pacific Cell, Field Trip, Oct. 12-14, 1984, p. 53-74, 1984.

Sieh, Kerry, and Marcus Bursik, Most recent eruption of the Mono Craters, eastern central California, J. Geophys. Res., 91:B12, 12539-12571, 1986.

Snow, Eleanour, and R.A. Yund, Origin of cryptoperthites in the Bishop Tuff and their bearing in its thermal history, J. Geophys. Res., 93:B8, 8975-8984, 1988.

Sorey, M.L., Evolution and the present state of the hydrothermal system in Long Valley caldera, J. Geophys. Res., 90:B13, 11219-11228, 1985.

Stine, Scott, Late Holocene lake level fluctuations and island volcanism at Mono Lake, California, in Stine, S., S. Wood, K. Sieh, and C.D. Miller, Holocene Paleoclimatology and Tephrachronology East and West of the Central Sierra Crest, Guidebook Friends Pleistocene, Field Trip, Oct. 12-14, 1984, p. 21-49, 1984.

Wood, S.H., Distribution, correlation, and radiocarbon dating of late Holocene tephra, Mono and Inyo Craters, eastern California, Geol. Soc. Am. Bull., 88, 89-95, 1977a.

______, Chronology of late Pleistocene and Holocene volcanics, Long Valley and Mono Basin geothermal areas, eastern California, US Geol. Surv. Final Tech. Rept., Geothermal Research Program, 1977b [reprinted as US Geol. Surv. Open-File Rept. 83-747, 76 pp., 1983].